SHIPPING AND LOGISTICS LAW

SHIPPING AND LOGISTICS LAW

Principles and Practice in Hong Kong

Second Edition

Felix W H Chan
Jimmy J M Ng
Sik Kwan Tai

HKU PRESS
香港大學出版社

Hong Kong University Press
The University of Hong Kong
Pokfulam Road
Hong Kong
www.hkupress.org

ISBN 978-988-8208-78-4 (*Hardback*)
ISBN 978-988-8208-79-1 (*Paperback*)

British Library Cataloguing-in-Publication Data
A catalogue record for this book is available from the British Library.

10 9 8 7 6 5 4 3 2 1

Printed and bound by Hang Tai Printing Co., Ltd. in Hong Kong, China

Contents

Preface to the Second Edition

Strategically located on the South China Sea with a deep, sheltered harbour, Hong Kong is among the top ten largest maritime centres in the world. For many years, Hong Kong has continued to be one of the world's major logistics hubs and international cargo handling centres. It is a most significant gateway connecting Mainland China with the rest of the world.

The international competitiveness of Hong Kong will increasingly depend on its transport and logistics. By sea, road, rail, or air, we depend heavily on transport to link individuals, businesses and cities. The maintenance of Hong Kong as a free port and global logistics centre is a crucial characteristic of its economic vitality. Rapidly expanding international trade with Mainland China and Hong Kong has generated significant changes to the shipping and logistics law in both jurisdictions. Therefore, it is essential that everyone involved in international trade with Hong Kong (and Mainland China via Hong Kong) understand the backbone of the legal issues which affect every aspect of the business.

We intend to provide a general introduction to the basic principles of shipping and logistics law in Hong Kong. This book contains many practical examples and illustrations from case law. Extracts of the relevant legislation and sample shipping documents are annexed in this book for reference.

In this second edition, we have reformatted some of the materials. The book has been reoriented to concentrate on the parts of the legal framework which are most directly relevant to the logistics and maritime industry of Hong Kong. This new edition takes account of a number of new cases, new international conventions (such as the Rotterdam Rules) and significant changes introduced by legislative amendments since the last edition.

The book is intended primarily for students and teachers of transport studies and business logistics management. However, we trust that it will also provide useful guidance to shipowners, carriers, shipping agents, traders, insurers, bankers, logistics managers, arbitrators, mediators and lawyers who need to acquire a clear understanding of the key principles in a practical context.

The division of labour associated with each of the chapters in this book is as follows: Chapters 4, 6.1, 6.2, 6.4, 11, 12, 13 (Felix Chan); Chapters 1, 6.3, 7.1, 7.2, 7.3, 7.4, 7.5, 7.6, 7.7, 7.8, 7.9, 7.10, 7.11, 10.1, 10.2, 10.3, 10.4, 10.5, 10.6, 10.7, 10.8, 10.9 (Jimmy Ng); and Chapters 2, 3, 5, 7, 8, 9 (S K Tai). The contribution of Mr Bobby Wong to the previous edition of this publication is acknowledged with gratitude.

This book has its origins in our lecture notes used in teaching. We are grateful to our students for their interest and enthusiasm. We also want to thank Hong Kong University Press for their efficient editorial work. Considering all of this help, any errors and shortcomings that remain must be ours.

Felix W H Chan
Jimmy J M Ng
Sik Kwan Tai
January 2015

Preface to the First Edition

The international competitiveness of Hong Kong will increasingly depend on its transport and logistics. By sea, air, road or rail, we rely on transport to connect individuals, businesses and cities. The maintenance of Hong Kong as a free port and international logistics hub is a crucial characteristic of its economic vitality. Rapidly expanding international trade with mainland China and Hong Kong has generated significant changes to the shipping and logistics law in both jurisdictions. Therefore, it is essential that everyone involved in international trade with Hong Kong and mainland China via Hong Kong understands the backbone of the legal issues which affect every aspect of the business.

We intend to provide a general introduction to the basic principles of shipping and logistics law in Hong Kong. This book contains many practical examples and illustrations from case law. Extracts of the relevant legislation and sample shipping and marine insurance documents are annexed in this book for reference.

The book is intended primarily for students and teachers of transport studies and business logistics management. However, we trust that it will also provide useful guidance to traders, insurers, bankers, logistics managers and lawyers who need to acquire a clear understanding of the key principles in a practical context.

The division of labour associated with each of the chapters in this book is as follows: Chapters 1, 2, 5, 7, 8, 9 and 19 (Felix W H Chan); Chapters 6, 14, 15, 16, 17 and 18 (Jimmy J M Ng); and Chapters 3, 4, 9, 10, 11, 12 and 13 (Bobby K Y Wong).

We are extremely grateful to Ms Mary Thomson, Partner of Koo and Partners in association with Paul Hastings Janofsky & Walker, and the anonymous reader for their comments on draft chapters. The errors and omissions that remain despite their help are our own responsibilities. In addition, we are grateful to our students for their interest and enthusiasm in shipping and logistics law. Finally, we would like to record our heartfelt thanks to the Department of Shipping and Transport Logistics of the Hong Kong Polytechnic University and the Department of Professional Legal Education of the University of Hong Kong for their support and encouragement.

The law is stated as known to us on 2 January 2002.

Felix W H Chan
Jimmy J M Ng
Bobby K Y Wong
January 2002

Acknowledgements

The authors wish to acknowledge with thanks the permission given by the following organizations to reprint material from the copyright sources indicated below:

Exits, Inc.
Sample Bill of Lading

International Chamber of Commerce
ICC Uniform Customs and Practice for Documentary Credits (UCP 600)
UNCTAD/ICC Rules for Multimodal Transport Documents
ICC Uniform Rules for a Combined Transport Document

Hong Kong Association of Freight Forwarding Agents Ltd.
HKFFA Standard Trading Conditions

The Government of Hong Kong Special Administrative Region
Bills of Lading and Analogous Shipping Document Ordinance
Carriage of Goods by Sea Ordinance
Carriage by Air Ordinance
High Court Ordinance
Marine Insurance Ordinance

Witherby & Co. Ltd.
Institute Time Clauses Hulls
Institute Cargo Clauses (A)
Institute Cargo Clauses (B)
Institute Cargo Clauses (C)

Table of Cases

Table of Statutes

STATUTORY INSTRUMENTS

OVERSEAS STATUTES

Australia

China

United Kingdom

International Conventions and Other Formulations of International Trade Law

PART I

Introduction

1

Introduction to the Hong Kong Shipping and Logistics Industry

1.1 BIRTH AND IMPORTANCE OF THE MARITIME CLUSTER IN HONG KONG

Hong Kong was surrounded by inhospitable and rocky terrain when British shipowners and traders arrived in the 1830s. The port of Hong Kong was used as a base for transshipment of goods traded in the East.[1] The Port of Hong Kong grew only because of the higher rate of utilization of the port by the shipping industry. The phenomenal rise of the Hong Kong shipping community was closely connected to the successful rehabilitation of Japan's economy after the Treaty of San Francisco in 1952 and the Korean War.[2]

The maritime cluster of Hong Kong was gradually established and has been widely recognized as an international maritime centre including ship management, ship owning, ship broking, ship registration, ship surveying services, maritime legal services, maritime arbitration, ship finance, marine insurance and maritime education.

The maritime industry directly contributed 1.9 per cent (equivalent to HK$30 billion or US$3.9 billion) to the GDP of Hong Kong and 1.7 per cent (equivalent to 60,800 jobs) of total employment in Hong Kong in 2010. About 700 shipping-related companies operate in Hong Kong providing a comprehensive range of maritime services. The port sector with the maritime industry accounts for 24 per cent (HK$373 billion or US$47.8 billion)

1. S Zarach, *Changing Places: The Remarkable Story of the Hong Kong Shipowners* (Hong Kong Shipowners Association 2007) 53.
2. ibid, 84.

of the GDP of Hong Kong and 23 per cent (788,000 jobs) of total employment in Hong Kong.[3]

According to the recent Consultancy Study on Enhancing Hong Kong's Position as an International Maritime Centre,[4] Hong Kong has advantages in soft power. To strengthen the institutional structure and dedicate human resources that proactively drive industrial development, the study proposed setting up a new statutory maritime body to propel the long-term development of the maritime industry in Hong Kong, including undertaking policy research, supporting manpower training, and conducting marketing and promotion regionally and internationally.[5]

1.2 LINER AND TRAMP SHIPPING

International liner shipping is a sophisticated network of regularly scheduled services of shipping that transports goods all over the world at lower cost and with greater energy efficiency than any other form of international transportation.[6] A deep sea tramp ship is prepared to carry any cargo to any port at any time, always providing that the venture is both legal and safe.[7] A common misconception in Hong Kong is that tramp shipping is not as important as liner shipping in which container cargo usually catches the attention of the general public on streets.

The United Nations Conference on Trade and Development (UNCTAD) divides all international seaborne trade in four categories, namely, crude oil and products which are in wet bulk, five major bulk (dry bulk), container (not in bulk) and other dry bulk. Wet and dry bulk cargo covers more than 75 per cent of the worldwide sea freight.[8] The majority of shipowners in Hong Kong operate or control dry or wet bulk carriers. A similar situation is observed in the shipping industry in China. Tramp shipping which is primarily involved in dry or wet bulk cargoes is equally important, if not more important, to the development of the shipping industry in Hong Kong.

3. Transport and Housing Bureau, HKSAR, *Key Messages and Questions & Answers of Hong Kong Shipping Delegation to Korea* (Transport and Housing Bureau, HKSAR 2011).
4. Transport and Housing Bureau, HKSAR, *Consultancy Study on Enhancing Hong Kong's Position as an International Maritime Centre* (Transport and Housing Bureau, HKSAR 2014).
5. Transport and Housing Bureau, HKSAR, *Press Release: Secretary of Transport and Housing Responds to 2014 Xinhua-Baltic Exchange International Shipping Centre Development Index Report*, 27 June 2014.
6. World Shipping Council <www.worldshipping.org> accessed 10 July 2014.
7. Institute of Chartered Shipbrokers, *Economics of Sea Transport and International Trade* (2010/2011 edn, Witherby Seamanship International Ltd) 85.
8. United Nations Conference on Trade and Development, *Review of Maritime Transport 2010* (UNCTAD 2010) 9.

1.3 SHIPPING INDUSTRY OF HONG KONG FACES COMPETITION

Shipowners in Hong Kong

The Hong Kong Shipowners Association (HKSOA) was incorporated in 1957 by 11 local shipowners with the purpose of creating a forum for shipowners resident in Hong Kong. Over the past 55 years, the Association has grown into one of the world's largest shipowner associations, its members owning, managing and operating a fleet with a combined carrying capacity of over 133 million deadweight tonnes. The HKSOA recognizes that an international shipping hub must provide the necessary ancillary services such as shipowners, operators and managers, ship finance banks, maritime lawyers, insurers and P&I Clubs, shipbrokers, classification societies and even journalists.[9] Hong Kong shipowners compete in the global shipping platform where competition is intense.

The Hong Kong Shipping Register

The Hong Kong Shipping Register (HKSR) has 88 million Gross Tonnage as of 23 May 2014. The Register is operated by the government of the HKSAR through the Marine Department which has over 150 years of experience in ship registration, inspection and survey. An international or local shipowner can register a ship with the HKSR subject to international standards for safety and protection of the marine environment. The benefits are that Hong Kong is one of the lowest tax regimes in the world with no profits' tax levied on overseas trade and double taxation relief arrangement with major trading partners; no nationality restrictions on manning; a clean, efficient and business friendly civil service; excellent ship management, financial, communication, legal and other support facilities; an independent, well-established common law system; and an important gateway to the mainland of China.[10]

Cargo throughput in the port of Hong Kong

The drop of 5.2 per cent (24.38 million Twenty-foot Equivalent Unit [TEU] in 2011 and 23.12 million TEU in 2012) in the container throughput in the port of Hong Kong in 2012 was unexpected under the negative impact of the

9. Hong Kong Shipowners Association, *Our History* (Hong Kong Shipowners Association 2011) <www.hksoa.org> accessed 17 July 2014.
10. Marine Department, HKSAR, *The Hong Kong Shipping Register* (Marine Department, HKSAR 2011) <www.mardep.gov.hk> accessed 17 July 2014.

global financial crisis and other factors. The port of Hong Kong was ranked as the third busiest container port in the world in container throughput in 2012.[11]

Freight forwarding

A freight forwarder is a company which specializes in providing a range of services to shippers including sea, air or land transport, consolidation, storage, inventory control, warehousing and physical distribution. Hongkong Association of Freight Forwarding and Logistics Limited (HAFFA) has several hundred members who move sea and/or sea freight in the region. HAFFA is responsible for setting industry standards and providing educational courses and business development programmes which enhance the professional levels of freight forwarders and logistics service providers in Hong Kong.[12] They usually avoid investment in assets like ships, aircraft or warehouses in order to maintain their flexibility in operation.

Maritime arbitration

Whether a place can be the maritime arbitration centre in a region largely depends on the preference and confidence of the contracting parties in the shipping business on that place. It is established through a long history of maritime arbitration activities in that place. Hong Kong Maritime Arbitration Group (HKMAG) is a division of Hong Kong International Arbitration Centre (HKIAC). Since 1985, HKIAC has processed a total of 645 maritime disputes.[13] In 2010, there are 131 appointments for the arbitrators listed in the HKMAG.[14] Those arbitrators in the list are experienced in shipping and are expert in handling maritime disputes. In recent years, Hong Kong has developed into a regional leader of maritime arbitration in the Asia-Pacific region. Singapore is another place which provides maritime arbitration in the region. From 2009, Singapore Chamber of Maritime Arbitration has handled 49 maritime cases.[15]

11. Marine Department, HKSAR, *Port of Hong Kong in Figures 2013 Edition* (Marine Department, HKSAR 2011).
12. Hongkong Association of Freight Forwarding and Logistics Limited (HAFFA) <www.haffa.com.hk> accessed 17 July 2014.
13. Source: HKIAC website and its Annual Reports.
14. The Speech of Arbitration in Hong Kong: Latest Trends and Developments made by Chiann Bao, Secretary-General of the HKIAC in British Chamber of Commerce in Hong Kong, 6 September 2011.
15. Statistics in the website of SCMA <www.scma.org.sg/pdf/casesummary.pdf> accessed 17 July 2014.

China is a major player in the international maritime business. Many international contracts that involve one party in Mainland China require arbitration outside of China. Administrative arbitration is more popular in Mainland China. In Hong Kong, both *ad hoc* arbitration and administrative arbitration are available in HKIAC.

In 2010, the Arbitration Institute of the Stockholm Chamber of Commerce surveyed the arbitration cost of major arbitration centres in the world for different levels of dispute value. The result shows that the average cost of arbitration in Hong Kong is quite competitive. The costs in the International Chamber of Commerce and Singapore International Arbitration Centre are the highest; it is followed by the Hong Kong International Arbitration Centre, Swiss Chambers of Commerce Association and Arbitration Institute of the Stockholm Chamber of Commerce. Compared with Hong Kong, the average arbitrators' fee in Singapore is higher, even with the inclusion of an administration fee in Hong Kong.[16]

16. M Luo and J Ng, 'Regional Competition for the International Shipping Center: The Development of Maritime Arbitration Center in Asia', *Journal of the Institute of Seatransport*, vol 100 (Winter 2012), 1–7.

2

Introduction to Contract Law, Agency Law and Negligence

2.1 INTRODUCTION

The law in relation to the shipping and logistics industry in Hong Kong is greatly based on contract law, agency law and negligence in tort law. Some of the contracts found in this industry are sale of goods contracts, sea carriage contracts, air carriage contracts, land carriage contracts, and marine insurance contracts among others. The relationship between a carrier and a shipper is contractual. Similarly, the relationship between a shipper and a consignee is also contractual. However, a shipper will usually form a carriage contract through the services provided by a freight forwarder. In this respect, the relationship between the shipper and the freight forwarder is based on an agency contract. Many cases of cargo damage or loss are due to the negligence of the carrier. Thus, many cargo claims are based on negligence of the tort law. Having basic knowledge of contract law, agency law and negligence in tort law is helpful for a better understanding of shipping and logistics law which will be discussed in the following chapters of this book.

2.2 CONTRACT LAW

A contract is a legally enforceable agreement between two or more parties. The court will enforce a contract or award remedies to the innocent party if a contract is breached. A contract may be formed orally or in writing, unless the law requires it to be in written form only. For a commercial contract, it is advisable to be in written form in order to have good evidence of its existence and the terms; for example, s 4 of the *Conveyancing and Property Ordinance* requires a transfer of the ownership of land to be in the form of a formal written contract which is known as a deed.

Three elements are required to form a contract:

(1) intention to create legal relations;
(2) agreement; and
(3) consideration.

Intention to create legal relations

To form a contract, the parties must have an intention to create legal relations at the time the agreement is formed. It is rare for them to explicitly declare whether they intend to create legal relations. To decide the issue, the law has developed two presumptions that apply unless the parties have indicated a contrary intention. Different presumptions apply to commercial and non-commercial (domestic or social) agreements.

For commercial agreements, the presumption is that the parties have an intention to create legal relations. This is a very strong presumption.

> *Esso Petroleum v Customs and Excise Commissioners*[1]
>
> E launched a sales promotion campaign linked to the 1970 World Cup. They produced 'coins' bearing the likenesses of the England squad. They advertised 'one coin given with every four gallons of petrol'. The ultimate issue for the court to decide was whether the 'coins' were subject to purchase tax. In the course of the hearing, E argued that the advertisement of the coins was not intended to create legal relations. The coins were of little intrinsic value. It was also unlikely that a customer would sue if E refused to give him the coins.
>
> However, the court held that E had the intention to create legal relations with its customers. The above-mentioned factors were irrelevant.

But if strong evidence shows the parties do not intend to be legally bound, the presumption may be rebutted. The best evidence is the agreement itself.

> *Rose and Frank Co v Crompton Bros Ltd*[2]
>
> D was a manufacturer of tissues for carbonising papers. By an agreement, D granted P the rights to distribute its products. The agreement

1. [1976] 1 All ER 117.
2. [1925] AC 445.

contained this 'Honourable Pledge Clause': 'This arrangement is not entered into nor is this memorandum written, as a formal or legal agreement, and shall not be subject to legal jurisdiction in the law courts either of the United States or England, but it is only a definite expression and record of the purpose and intention of the parties concerned, to which they each honourably pledge themselves.'

In deciding a dispute between P and D, the court held, among other things, that the terms of the agreement made it beyond doubt that the parties had no intention to create legal relations. P's claim under the agreement therefore failed.

On the other hand, the presumption for non-commercial agreements is that the parties have no intention to form legal relations.

Balfour v Balfour[3]
D was ordered to station in Ceylon by the UK government. Before he went abroad, it became clear that his wife, P, could not go with him due to poor health. D agreed to pay P a monthly maintenance during the time they lived apart. D broke his promise to pay. P sued D for the maintenance. It was held that the parties had not intended to create legal relations.

It is also a very strong presumption and may only be rebutted by strong evidence showing the parties' intention to create a legally binding agreement. One example of such evidence is an explicit statement in a written agreement that the parties intend the agreement to be legally enforceable.

Agreement

A contract is first an agreement. There can be no contract without agreement. In determining whether an agreement has been formed, the 'offer and acceptance' analysis is a useful tool. But it should not be viewed as the only way of determining whether an agreement has been formed. For example, where there is a formal written agreement, there is generally no need to ask who made the offer and who accepted it unless it is necessary to know when or where the contract was made.

Offer

Article 14(1) of the *United Nations Convention on Contracts for the International Sale of Goods 1980* defines offer as:

3. [1919] 2 KB 571.

> a proposal for concluding a contract addressed to one or more specific persons . . . [which] is sufficiently definite and indicates the intention of the offeror to be bound in case of acceptance

A party who makes an offer is the offeror. An offeree is a party to whom the offer is made.

It is important to distinguish 'offers' from mere indications a party wishes to negotiate. Such an indication is called an 'invitation to treat'. An invitation to treat is not an offer. Similarly, the mere supply of information is, of itself, not an offer. Sometimes it may not be easy to decide whether an offer has been made. Over the years, common law guidelines and statutory rules have been developed for certain frequently encountered situations. For example:

- A display of goods for sale is only an invitation to treat.[4]
- Putting goods up for sale by auction is not an offer.[5]
- An advertisement is an invitation to treat.[6]

The common law guidelines must not be over-generalized and treated as strict rules. For example, an advertisement may in certain circumstances constitute an offer.

> *Carlill v Carbolic Smoke Ball*[7]
>
> D, the manufacturer of 'The Carbolic Smoke Ball', advertised that the smoke ball could prevent people from catching influenza. It was said in the advertisement that if any purchaser had used the smoke ball in accordance with the instructions for a certain period but nevertheless succumbed to influenza, D would pay him £100. D also stated that £1,000 had been deposited with a bank for the purpose. Relying on the advertisement, P had bought the product and used it as instructed by the manufacturer. P caught influenza. P sued D for the £100. D argued, among other things, that the advertisement was 'mere puff' and that there was no offer to any particular person.
>
> The court held that the advertisement amounted to an offer which P accepted by conduct. P was entitled to the £100.

An offer may be made to one person, a group of people or the world at large. Only the offeree may accept the offer.

---

4. Section 40 of the Sale of Goods Ordinance.
5. *Pharmaceutical Society v Boots* [1952] 2 QB 795; affirmed [1953] 1 QB 401.
6. *Partridge v Crittenden* [1968] 1 WLR 1204.
7. [1893] 1 QB 256.

Boulton v Jones[8]

D had been accustomed to dealing with X, who supplied goods to D. D sent an order to X for some goods. Unknown to D, X had sold the business to P. P delivered the goods to D. D used the goods in the mistaken belief that they had been supplied by X. When P billed D, D refused to pay, arguing that he had intended to do business with only X.

The court held that D was not liable for the price of the goods.

An offer is not effective until it is communicated to the offeree. It may end in many ways:

- lapse of time (specified or 'reasonable');
- revocation (withdrawal);
- rejection (which terminates the offer);
- counter-offer (which has the same effect as a rejection);
- acceptance (a contract is formed).

If an offer states the time it will end, it ends at the specified time. If no time is mentioned, it ends after a reasonable time. Besides, the offeror may revoke it at will. After it is revoked, it can no longer be accepted. But revocation is not allowed if the offer has been accepted. The offeree is, on the other hand, under no obligation to accept the offer. It ends when the offeree rejects it. Offerees can no longer accept it even if they later change their minds. Often, the offeree may, based on the terms of the offer, make a different proposal by introducing some new terms. This proposal is a counter-offer which ends the original offer. If the offeree accepts the offer (or the offeror accepts the counter-offer), a contract is formed and the offer (or counter-offer) ends. The offeree can accept an offer only once. Otherwise, the offeror who makes only one offer might be bound by many contracts.

Acceptance

The offeree accepts the offer when expressing unqualified assent to all the terms of it. In general, acceptance is not effective until it is communicated to the offeror. As the maker of the offer, the offeror may specify a particular method of acceptance. The circumstances of the case and commercial reality determine what may be acceptable means of communication.

8. (1857) 27 LJ Ex 117.

> *Yung Zeng Industrial Co (HK) Ltd v Zhong Shan Native Produce & Animal By-products Co Ltd*[9]
>
> On 30 September, D sent P a fax containing an offer to sell oil. The offer remained open until 4 October. On 1 October, D made a second offer, in a signed form of contract, to P. The content of the second offer was the same as the first. The second offer contained this statement: 'Kindly sign and return one original of the sales contract to us immediately upon the receipt hereof.' P told D that it accepted the offer in subsequent telephone conversation. P signed the form but did not fax it back to D until 27 October, arguing that it had not faxed it back earlier because it was pointless to do so since there was no 'original'. P later claimed against D under the contract.
>
> The court held that P's claim failed. The circumstances of the case and modern commercial reality determined how the words 'sign and return original . . . immediately' were construed. The fax must be an acceptable means of acceptance, as implied by the fluctuating market, the urgency and the fact that D elected to use the fax in making the offer. The word 'original' was therefore superfluous. However, the words 'sign and return immediately' could not be ignored. Prompt written acceptance was required and oral acceptance excluded. There was no communication of written acceptance from P until 27 October. By that time the offer had lapsed.

If no means of communication is specified, the communication may be by any reasonable means. For example, acceptance by conduct is effective when it comes to the attention of the offeror.

The requirement for communication of the acceptance may be waived by the offeror. If the offer expressly or implicitly allows acceptance by performance, the acceptance needs not be communicated. The offeree may instead accept by actual performance of the contract; see *Carlill v Carbolic Smoke Ball* above.

The postal rule is another exception to the general rule that an acceptance must be communicated. If the postal rule applies, acceptance takes effect when the letter is put in the control of the post office, ie, when it is posted. The rule does not apply in every acceptance by post. There are two conditions for it to apply:

(1) It is reasonable to accept by post, and
(2) it is not required that acceptance must be communicated to the offeror.

It is not reasonable to accept by post when the offer requires the offeror's personal acknowledgment of the acceptance or when it would lead to great

9. [1997] HKLRD J19.

inconvenience; see, for example, *Holwell Securities v Hughes*.[10] In this case, the postal rule did not apply because the offer was required 'to be exercisable by notice in writing to the [offeror]'. Where the offeror does not specify how the acceptance is to be communicated, it is generally reasonable to accept by post if the offer is made in a letter posted to the offeree.

The postal rule also applies to telegrams. It does not apply to instantaneous means of communication, such as telex or facsimile:[11]

Consideration

Consideration is the 'price for which the promise [or act] of the other is bought'.[12] It must be 'sufficient at law'. In other words, it must be 'something of value' as viewed by the law. It may be in the form of rights or benefit accruing to the promisor, or detriment suffered or undertaking given by the promise.[13] In general, anything tangible or capable of being expressed in terms of money will do. Emotional benefits such as 'natural love and affection' or 'promise of refraining from boring you' are, however, not sufficient.[14]

If something is sufficient for consideration, its value is irrelevant. For example, $1 or a peppercorn may be good consideration for a house or a car.[15]

Past consideration

Past consideration refers to acts done by one party before the promise or act of the other party (which is 'to be bought') is made. The general rule is that past consideration is not good consideration. It cannot support a contract since the element of bargain is missing.

> *Re McArdle*[16]
> A and B were entitled to their family house after the mother's death. During their mother's life, A made various improvements to the house. At a later date, B wrote to A, stating that 'in consideration of your carrying out certain alteration and improvement to the property', A would be given the amount incurred for the works. An agreement was formed. When B refused to pay, A sued B for the amount.

10. [1974] 1 WLR 155.
11. *Susanto Wing Sun and Yung Chi Hardware* (1989) HKLY 134.
12. *Dunlop Pneumatic Tyre Co Ltd v Selfridge & Co Ltd* [1915] AC 847.
13. (1875) LR 10 Ex 153.
14. *White v Bluett* (1853) 23 LJ Ex 36.
15. *Thomas v Thomas* (1842) 2 QB 851.
16. [1951] Ch 669.

> The court held that since the improvements had all been done before the agreement was concluded, this was a case of past consideration and as a result, no contract had been formed.

There is one exception. If X did something upon the request of Y and the understanding was that Y would pay X in return, the service provided by X is good consideration for Y's promise (made after X's service was provided) to pay for X's service.[17] The rationale is the implied promise to pay when Y made the request. The later promise merely fixes the price to be paid.

Terms of contract

During the course of negotiation the parties may make many statements. A statement made before the contract was formed may or may not have legal effect. It may be 'mere puff' which has no legal effect. If it is a term or misrepresentation or both, it carries certain legal consequences. The law relating to terms is considered first.

The terms of a contract set the limits of the contractual rights and obligations of the parties. There are two types of terms: express and implied terms.

Express terms

Express terms are terms the parties expressly agree on either orally or in writing.

The test for deciding whether or not a statement is a term, is the intention of the parties at the time the contract was formed as evidenced by their conduct. All the relevant circumstances should be considered to determine the intent of the parties. Common factors to consider are, the importance of the statement, the time at which the statement was made, failure to include it in the written agreement, the expertise of the statement maker, and other special circumstances.

A statement that is important to one or both parties is likely to be a term.

> *Bannerman v White*[18]
>
> During negotiation, the buyer of hops asked the seller whether sulphur had been used in the treatment of the hops. He made it clear that, if it had, he would not even bother to ask the price. The seller replied that it had not. It was discovered that sulphur had in fact been used. The buyer refused to pay the price.

17. *Lampleigh v Braithwait* (1615) Hob 105.
18. (1861) 10 CB(NS) 844.

The court held that the buyer had placed great importance on the absence of sulphur, and the seller had known that. The seller's reply was a term of the contract. In the circumstances, the buyer was entitled to refuse payment.

A statement made or repeated immediately before the contract was formed is more likely to be a term, compared with one made early in the negotiations and not repeated later.

Routledge v McKay[19]
On 23 October, the seller of a car told the buyer that the car was a 1942 model. On 30 October, the contract was formed, without referring to the date of the model. The car was in fact a 1930 model. The buyer sued the seller for breach of contract.

The court held that, since the statement had been made several days before the contract was concluded and had never been mentioned again, the statement about the car's year of manufacture was not a term of the contract.

Where there is a written agreement, a statement that is not included in it is less likely to be a term.

If one party is an expert and the other party is not, what the expert said is more likely to be a term of the contract. Contrast the following two cases.

Dick Bentley Productions Ltd v Harold Smith Motors Ltd[20]
P went to D's shop and told D that he wanted to buy a 'well-vetted' car. D showed P a car, saying that it had done only 20,000 miles since the engine and gearbox had been replaced. P tried the car and bought it. The statement about mileage was untrue.

The court held that the statement was a term of the contract.

Oscar Chess Ltd v Williams[21]
D wanted to buy a car from P, a car dealer. D would trade in his old car for part of the price. The registration book indicated that the date of the old car was 1948. D confirmed that the information was true. A trade-in price was agreed. The car was in fact a 1939 model with a much lower trade-in price. P claimed against D for the difference between the trade-in prices.

The court held that the statement about the age of the car was not a term. One relevant factor was that D was a car dealer who should have had special knowledge and skill.

19. [1954] 1 WLR 615.
20. [1965] 1 WLR 623.
21. [1957] 1 WLR 370.

Any special circumstances of the case should always be considered. Often, they are decisive. A comparison of the following cases makes the point clear.

> *Schawel v Reade*[22]
>
> D advertised the sale of a horse. P wanted to buy it. When P started to examine the horse, D interrupted and said 'You need not look for anything: the horse is perfectly sound'. P then stopped the examination. P bought the horse from D. The horse was in fact unsound. P sued D for breach of contract.
>
> The court held that D's statement as to the soundness of the horse was a term of the contract.
>
> *Hopkins v Tanguerary*[23]
>
> D was offering his horse for sale by auction and gave an assurance as to the perfect soundness of the horse on the day before the auction. The auction was held at a place where, by custom, all sales were without warranty. P bought the horse. The horse was in fact unsound. P sued D for breach of contract.
>
> The court held that the assurance was not a term of the contract.

Implied terms

Implied terms are terms on which the parties do not explicitly agree. Terms may be implied 'in fact' (ie, by the facts of the case), by custom or by law.

The test for implying terms 'in fact' is strict. For a term to be implied in this way, it must be:

- reasonable and equitable;
- necessary to give business efficacy to the contract;
- so obvious that it 'goes without saying';
- capable of clear expression; and
- not contradictory to any express term of the contract.

> *Gardner v Coutts & Co*[24]
>
> A landlord sold a piece of land to a purchaser. The landlord undertook that the purchaser had 'the right of first refusal' if the adjoining land was sold. The landlord later disposed of the adjoining land by way of gift.
>
> The court held that a term prohibiting the landlord from doing so was implied 'in fact'.

22. [1913] 2 IR 64.
23. (1854) 15 CB 130.
24. [1968] 1 WLR 173.

A contract is deemed to incorporate the relevant local custom, market practice or trade usage. Terms implied in this way may govern matters such as delivery time or time and method of payment. They are subject to terms agreed by the parties.

Terms implied by law are usually found in the relevant legislation. For example, the Sale of Goods Ordinance (Cap 26) implies certain terms in every sale of goods contract. By virtue of the Ordinance, the seller impliedly guarantees the right to sell the goods, the goods are of merchantable quality and reasonably fit for the purpose known to the seller, and they correspond with their description. Terms may also be implied by the common law. For example, it was held in *Liverpool City Council v Irwin*[25] that, in a landlord and tenant agreement, the landlord undertook impliedly to maintain the common parts, such as the stairs and lift, in a state of reasonable repair.

Relative importance of terms

Terms of contracts may be classified as conditions, warranties or innominate (or intermediate) terms. This classification is necessary to determine what kinds of remedies are available when a term is breached. In the past, a term could only be a condition or a warranty. A third type of term, innominate, has recently been recognized.

Every term of a contract is important in the sense that remedies are available to the innocent party if any term is breached. However, some terms are more important than others. Such terms are called conditions. Breach of a condition entitles the innocent party to discharge the contract and sue for damages. That party may, however, treat the contract as still binding and sue for damages. Terms of comparatively less importance are called warranties. For breach of a warranty, the innocent party is entitled only to damages. The contract is still valid and the parties are required to perform. Innominate terms refer to those terms that may be breached in many ways, some serious and some trivial. What the innocent party may do depends on the consequences of the breach. If it is serious, the breach is treated as a breach of condition and the innocent party may terminate the contract. If it is not serious, the breach is regarded as a breach of warranty and that party may only sue for damages. Innominate terms were first recognized in *Hong Kong Fir Shipping Co Ltd v Kawasaki Kisen Kaisha*.[26] It is recognized in this case that the carrier's undertaking to provide a seaworthy ship cannot be simply classified as a condition or a warranty, because 'seaworthiness' is a very

25. [1977] AC 239.
26. [1962] 2 QB 26.

complex concept. Some breaches of the undertaking give rise to an event that will deprive the innocent party of 'substantially the whole benefit which it was intended that he should obtain from the contract'. But breach of the undertaking may be very trivial. The nature of the breach must, therefore, be taken into account in deciding the remedies available to the innocent party. The court will look at the contract as a whole to determine whether a term is a condition, warranty or innominate term. The label the parties give to it is not decisive.

Wickman Machine Tool Sales v L Schuler AG[27]
In a four-year distributorship agreement, it was a 'condition' that the distributor should visit six named customers once a week. In a total of about 1,400 visits, the distributor had breached the obligation once. The manufacturer argued that the breach of the condition terminated the contract.

The court held that the contract was not terminated by a mere breach of the term. The parties could not have contemplated that one breach out of 1,400 visits would have such drastic consequences.

City Famous Ltd v Profile Property Ltd[28]
P bought a number of shops from X. P sold some of them to D. One clause of the contract between P and D provided that P 'shall not without the prior written consent of [D] make or agree to any modification or amendment of the [agreement between P and X] or agree to waive or to release any of the obligations of [X] under the [agreement between P and X]'. P and X entered into a supplementary agreement to change the date of the first instalment without D's consent. The change was in fact trivial. D claimed that the clause was a condition and the contract was repudiated by P's breach.

The court held that the contract was not terminated. D was given wide power under the clause to direct P in dealing with X in different situations. The wide scope of the clause meant that P and D could not intend that the contract should be terminated by a trivial breach. A clause which might be breached in many, some fundamental and some trivial, ways was most likely to be classified as innominate.

Exemption clauses

Exemption clauses are terms limiting the liability of a party when in breach of the contract. If the parties are of equal bargaining power, none of them can force the other to accept unfair terms. In such a case, exemption clauses

27. [1974] AC 235.
28. [1999] 3 HKLRD 15.

create no problem and would be regarded as valid. In consumer situations, exemption clauses may, however, cause hardships to the consumer, who often has no other alternative but to accept the standard terms of the seller or service provider. To protect the interests of consumers, the law limits the operation of exemption clauses.

The first point to note is that, for an exemption clause to have any effect, it must be incorporated into the contract.

> *Thornton v Shoe Lane Parking*[29]
>
> P was a driver and had not been in D's automatic car park before. Outside the park, there was a notice containing this statement: 'All cars parked at owners' risk'. Nobody was in attendance at the entrance. P put coins into the slot of a machine. When sufficient money was given, a ticket was pushed out from the machine and the plaintiff could then enter the car park. The notice contained some words but P did not read them. The words stated that the ticket was issued subject to terms and conditions displayed inside the car park. The terms and conditions could only be seen after P entered the car park. One condition purported to exempt D from liability for injury to customers. P was injured inside the park and sued D for breach of contract. D tried to rely on the exemption clause.
>
> The court held there was an offer when D held itself out as being ready to receive money through the machine. P accepted the offer by putting his money into the slot. The ticket was pushed out after the contract had been formed. The contents of the notice outside the car park were terms of the contract. But the exemption clause displayed inside the car park was not a term of the contract. Thus, D could not rely on it.

If the exemption clause is a term of the contract, its validity becomes the issue. The law is governed by the Control of Exemption Clauses Ordinance ('the CECO'). It applies generally to business liability. Under the CECO, certain types of exemption clauses are declared invalid. Some other types are subject to the reasonableness test.

By virtue of the CECO, the following liabilities cannot be excluded or limited:

- terms which exclude or limit liability in respect of personal injury or death resulting from negligence;
- terms which exclude or limit liability for breach of the obligations arising from section 14 of the Sale of Goods Ordinance (Cap 26) (seller's right to sell the goods);

29. [1971] 2 QB 163.

- terms which exclude or restrict liability for loss or damage arising from defects in goods (of a type ordinarily supplied for private use or consumption) resulting from the negligence of a person concerned in the manufacture or distribution of the goods.

The CECO gives the strongest protection to consumers. A person 'deals as consumer' if the following three conditions are satisfied:

- the contract is neither made in the course of a business nor held out as being done so;
- the other party makes the contract in the course of a business; and
- if it is a sale of goods contract, the goods sold are of a type ordinarily supplied for private use or consumption.

Against a consumer, an exemption clause may be invalid or subject to the reasonableness test.

Terms excluding or limiting liability for breach of the obligations arising from section 15, 16 or 17 of the Sale of Goods Ordinance, with respect to the seller's implied undertaking as to conformity of the goods with their description or sample, or as to the quality or fitness of the goods for particular purposes made known to the seller, are invalid.

The following types of exemption clauses are subject to the reasonableness test when against a consumer:

- terms which exclude or limit the other party's liability in respect of any breach;
- terms which allow the other party to render a contractual performance substantially different from what was reasonably expected of him;
- terms which allow the other party to render no performance at all;
- terms which require a consumer to indemnify another person in respect of liability that may be incurred by the other for negligence or breach of contract.

The test of reasonableness is also applied in some other situations. The following terms are subject to the test when against a person dealing on the standard terms of the other party:

- terms which exclude or limit the other party's liability in respect of any breach;
- terms which allow the other party to render a contractual performance substantially different from what was reasonably expected of him;
- terms which allow the other party to render no performance at all.

Further, the following exemption clauses are subject to the reasonableness test in the prescribed situations:

- in any case, terms which exclude or limit liability for negligence causing loss or damage other than personal injury or death;
- against a person dealing otherwise than as consumer, terms which exclude or limit liability for breach of the obligations arising from section 15, 16 or 17 of the Sale of Goods Ordinance (seller's implied undertaking as to conformity of goods with description or sample, or as to their quality or fitness for a particular purpose).

The reasonableness test is objective. A term is reasonable if it is a fair and reasonable one having regard to all the circumstances which were, or ought reasonably to have been known to or in the contemplation of the parties when the contract was made.

In particular, with respect to contracts for the sale of goods, Schedule 2 to the CECO provides that, in determining the reasonableness of an exemption clause, the following matters should be considered:

- the strength of the bargaining positions of the parties relative to each other;
- any inducement received by the customer, or other opportunity of entering into a similar contract without such terms with another party;
- the knowledge that the customer had or should have reasonably had of the existence of the terms;
- Where liability is excluded or restricted if some conditions are not complied with, the reasonableness of the expectation that compliance with such conditions would be practicable;
- any special order of the customer in relation to the manufacturing of the goods.

Exemption clauses not mentioned in the CECO are not subject to the ordinance. For example, the validity of a term excluding a seller's liability for loss suffered by a buyer as a result of the improper use of the goods is determined by the common law.

Privity

The doctrine of privity of contract states that a person who is not a party to a contract cannot sue or be sued under it.

Dunlop Pneumatic Tyre Co Ltd v Selfridge & Co Ltd[30]

D was a motor manufacturer and sold tyres to X, who was a dealer. The contract stated that X must not sell the tyres below D's listed price. S bought tyres from X and sold them to customers below the listed price. D sought an injunction and sued S for damages for breach of the agreement between D and X.

The court held that the restriction was valid against only X. D could not enforce the contract against S for want of privity between D and S.

Some exceptions to the rule of privity are:

- Agency—an agent is entitled to make contracts on behalf of his principal. Such contracts are binding upon principals even though they do not sign the agreement.
- Assignment—(1) by operation of law—on a person's death, their rights and liabilities under contracts not of a personal nature devolve upon the person's personal representative; (2) by agreement—in equity, it is possible for B to assign to C certain rights or benefits under a contract between A and B which C may then enforce against A. But no burdens or obligations may be assigned without A's consent.
- Collateral contract—the third party may be liable under another contract collateral to the main contract. For example, A, a car dealer, says to B, a customer, that 'It is a good car. I'd stake my life on it'. Encouraged by A's statement, B enters into a hire-purchase arrangement, by which (a) A sells the car to a bank, and (b) the bank contracts with B that B is allowed to use the car and the car will become B's property after B makes all the payments. On the face of it, there is no contract between A and B. But there is a collateral contract between them. The consideration is A's promise that the car is sound in return for B's promise to enter into the contract with the bank. Therefore, if the car is defective, B may sue A under the collateral contract.[31]

Vitiating factors

Vitiating factors are those elements which may affect the validity of a contract. The following are vitiating factors: misrepresentation, mistake, duress, undue influence, incapacity, illegality and unconscionability.

30. [1915] AC 847.
31. See *Brown v Sheen & Richmond Car Sales* [1950] 1 All ER 1102; *Andrews v Hopkinson* [1957] 1 QB 229.

Misrepresentation

A misrepresentation is an untrue statement of fact made by the representor to the representee inducing the representee to enter into a contract with the representor.

A statement as to intention may be a misrepresentation, if the statement maker does not, at the time of making it, have that intention. An opinion cannot be a misrepresentation, unless the person making it does not in fact hold the opinion or it cannot be held by a reasonable man.

> *Bissett v Wilkinson*[32]
> The vendor told the purchaser that the land could, in his judgment, carry 2,000 sheep. The land had never been used as a sheep farm.
> The court held that what the vendor said was an honest expression of opinion, not a representation of the actual capacity of the land.

> *Smith v Land and House Property Co*[33]
> Before the sale of a house, the vendor wrote to the purchaser that the property was being 'let to X (a most desirable tenant) . . . thus offering a first-class investment'. In fact X had long been in arrears with his rent. As to the rents that X had paid, the payment had been made by instalments under pressure from the vendor.
> The court held that the vendor's description of X as 'a most desirable tenant' was not opinion. It was an untrue assertion of fact that there had been no facts indicating that X was an undesirable tenant.

There is no misrepresentation if the representee is not induced by the untrue statement of fact. No one can be induced by a statement they do not know or know to be untrue. Thus, if the representee, at the time of making the contract, realizes the statement is untrue or is not aware of it, there is no misrepresentation.

There are three types of misrepresentation: fraudulent, negligent and innocent.

Where the representor makes the misrepresentation either (a) knowingly, (b) without belief in its truth, or (c) recklessly without caring whether it is true or false, it is a fraudulent misrepresentation.

> *Derry v Peek*[34]
> A company was set up to run trams by steam power. The consent of the Board of Trade was required for the business. The director, in the absence of any signs of objection, honestly believed that consent would

32. [1927] AC 177.
33. (1884) 28 ChD 7.
34. (1889) 14 App Cas 337.

be given. They issued shares to the public, stating that the company had the right to run steam-powered trams. The consent was never given and the company was ultimately wound up.

The court held that the directors had not made a fraudulent statement because they honestly believed that what they asserted was true.

A negligent misrepresentation is one made without the standard of care owed by the representor to the representee. If a reasonable man would not make the misrepresentation, the misrepresentation is a negligent misrepresentation. Negligent misrepresentation is first recognized in *Hedley Byrne v Heller*.[35]

If the misrepresentation is not fraudulently or negligently made, it is an innocent misrepresentation. In other words, it is an innocent misrepresentation if there are reasonable grounds for the representor to believe, at the time of making the statement, that it is true.

A misrepresentation renders the contract voidable at the option of the representee. If the representee chooses to rescind the contract, it is void *ab initio*, ie, as if the contract had never existed. Anything done under the contract is of no legal effect upon the rescission. However, rescission is not allowed in the following situations:

- Where the injured party affirms the contract, expressly or implicitly, with knowledge of the misrepresentation;[36]
- Where the representee does not seek to rescind within a reasonable time after knowing the statement was untrue;[37]
- Where the parties cannot be restored to their previous position;
- Where a third party would suffer if rescission is ordered, eg, the party has acquired in good faith and for value an interest in the subject matter;
- Where restitution is impossible, eg, the goods have been consumed or destroyed.

The innocent party may also sue for damages. For fraudulent misrepresentation, representees would be put in the position in which they would have been if the misrepresentation had never been made. In *Doyle v Olby (Ironmongers)*,[38] it was held that the representor was liable 'for all the actual damages directly flowing from the fraudulent inducement'. For negligent misrepresentation, the representor is only liable for losses which were reasonably foreseeable when the misrepresentation was made. In the case of

35. [1964] AC 465.
36. *Long v Lloyd* [1958] 2 All ER 402.
37. *Leaf v International Galleries* [1950] 2 KB 86.
38. [1969] 2 QB 158.

innocent misrepresentation, the innocent party is not entitled to damages at common law. But the court may order damages in lieu of rescission under the Misrepresentation Ordinance.

Mistake

As a general rule, a mistake does not affect the validity of a contract. We sometimes say that one party made a 'mistake' in a bad deal. This is, however, not an operative mistake which renders the contract void. Actually, only in very rare circumstances may a party plead an operative mistake at law. To be operative, the mistake must be one of fact. Mistake or ignorance of the law is not an operative mistake.

> *Citilite Properties Ltd v Innovative Development Co Ltd*[39]
> P bought a commercial property from D. In the contract, D warranted that the 'saleable area' was 7,864 square feet. The actual 'saleable area' was 6,596.5 square feet. D sought rectification of the clause, saying that it made a mistake of fact on the meanings of 'saleable area' and 'gross floor area'.
> The court held that the doctrine of rectification did not apply because the mistake was of law. D knew that the warranty was given for 'saleable area' but not 'gross area'. D made a mistake of law that the two terms referred to the same area.

There are three types of mistake: common, mutual and unilateral mistake.

The parties make a common mistake if they reach the agreement on the same mistake on certain fundamental facts. At common law, the only established operative common mistake is non-existence of the subject matter.

> *Couturier v Hastie*[40]
> A and B formed a contract for the sale of a cargo of corn. At the time the contract was formed, A and B thought that the cargo was being carried from port X to port Y. Unknown to them, the cargo had been sold by the captain because it had become fermented.
> The court held that the contract was void for common mistake.

If one party guarantees the existence of the subject matter, the situation is completely different.

39. [1998] 2 HKLRD 705.
40. (1856) 5 HL Cas 673.

McRae v Commonwealth Disposals Commission[41]

The defendant invited tenders 'for the purchase of an oil tanker lying on [X]'. The plaintiff bid and won the contract. There was in fact no such place known as X.

The court rejected the argument that the contract was void for mistake, holding that the defendant had implicitly warranted the existence of the tanker. Non-existence of it amounted to a breach of contract.

In principle, mistake as to the quality of the subject matter of the contract could be operative. But the test laid down in *Bell v Lever Brothers Ltd* [1932] AC 161 is extremely difficult to pass, although the possibility for a contract void for this kind of mistake is not completely ruled out.

Bell v Lever Brothers Ltd[42]

B was managing director of a subsidiary company of L. L requested B to resign since he had become redundant due to the amalgamation of the subsidiary with another company. The parties thought that B was, in accordance with the employment contract, entitled to compensation. L paid B a large sum as compensation. It was later found out that B had previously committed several serious breaches of duty. As a result, B had no right to compensation. Pleading common mistake, L claimed for the recovery of the compensation.

In delivering his judgment, Lord Atkin said that 'mistake as to quality of the thing contracted for raises more difficult questions. In such a case a mistake will not affect assent unless it is the mistake of both parties, and is as to the existence of some quality which makes the thing without the quality essentially different from the thing as it was believed to be.' On the facts, there was no operative mistake.

Where a contract is void for operative mistake at common law, what has been done under it is of no legal effect. This is important when a third party is involved. Take this example: X sold goods to Y. Y then resold them to T. If the contract between X and Y is void for operative mistake, T does not become the owner of the goods since Y never had the right to sell. X can take the goods back from T. If the contract is not void for mistake but is only voidable for misrepresentation, the court may not allow X to rescind the contract. T becomes the owner and X cannot take back the goods.

Equity generally follows common law. It is also very difficult to find an operative mistake in equity. But in equity, the test for an operative mistake is less stringent. It is an operative mistake in equity where A sells B's own

41. (1951) 84 CLR 377.
42. [1932] AC 161.

goods to B.[43] The court has the discretion to set aside the contract or order the parties to enter into a just and fair contract.

In case of mutual mistake, the two parties are at cross-purposes. For example, A owns two companies. A intends to sell one to B, while B wants to buy the other. An objective test is used to decide whether there is a contract and, if there is one, its terms. The questions to ask are:

(1) Would a reasonable man, after considering all the relevant circumstances, conclude the parties formed a contract?
(2) If the answer to the first question is 'yes', what are the terms of the contract as seen by the reasonable man?

In fact, the two questions are closely related. If a reasonable man concludes the parties did not form a contract or he cannot otherwise determine the terms, there is no contract.

> *Raffles v Wichelhaus*[44]
> A contract was made to sell a shipment of cotton 'ex Peerless from Bombay'. It happened that two 'Peerless' were to sail from Bombay (one departed in October and the other in December) with cotton as cargo on board. The buyer was sued for non-acceptance.
> The court held there was no contract because it was not possible to determine to which ship the contract referred.

In a case of unilateral mistake, one party has made a mistake which is known to the other. The test for an operative mistake is again objective. One common example of a unilateral mistake is 'document mistakenly signed'. The basic rule is that once a person signs a document, that person is bound by its terms. A party is bound even if that person has not read the document before signing. There is one exception to the rule. If an innocent party was induced by the fraud of a rogue to sign a document altogether different from what was believed by the innocent party to be signing, the innocent party may plead 'non est factum' (ie, not my deed). This defence is not available if the person was negligent in signing the document. If the defence is established, the document is void and the innocent party not bound by it.

The following pairs of documents are fundamentally different:

- a bill of exchange and guarantee;
- a guarantee of overdraft and a proposal for insurance;
- a mortgage agreement and a power of attorney.

43. *Cooper v Phibbs* (1867) LR 2 HL 149.
44. (1864) 2 H & C 906.

Duress and undue influence

The basic principle is that a party is bound by a contract only if entering into it wilfully. A contract a party formed under duress (ie, actual violence or threats of violence) is voidable at the option of that party.[45] Economic duress may also affect the validity of a contract. It refers to illegitimate threat of damage to a party's economic interest.[46] A threat to breach an existing contract with the same party may be economic duress. However, mere commercial pressure is not enough.

Undue influence is an equitable doctrine. In a case of undue influence, the innocent party is usually not under any duress. A party enters a contract under undue influence if that party has, at the time of forming the contract, not been allowed to exercise a free and deliberate judgment on the matter. A contract tainted by undue influence may be set aside by the innocent party. Undue influence usually arises in cases where one party owes the other a fiduciary duty. Examples of relations giving rise to a fiduciary duty are:

- lawyer and client
- doctor and patient
- trustee and beneficiary

However, there can be undue influence without a fiduciary relationship.

> *Lloyds Bank v Bundy*[47]
> D was an old farmer. He owned a farmhouse, the value of which was about £10,000. Initially, D offered the farmhouse as security to guarantee his son's overdraft (up to £7,500) to a bank. D, his son and the son's business all had their own accounts with the bank. The son was in financial difficulty. He told the bank that D would help. The bank's manager came to D asking for an increase of the guarantee to £11,000. He said that the bank would continue to support the son only if D executed the guarantee. D agreed to increase the guarantee. Later, the bank wanted to sell the farmhouse under the guarantee.
>
> The court set aside the guarantee. D looked to the bank for financial advice and placed confidence in it. The bank could not give independent advice to D because it had interests in the matter. D entered into the contract under undue influence.

---

45. *Barton v Armstrong* [1976] AC 104.
46. *B & S Contracts & Design Ltd v Victor Green Publications Ltd* [1984] ICR 419.
47. [1975] QB 326.

In *Diner's Club International v Ng Chi-sing*,[48] it was held that improper pressure put on a father by an officer of a financial company could amount to undue influence. The improper pressure in this case is the threat to report the son's wrongdoing, which would lead to lawful arrest.

In the absence of a fiduciary relationship, a party can avoid a contract on the ground of undue influence only by showing:

- the contract was manifestly disadvantageous; and
- the other party took unfair advantage.

A contract formed under undue influence can be affirmed by the influenced party after becoming aware of the undue influence. Once affirmed, the contract cannot be avoided.

Incapacity

The validity of a contract may be affected by the minority, drunkenness or mental state of a party.

The age of majority is 18. A person under 18 is a minor. A contract made by a minor may be valid, voidable or unenforceable.

An executed contract for necessaries is valid. Necessaries are 'goods suitable to the condition in life of the minor and to his actual requirements at the time of sale and delivery'. Whether the goods are necessaries is a mixed question of law and fact. Food, clothing, lodging, medicines and means of transport may be necessaries. Although the contract is valid, the minor may not be required to pay the contract price. If the contract price is too high, the minor is only required to pay a reasonable price.

Contracts for the benefit of minors are also valid. Examples are contracts of service, apprenticeship and education. The contract must, as a whole, be beneficial to the minor.

> *De Francesco v Barnum*[49]
> D was a girl of 14. She concluded an apprenticeship contract with P for seven years to be a dancer. D agreed that she would neither marry during the apprenticeship nor accept professional engagements without P's permission. P would not provide D with engagements or maintain her while unemployed. P would pay D a small amount when she was in employment. P would be entitled to terminate the contract if, after a fair trial, D was found unfit to be a dancer.

48. [1987] 1 HKC 78.
49. (1890) 45 ChD 430.

> The court held that the contract as a whole was unreasonable and unenforceable. D was at the absolute disposal of P. She received no pay and no maintenance except when employed. P had no obligation to find employment for her. P alone could terminate the contract and thus destroy D's chance of success.

A contract from which a minor acquires an interest in a subject matter of a permanent nature is valid, unless the minor repudiates it during the minority period or within a reasonable time after the attainment to the majority. Examples of such contracts are leases of land and agreements for the purchase of shares.

No other contracts are enforceable against a minor. However, a contract becomes valid if the minor ratifies it upon reaching the majority.

For a contract formed by a drunken or insane person, the general rule is that the contract is valid, unless that party can prove that:

- when the party formed the contract, its nature was not understood; and
- the other party knew this to be the case.

Further, the contract is binding if later, when having a better understanding, the party ratifies it.

If the contract is for necessaries, the mentally disordered or drunken person who buys the goods is required to pay a reasonable price.

Unconscionable contract

The relevant legislation is the Unconscionable Contract Ordinance ('the UCO'). The UCO applies to 'a contract for the sale of goods or supply of services in which one of the parties deals as consumer'. In other words, the UCO does not cover a contract formed by, for example, two carriers, since neither of them deals as consumer.

To determine whether the contract is unconscionable, the court will consider all the relevant circumstances. It may in particular take into consideration the relative strengths of the bargaining positions of the parties, the necessity of protecting the non-consumer's legitimate interests, the consumer's ability to understand the contractual document, the presence of any undue influence, pressure or unfair tactics, and the amount for acquiring the same goods or service from another.

If the contract is unconscionable, the court has wide powers to deal with the case. It may refuse to enforce the whole or the unconscionable part of the contract. It may also limit the application of, revise or alter the unconscionable parts of the contract.

Illegality

As a legally enforceable agreement, a contract gives the parties the right to enforce their private rights in court. It is also in the interests of the public that promises are kept. However, to invoke the machinery of the court, a party must act within the approved framework of the law. The court will not enforce an illegal contract. Illegality in this context encompasses not just criminal acts but also breaches of public morality.

A contract may be illegal as formed or as performed. A contract to murder for money is an example of an illegal contract as formed, since the subject matter of the contract is illegal. A contract to hire a car to carry smuggled goods is illegal as performed. The subject matter of the contract, carrying goods, is not illegal. But the contract is performed for an illegal purpose. For a contract illegal as formed, the court will not entertain any claim relating to it. For contracts illegal as performed, innocent parties may claim for things done before the illegal elements are known to them.

Discharge of contract

A contract is discharged when no further rights and obligations exist under it. It may be discharged in several ways: by performance, agreement, breach or frustration. A contract is discharged by performance if both parties have performed the contract. If the parties agree later that no further performance is required, the contract is discharged by agreement. A breach of condition may allow the innocent party to treat the contract as discharged and sue for damages. A contract will be automatically discharged on the occurrence of a frustrating event.

The law does not require exact and precise performance to discharge a contract. Substantial performance is enough.

> *Hoenig v Isaacs*[50]
> P contracted with D for the redecoration of D's flat. Under the contract, P was required to pay D $750 for the work. D's work was defective but could be made good for $55. D was prepared to pay P reasonable remuneration for the work done, but refused to pay him according to the terms of the contract.
>
> The court held that P was entitled to be paid at the contract rate, less the cost of making good the defects, because he had substantially completed the contract.

50. [1952] 2 All ER 176.

The parties may reach a subsequent agreement to discharge the contract. The contract may also state some events upon the occurrence of which the parties agree to discharge it.

Only breach of condition entitles the innocent party to terminate the contract. Sometimes a party shows an intention not to perform the contract before performance is due. This is a case of anticipatory breach and the innocent party has several choices:

- Do nothing and wait for the time fixed for performance. The situation is treated as if nothing had happened. Until the time for performance, the other party may decide to perform the contract.
- He may 'accept' the repudiation and seek remedies immediately.
- Perform the contract and claim the contract price. This option is possible only if the innocent party is able to complete the contractual obligation without further assistance from the first party.

A frustrating event is an intervening event or change of circumstances which:

- happens after the formation but before complete performance of the contract;
- makes further performance of the contract impossible;
- is not caused by the parties; and
- is beyond the parties' contemplation at the time of the contract.

Some examples of frustrating events are:

- Destruction of a specific thing on which the contract depends. In *Taylor v Caldwell*,[51] a contract for performance in a concert was frustrated when the hall in which the concert was to be held was destroyed by fire.
- Cancellation of an event which is central to the contract. In *Krell v Henry*,[52] a contract to rent a room for the purpose of viewing a procession was frustrated when the procession was cancelled.
- Death or incapacity of a person whose personal service is required under the contract.
- Supervening illegality, ie, performance becomes illegal after the contract has been formed.

An event which makes the performance of the contract more expensive or difficult is not a frustrating event.

51. (1863) 3 B & S 826.
52. [1903] 2 KB 740.

Tsakiroglou v Noblee[53]

The seller agreed to sell and deliver goods from Sudan to Germany. Although not stated in the contract, the parties at the time of the contract expected shipment via the Suez Canal. Due to the outbreak of war, the Suez Canal was closed. The shipment had to be made via the Cape of Good Hope. That made the shipment two and a half times longer than the voyage through the Suez Canal. The seller argued that the contract was frustrated.

The court held that the closure of the Suez Canal merely made the performance of the contract more expensive. The contract was not frustrated.

If a contract is discharged by frustration,

- no further performance is required from the parties;
- money paid before frustration is recoverable;
- money payable before frustration ceases to be payable;
- at the discretion of the court, the recipient of money paid before frustration may retain an amount not exceeding the expenses incurred for the contract;
- the court has discretion to order a party to pay for a 'valuable benefit' received before frustration.

Remedies for breach of contract

When a contract is breached, the innocent party's rights are infringed. That party may ask the court for remedies. By awarding remedies to the wronged party, the court tries to put that party into a position as if the breach had never happened. There are two categories of remedies: one at common law, the other in equity.

Damages

The usual common law remedy is damages. Damages are money paid by the party in breach of the contract to the innocent party. Damages are compensatory, but not punitive in nature. Only losses suffered by the wronged party as a result of the breach will be recoverable. It follows that if the injured party does not in fact suffer any loss, the court will award only 'nominal damages' of, say, $1.

The innocent party may not be awarded damages for all the losses suffered. Generally, only losses expressed in terms of money may be recoverable.

53. [1962] AC 93.

For example, no damages will be awarded for inconvenience or disappointment under an ordinary commercial contract. The particular nature of a contract may, however, make the award of damages for mental suffering possible.

> *Jarvis v Swan Tours*[54]
>
> D contracted to provide P with a 15-day Christmas winter sports holiday at a hotel in Switzerland. D described the holiday as a 'house-party' and promised to provide a variety of entertainment, such as skiing, a yodeller evening, a bar, etc. In the first week, there were only 13 people at the hotel. In the second week, only D remained. The promised entertainment was wholly inferior in quality to that described in the advertisement.
>
> The court held that D was entitled to damages in compensation for his disappointment.

Even if the loss is generally recoverable, other conditions must be satisfied for the innocent to be compensated. The loss must:

- have been caused by the breach; and
- be not too remote.

Causation is a matter of fact.

> *Monarch SS Co Ltd v Karlshamns Oljefabriker* (A/B)[55]
>
> A, a shipowner, agreed to carry goods for B. By the time they signed the contract, war had broken out between Britain and Germany. The ship was also unseaworthy due to the defective state of the boilers. As a result of the war and the poor condition of the ship, A could not deliver the goods in accordance with the terms of the contract.
>
> The court held that B was entitled to damages. The question must always be 'what reasonable businessmen must be taken to have contemplated as the natural or probable result if the contract was broken. As reasonable businessmen, each must be taken to understand the ordinary practices and exigencies of the other's trade or business.' The possibility of war must have been present in the minds of the parties. And they must have contemplated the likely delay of the voyage due to the war.

For a loss to be recoverable, it cannot be too remote. The basic principle is laid down in *Hadley v Baxendale*.

54. [1973] QB 233.
55. [1949] AC 196.

Hadley v Baxendale[56]

A mill ceased to be productive due to a broken crankshaft. A, the owner of the mill, concluded a contract with B, a common carrier, to bring the broken shaft to another city for repair. B breached the contract for undue delay in delivery. The mill could not work for a long time. A sued B for, among other things, the loss of profit caused by the delay.

The court held that 'where two parties have made a contract which one of them has broken, the damages which the other party ought to receive in respect of such breach of contract should be such as may fairly and reasonably be considered either arising naturally, i.e., according to the usual course of things, from such breach of contract itself, or such as may reasonably be supposed to have been in the contemplation of both parties, at the time they made the contract, as the probable result of the breach of it'. A's claim for the loss of profit failed, because (1) the want of the shaft would normally not cause the mill to cease to be operative. A reasonable miller should have a spare shaft; and (2) B had not been informed of the possible consequences of his breach, i.e., the loss of profit was not contemplated.

Even though the innocent party is entitled to damages, he cannot just wait for compensation, must take reasonable steps to reduce the loss. The court will not award damages for loss which could have reasonably been prevented by the innocent party.

Payzu v Saunders[57]

A agreed to buy goods from B. Delivery was to be made by instalments. Payments were to be made within one month of each delivery, less 2.5% discount. A failed to pay the first instalment. B treated this as repudiation. B offered to continue deliveries at the contract prices for cash sale. A refused the offer. The price of the goods then went up. A sued B for the breach of contract.

The court held that B was in breach of contract because A had not repudiated the contract. A was entitled to damages. However, to reduce the loss, A should have accepted B's offer. As a result, the damages awarded would not be the difference between the contract sum and the market price, but the loss that would have been suffered had the offer been accepted.

Equitable remedies

The general rule is that no one has an absolute right to equitable remedies. The grant of such remedies is always subject to the court's discretion. The

56. (1854) 9 Exch 341.
57. [1919] 2 KB 581.

court will do so if it is just and fair in the circumstances. Examples of equitable remedies are specific performance, injunction and rectification.

An order for specific performance forces the party against whom the order is made to perform the contract. The court only orders specific performance if an award of damages is inadequate (eg, if the buyer buys a unique painting to enjoy and appreciate) and it is just and equitable to do so. The court will not order specific performance if

- it is impossible for the order to be complied with,
- it would cause severe hardship, or
- it requires personal services.

An injunction is a court's order to prohibit a party from breaching a contract. For example, the party may be ordered not to sell the goods to third parties or carry on business with certain people.

Where it is equitable in the circumstances, the court may rectify the terms of the contract to reflect the true intent of the parties.

2.3 AGENCY LAW

Introduction

The term 'agency' refers to the legal relations between two parties. One is called the 'principal' and the other the 'agent'. On behalf of the principal the agent is authorized to do certain things. The principal is liable for the authorized acts done by the agent. Generally, what the agent may do on the principal's behalf depends on the agency agreement. In the shipping and logistics industries, agency may, for example, be found between the following parties: forwarder and consignor, carrier and consignee, and employer and employee. In addition to other special rules, the law of contract generally applies to the agency agreement. With this understanding, the following will be considered below: the formation of agency; authority, rights and obligations of agents; agency and third parties; and termination of agency.

Formation

Agency may be created by agreement, estoppel, ratification or necessity.

If there is an agency agreement, the law of contract generally applies. The agreement may be express (written or oral) or implied (by conduct or past dealing). The facts of the case must be carefully examined to decide whether there is an agency.

Newsholme v Road Transport & General Insurance[58]

N gave untrue answers to X, an agent of an insurance company. N permitted X to fill in the information in an insurance proposal form signed by N. The insurance company later refused to pay N on the grounds of the untrue answers. N argued that the insurance policies were valid because the answers were given by an agent of the company.

The court held that X was, in filling in the answers, N's agent.

The doctrine of estoppel operates to protect the interests of innocent parties. It is relevant in the following situation. A is not the agent of P. P, however, tells T that A is P's agent. Relying on what P said, T forms a contract with A in the belief the contract is formed with P. In such a case, P is not allowed to deny being the principal of A.

Spiro v Lintern[59]

H was the owner of a house. He asked his wife to arrange for the sale of the house with an estate agent. The agent informed the wife of an offer over the phone. Although the wife had no authority to sell the property, she authorised the agent to sign the sale and purchase contract. H allowed the purchaser to visit and repair the house. H later gave a power of attorney to transfer the house. The wife then conveyed the house to another party. The purchaser claimed ownership of the house.

The court held that H was estopped from denying that his wife was his agent in arranging for the sale with the estate agent.

A person may choose to ratify the act of another person who is not an agent. After ratification, the person is liable as principal for the acts of that other person. If A forms a contract purportedly on behalf of B, the contract is generally not binding on B because A is in fact not B's agent. But B is bound if B later, with full knowledge of the material circumstance in which the contract was formed, approves of the contract. The approval may be implied by conduct.

Hunter v Parker[60]

A master sold the ship without the authority of the shipowner. With knowledge of the circumstances of the sale, the shipowner received the proceeds.

The court held that the receipt of the proceeds amounted to ratification.

An agency by necessity may be created in an emergency. If the situation is urgent, a party may do anything reasonable in the circumstances

58. [1929] 2 KB 356.
59. [1973] 1 WLR 1002.
60. (1840) 7 M & W 322.

to preserve another party's property or further that party's interests. In so doing, that party is deemed to act as his agent even though he has never been appointed as such. For an agency by necessity to be created, it must be practically impossible for the agent to obtain instructions from the principal.

> *The Choko Star*[61]
> The master formed a salvage contract with a Greek salvor when the ship was stranded in Argentina. The cargo owner had not been consulted. In the circumstances, the contract was binding on the cargo owner only if the master was an agent of necessity.
> The court held that the master was not an agent of necessity because of the lack of communication with the cargo owner as well as the unreasonableness of engaging the Greek salvor even though local salvors were available.

Authority of an agent

The authority of an agent may be categorized as actual, implied or apparent.

Actual authority is the authority expressly given to the agent by the principal. In determining the extent of the actual authority, the general principles of construction, trade usage and past dealings are relevant.

An agent also has implied authority to do what are 'reasonably incidental to and necessary' for the proper discharge of duties. Such authorities need not be stated in the agency agreement and are generally given to the 'type', trade, and profession of agent (eg, forwarder, carrier, consolidator, chairman of a company, lawyers, etc) involved in the particular case. The existence and scope of implied authority are basically determined by the facts of each case.

> *The Yuta Bondarovskaya*[62]
> M let a ship to S under a charter by demise. S sub-let the ship under a time charter to E. I supplied bunkers to E. The contract between I and E stated that sales of fuel were made on the credit of the ship as well as of the buyer. The master employed by S acknowledged receipt of bunkers supplied by I to E. The ship was arrested when E failed to pay I. When S applied to set aside the arrest, I claimed that S was responsible because, among other things, E, as a charterer, had implied authority to contract on behalf of S.
> The court held that the argument of I could not stand. There was no support for the suggestion that a time charterer had implied authority to contract on behalf of the shipowner or a demise charterer.

61. [1990] 1 Lloyd's Rep 516.
62. [1998] 2 Lloyd's Rep 357.

Apparent authority is relevant to the unauthorized acts of an agent. Such acts are binding on the principal in the following situation: a principal makes a representation to a third party that the agent has certain authorities which his agent does not have. In making the representation, the principal intends that the third party act upon it. The third party, in fact, acts on the representation. In such a case, the principal is bound by the acts of the agent. In other words, the principal is not allowed to retract words. For example, a principal tells a third party the agent has the authority to form contracts up to $1 million. The principal intends that the third party would enter into a contract of $0.8 million with the agent. Relying on what the principal said, the third party forms the $0.8 million contract with the agent. In such a case, the agent is said to have apparent authority to form a contract on the principal's behalf up to $1 million. The principal is bound by the contract, even though the agent is in fact authorized to form contracts up to only $0.5 million.

> *The Unique Mariner*[63]
> The shipowner notified the master that a tug was on its way to rescue the stranded vessel. The captain of another tug who had completed another salvage operation nearby offered his services. The master mistakenly thought that this tug was sent by the shipowner and formed a salvage contract with the captain.
>
> The court held that the master had apparent authority to form the contract.

Duty of agent

The agent has the following duties:

(1) To perform the undertaking;
(2) To follow further and reasonable instructions of the principal;
(3) To act with reasonable skill and care.

> *Gomer v Pitt & Scott*[64]
> X asked Y, a shipping agent, to 'see the policy' to ensure X's goods were properly covered by a marine insurance policy. Y agreed to do so gratuitously. Y obtained an oral assurance from X's bank without looking at the policy itself. X later suffered loss for getting no indemnity under the policy.
>
> The court held that Y had not exercised reasonable care and was liable for the loss of X.

63. [1978] 1 Lloyd's Rep 438.
64. (1922) 12 Ll Rep 115.

(4) To be the actual person providing the service, unless
- there is a usual practice for the appointment of a sub-agent;
- the principal does not object to the sub-agency;
- it is necessary to obtain assistance from other parties;
- only administrative work is delegated; or
- it is authorized by the principal.

(5) To avoid conflicts of interest: If confidence is imposed in the agent by the principal, the two parties are in a fiduciary relationship. The agent, therefore, owes the principal certain fiduciary duties. Generally, the principal's interests must be put before the agent's. Avoidance of conflicts of interest is part of the fiduciary duties. The agent is in breach of this duty if allowing to be in a situation in which the principal's interests and the agent's are in conflict. Examples of conflicts of interest are: an agent acting for both the principal and the other party to the contract, an agent forming a contract with the principal, and an agent having direct or indirect interests in the contract formed on behalf of the principal.

Talent-sign Properties Ltd v Tang Chi Ming[65]

X was an estate agent and formed an agency agreement with Y. Under the agreement, X was to find a suitable flat and conduct negotiations on Y's behalf with the vendor. X informed Y that it would receive double commission, i.e., commission from both Y and the vendor. Y was interested in one flat found by X. No deal was concluded through X. Y later bought the flat through another estate agent. X claimed commission in accordance with the agency agreement. Y pleaded, among other things, the defence of double agency.

The court held that the defence of double agency was established and X's claim failed. X, as the agent of Y, could not be the agent of another principal if that required X to act in a manner inconsistent with its duty to Y, unless X disclosed its interest fully to both principal and obtained their consents. It was not enough for X to merely give notice to Y that he would receive commission from the vendor.

Reid-Newfoundland Co v Anglo-American Telegraph Co Ltd[66]

A was a telegraph company and the owner of 'special wire' erected alongside a railway line. B, the railway company, was under a contract entitled to use the wire for its own business. B was not allowed to use the wire to transmit commercial messages except to A. B, however, used the wire for its own shipping and other business.

65. [2000] 1 HKLRD B1.
66. [1912] AC 555.

The court held that B was the agent of A. Since B obtained profits by using A's property, it was accountable to the principal for all the profits.

Henry v Hammond[67]
C was an average adjuster and employed D to sell cargo from a wrecked ship. A certain sum was left in D's hands after deducting the amount due to C for salvage from the proceeds.

The court held that D was not in a fiduciary relationship with C. With respect to the money, D was only a debtor to C.

(6) To account to the principal: Money or other properties obtained by the agent on behalf of the principal must be separate from the agent's own. The agent must return them to the principal upon the latter's request.
(7) Not to make secret profit: This is another fiduciary duty. If the agent obtains a profit in the course of carrying out duties without the consent of the principal, there is liability to account for the secret profit to the principal, ie, the principal is entitled to get all the profit.

W.A. Phillips, Anderson & Co v Euxine Shipping Co[68]
J hired K to negotiate with L, a ship builder, for the building of a vessel. L paid commission to M, a shipbroker, for his service in negotiating on L's behalf. Under a secret agreement between M and K, half of M's commission went to K.

The court held that J was entitled to the amount obtained by K from M.

Rights of agent

An agent is entitled to be paid according to the agency agreement and claim indemnity against all reasonable expenses or losses incurred for discharging the duties. If the agreement does not mention the agent's fee, the agent is entitled to a reasonable remuneration.

If the principal does not pay the amount due to the agent, the agent may sue the principal under the agency agreement. The principal's property may also be retained until payment is made.

Relationship between principal and third party

A principal may be disclosed or undisclosed. A principal is disclosed if the agent makes known to the third party at the time of contract the agent is

67. [1913] 2 KB 515.
68. [1955] 2 Lloyd's Rep 512.

acting on behalf of a principal. Whether or not the identity of the principal is also disclosed is not relevant. A principal is undisclosed if, at the time of contract, the third party does not know the agent is acting on behalf of another party.

Disclosed principal

Generally, disclosed principals can sue or be sued under a contract formed by agents on their behalf with actual or implied authority.

If the contract is formed on the basis of apparent authority, the principal may be sued by the third party. But the third party cannot enforce the contract.

> *Gurtner v Beaton*[69]
> A company (A) employed B to service its aircraft and teach its employees to fly. A permitted B to carry on an air taxi business even though none of them was licensed to do so. The trade name of the business gave the impression that it was carried on by A.
> The court held that A was liable for an airplane crash which happened while B was providing the service of air taxi.

If the agent enters into a contract without any authority, the principal cannot be sued without ratifying it. But the agent is liable to the third party for breach of warranty of authority.

Undisclosed principal

If the agent has actual authority to form a contract, the undisclosed principal may sue or be sued under the contract.

> *The Barquentine Osprey*[70]
> A shipping company (A) instructed shipping agents to procure workmen's compensation insurance on the crew of its ship. The insurance policy was taken out in the shipping agents' name and the insurance company did not know that the agents were not the employers of the crew. In a claim in relation to the deaths of two crewmen, the insurance company argued that it was not liable because A did not take out the policy itself.
> The court held that A was an undisclosed principal. A might sue or be sued under the contract because it was formed within the agents' actual authority and the agents entered the contract intending to act

69. [1993] 2 Lloyd's Rep 369.
70. [1994] 2 AC 199.

on A's behalf. The contract was not a 'personal' contract because the identity of the employer was a matter of indifference to the insurance company.

An undisclosed principal is not allowed in the following situations:

- where the contract states that the agent is the sole principal; or
- where the contract is for personal performance; or
- where the terms of the contract imply the agent is the principal.

Humble v Hunter[71]
The agent of a shipowner executed a charterparty in his own name. The charter stated that the agent was the owner of the ship.

The court did not allow the shipowner to adduce evidence to prove that the contract was formed on his behalf, on the ground that it was inconsistent with the statement that the agent was the shipowner.

Relationship between agent and third party

The general rule is that an agent is not a party to the contract formed on behalf of the principal. The following points should, however, be noted:

(1) In a case of a disclosed principal, an agent may be sued by the third party for breach of warranty of authority if entering into a contract beyond the agent's actual or apparent authority.

Suart v Haigh[72]
A, a shipbroker, was the agent of B. A purportedly made a charterparty on behalf of B. A signed the contract, stating 'by telegraphic authority of B, A as agent'. B did not authorise A to form the charterparty.

The court held that A was liable for breach of warranty of authority. The fact that A acted in good faith and believed that B's telegram would give him the authority was no defence.

(2) In a case of undisclosed principal, the agent may sue or be sued under the contract before the intervention of the principal. After the principal is revealed, the agent can no longer sue under the contract. But the third party may now enforce the contract against both the agent and the principal.
(3) If agents undertake personal liability, the third party may sue them. Agents undertake personal liability if they, for example, execute a

71. (1848) 12 QB 310.
72. (1893) 9 TLR 488.

document under their own seal, sign a cheque or bill of exchange in their own name, or enter into a contract in their own name without indicating they are acting as agents.

Parker v Winlow[73]

A charterparty expressly stated that it was made between A and B, agent for C. B signed the contract without any qualification.

The court held that B was personally liable. The terms of the contract contained nothing inconsistent with an intention to contract personally on the part of B, although the principal was named.

Lark International Finance Ltd v Lam Kim[74]

P, a foreign exchange company, discounted cheques for X. Many of X's cheques were dishonoured and X as a result owed P $10 million. Y was X's husband. He took over the negotiations with P and repaid some of the debt out of his own pocket. He also gave his own property as security for the debt. Y then signed an agreement in his name for a repayment schedule. Y signed and gave P 15 post-dated cheques drawn on the account of X. Some of the cheques were dishonoured. P claimed against Y under the agreement and alternatively for the dishonoured cheques. Y argued that he was only an agent of X.

The court held that Y was liable to P. The facts were that Y, without X's approval, (1) actively involved in a negotiation with P in which X played no part, (2) provided security for the debt, and (3) repaid some of the debt with his own money. Objectively, Y assumed personal liability in signing the agreement as a principal. Besides, 'as a matter of law, if a person signed a contract without making it clear at the time of signing that he was signing it not on his own behalf, but for and on behalf of a principal, either disclosed or not disclosed, he cannot escape personal liability on the contract'.

Termination of agency

An agency may be terminated by act or agreement of the parties or by operation of law.

By act or agreement

If the agent is appointed for a special task, the agency is terminated when the job is done. Besides, the parties may at any time agree to terminate an agency.

73. (1857) 7 E & B 942.
74. [2000] 1 HKLRD C4.

By operation of law

The law states that an agency ends in any of the following situations:

- death of the agent or principal;
- insanity of the agent or principal;
- bankruptcy of the principal which affects any property dealt with by the agent;
- bankruptcy of the agent which makes him unfit to act;
- frustration.

Effect of termination

Upon termination of the agency, the agent has no more actual authority. But if third parties are not informed of the termination, the principal may, by virtue of the doctrine of apparent authority, be liable for acts done by the agent after the termination. The agent should return any of the principal's property and documents. Termination of the agency does not affect any accrued rights and obligations. For example, the agent is entitled to sue for remuneration already due and the principal may sue a third party under a contract formed by the agent before the termination.

2.4 LAW OF NEGLIGENCE

Introduction to the law of negligence

In the absence of a contract, if a party (P) wants to sue another (D) for loss suffered as a result of D's negligence, P must prove the following:

(1) D owed P a duty of care;
(2) D was in breach of that duty; and
(3) P suffered some legally recognized loss as a result of D's breach of duty.

If P fails to prove any of the three elements, he no compensation can be obtained from D. Even if P can prove all three elements, P also gets no compensation if D has a defence.

Duty of care: General principles

The duty of care is in effect a control mechanism to limit the liability of a party whose fault has caused losses to other parties. In the event that D does not owe P a duty of care, D is not liable for any loss suffered by P.

To determine whether D owes P a duty of care, the starting point is the neighbour principle laid down in the landmark case *Donoghue v Stevenson*.[75]

> *Donoghue v Stevenson*[76]
>
> D manufactured ginger beer in "opaque" bottles. The beer was sold to X, the owner of a cafe. T, a customer, bought it from X and then gave it to P. P drank some and then discovered the remains of a snail in the bottle. P became sick.
>
> The court held that D had breached his duty of care to P. P was a foreseeable victim. Lord Atkin stated the neighbour principle: 'you must take reasonable care to avoid acts or omissions which you can reasonably foresee would be likely to injure your neighbour . . . [i.e.] . . . persons who are so closely and directly affected by my act that I ought reasonably to have them in contemplation as being so affected when I am directing my mind to the acts or omissions which are being called in question'.

The modern test for duty of care is found in *Smith v Bush*.[77]

> *Smith v Bush*[78]
>
> A surveyor was instructed by a building society to prepare a report on a house. The surveyor failed to discover a serious defect (inadequate support for upstairs chimneys) and the report stated that the house needed no essential repairs and gave the house a value more than the actual value. The report was shown to a prospective purchaser. The purchaser bought the house and suffered loss.
>
> The court held that, although instructed by the building society, the surveyor owed a duty to exercise reasonable care in inspecting and reporting to the purchaser, because he knew that the report would be shown to the purchaser, who would likely rely on it.

The modern test to determine whether there is a duty of care consists of three parts. Where D has done an act or made a statement to P, D owes P a duty of care if:

(1) It is reasonably foreseeable that if D's act or statement is negligently done or made, P is likely to suffer damages.
(2) P and D are in a sufficiently proximate relationship, ie, D knows that reliance is to be placed on D's advice by P who belongs to an identified or identifiable third party group.
(3) It is just and reasonable to impose the duty.

75. [1932] AC 562.
76. [1932] AC 562.
77. [1990] 1 AC 831.
78. [1990] 1 AC 831.

Melissa (HK) Ltd v P & O Nedlloyd (HK) Ltd[79]

P sold white garlic to X in Trinidad. The goods were carried to X by P's carrier. D was the agent of the carrier and signed the non-negotiable bills of lading issued by the carrier. The bills of lading stated that 'one original bill of lading must be surrendered . . . in exchange for the goods'. The goods were handed over to the agent of the carrier in Trinidad, M. The bill of lading that P had intended to send by post to X remained in P's office. X did not receive the bill. M however gave the goods to X without presentment of any bill. P as a result suffered loss. P sued D in the law of negligence.

The court held that D owed P no duty of care. There was no sufficient proximity between the parties, D being only the agent of the carrier.

Yim Yai-fai v AG[80]

P was 19 and in police custody. His life was saved by the duty officer when he was attempting suicide in his cell. However, P suffered irreparable brain damage. His father said that the Commissioner of Police was aware of P's suicidal tendencies and sued him for his negligence in failing to prevent P from attempting suicide.

The court held, among other things, that the claim failed on the ground of public policy. 'A person who wilfully injures himself whilst of years of discretion and in his senses, by any means should not be able to lay the blame at the door of another.'

(The law has probably been changed by *Reeves v Commissioners of Police for the Metropolis* [2000] 1 AC 360. The facts of the two cases are similar. The House of Lords held in *Reeves* that, in similar situation, the Commissioners owed a detained person who had committed suicide a duty of care. The court admitted that the duty 'is a very unusual one, arising from the complete control which the police or prison authorities have over the prisoner, combined with the special danger of people in prison taking their own lives'.)

Economic loss and negligent statement

If P suffers personal injury or P's property is damaged by the negligent act of D, it is no problem for P to be awarded damages for personal suffering or the value of the property based on the neighbour principle.

It is more difficult for P to get compensation in the suffering economic loss, such as loss of profit. The categorization of damage to property and economic loss is in a legal sense only. In fact, all damages to property are

79. [1999] 3 HKLRD 674.
80. [1986] HKLR 873.

'economic losses'. If P's car is destroyed, the destruction of the car is classified as damage to property. What P suffers is actually the value of the car. The loss is 'economic' as seen by most people. But in considering whether P can recover the loss in the law of negligence, the classification is crucial.

There are two types of economic loss:

(1) Loss of money resulting from (or is consequential to) personal injury or physical damage of property: This type of loss is generally recoverable by applying the neighbour principle, even if the loss is unquestionably financial in nature. For example, the victim in *Donoghue v Stevenson* might recover loss of earnings and medical expenses.
(2) Pure economic loss: Loss of money which is not consequential to personal injury or physical damage to property. The test is whether the link of physical harm is absent. This type of loss is generally not recoverable, unless there are some special factors.

Sometimes it is not easy to distinguish the two types of loss.

> *Spartan Steel v Martin*[81]
> P had a factory manufacturing stainless steel. Continuous supply of electricity was crucial to the production. Due to D's negligence, the electricity supply to the factory was cut off. Metal solidified in the factory's furnace and no production was possible before electricity supply resumed. P claimed: (1) the reduction in value of the solidified 'melt' and the loss of profit associated with that 'melt', and (2) the loss of profits on four further 'melts' which could have been processed before the electricity was restored.
>
> The court held that item (1) was recoverable but not item (2). The reasons were that item (1) is consequential on the physical damage to P's property, i.e., the solidified 'melt', but item (2) is not a consequence of any damage to P's property — it is only caused by the interruption of the electricity supply.

In the past, only fraudulent statements carried liability for deceit and there was no liability for negligent statements. The law has been changed and a person making a statement negligently may be liable. Losses caused by negligent statements (eg, loss of profit) are usually not consequential to any personal injury or physical damage to property. If the rule that pure economic loss is irrecoverable were applied to the letter, there would be no liability for a negligent statement. The situation would not be satisfactory if that were the case. On the other hand, the court is cautious and takes a restrictive approach in imposing liability for negligent statements because

81. [1972] 3 All ER 557.

words can fly. The law has to strike a balance in setting the extent of liability for negligent statements. The landmark case in this area is *Hedley Byrne v Heller*.[82]

> *Hedley Byrne v Heller*[83]
>
> P was an advertising agent and concerned with the financial standing of E. P asked its banker to obtain a report from D, E's banker. D replied 'without responsibility' that E was 'good' for its business liabilities. The statement was negligently made. P relied on this statement and suffered loss when E went out of business. P sued in tort for D's negligent statement.
>
> The court held that D could be liable for the negligent statement, but in this case owed no duty to P because of the disclaimer. (Note that the Control of Exemption Clauses Ordinance may render similar disclaimers invalid now.)

The modern test is given in *Caparo v Dickman*.[84] In this case, the test laid down in *Hedley Byrne v Heller* is refined. To hold D liable for negligent statements, it is necessary for P to prove that at the time the statement was made:

(1) D was aware of the nature of the transaction contemplated by P,
(2) D understood the statement would be communicated to P,
(3) D knew P would rely on it in deciding whether to carry out the contemplated transaction.

Breach of duty (standard of care)

Liability for negligence is based on fault. D may owe P a duty of care and injure P, but if P cannot prove D was careless (ie, negligent), P will fail in the action against D. In each case, P must prove D was in some way 'at fault'. In other words, P must show D was in breach of the duty of care. The issue of breach of duty refers to D's failure to achieve the standard of care required by law.

If a reasonable man would not do what D actually does in the circumstances, D is in breach of the duty of care. Similarly, if D does not do what a reasonable man would do, D is also in breach of the duty of care.

The 'reasonable man' test is used to determine whether D is in breach of the duty of care. Reasonable man refers to a reasonable person of ordinary intelligence and experience. The test is objective. What D thinks or

82. [1963] 2 All ER 575.
83. [1963] 2 All ER 575.
84. [1990] 2 AC 605.

believes is irrelevant. In the end, the question is: 'Did D act reasonably in all the circumstances?' To answer this question, all relevant factors should be considered. The following five factors are usually relevant:

(1) Foresight of harm
(2) Degree of risk of harm
(3) Seriousness of consequences for P if harm occurs
(4) 'Utility' or usefulness of D's conduct
(5) Cost of taking preventive measures

All the relevant circumstances have to be balanced to determine whether D was negligent.

> *Watt v Hereford City Council*[85]
> W was a fireman employed by HCC. HCC received an emergency call reporting that a woman was trapped under a heavy vehicle. To rescue her, a lifting jack was put on the only truck available at the time. The truck was not specially fitted for carrying the jack. On the way to the scene of accident, the jack shifted and injured W. W sued HCC for his injury in the law of negligence. The issue was whether HCC was in breach of its duty of care in exposing W to the risk of injury by transporting the jack in this way.
> The court held that HCC was not negligent. The risk must be balanced against the end to be achieved. In this case, the saving of life or limb justified taking a considerable risk.

Causation and remoteness of damage

Suppose D owes P a duty of care, the duty of care is breached, and P suffers loss. Does it mean P is liable for the loss? Not necessarily so. The issue now is the connection between the negligent act and the loss. To hold D liable for the loss, D's act must cause P's loss (the issue of causation) and P's loss must not be too remote (the issue of remoteness).

Causation is a matter of fact. In deciding the issue, common sense should help.

> *Wong Lai Kai v Wu Chan Choi*[86]
> P was walking past a shop when he was injured by a falling awning attached to the external wall of the shop. P sued the tenant and the owners of the shop for his injury. The tenant argued that the collapse

85. [1954] 1 WLR 835.
86. [1999] HKLRD (Yrbk) 376

was an Act of God since there had been unusually heavy rain for over ten hours beforehand.

The court held that the injury of P was caused partly by the negligence of the tenant. The tenant should have been aware of the potential danger caused by the heavy rainfall and taken proper precautions against it. The defence of Act of God was not available to the tenant.

The 'but-for' test is used to determine the issue of causation. If a result would not happen but for a certain event, the event is the cause of the result. If it would happen anyway, the event is not a cause.

Barnett v Chelsea and Kensington HMC[87]

P's husband became ill after drinking tea at work and went to D's hospital. But the doctor told him to go home and contact his family doctor. P's husband died shortly afterwards. It was found that the cause of the death was poisoning. D was found negligent for not examining P's husband. The issue was whether D's negligence caused P's husband's death.

The court held that D was not liable, because P had to prove, on a balance of probability, that D's negligence caused her husband's death. However, there were medical expert witnesses who testified that *even if* the husband had been properly examined and treated, it would not have made any difference because the poison would have killed him anyway.

If D's negligence in fact caused P's loss, D is still not liable if the loss is too remote. The test for remoteness is whether the kind of loss is reasonably foreseeable. The test is again objective.

Hughes v Lord Advocate[88]

Workers employed by D dug a hole in the road. The workers covered the manhole with canvas shelter and surrounded it by warning paraffin lamps. P, a boy of 8, played with a lamp and caused it to fall into the manhole. A violent explosion followed and P fell into the hole, sustaining terrible injuries from burns. It was found that D's workers were in breach of their duty of care in leaving the manhole unattended because they should have appreciated that a boy might take a lamp into the shelter and suffer serious injury from burning, even though an explosion was unpredictable. Was D liable for P's injury consequential to the explosion?

The court held that D was liable. As the type of injury, burns, was reasonably foreseeable, the fact that the manner in which the injuries were caused (i.e., explosion) was not foreseeable did not matter.

87. [1969] 1 QB 428.
88. [1963] AC 837.

Smith v Leech Brain[89]

P worked at D's iron works. P's job involved dipping metal into a tank containing chemicals. Some liquid splashed onto P's lip, causing a burn. The burn later developed into cancer because P's skin tissues were in a pre-malignant condition and the burn acted as a catalyst. P died. It was found that D was negligent in not providing adequate protection to workers. The issue in this case was whether P's cancer was too remote.

The court held that it was not. 'The test is not whether D could reasonably have foreseen that a burn would cause cancer and that P would die. The question is whether D could reasonably foresee the type of injury which he suffered, namely, the burn.' It is no answer to P's claim for damages that he would have suffered less injury if he had not had an unusually thin skull or an unusually weak heart.

Defence

Where P can prove that D owed a duty of care to P, that D was in breach of that duty, and that the breach caused P loss which was not remote, P may nevertheless be awarded no damages if D has a defence.

The common defences are contributory negligence, consent and exclusion of liability.

Contributory negligence

The defence of contributory negligence ('CN') is now governed by section 21 of the Law Amendment and Reform (Consolidation) Ordinance ('LARCO') (Cap 23). If P's negligence contributes to the loss or injury, the damages recoverable may be reduced, having regard to P's share in the responsibility for the loss.

Consent (volenti non fit injuria)

In order to prove consent, D must show that P, with full knowledge of the nature and extent of the risk, has voluntarily agreed that P will run the risk of injury due to D's negligence. It is not easy to establish this defence.

Slater v Clay Cross[90]

P was lawfully walking along a narrow tunnel on a railway track owned and occupied by D when she was struck and injured by a train

89. [1961] 3 All ER 1159.
90. [1956] 2 QB 264.

negligently operated by D's employee. D contended that by walking in the tunnel, P impliedly consents to the risk of being struck.

The court held that the defence of consent was not established. 'When P walked in the tunnel, although it may be said that she voluntarily took the risk of danger from the running of the railway in the ordinary and accustomed way, nevertheless she did not take the risk of negligence by the driver.'

Although the defence of consent is very hard to establish, the conduct of P may show that P must have consented to the act of D.

Imperial Chemical Industries v Shatwell[91]

P worked in D's quarry. P and another worker, X, in disregard of D's instructions and the statutory regulations, tested detonators without taking proper precautions. An explosion was caused and P injured. P sued D for his injury, contending that D was vicarious liability for X's negligence.

The court held that P must have fully appreciated the risk of injury that might be caused by explosion. Therefore, he could not argue that he had not consented to the very conduct which had caused his injury. Since the defence of consent would have been available to X if P had sued X, D was not liable in this case.

Exclusion of liability

For business liability, this defence is governed by the Control of Exemption Clauses Ordinance Cap 71. The rules are:

(1) Liability for death or personal injury resulting from negligence cannot be exempted or restricted; and
(2) Liability for other loss or damage caused by negligence can be excluded or restricted, provided the exclusion clause satisfies the reasonableness test.

In deciding whether or not an exemption clause is reasonable, all the relevant circumstances should be taken into account. In *Smith v Bush*, it was held that the following factors should be considered:

- Practicability of the victim's obtaining a second opinion;
- Difficulty of the task undertaken by D;
- Practical consequences of the decision on the question of reasonableness, eg, whether the party liable may cover the risk by taking out insurance policies.

91. [1965] AC 656.

3

Laws of Conversion, Lien and Bailment

3.1 INTRODUCTION

This chapter covers carrier's rights and obligations at common law or, the areas of law relating to goods in transit, ie, the laws of conversion, lien and bailment.

Since carriers provide services for the transport of cargo, they are basically not different from other service providers. But what makes carriers different from other kinds of service providers is that they are always in possession of the goods in transit. To discharge their contractual obligations, they are required to deliver the goods in accordance with the terms of the carriage of goods contract. If shippers fail to pay the freight, they may exercise their right of lien to retain the goods until payment is made. However, if they unlawfully interfere with the rights of the cargo owner, they commit an act of conversion. A carrier in possession of other's goods is also liable as a bailee. Other intermediaries in the logistics industry may similarly be liable because they often deal with other parties' goods. Besides, they may inadvertently act as carriers and be subject to various international conventions. These areas of law are thus particularly relevant to practitioners in the transport and logistics industry.

3.2 TYPES OF CARRIER

Two types of carriers are common carrier and private carrier. The rights and obligations of a common carrier and a private carrier are different.

Common carrier

If a carrier is prepared to offer carriage to anyone's goods for reward, he is a common carrier. A carrier who has no intention to carry anyone's goods is still a common carrier if something has been said or done which makes others think carriage would generally not be refused. In short, the common carrier must carry the goods of any person who wishes to employ his service.[1] If carriers reserve the right of refusal, they are not a common carrier. The test for common carrier is given in *Belfast Rope Work v Bushell.*[2]

> did [the carrier] while inviting all and sundry to employ him, reserve to himself the right of accepting or rejecting their offers of goods for carriage whether his [vehicles] were full or empty, being guided in his decision by the attractiveness or otherwise of the particular offer and not by his ability or inability to carry having regard to his other engagements?

If carriage is not for reward, the carrier cannot be a common carrier. One reason for this rule is that it would be unfair and unreasonable to impose the onerous obligations of common carrier upon a carrier who provides services for free. A gratuitous carrier is, therefore, not a common carrier.[3]

It is not true that a carrier can only be either a common carrier or a private carrier at any one time. Each carriage must be considered separately. The carrier's status may be different for different kinds of carriage. Further, it may also be different for the same type of carriage at different times. Thus, a carrier may be a common carrier for certain types of goods or a fixed route, whilst a private carrier may be so for others. The test is his right of refusal in the particular carriage at issue. It was held in *Johnson v The Midland Railway Company*[4] that:

> At common law a [common] carrier is not bound to carry for every person tendering goods of any description, but his obligation is to carry according to his public profession . . . A person may profess to carry a particular description of goods only, for instance, cattle or dry goods, in which case he could not be compelled to carry any other kind of goods; or he may limit his obligation to carrying from one place to another . . . and then he would not be bound to carry to or from the intermediate places.

1. *Belfast Rope Work Co v Bushell* [1918] 1 KB 210.
2. ibid.
3. *Tyly v Morrice* (1699) Carth 485; 90 ER 879.
4. (1849) 4 Exch 367.

On the other hand, the carrier is a common carrier with respect to a particular carriage if all the conditions are satisfied.[5] For example, a carrier may be a common carrier for the transport of fruit for the route from Mong Kok to Tai Po but a private carrier for other kinds of cargo or routes.

It is sometimes difficult to determine whether a carrier is a common carrier or private carrier. The general rule is that all the relevant circumstances should be considered. The label chosen by carriers for themselves is merely a factor to be taken into account.

Liabilities of common carrier

Generally, common carriers have three special obligations:

(1) To accept for carriage goods delivered to them
(2) To charge only a reasonable freight
(3) To be strictly liable for all loss of or damage to the goods during the carriage

Duty to accept

Common carriers refusing to accept goods tendered to them for carriage are generally liable for damages for unlawful refusal.[6] They may, however, refuse to carry the goods if:

- the vehicle is full;
- the goods are not of the kind they usually carry;
- the goods are dangerous;
- the goods are not properly packed;
- the consignor refuses, upon request by the carrier, to disclose the nature of the goods;
- the tender of the goods is done in an unreasonable time; or
- The consignor refuses to pay the freight when the goods are tendered to the carrier.

Duty to charge only reasonable freight

Common carriers may only charge a reasonable freight. They cannot ask for an unreasonably high rate. This rule is a variation of the duty to accept goods. The rationale is that, if common carriers were allowed to charge any

5. *Robinson v Dunmore* (1801) 2 B & P 416, 419.
6. *Jackson v Rogers* (1683) 2 Show KB 327; 89 ER 968.

freight, they might indirectly refuse to carry goods by charging an unreasonably high freight. If unreasonable freight is charged, the common carrier is liable accordingly.[7] Whether the freight charged is reasonable depends on all the relevant circumstances of the case. However, the carrier is not bound to charge all consignors the same rate.

Duty to answer for loss or damage

It is often said common carriers are liable as an insurer. Generally, their duty for loss of or damage to goods is strict, in the sense they are liable even though they are not at fault or negligent. For example, they may be liable for loss of or damage to the goods occurring in the carriage even though it was caused by robbery, fire, riot or inevitable accident.

However, if the carrier can prove any of the following excepted perils, there is exemption from liability. The excepted perils are:

- Act of God;
- act of the government's enemies;
- inherent vice; and
- fault or fraud of the owner or consignor of the goods.

Act of God means any operation of natural forces such as lightning, heart attack, frost, storm and flood. It is not reasonably possible for human beings to guard against any of these. In order to establish the defence of Act of God, carriers must prove the occurrence was unusual or unpredictable, and that it would be unreasonable to expect them to take precautions against it. With the advent of technology, the scope of this defence is diminishing.

An act of the government's enemies means any act of the armed forces of a state at war with the sovereign state of the carrier. For a Hong Kong carrier, the sovereign state is the PRC.

Inherent vice refers to the inherent defect, nature or deficiency of the goods as well as containers supplied by the consignor. Examples of inherent vice are:

- animals attacking one another;
- damp rags becoming overheated;
- fruit rotting by virtue of a latent defect;
- explosion of fermented wine;
- inadequate strength of packing.

7. *Harris v Packwood* (1810) 3 Taunt 264, 272; 128 ER 105, 108.

Fault or fraud of the owner or consignor of the goods refers to anything done with elements of blameworthiness or dishonesty. Examples are:

- fraudulent or negligent false statements;
- defective packing;
- misleading labelling; and
- insufficient addressing.

Even if one or more excepted perils are proved, the common carrier is still liable if negligence also contributed to the loss of or damage to the goods.

Private carrier

A carrier who is not a common carrier must be a private carrier. Because the duties imposed on common carriers are onerous, nearly all carriers now reserve a right of refusal in their standard forms of contract. The traditional way to describe the situation is that the parties form a 'special contract'. It is safe to presume that, in today's business world, all carriers are private carriers. Even the ordinance establishing the KCR states explicitly the KCR is not a common carrier.[8]

It should be noted only the reservation of the right is important. It is irrelevant whether or not it is actually exercised. For example, a carrier who includes a clause of refusal such as 'the carrier shall not be under an obligation to accept any goods for shipment' is a private carrier, even though rarely refusing any tendered goods.

The private carrier's rights and obligations are governed by the 'special contract' formed with his clients. The rule is that only parties to a contract may sue or be sued under it. It is thus important to know who the parties are to the carriage of goods contract.

Subject to the terms of the contract, a private carrier impliedly undertakes to exercise reasonable skill and care. If negligence causes loss to other parties, carriers may be liable in the law of negligence. They are not, however, responsible for the acts of third parties over whom they have no control.

Parties to the contract of carriage

It is important to identify the parties to the carriage of goods contract for:

8. See section 4 of the Kowloon-Canton Railway Corporation Ordinance (Cap 372).

- identifying the proper person to sue the carrier in the event of loss, damage or delay;
- identifying the person liable to pay the freight when it is unpaid.

If the carriage is done in pursuance of a sale of goods contract, the seller often concludes the carriage of goods contract on behalf of the buyer. If that is the case, the seller becomes the consignor. Often, the buyer is named as the consignee. This arrangement is common if the property in the goods has been passed to the buyer before shipment. In fact, section 34 of the Sale of Goods Ordinance (Cap 26) generally requires a seller to make a contract with the carrier on behalf of the buyer. In such cases, the consignee (the buyer), as the principal of the consignor (the seller), may sue the carrier for loss of or damage to the goods. Similarly, the consignee has, as a party to the contract, a duty to pay the freight.

Where goods are to be delivered to a party other than the consignor, it seems that the carrier may presume the goods are delivered in pursuance of a sale of goods contract. Without knowledge of the terms of the sale of goods contract, however, the carrier cannot know whether or not the seller/consignor forms the carriage of goods contract on behalf of the buyer/consignee.

General obligations of private carrier

The private carrier's obligations and liabilities are generally governed by the carriage of goods contract. Generally, there is an agreement in writing. If there are no express terms on matters relating to delay or deviation, certain terms are implied by the common law. The Supply of Services (Implied Terms) Ordinance (Cap 457) also implies some terms in the contract.

Delivery

The primary obligation of a carrier under a carriage of goods contract is to deliver the goods. The general rule is no delivery until the carrier passes custody and control of the goods to the consignee (or a party entitled to take delivery) or his agent.[9] The contract may impose a duty upon the carrier to hold the goods until the consignee comes and collects them. If the time and place of delivery are specified in the contract, the carrier must deliver the goods in accordance with the terms. If the contract does not state how delivery is to be made, the carrier may deliver the goods to the consignee

9. *Marten v Nippon Sea & Land Ins Co Ltd* (1898) 3 Com Cas 164.

at a reasonable hour of the day to the usual place of delivery. In so doing, the carrier discharges his obligation to deliver under the carriage of goods contract. This rule applies even though the consignee does not turn up and take delivery.[10]

Loading and unloading

The carriage of goods contract may state whether the carrier or consignor has the duty to load or unload the goods. If not required by the contract, the carrier has no duty to provide special equipment for loading.

Delay

At common law, a private carrier is under a duty to exercise reasonable skill and care to deliver the goods at the agreed time or, if the contract does not state the time for delivery, within a reasonable time. In the case the contract does not specify liability for delay, the carrier is not liable unless the delay is caused by his fault. What is a reasonable time is a matter of fact depending on the particular circumstances of the case.

> *Panalpina International Transport Ltd v Densil Underwear Ltd*[11]
> The consignors formed a carriage of goods by air contract with the carrier for the consignments of shirts to Nigeria for the Christmas market. The consignors told the carrier in a telephone conversation that the Nigerian customers wanted the goods before Christmas. The first consignment was intended to leave on 2 December. In fact, the goods did not arrive in Nigeria until 18 December. The consignors sued the carrier for the delay.
>
> The court held that the carrier was liable. The carrier was under an obligation to deliver within a reasonable time. The fact that carriage by air was chosen indicated that the consignors intended to enjoy the advantages of speed that air transport offered. A delay of 16 days was undue and unreasonable in the circumstances.

Deviation

In a general sense, deviation means any fundamental departure from the agreed method of performing the carriage of goods contract. Usually, deviation refers, in the context of carriage of goods, to departure from the

10. *Heugh v L.N.W. Ry* (1870) LR 5 Ex 51.
11. [1981] 1 Lloyd's Rep 187.

agreed route.[12] The parties may agree on the consequences of deviation. The common law rules apply if there is no agreement on the matter.

If there is no agreed route, the carrier should take the usual route. If there are several alternative routes, the carrier may choose any of them. The law also takes into account commercial reality. If the carrier loads the goods of two consignors onto the same vehicle, delivering the goods to one consignor first and then the other, there is no deviation in relation to the carriage of the second contract.

> *Mayfair Photographic Supplies (London) Ltd v Baxter Hoare & Co Ltd*[13]
> The owners of a consignment of cameras formed a contract with a forwarder to arrange for the carriage of the goods to X. The forwarder contracted a carrier for the carriage. The contract for the carriage of goods contained this term: 'Subject to express instructions in writing given by the Customer the Company reserves to itself complete liberty in respect of means, route and procedure to be followed in the handling and transportation of goods.' The carrier also had some sewing machines to be delivered to another place, Y. The carrier put the cameras and sewing machines on the same lorry. The driver was instructed to deliver the sewing machines to Y first and then the cameras to X. When the lorry was on the way to Y, it was hijacked and the cameras stolen. The owners of the cameras sued the forwarder and carrier for the loss. One of the grounds was that there was an act of deviation.
>
> The court held that there was no deviation. The contract permitted the carrier to carry goods of different customers in a lorry. It was in fact normal practice to use one lorry to carry different goods belonging to different customers. As to the route, it was on the facts reasonable to send the lorry to Y first.

An act of deviation is one of the most serious breaches of contract unless the act is justifiable. It is likely a deviation is justified if it is necessary for the purposes of saving life or avoiding danger.

Once the carrier has committed the act of deviation, the consignor may, subject to the terms of the contract, terminate the carriage of goods contract. If the contract is terminated on this ground, the carrier loses his contractual right to freight. That does not mean the carrier will get nothing. The rule against unjust enrichment applies in such circumstances and the carrier is entitled to reasonable remuneration for the services provided. Besides, the carrier is liable as an insurer, ie, the liability becomes strict. If the goods are lost or damaged, the carrier is not entitled to rely on any defence provided in the contract, even though the deviation is not the cause of the loss or

12. See, for example, *L. & N. W. Ry v Neilson* [1922] 2 AC 263.
13. [1972] 1 Lloyd's Rep 410.

damage. Another consequence is that the carrier is no longer protected by any exemption clauses in the contract of carriage. Besides, the consignor or consignee is entitled to take delivery immediately. In other words, instead of waiting for the goods to be delivered, the consignor may demand the immediate return of the goods.

Supply of Services (Implied Terms) Ordinance

Because a contract of carriage of goods is 'a contract for the supply of a service', it is governed by the Supply of Services (Implied Terms) Ordinance (SOSITO). The main purposes of the SOSITO are to codify the relevant common law rules and to invalidate, in certain circumstances, terms in the contract which purport to exclude the liability of the service provider.

If the carrier performs the carriage of goods in the course of a business, section 5 of the SOSITO implies a term to the contract that the carrier 'will carry out the service with reasonable care and skill'. Further, section 6 also implies the carriage of goods contract must be performed within a reasonable time. No such term is implied if the parties agree on the time for performance or the time is determined by the course of dealings between them. A term that the carrier may only charge reasonable freight is implied by section 7. As mentioned above, similar terms are also implied by common law.

In general, the rights and liabilities arising under the SOSITO may be modified by express agreement, the course of dealings between the parties, or such usage binding the parties. However, one important exception occurs in section 8 which states that such rights, duties or liabilities may not be negated against a party who 'deals as consumer'.

The SOSITO only lays down the general statutory requirements for service providers, eg, carriers and forwarders. If other law imposes more onerous obligations or the contract provides stricter duties, the carrier or forwarder is subject to the more demanding obligations.[14]

Freight

Unless the contract provides otherwise, a carrier is entitled to his freight only on delivery of the goods.[15] Usually, the consignor is liable to pay the freight. Where an underlying sale of goods contract exists, the case is more complicated. As mentioned above, the consignor often makes the carriage of goods contract on behalf of the consignee, who is the buyer under the sale

14. See section 9 of the SOSITO.
15. *Krall v Burnett* (1877) 25 WR 305.

of goods contract. If so, the consignee is liable to pay the freight.[16] The consignor, as merely the agent of the consignee, is not liable for the freight.[17]

In a case of short delivery, the consignor often wants to set off in respect of loss against the freight. For carriage of goods by sea, the law is that no set off is allowed.[18] The consignor must first pay the whole freight and then sue for the short delivery. Recently, the court held this rule also applied to domestic carriage of goods by land.

> *United Carriers Ltd v Heritage Food Group (UK) Ltd*[19]
> The defendant carried on the business of selling Christmas hampers. They formed a contract for the carriage of 120,000 hampers for the coming Christmas with the plaintiff. It was found that some hampers were missing on delivery. The defendant refused to pay the freight. When the plaintiff sued the defendant for the freight, the defendant attempted to deduct from the freight the loss caused by the short delivery.
>
> The court held that the rule in maritime shipping law that 'a claim in respect of cargo cannot be asserted by way of deduction from freight' applied in this case. However, the judge made it clear that, but for 'the cumulative weight of relevant authorities', he would hold that this rule only applied to carriage by sea and international carriage by land and air, but not domestic carriage by land.

3.3 CARRIER'S LIABILITY IN CONVERSION

The essence of the tort of conversion is the infringement of the owner's rights.[20] An act of conversion is an act of wilful interference with the goods of another person. It must be done without lawful justification and in a manner inconsistent with the rights of the owner who is thereby deprived of the use and possession of the goods. The tort is committed once an act of conversion occurs. The mental state or motive of the person who commits the act is irrelevant. It is, therefore, said that liability for conversion is strict, in the sense that a person may be liable even though not at fault or negligent. It was held in *Willis v British Car Auctions*[21] that the law

16. cf *Dickenson v Lano* (1860) 2 F & F 188; 175 ER 1017.
17. *Great Eastern Rly v Nix* (1895) 39 Sol Jo 709.
18. *The Brede* [1974] QB 233; *Aries Tanker Corporation v Total Transport Ltd* [1977] 1 WLR 185.
19. [1996] 1 WLR 371.
20. See, for example, *Wo Loong Hing v Zung Fu Company* [1953] HKLR 213 (godown-keepers) and *Chan Kin Har v Mansion House Securities Ltd* [1986] HKLY 26 (undisclosed principal situation).
21. [1978] 2 All ER 392, 395.

> has protected the property rights of the true owner. It has enforced them strictly as against anyone who deals with the goods inconsistently with the dominion of the true owner. Even though the true owner may have been very negligent and the defendant may have acted in complete innocence, nevertheless the common law held him liable in conversion.

Carriers and other practitioners in the shipping and logistics industry, such as forwarders and consolidators, are often in possession of their clients' properties. The law of conversion is, therefore, particularly relevant to them. Generally, a carrier is liable for conversion if delivery of the goods was made to a third party who is not the owner and has no right to take delivery. The carrier's innocence or absence of knowledge provides no defence. Lord Diplock stated in *Marfani & Co v Midland Bank*[22] that:

> At common law one's duty to one's neighbour who is the owner, or entitled to possession, of any goods is to refrain from doing any voluntary act in relation to his proprietary or possessory right in them. Subject to some exceptions . . . it matters not that the doer of the act of usurpation did not know, and could not by the exercise of any reasonable care have known, of his neighbour's interest in the goods. This duty is absolute; he acts at his peril.

Similarly, a carrier commits an act of conversion by refusing to deliver the goods to the person entitled to take delivery. Non-delivery may also be an act of conversion.

> *The Antwerpen*[23]
>
> Two containers of whisky were stolen from a container terminal. Since the containers left the terminal through the main gate, it was highly likely that at least one of the employees of the terminal operator was involved in the theft. The consignee sued the carrier, who was the principal of the terminal operator, for conversion.
>
> The court held that, on the basis that the employees of the operator assisted the theft and delivery of the cargo to a party other than the consignee, the operator had made an unauthorised delivery. The carrier, as the principal, was vicariously liable for the act of conversion committed by the operator.

However, a carrier who refuses to part with the cargo is not liable for conversion if:

22. [1968] 1 Lloyd's Rep 411, 421.
23. [1994] 1 Lloyd's Rep 213.

- exercising the right of lien; or
- reasonable doubt exists the party demanding delivery is not so entitled; but the party may become liable after a reasonable time.

Not every act without the consent of the owner amounts to conversion. Carriers are liable for conversion only if their acts are contrary to the rights of the owner. Their fault in other aspects is immaterial.

> *Peereboom v World Transport Agency*[24]
> The defendant was entrusted with 500 cases of goods to be forwarded. The seller delivered the goods to a warehouse. He gave instructions that the goods were held to the order of the defendant. As usual practice, the defendant sent a bill of lading to the carrier. The bill of lading signed by the carrier covered only 339 cases, the other 161 cases being shut out and returned to the warehouse. The seller then wanted to stop shipment. The defendant gave the seller a delivery order for the 161 cases. After some months, when the seller presented the delivery order to the warehouse, the 161 cases had been shipped to an unknown consignee and could not be recovered.
>
> The court held that the defendant was not liable for conversion, because he had not dealt with the goods contrary to the seller's title or rights in respect of the goods. The defendant was negligent in not informing the seller that the carrier had been in dispute with the warehouse owner as to the whereabouts of the goods. However, this fact was not relevant.

Common acts of conversion

The two most common acts of conversion are:

- wrongful delivery of goods; and
- retention of goods with the intention of keeping them in defiance of the owner.

Wrongful delivery of goods

Wrongful delivery of goods or documents of title is an act of conversion. Where the goods or the bills of lading are delivered to a party not entitled to take delivery, a wrongful delivery of goods occurs. Since liability for conversion is absolute, the carrier is liable even though the delivery was made only as a result of misrepresentation.

24. (1921) 6 Ll L Rep 170.

> *Kolbin v United Shipping Co Ltd*[25]
> A shipping company formed a contract to arrange for the carriage of the consignor's goods. In the circumstances it was impossible to deliver the goods as instructed, the shipping company decided to sell the goods 'as reasonably careful businessmen' would do. The bill of lading was given to R, who misrepresented himself as the agent of the consignor. R sold the goods and obtained the money, and subsequently became bankrupt. The consignor sued the shipping company for conversion.
>
> The court held that the shipping company was liable. By handing over the bill of lading to R and taking no further responsibility for them, the company abandoned the charge of the goods and was therefore liable for conversion. R's misrepresentation provided no defence.

In a case of wrongful delivery, the knowledge of the carrier is crucial in determining liability. If the goods are delivered to the consignee or to the consignor's order without knowledge a third party has a claim on the goods, the carrier is not liable for conversion.[26] It was held in this case that:

> On principle, one who deals with goods at the request of the person who has the actual custody of them in the bona fide belief that the custodier is the true owner, or has the authority of the true owner, should be excused for what he does if the act is of such a nature as would be excused if done by the authority of the person in possession, if he was . . . entrusted with their custody'.

In short, if the carrier delivers the goods without knowledge it is in pursuance of a sale or other disposition of title on the goods, there is no liability for conversion. In such cases, the carrier who makes the delivery in accordance with the instructions of the consignor is not liable even though the consignor's transaction amounts to conversion. In many cases, the carrier should know the goods had been sold to the consignee. Strictly speaking, the carrier may be liable for conversion. Nevertheless, the general view is it would be unreasonable to impose liability on the carrier in such a case.

The law on this point is not completely settled. According to *Hollins v Fowler*, if a carrier or forwarder disposes of the goods with knowledge the act is part of a transaction affecting the title to the goods, and not merely their possession, they are liable for conversion. In *National Mercantile Bank v Rymill*,[27] it was, however, held that an agent who delivered goods to a third party with knowledge the delivery was done in pursuance of a sale of goods contract was not liable for conversion. It should be noted the correctness of

---

25. (1931) 40 Ll L Rep 241.
26. *Hollins v Fowler* (1874) L R 7 HL 757.
27. (1881) 44 LT 767.

National Mercantile Bank v Rymill is doubted in *Willis v British Car Auctions.*[28] It is submitted that carriers who merely transfer the possession of the goods as innocent handlers should not be liable in conversion, provided they have no knowledge of the true owner's title to the goods. On the other hand, they should be liable if they are involved in a sale or other transaction which affects the title to the goods.

If the carrier is acting as an agent of the consignor, it is not liable for conversion for ministerial acts which are done *bona fide.*[29]

Retaining the goods

Goods are delivered to carriers for the purpose of carriage. A carrier may only retain the goods for this purpose. Retention of goods for other purposes is unjustifiable. If a carrier unjustifiably retains the goods, an act of conversion is committed. Similarly, a refusal to deliver the goods to the named consignee would also be conversion.

> *Perry & Co v British Railways Board*[30]
> A carrier refused to deliver a consignment of steel to the consignee when the goods were being carried to the consignee. The decision was made because the defendant feared that delivery of the steel to the consignee would cause industrial action by railway workers who supported the industrial action by steelworkers.
>
> The court held that it was a clear case of conversion in that the defendant was denying the plaintiff most of the rights of ownership, including the right to possession, for an indefinte period.

What should a carrier do to protect himself?

To avoid committing acts of conversion innocently, a carrier should, if possible, make sure the client is the owner of the goods or has the authority to deal with them. In practice, a carrier should check whether the client has the relevant document of title, such as a bill of lading covering the goods. Sometimes the client is not the owner. The problem, in such cases, is that a carrier who properly follows the instructions of his client may still be liable for conversion. If the consignor subsequently changes his mind and gives instructions for the delivery of the goods to a third party, the carrier should act accordingly. In doing so, the carrier may be liable for conversion to the

28. [1978] 1 WLR 438.
29. *Re Samuel (No. 2)* [1945] Ch 408.
30. [1980] 2 ALR 579.

named consignee who has become the owner by, say, obtaining the bill of lading from the consignor.

In practice, carriers would cover their potential liability for conversion by appropriate insurance policies.

Indemnity for conversion

It is common for carriers or forwarders to insert an indemnity clause in their standard forms of contract. The clause usually states the client is under an obligation to indemnify the carrier or forwarder who, acting in accordance with the instruction of the client, is subsequently held liable for conversion.

3.4 CARRIER'S LIEN

Generally, the right of lien refers to the right of a creditor to retain properties of the debtor as security for the payment of the debt. The exercise of the right of lien deprives the debtor of the use and possession of the property. It is a forceful means of compelling the debtor to pay or provide other forms of security. Non-payment of freight is not uncommon to carriers. In such a case, the exercise of lien on the goods may be the best way to force the consignor to pay.

There are two types of lien:

- particular lien; and
- general lien.

A carrier's lien is a possessory lien. It is not exercisable on goods not in possession. Loss of possession of the goods means loss of the right of lien. The lien is lost if the goods are returned to the owner. But if the goods are returned to the owner for other purposes and on the understanding they are to be returned afterward, the carrier does not lose the lien.[31] Where the owner gets the goods back in an improper or unlawful manner, the carrier's lien is not lost. For example, if the owner takes the goods without permission or by deception induces the carrier to return the goods, the lien is not affected.[32]

If a carrier exercises a particular lien for the freight payable under a contract, only the goods carried under the contract may be retained. A carrier cannot retain other goods of the same consignor. On the other hand,

31. cf *Albemarle Supply Co Ltd v Hind & Co Ltd* [1928] 1 KB 307.
32. *Wallace v Woodgate* (1824) 1 C&P 575; *Earl of Bristow v Wilsmore* (1823) 1 B&C 514.

a general lien is exercisable on any property of the consignor which is, for whatever reasons, in the possession of the carrier.

At common law, a common carrier may exercise a particular lien for freight on the client's goods. Such a common law lien is created independent of the carriage of goods contract. It is good against the whole world. If the owner (who is not the client) wants to have the goods returned, payment of freight must be made or otherwise with the consent of the carrier. It is not clear whether such a common law lien is available to a private carrier. In practice, a private carrier often has a contractual lien with generally no need to exercise the common law lien.

Generally, a carrier does not have a general lien on the goods of his clients. A general lien may only be created by contract, local custom or past dealings between the parties. The common practice is for carriers to insert a lien clause in their standard forms of contract. This clause often gives the carrier a right of general lien. In a case of non-payment of freight, the carrier may retain any goods of the consignor happening to be in his possession. Since the lien is contractual in nature, it is exercisable only against a party to the contract, eg, the consignor. If the client is not the owner of the goods, the carrier must return the goods upon the request of the owner.

> *The China State Bank v The Dairy Farm*[33]
>
> X imported and stored large quantities of meat in D's cold store. The import of the meat was financed by several banks, including P. The arrangement was such that P was the owner of the meat until X paid off the amount advanced. Such meat was to be delivered to P's order. A term in the contract between X and D stated, 'Goods are received subject to [D]'s right to a general lien thereon for all charges accrued or accruing against the storer and/or owner of the goods and for all other moneys, accounts and liabilities of the storer and/or owner of whatsoever kind or nature and if such lien is not satisfied within seven days of notice in that behalf being given to the storer and/or owner or sent by post to his last known address, the goods or any part thereof may be sold by auction or by private treaty to defray such lien and all expenses incurred'. Most of the meat was stolen. P agreed to pay for the storage charges for the remaining meat in exchange for the delivery. D refused and sold the remaining meat because it asked for all the outstanding storage fees. D claimed that it had a general lien.
>
> The court held that D did not have a general lien on the meat against P. It was a breach of the contract of storage by not accepting P's payment.

33. [1967] HKLR 95.

Similarly, carriers who act as an agent of the consignor/seller or consignee/buyer have a particular lien for outstanding charges against their principal. If their principal is not the owner of the goods, the carrier's right of lien provides no security. This is the case if the principal is the seller but the property in the goods has passed to the buyer; or, the principal is the buyer and the property of the goods has not been delivered.

Lien and conversion

As explained above, a private carrier's lien is good against only the client who has not paid the freight when due. If the client is not the owner of the goods, the lien is not exercisable against the owner. The carrier commits an act of conversion if wrongfully retaining the goods against the owner. Besides, the carrier is, in exercising the right of lien, under a duty to exercise reasonable skill and care for the safety of the goods.

Lien and seller's stoppage in transit

Section 46 of the Sale of Goods Ordinance (Cap 26) (SoGO) gives the seller a right of stoppage in transit. The section reads:

> Subject to the provisions of this Ordinance, when the buyer of goods becomes insolvent, the unpaid seller who has parted with the possession of the goods has the right of stopping them in transitu, that is to say, he may resume possession of the goods as long as they are in course of transit, and may retain them until payment or tender of the price.

The carrier's common law lien takes priority over the seller's right of stoppage in transit. A contractual lien has no such priority unless the carriage of goods contract gives priority to the lien.

A carrier may enter, on behalf of the seller, into a contract of carriage with another carrier. For example, a land carrier may conclude on behalf of the consignor a carriage of goods by sea contract with a sea carrier. If the carrier is instructed to exercise the right of stoppage in transit, the stoppage is subject to the second carrier's common law lien.

Section 48 of the SoGO states:

> (1) The unpaid seller may exercise his right of stopping in transitu either by taking actual possession of the goods or by giving notice of his claim to the carrier or other bailer in whose possession the goods are. Such notice may be given either to the person in actual possession of the goods or to his principal. In the latter case the

notice, to be effectual, must be given at such time and in such circumstances that the principal, by the exercise of reasonable diligence, may communicate it to his servant or agent in time to prevent a delivery to the buyer.

(2) When notice of stoppage in transitu is given by the seller to the carrier or other bailee in possession of the goods, he must re-deliver the goods to, or according to the directions of, the seller. The expenses of such re-delivery must be borne by the seller.

A carrier who has received a notice of stoppage but wrongfully delivers the goods to the buyer is liable to the owner for conversion.

An unpaid seller who has exercised the right of stoppage must give proper instructions for the disposal of the goods. The carrier is not liable for any loss if the seller fails to do so.[34]

Disposal of goods by carrier

The duty of the carrier is to deliver the goods in accordance with the carriage of goods contract. Generally, the carrier has no right to dispose of the goods during the carriage. A carrier who makes an unauthorized disposal is in breach of the contract. The disposal is also an act of conversion. The carrier may, however, sell perishable goods with the permission of the consignor or as an agent of necessity.

The following is an example of agent of necessity.

Sims v Midland Ry[35]

The railway system was paralysed by the national strike of railway workers. This caused delay to the carriage of some butter in transit. The butter began to deteriorate. Without the consent of the consignor, the carrier sold the goods at a price lower than the market price. The consignor sued the carrier for the loss.

The court held that the carrier was not liable. In selling the goods, the carrier was acting as an agent of necessity.

If the goods are noxious or dangerous, the carrier may dispose of them for the safety of the vessel or other cargo (see below).

Dangerous goods

At common law, there is an implied warranty of fitness for carriage. This is a strict liability. The consignor is liable for the loss of the carrier caused by

34. *Booth SS Co Ltd v Cargo Fleet Iron Co Ltd* [1916] 2 KB 570.
35. [1913] 1 KB 103.

the dangerous goods, even though the consignor was not negligent or had no knowledge of the unfitness of the goods.[36]

If the goods are dangerous, the consignor has an obligation to give notice to the carrier as to their nature and any precaution that should be taken. This obligation is strict. The consignor is liable for the loss suffered by the carrier as a result of carrying the dangerous goods. But the consignor is not liable if the risk is discoverable on reasonable inspection. The carrier is not required to conduct a detailed inspection to find out whether the goods are dangerous. The nature of the goods and the packing are relevant to the issue of discoverability. For example, the dangerous nature of the goods is not reasonably discoverable if the goods are packed in such a way that the dangerous nature of the goods cannot be discovered on visual or other usual inspections. If the consignor fails to give notice and the goods appear to be harmless, they are liable for the loss suffered by the carrier as a result of carrying the goods.

> *The Great Northern Railway Company v L.E.P. Transport & Depository Ltd*[37]
>
> D was a forwarding agent. D sent a quantity of liquid which was called 'oxygen water' to P, a carrier. It was common knowledge that ordinary oxygen water was harmless. P carried the goods in a van which also contained felt hat bodies. Some felt hat bodies were damaged by the 'oxygen water' spilt during the journey. It was found that sulphuric acid had been added to the 'oxygen water' which had become highly corrosive and dangerous. P sued D for the compensation that P paid for the damaged felt hat bodies.
>
> The court held that D was liable. By describing the goods as 'oxygen water', D impliedly gave the warranty that the goods were not dangerous. Since the goods were in fact dangerous, D was in breach of the warranty.

In short, a warranty is implied that the goods are not dangerous and may be safely carried. But the carrier will bear the loss if there is actual knowledge or anticipation of the risk.

As indicated by *The Great Northern Railway Company v L.E.P. Transport & Depository Ltd*, a contracting carrier who forms a sub-contract for carriage of the goods with the actual carrier is similarly under a duty to give notice as to the dangerous nature of the goods to the actual carrier. Failing to do so, the carrier impliedly gives an absolute warranty that the goods are fit for carriage, whether or not there is actual knowledge of the dangerous nature.

36. *Bamfield v Goole and Sheffield Transport Co* [1910] 2 KB 94.
37. (1922) 11 Ll L Rep 133.

If the goods are not so fit, the carrier is liable for any loss suffered by the actual carrier as a result of the unfitness of the goods.

3.5 BAILMENT

Basically, a bailment is a delivery of goods to the bailee (the one who holds the goods on bailment) on condition they are to be returned to the bailor (the one who delivers the goods to the bailee or creates the bailment) or to his order as soon as the purpose for which the goods were delivered has been completed. A more elaborated definition is given in *Cher Singh v Forja Singh*:[38]

> A bailment is the transfer of the actual or constructive possession of a specific chattel by one person to another, in order that that other person may perform some act, for which such possession is necessary in connection therewith, and upon the understanding, either expressed or implied, that, when the act is performed, the recipient shall redeliver the chattel to its owner or to his nominee.

Although a bailment is usually created by a contract, bailment is possible without agreement.

Common examples of bailment are:

- borrowing a book from a library;
- hiring of roller-skates;
- giving a car for parking service.

The law of bailment is particularly relevant to carriers and other practitioners in the shipping and logistics industry since they are always in possession of their clients' goods. The law of bailment determines the rights, duties and powers of the bailee and bailor. Where there is no contract between the owner and the carrier, the law of bailment governs. Even though there is a contract, it still applies if the contract is silent on the relevant points.

Elements of bailment

The recent case of *The Pioneer Container*[39] states the elements of bailment:

(1) Possession by the bailee; and
(2) The bailee's consent.

38. [1922] HKLR 49.
39. [1994] 2 AC 324.

Generally, the bailor's consent is not a necessary element in holding the bailee liable. For the creation of sub-bailment, the consent of the bailor becomes an issue (discussed later). The consent of the bailor is also relevant in that the bailee's possession of the goods without the consent of the bailor amounts to an act of conversion.

Possession

The bailee's possession is essential to create a bailment. The bailee may hold the goods in possession, or may be in possession through his agent. Basically, possession has two elements: control and intention. Control refers to the existence of some power over the property. Intention means the possessor's attitude to deny the other person use of or access to the property without his permission. The two elements must co-exist to effect possession. Whether they co-exist depends largely on the circumstances.

> *Spectra International Plc v Hayesoak Ltd*[40]
>
> D carried on the business of freight forwarding. P formed a contract with D to arrange for the receipt of certain goods shipped from Hong Kong to the UK. The contract stated that the goods were 'taken into [D's] charge'. D sub-contracted the haulage of the goods to other parties. P gave instructions to D for the delivery of the goods. D then repeated the instructions to the sub-contractor. Some of the goods were stolen when they were being delivered by the sub-contractor. P sued D for the stolen goods. One of the issues was whether D was a bailee of the goods.
>
> The court held that D was in possession of the goods when they were stolen. D was a bailee of the goods. The fact that D had never been in physical possession of the goods was immaterial. There was no rule that D could only be liable as bailee 'if at some stage [D had] had physical charge of the goods. In any event . . . [the contract] plainly stipulated that the goods were 'taken into [D's] charge, whether held personally or by a sub-contractor'.

The degree of control a person has over the property has a bearing on the element of intention. A person having very firm control over the property is usually in possession of it, even though nothing is done to show intention to deny others access to it.[41] A householder who holds the door key and has the power of not allowing others to enter the premises is in possession of anything found in the house, even though nothing is done

40. [1997] 1 Lloyd's Rep 153.
41. *Parker v British Airways Board* [1982] 1 QB 1004.

to show intention of denying others to the access of the thing found in the premises.

However, a party may merely obtain a license to use the premises but have no possession of the goods there. For example, if a man parks his car at his own risk in a car park, only a license to park is obtained. The car is still in his possession and the car park owner is not liable as a bailee.[42] Similarly, in the event the consignor ships goods at personal risk (eg, in a gratuitous carriage), the carrier is not liable as a bailee.

Since possession is essential for the creation of bailment, a carrier or freight forwarder who has provided some services in relation to the goods is not liable as bailee if the goods are not in possession.[43]

Bailee's consent

A person in possession of the property of another may not be a bailee. Generally, the bailee must be willing to hold the goods in possession. In the past, the term 'involuntary bailment' was often heard. The carrier is an 'involuntary bailee' if the consignee refuses to take delivery of the goods at the destination and the carrier (assuming that all the freight has been paid) is still holding the goods in possession. The traditional view is that an 'involuntary bailee' owes a lower standard of care to the owner of the goods. The law seems to have been changed by *The Pioneer Container*.[44] According to this case, a bailment arises only if a person is voluntarily in possession of the goods. The case is about a sub-bailment. However, it is likely that the new definition will be accepted generally.

Types of bailment

If the bailee receives no benefit for having the goods in possession, there is a gratuitous bailment. In the normal case where the bailee gets some benefit in return, there is a bailment for reward.

Duration of bailment

A bailment is generally at will unless the parties agree the bailee holds the goods for a fixed period of time. The bailor and bailee's rights and obligations under the bailment end with the termination of the bailment. There is

42. *Ashby v Tolhurst* [1937] 2 KB 242.
43. *Jones v European and General Express Co Ltd* (1920) 4 Lloyd's Rep 127.
44. [1994] 2 AC 324.

an implied term in a contract of bailment (eg, under which a party obtains possession of goods of the other) that the bailor will take back the goods within a reasonable time after the termination of the bailment. If the bailor fails to do so, the bailee is relieved of his obligations.[45]

Rights and obligations of bailee

By virtue of his possessory title, a bailee may sue third parties for interference with the goods, trespass, negligence or conversion during the term of the bailment. A bailee is also entitled to remuneration for or expenses incurred in possession of the goods. Also, a bailee has a right to sale in case of default to repay or collect.

In case of contractual bailment, the bailor and bailee must observe the terms of the contract. For non-contractual bailment, a bailee is under an obligation to exercise reasonable care to protect the goods; the goods should be protected against the wrongdoing of third parties. Generally, the standard of care required in the law of negligence applies.[46]

> *Dense Billion Ltd v Hui Ting-sung*[47]
> L was an employee of D, a transportation company. D formed a contract with P for the carriage of a load of silk fabrics from China to HK. The contract contained the following term: 'During transport, all risk caused by wind or fire or water and other accident shall be the responsibility of the consignor'. When the lorry had arrived in HK, L parked it near his home. The silk fabrics on the lorry were stolen. P sued D for damages, arguing that the theft was not foreseeable as well as relying on the exemption clause.
>
> The court held that D was liable. As a bailee for reward, D owed P a duty of care for the safety of the fabrics. The risk of theft in unattended open parking was foreseeable. D could not say that no precautions were needed, taking into account the value and ready disposability of the goods.

It seems that the standard of care for a gratuitous bailee is lower. It was held in *Tung Chi Cheung v Jamsons Shipping Co (H.K.) Ltd*[48] that a gratuitous bailee is only liable for gross negligence.

A bailee is under an obligation not to convert by selling or appropriating the goods for personal gain.[49] If the bailee deals with the goods contrary

45. *Jerry Juhan Developments SA v Avon Tyres Ltd* [1999] CLC 702.
46. *Houghland v Low* [1962] 1 QB 694.
47. [1996] 2 HKLR 107.
48. [1961] HKDCLR 124.
49. *Morris v C W Martin & Sons Ltd* [1966] 1 QB 716.

to the terms of the bailment, liability as an insurer is incurred until the goods are returned.[50] The bailee is, however, not liable if the damage to the goods is not connected to wrongful dealing with the goods.[51]

The duty not to deviate also applies to a bailee. Deviation here means serious breaches of the terms of bailment. For example, the duty not to deviate is breached by a carrier who stores the goods in a place different from the agreed location;[52] or entrusts the goods to a third party without permission.[53]

Depending on the nature of the bailment, the carrier may be required to redeliver the goods to the consignor or to his order. The time and place for the redelivery should be agreed. Generally, it is unclear whether a carrier is under an obligation to seek out the party to whom delivery is to be made. But if the bailment agreement imposes such a duty, the bailee must seek out the party entitled to delivery. Delivery to a wrong party constitutes a breach of bailment as well as an act of conversion.[54]

A bailee is vicariously responsible for the wrong done by an employee or agent.[55]

Rights and obligations of bailor

A bailment does not affect ownership of the goods.[56] The owner's rights are, however, affected. During the term of the bailment, the owner may not enjoy the goods or sue a third party for interference with the goods. The owner loses the right of possession and may only regain possession of the goods after the end of the bailment.

The bailor guarantees the owner has the right to bail the goods, ie, that he is the owner or has rights delegated by the owner. If the guarantee is breached, the bailor is liable to compensate any loss suffered by the bailee. If the bailment is for some particular purpose, the bailor guarantees the goods are safe and suitable for the purpose.[57] For unsafe goods entrusted to a bailee, the bailor is strictly liable for the loss suffered by the bailee.[58]

50. *Jackson v Cochrane* [1989] 2 Qd R 23.
51. *James Morrison v Shaw Savill and Albion Ltd* [1916] 2 KB 783.
52. *Lilley v Doubleday* (1881) 7 QBD 510.
53. *Edwards v Newland & Co* [1950] 2 KB 534.
54. *Devereux v Barclay* (1819) 2 B & Ald 702.
55. *Morris v C W Martin & Sons Ltd* [1966] 1 QB 716.
56. *Franklin v Neate* (1844) 13 M & W 481.
57. *White v Steadman* [1913] 3 KB 340.
58. *Bamfield v Goole & Sheffield Transport Co Ltd* [1910] 2 KB 94; *Burley v Stepney Corp* [1947] 1 All ER 507.

Unless the parties agree the bailor should repossess the goods at a certain time, the bailor is not liable for delay in repossession. The bailor's delay in repossession renders the bailment gratuitous or, in the past, 'involuntary'. The bailee's duties may, therefore, become less onerous.

The bailee's estoppel

A bailor may not be the true owner of the goods. For example, the goods may have been bought from a thief who had stolen them from the true owner. At law, the true owner may, despite what happened after the theft, still be the owner of the goods. In such a case, the bailee may commit an act of conversion by delivering the goods to the bailor's order.

Generally, the bailee cannot refuse to follow the bailor's order on the ground the true owner has a better title. The bailee is at common law estopped from denying the bailor's title.[59] Exceptions to this rule are:

- where the bailee has to surrender the goods;[60]
- where the bailee is acting on behalf of the party having the better title;[61] and
- where the bailee has obtained the rights of the party having the better title.[62]

If none of the exceptions applies, it is advisable for the bailee to ask for an indemnity from the bailor against liability for conversion.

Sub-bailment

The traditional view of whether the bailee may sub-bail depends on the terms of the principal bailment (head bailment) agreement. In the usual case that a carrier reserves the right to sub-contract, sub-contracting an actual carrier to perform the carriage creates a sub-bailment. If the contract is silent on the right of sub-bailment, sub-bailment is normally allowed. However, if the goods being shipped are of exceptionally high value or unusual delicacy, sub-contracting is not permitted without the bailor's consent.

59. *Ross v Edwards* (1895) 73 LT 100.
60. *Biddle v Bond* (1865) 6 B & S 225.
61. *Rogers, Sons & Co v Lambert & Co* [1891] 1 QB 318.
62. *Webb and Webb v Ireland and the Attorney General* [1988] ILRM 565.

> *Garnham, Harris and Elton Ltd v Alfred W. Ellis (Transport) Ltd*[63]
>
> P frequently employed D for the carriage of copper wire from London to Glasgow. The contracts were made orally. It was common knowledge that copper wire was valuable and easy to steal. Further, stolen copper wire was untraceable. For the carriage in issue, D arranged with X for the delivery. P did not know X and was not informed of the arrangement. In making the arrangement with X, D knew that P would not have consented to any sub-contracting without making enquiries. Eventually, X fled with the copper wire. P sued D for the loss.
>
> The court held that D was liable. Although a contract of carriage might normally be sub-contracted, the right to sub-contract could, however, be denied if the goods were the sort of commodity that every lorry thief would strive to steal. Express words were necessary to give the carrier a right to sub-contract. In this case, D was liable under the contract. Besides, delivery of the wire to X without P's consent was an act of conversion and D was liable accordingly.

The rule stated in *The Pioneer Container*[64] is that the sub-bailee must be aware of the bailor. In the words of Lord Goff,

> a sub-bailee can only be said for these purposes to have voluntarily taken into his possession the goods of another if he has sufficient notice that a person other than the bailee is interested in the goods so that it can properly be said that (in addition to his duties to the bailee) he has, by taking the goods into his custody, assumed towards that other person the responsibility for the goods which is characteristic of a bailee.

It is now clear no sub-bailment can be permitted if the bailor does not agree to the bailee's possession. If the bailor has generally agreed to the bailee's possession, consent to the sub-bailment is not required even though the terms or details of the bailee's possession have not been agreed on.[65] The judges in that case refused to consider the possibility of a wholly unauthorized sub-bailment. So the law is unclear on this point.

The bailee is generally under a duty to exercise reasonable skill and care in the selection of the sub-bailee. Subject to the bailment agreement, the bailee is also liable for the sub-bailee's failure to discharge his obligations. As explained in *BRS v Arthur V. Crutchley & Co*:[66]

63. [1967] 2 All ER 940.
64. [1994] 2 AC 324.
65. *The Pioneer Container* [1994] 2 AC 324.
66. [1968] 1 All ER 811, 820.

> the obligation of the bailee can be formulated as an implied term of contract . . . The bailor could not reasonably be expected to be content with a contractual promise of the bailee to take proper care of the goods or engage a competent contractor to do so. If that were the contractual promise, then in the event of default by a competent contractor duly selected by the bailee, the bailor would have no remedy against the bailee and would have to rely on the possibility of an action of tort against the contractor. To give business efficacy to the contract, the bailee's implied promise should be that he will himself or through his servants or agents take proper care of the goods.

In general, the bailee and sub-bailee owe the same duty to the bailor.[67] The sub-bailee is, however, not entitled to obtain remuneration from the bailor. Only the bailee is liable to pay the sub-bailee under the sub-bailment agreement. A slightly different rule applies for expenses. Although there is no general right to reimbursement, *The Winson*[68] makes it clear that a bailee or sub-bailee has a correlative right to reimbursement if:

(1) The contract of bailment is a commercial salvage contract;
(2) The service of the bailee or sub-bailee (ie, the salvor) has preserved the goods;
(3) After the contract of bailment ends, the bailee or sub-bailee continues in possession of the goods as a gratuitous bailee;
(4) The bailee or sub-bailee incurs reasonable expenses in safeguarding and preserving the goods to the benefit of the bailor; and
(5) The bailor stands by, knowing that the bailee or sub-bailee is so acting to his benefit.

The legal relations among a bailor, bailee and sub-bailee are explained by way of an example. A carrier (A) forms a carriage of goods contract with his client (C). A then sub-contracts with another carrier (B) for the performance of the carriage. B gets reward from A and is liable to A under the sub-contract and also as a bailee. A is similarly liable to C under the main carriage of goods contract and as a bailee. Further, B is liable as sub-bailee to C. If the goods are damaged in the carriage due to B's fault, C may sue A under contract or B as a sub-bailee. The general practice is for C to sue A under the main contract. A would then join B as defendant or turn to B for compensation.

67. *Morris v Martin* [1966] 1 QB 716.
68. [1982] 1 Lloyd's Rep 117.

Exemption clauses

Subject to the Control of Exemption Clauses Ordinance (Cap 71) and the relevant common law rules, a bailee may exempt or limit liability.

In case of sub-bailment, if the sub-bailment agreement contains exemption clauses, the sub-bailee is also protected against the bailor.[69] However, the sub-bailee may claim the protection of the clauses only if the bailor has generally consented to sub-bailment.[70]

If a sub-bailee would be liable under the sub-bailment, relying on the exemption clauses of the agreement of the head bailment would not be an option.

> *Bewise Motors Co Ltd v Hoi Kong Container Services Ltd*[71]
>
> F was a freight forwarder and contracted with P to arrange for the carriage of P's cars in containers to China. Under a contract between F and D, a container depot operator, D was responsible for putting the cars in containers and aboard a ship. P sent the cars directly to D. Because of D's negligence, the cars were stolen from the depot. D would be liable under its own contract. Liability for the theft was, however, exempted by a term in the contract (also an agreement of bailment) between F and P. D argued that it, as a sub-bailee, was protected by the exemption clause.
>
> The court held that D was liable. Subject to any modifying terms, 'the duty of a bailee is to take reasonable care of the article bailed to it. It follows logically that if a bailee negligently allows the article to be stolen it is liable whether the theft is committed by a stranger or by a servant or agent whether entrusted with the article or otherwise.' The exemption clause in the contract between F and P could not 'supervene over the actual terms of a sub-contract or sub-bailment'.

69. *Morris v Martin* [1966] 1 QB 716, 729.
70. *The Pioneer Container* [1994] 2 AC 324.
71. [1998] 2 HKLRD 645.

4

Carriage and International Trade Finance

4.1 INTRODUCTION

Consider the following situation: the Indian buyer 'Indian Import Company Limited' intends to purchase a consignment of cargo from the Hong Kong seller 'HK Export Company Limited'.

The managers of the Hong Kong company may have the following questions in their minds:

- How can we minimize the risk of non-payment? How do we know whether the Indian company is able to pay on time once the goods have been shipped?
- How can the bank help us in the practical arrangements for the transaction? We may need assistance from the banks with all the necessary documentation, particularly with the specific procedure to be followed. We may also need advice from the banks concerning matters such as currency restrictions and import/export licences.
- How can we prevent the Indian buyer from discovering and contracting directly with our supplier? Our trading company does not have any manufacturing operations. After entering into the contract of sale and purchase with the Indian buyer, we will have to purchase the requested goods from a manufacturer in Mainland China and export the goods to India for profit. We would like to make sure that the confidentiality of the identity of the Mainland Chinese manufacturer is preserved.

On the other hand, the buyer in India may have the following concerns:

- Can we be sure that the Hong Kong seller will deliver the goods on time?
- Before we pay, how can we check that the goods are exactly those we ordered? We want to be absolutely sure that the seller in Hong Kong has shipped the goods of the right quality, quantity and packing requirements before we pay for the goods.
- We would prefer to delay paying for the goods until we have sold them. Is this possible?

Basically, three payment options are available: payment after delivery of the goods in India, payment of the goods after shipment from Hong Kong, and payment in advance by the Indian buyer before shipment. From the point of view of the Indian buyer, payment after delivery is the best option. After delivery of the goods at the port of discharge in India, the buyer has the opportunity to inspect the cargo, and then makes payment only if the cargo is in compliance with the contractual requirements. In contrast, the Hong Kong seller will be very reluctant to accept this payment option. After taking delivery of the cargo, the Indian buyer may refuse to pay or even go into liquidation. It will be very inconvenient and expensive for the Hong Kong seller to sue the Indian company for non-payment. Hence, the option of payment in advance before shipment is likely to be preferred by the Hong Kong seller.

As shown in the above example, a commercially acceptable compromise is needed between the buyer and the seller. Otherwise, there would be no trade between nations. This is facilitated by use of a letter of credit as a mechanism of international payment. Lord Wright said: 'The general course of international commerce involves the practice of raising money on the documents so as to bridge the period between the shipment and the time of obtaining payment against documents.'[1]

Definition of letter of credit

Documentary credits are most commonly used in financing international sales transactions. A letter of credit may be described as an undertaking or promise by a bank to pay money to or on behalf of the customer for whom it has issued the credit. Expressed more fully, it is a written undertaking by a bank (*issuing bank*) given to the seller (*beneficiary*) at the request, and in accordance with the instructions, of the buyer (*applicant*) to effect payment

1. *TD Bailey v Ross T Smith & Co Ltd* (1940) 56 TLR 828.

up to a stated sum of money, within a prescribed time limit and against stipulated documents.

International trade and financing procedure: The five basic steps

1. The seller and the overseas buyer conclude a contract of sale providing for payment by a letter of credit

In international commercial practice, CIF and FOB contracts are the most common forms of international sales, although the parties are absolutely free to decide between themselves how the contract shall be fulfilled.

CIF stands for 'Costs, Insurance and Freight'. The seller must pay the costs and freight necessary to bring the goods to the named destination. In addition, the seller has to procure marine insurance against the risk of loss or damage to the goods during the carriage. The seller contracts with the insurer and pays the insurance premium.

FOB means 'Free On Board'. The goods are placed on board a ship by the seller at a port of shipment named in the sales contract. The buyer has to nominate a carrier and contract for the carriage and pay the freight. The risk of loss of or damage to the goods is transferred from the seller to the buyer when the goods pass the ship's rail. The marine insurance is normally arranged by the buyer directly, though sometimes the sales contract may require the seller to arrange it for the account of the buyer.

2. The application and the issuance of the credit

The overseas buyer (the '*applicant* for the credit') instructs a bank at his place of business (the '*issuing bank*') to issue a letter of credit in favour of the seller (the '*beneficiary*') on the terms specified by the buyer in his instructions to the issuing bank. The instructions to be given by the applicant to the issuing bank will cover items such as:

- the full name and address of the beneficiary;
- the amount and type of credit;
- the details of the documents required;
- the date and expiry of the credit.

Usually, a prudent buyer will fax a draft of the credit application to the seller for his confirmation and comment before actually instructing the bank to issue the credit. The issuing bank (commonly in the country of the buyer) arranges with another bank (usually in the country of the seller) to act as the *nominated/advising/confirming* bank.

- The *nominated bank* is authorized to pay or accept bills of exchange against presentation of documents by the seller.
- If the second bank is simply 'advising the credit' (the *advising* bank), it will mention this fact when it forwards the credit to the seller. Such a bank is under no commitment/engagement to pay the seller. It is only responsible for advising (ie, informing or notifying) the seller that the credit has been issued.
- If the nominated/advising bank is also 'confirming the credit' (the *confirming* bank), it will so state. This means that the confirming bank adds its own undertaking to pay or accept bills of exchange, provided all the documents are in order and the credit requirements are met.

3. Presentation of documents required by the letter of credit

As soon as the seller receives the credit, and provided the correct documents conforming with the terms of the credit are tendered by the seller before the expiry of the credit, the issuing bank (and also of the confirming bank if it is confirmed) must pay the purchase price to the seller. The advising/nominated/confirming bank has to check the documents carefully against the credit. If there is no nominated bank, the seller has to present the documents directly to the issuing bank. The documents that are usually stipulated in a credit include, *inter alia*:

- Ocean bill of lading: This is a document issued by the carrier. It is evidence of a contract of carriage, a receipt for the goods, and a document of title to the goods.
- Charterparty bill of lading: This is a bill which incorporates, by reference, some of the terms of the charterparty. Unless otherwise stipulated in the credit, banks will reject a document which 'indicates that it is subject to a charterparty'.
- Combined transport (multi-modal transport) document: This is also called a 'through bill of lading'. Where the ocean shipment forms only part of the complete journey, and the goods have to be carried by other land/air/sea carriers, it is more convenient for the cargo consignor to contract with a multi-modal transport operator (MTO). The MTO will then sub-contract with the various carriers who have to carry the goods at the consecutive stages of the journey.
- Air waybill.

- Commercial invoice: This normally includes the names and addresses of the seller and the buyer, the date, a description of the goods, the weight/quantity of the goods, the number of packages, the shipping marks, the unit price and the total price. The particulars of the invoice must agree with those stated in the credit.
- Packing list: This sets out the contents of each package without prices. The packing list is used for various purposes, eg, as the basis for freight calculation or to satisfy customs requirements.
- Insurance certificate: The credit shall specify the type of insurance required and, if any, the additional risks which are to be covered.
- Certificate of origin/inspection/quality/quantity.

4. Settlement or payment under the letter of credit

After presenting the documents that comply with the terms and requirements of the letter of credit, the beneficiary is entitled to receive payment from the bank. A 'payment at sight credit' requires a payment immediately at sight against presentation of the conforming documents. If the parties have arranged a 'deferred payment credit', the advising/nominated/confirming bank will make payment at a future date. For example, the letter of credit may provide for 'payment 60 days from the date of the bill of lading'.

If the credit is an '*acceptance credit*', the seller presents the documents required by the letter of credit, accompanied by a bill of exchange (also called a 'draft'). The bill will normally be a time draft (post-dated). If the advising/nominated/confirming bank accepts the bill of exchange, it commits to pay the face value on maturity to the payee/indorsee presenting the bill. The seller may not wish to wait until the bill matures, and he can discount it to his own bank or sell it. If the credit is a '*negotiation credit*', the seller presents the documents required by the letter of credit, accompanied by a bill of exchange. If the advising/nominated/confirming bank is satisfied that the documents presented are in order, it will purchase (ie, negotiate) the bill of exchange (usually subject to deduction of discount or commission). The advising/nominated/confirming bank also becomes the negotiating bank. The seller receives payment for the bill of exchange, not payment under the credit. The negotiating bank will deliver the documents to the drawee. The drawee will honour the bill of exchange if the documents are in order.

5. Reimbursement

The advising/nominated/confirming bank sends the documents to the issuing bank. The issuing bank checks the documents. If the documents meet the credit requirements, the issuing bank will reimburse the advising/nominated/confirming bank the sum that has been paid to the seller or the value of the bill of exchange accepted. The issuing bank then releases the documents to the buyer upon payment of the amount due, or upon other terms agreed between the buyer and the issuing bank. The buyer presents the transport document to the carrier and obtains delivery of the cargo at the port of discharge.

Common types of credits (as defined by International Chamber of Commerce)

Revocable and irrevocable credit

From the seller's perspective, a revocable credit involves risk. The credit may be altered or cancelled while the goods are in transit and before the documents are presented, or, though presented, before payment has been made. However, it gives the buyer maximum flexibility as it can be altered or cancelled without prior notice to the seller up to the moment of payment by the advising/confirming bank. The credit should expressly indicate whether it is revocable or irrevocable. In the absence of such indication, the credit shall be deemed to be irrevocable.

Confirmed and unconfirmed credit

A confirmed credit gives the seller a double assurance of payment since a bank in the seller's country has added its own promise to honour the credit. However, it represents an additional requirement on the part of the buyer and is more costly.

Revolving credit

If the buyer is a regular customer of the exporter, the parties may wish to arrange for a revolving credit for an agreed maximum amount. It still has an expiry date, but renewal is not required before the credit expires.

Transferable credit

A transferable credit is one that can be transferred by the first beneficiary to a second beneficiary. It is normally used when the first beneficiary does not supply the cargo himself, but is an agent and therefore intends to transfer part, or all, of his rights and obligations to the actual supplier as the second beneficiary.

Back-to-back credit

Under the back-to-back concept, the seller as the beneficiary of the first credit, offers it as 'security' to the advising/nominated/confirming bank for the issuance of the second credit. The seller is the beneficiary of the first credit but is the applicant of the second credit. The advising/nominated/confirming bank of the first credit is also the issuing bank of the second credit. The second credit must be worded so as to produce the documents (apart from the commercial invoice) required by the first credit, and to produce them within the time limits set by the first credit.

Standby credit

This is triggered by the submission of documents in compliance with the credit's requirements. Although a common letter of credit usually obliges the beneficiary to submit the transport documents (and other specified documents), a standby credit may be activated by a document of any description, eg, a demand by the beneficiary, or a statement by the beneficiary that he has shipped the cargo. Its function is similar to that of a bank guarantee.

Trust receipts

Having effected payment and received the documents from the advising/nominated/confirming bank or the seller, the issuing bank will seek reimbursement from the buyer before releasing the documents to the buyer. However, the buyer will need the documents in order to obtain the cargo from the shipping company for the purpose of selling them, for it is out of the proceeds of sale that he will repay the bank. A trust receipt (or a trust letter) is a document by which the buyer undertakes that in consideration of the release of the documents to him, he will hold the documents on trust for the bank, will use the documents to obtain the cargo and sell the cargo as the bank's agent, and will hold the cargo until sale, and the proceeds after sale, on trust for the bank. Even if the buyer is a company, it is held that the trust receipt does not amount to a charge so as to attract registration under the Company Ordinance.[2] Sometimes the issuing bank may require the bill

2. *Re David Allester Ltd* [1922] 2 Ch 211.

of lading be issued 'to the order of the bank' as the consignee. The bank may also insist that the cargo be warehoused in the bank's name. This will enable the issuing bank to retain the cargo as against the buyer until the sum secured by the cargo has been paid.

4.2 UNIFORM CUSTOMS AND PRACTICE FOR COMMERCE (UCP)

International letter of credit practice has been standardized by the International Chamber of Commerce ('ICC'). The latest revision of the Uniform Customs and Practice for Documentary Credits ('UCP 600') came into force on 1 July 2007.[3]

UCP is the universally recognized set of rules governing letters of credit and is widely regarded as the most successful act of commercial harmonization in the history of world trade. The UCP is indispensable to letter of credit operation. Most of the banks adopt the UCP in the letters of credit that they issue. In recent years, the judicial guidelines pertinent to the interpretation of the UCP have been developed and further refined by judges in a number of major lawsuits involving letters of credit. With the kind permission of the International Chamber of Commerce, a relevant extract of the UCP is now produced at the end of this chapter as Annex.

4.3 STRICT COMPLIANCE

Articles 14, 15 and 16 of the UCP 600 state that upon receipt of the documents, the issuing bank and/or correspondent bank must determine on the basis of the documents alone whether or not they appear on their face to be in compliance with the terms and conditions of the credit. The bank is not required to make inquiries as to any ambiguity or discrepancy. The 'strict compliance' rule has long prevailed in English law. Lord Sumner's classic statement in *Equitable Trust Co of New York v Dawson Partner*[4] is often cited: 'There is no room for documents which are almost the same, or which will do just as well. Business could not proceed securely on any other lines.' An American court (*Fidelity National Bank v Dade County*)[5] has paraphrased the gist of Lord Sumner's statement: 'Compliance with the terms of a letter of credit is not like pitching horseshoes. No points are awarded for being

3. The first adopted version of the UCP was published in 1929. Subsequent revisions of the UCP were made in 1933, 1962, 1974, 1983, 1994, and 2006. The 1994 version of the UCP is also known as 'UCP 500'.
4. (1927) 27 Lloyd's Rep 49.
5. 371 So 2d 545 at 546 (1979).

close.' The doctrine of strict compliance is further illustrated in the classic case *JH Rayner & Co Ltd v Hambro's Bank Ltd.*[6] The credit described the goods as 'Coromandel groundnuts'. The sellers had tendered a bill of lading referring to the goods as 'machine-shelled kernels'. It was held that the bank was entitled to reject the documents and refuse payment, even though it was well known in the trade that the two terms are one and the same: 'It was quite impossible to suggest that a banker is to be affected with knowledge of the customs and customary terms of every one of the thousands of trades for whose dealings he may issue a letter of credit.'

In *China New Era International Limited v Bank of China (HK) Limited,*[7] when opening the credit, *China New Era* provided *Bank of China* with a specimen copy of its authorized signature and chop. The specimen chop gave *China New Era*'s Chinese name in traditional Chinese characters. *Bank of China* was instructed that the signature and chop on any cargo receipt submitted under the credit had to be verified against the specimen. However, the chop imprint on the cargo receipt subsequently tendered under the credit gave *China New Era*'s name in simplified (not traditional) Chinese characters. It was held that the cargo receipt failed to conform with the letter of credit.

There is no room for a *de minimis* effect, as decided by the English Court of Appeal in *Seaconsar Far East v Bank Markazi Jomhouri Islami Iran.*[8] The plaintiff agreed to sell a large quantity of artillery shells to the Iranian Ministry of Defence for US$193m. Payment was to be by letter of credit. The letter of credit stipulated, *inter alia*, that all documents presented to the bank should carry the credit number and the buyer's name. One of the documents tendered omitted to state the credit number and the buyer's name. It was held that the tender was bad. The bank was right to reject the documents even though the discrepancy might be trivial. Lloyd LJ stated:

> [The plaintiff's counsel] argues that the absence of the letter of credit number and the buyer's name was an entirely trivial feature of the document. I do not agree. I cannot regard as trivial something which, whatever may be the reason, the credit specifically requires. It would not help, I think, to attempt to define the sort of discrepancy which can properly be regarded as trivial. But one might take, by way of example, *Bankers Trust Co v State Bank of India* [1991] 2 Lloyd's Rep 443 where one of the documents gave the buyer's telex number as 931310 instead of 981310. The discrepancy in the present case is not of that order.

6. [1943] KB 37.
7. [2009] HKEC 2020.
8. [1993] 1 Lloyd's Rep 236.

Previously, in *Hing Yip Hing Fat Co Ltd v The Daiwa Bank Ltd*,[9] Kaplan J of the Hong Kong Court held that the use of the word 'industrial' in the documents presented to the advising bank was an obvious typographical error from the word 'industries'. It was not a discrepancy upon which the defendant can rely. Such an error could easily occur in a society where English is not the first language for 98 per cent of the population. With due respect, it is submitted that in light of *Seaconsar*, bankers and their legal advisors must exercise the utmost care if they discover some possible 'typographical errors' contained in the documents required by the credit. An omission or mistake caused by a typographical error may possibly be considered to be fatal to the beneficiary's entitlement under the credit if the omission or error is in connection with something which the credit specifically requires. The court may not easily regard a typographical error as 'trivial' on every occasion,[10] bearing in mind that the banks are not expected to evaluate the importance of the information or particulars required under the credit.

The plaintiff's counsel in *Seaconsar* further argued that the discrepancy, even if it cannot be regarded as trivial, can be cured by reference to the other documents, such as the certificate of inspection and the certificate of quality. He relied on the opinion expressed by Parker J in *Banque de l'Indochine et de Suez SA v JH Rayner Ltd*,[11] where he said:

> I have no doubt that so long as the documents can be plainly seen to be linked with each other, are not inconsistent with each other or with the terms of the credit, do not call for inquiry and between them state all that is required in the credit, the beneficiary is entitled to be paid.

The Court of Appeal, however, took the view that Parker J was not saying that a deficiency in one document could be cured by reference to another. There was an express requirement that the documents should be linked in the sense that each of them should contain the letter of credit number and the name of the buyer. Whatever the reason for that, the requirement was clear. Lloyd LJ continued:

> I do not see how Bank Melli could ignore that requirement. It may be that the procès-verbal in fact related to the same goods, and that one can see this by inference from the other documents. But the absence of the letter of credit number and the name of [the buyer] on the procès-verbal called for some explanation. The bank was therefore entitled to reject . . . To hold that it is even arguable that the documents as

9. [1991] 2 HKLR 35.

10. Unless the typos are minor errors in respect of the credit applicant's address, telephone number, email, etc (UCP 600 Art 14(j)).

11. [1982] 2 Lloyd's Rep 476.

> presented were a valid tender under the credit would, I suspect, cause surprise among bankers, and risk upsetting the ordinary course of business. In my judgment the tender was clearly bad.

What is the precise scope of banks' duties in connection with the examination of documents stipulated in a letter of credit? In what circumstances can the documents be regarded as sufficiently linked with each other? According to Articles 14(a) and 15 of UCP, on receipt of the documents, the bank(s) concerned must determine *on the basis of documents alone* whether they appear on their face to be in compliance with the terms and conditions of the credit.

The autonomy of the credit: The 'fraud exception'

UCP Article 4 states that: 'A credit by its nature is a separate transaction from the sale or other contract on which it may be based. Banks are in no way concerned with or bound by such contract, even if any reference whatsoever to it is included in the credit.' Article 5 states that: 'Banks deal with documents and not with goods, services or performance to which the documents may relate.' The letter of credit transaction is thus a paper transaction. However, there is a very important exception to the principle of autonomy of the letter of credit: the 'fraud exception'. The allegation of fraud is normally raised by the buyer. The buyer will attempt to prevent the bank from honouring the credit. The bank should refuse to pay under the credit if it is proved to its satisfaction that the documents, though apparently in order on their face, are fraudulent and the beneficiary/seller was involved in the fraud.

In *Unicredito Italiano S.P.A. Hong Kong Branch v Guang Xin Enterprises Ltd*,[12] the Hong Kong Court held: 'It is a well-established law that the test is whether, standing in the shoes of the paying bank at the time of payment, the fraud was clear and obvious to it . . . If fraud was clear and obvious, then the bank pays the beneficiary at its own peril and it is not entitled to reimbursement . . .' In *United City Merchants (Investments) Ltd v Royal Bank of Canada*,[13] it was held that even if there is unambiguous evidence before the bank that the documents are fraudulent or forged, the bank must pay if there is no evidence before the bank which shows that the beneficiary/seller knew of the fraud. In this case, the bill of lading presented by the seller was backdated. The goods were actually loaded after the latest date of shipment required by the credit. The bill of lading complied with the credit on its face.

12. [2002] 339 HKCU 1.
13. [1983] 1 AC 168.

The false date was inserted by an employee of the loading brokers and the seller/beneficiary knew nothing about it. The court held that the bank was obliged to pay the seller/beneficiary. The bank's duty is limited to verifying the apparent good order of the documents: *Gian Singh & Co. Ltd v Banque de l'Indochine.*[14] It is not responsible for ensuring the accuracy or authenticity of the documents, but must exercise reasonable care that the documents appear to be in order.

Must a notice of rejection be given by telecommunication?

In *Seaconsar*, the House of Lords held that the plaintiff should have leave to serve the proceedings out of the jurisdiction. As discussed, the documentation presented to the defendant bank on behalf of the plaintiff (the beneficiary of the credit) contained discrepancies. The bank sought to reject the documents orally, a rejection that the plaintiff argued was invalid under UCP (see UCP 600 Art 16 (d)). It stipulates that if a bank decides to refuse the documents, it must give notice to that effect by telecommunication or, if not possible, by other expeditious means. The plaintiff's contention was rejected by the trial judge below. The plaintiff appealed, arguing that the use of some other expeditious means was only permissible if '*telecommunication*' was impossible. It was not impossible in the instant case as the defendant bank had been given a telephone number and two telex numbers of the plaintiff. The Court of Appeal dismissed the appeal.[15] The trial judge's conclusion was upheld on the basis that notice was not required to be given by telecommunication if a senior official of the beneficiary was present at the bank to receive notice. Sir Christopher Staughton said:

> As a matter of construction of the Uniform Customs, we do not consider that they require notice to be given by telecommunication if a senior official of the beneficiary (or the remitting bank, as the case may be), under whose aegis the documents were presented, is present at the bank to receive notice. It must surely be an implied term that notice can then and there be given *viva voce*, rather than to another person who is some distance away.

Several practical difficulties in applying this test can be identified. Who is a 'senior official' of a corporate beneficiary of a letter of credit? In some companies, all the directors devote their whole time and attention to the company's affairs. In other companies, the directors divide the various sectors of management between themselves. For instance, the full-time

14. [1974] 2 All ER 754.
15. [1999] 1 Lloyd's Rep 36.

executive directors may be responsible for the day-to-day management of the company, while other part-time non-executive directors may only contribute their skill and experience in some of the major decisions made by the board of directors. In addition, the senior managers or consultants of a company may be given wide powers by the board of directors to run the company, even though they are only employees under service contracts with the company. The uncertainty as to whether a person is a 'senior official' is more apparent in the situation of partnership and sole proprietorship.

ANNEX

UCP 600 Articles 19–23

Article 19: Transport Document Covering at Least Two Different Modes of Transport

a. A transport document covering at least two different modes of transport (multimodal or combined transport document), however named, must appear to:

   i. indicate the name of the carrier and be signed by:
      - the carrier or a named agent for or on behalf of the carrier, or
      - the master or a named agent for or on behalf of the master.

      Any signature by the carrier, master or agent must be identified as that of the carrier, master or agent.

      Any signature by an agent must indicate whether the agent has signed for or on behalf of the carrier or for or on behalf of the master.

   ii. indicate that the goods have been dispatched, taken in charge or shipped on board at the place stated in the credit, by:
      - pre-printed wording, or
      - a stamp or notation indicating the date on which the goods have been dispatched, taken in charge or shipped on board.

      The date of issuance of the transport document will be deemed to be the date of dispatch, taking in charge or shipped on board, and the date of shipment. However, if the transport document indicates, by stamp or notation, a date of dispatch, taking in charge or shipped on board, this date will be deemed to be the date of shipment.

   iii. indicate the place of dispatch, taking in charge or shipment and the place of final destination stated in the credit, even if:
      a) the transport document states, in addition, a different place of dispatch, taking in charge or shipment or place of final destination, or
      b) the transport document contains the indication '"intended"' or similar qualification in relation to the vessel, port of loading or port of discharge.

   iv. be the sole original transport document or, if issued in more than one original, be the full set as indicated on the transport document.

   v. contain terms and conditions of carriage or make reference to another source containing the terms and conditions of carriage (short form or blank back transport document). Contents of terms and conditions of carriage will not be examined.

   vi. contain no indication that it is subject to a charter party.

b. For the purpose of this article, transhipment means unloading from one means of conveyance and reloading to another means of conveyance (whether or not in different modes of transport) during the carriage from the place of dispatch, taking in charge or shipment to the place of final destination stated in the credit.

c. i. A transport document may indicate that the goods will or may be transhipped provided that the entire carriage is covered by one and the same transport document.

ii. A transport document indicating that transhipment will or may take place is acceptable, even if the credit prohibits transhipment.

Article 20: Bill of Lading

a. A bill of lading, however named, must appear to:

i. indicate the name of the carrier and be signed by:
- the carrier or a named agent for or on behalf of the carrier, or
- the master or a named agent for or on behalf of the master.

Any signature by the carrier, master or agent must be identified as that of the carrier, master or agent.

Any signature by an agent must indicate whether the agent has signed for or on behalf of the carrier or for or on behalf of the master.

ii. indicate that the goods have been shipped on board a named vessel at the port of loading stated in the credit by:
- pre-printed wording, or
- an on-board notation indicating the date on which the goods have been shipped on board.

The date of issuance of the bill of lading will be deemed to be the date of shipment unless the bill of lading contains an on board notation indicating the date of shipment, in which case the date stated in the on board notation will be deemed to be the date of shipment.

If the bill of lading contains the indication "intended vessel" or similar qualification in relation to the name of the vessel, an on board notation indicating the date of shipment and the name of the actual vessel is required.

iii. indicate shipment from the port of loading to the port of discharge stated in the credit.

If the bill of lading does not indicate the port of loading stated in the credit as the port of loading, or if it contains the indication "intended" or similar qualification in relation to the port of loading, an on board notation indicating the port of loading as stated in the credit, the date of shipment and the name of the vessel is required. This provision applies even when loading on board or shipment on a named vessel is indicated by pre-printed wording on the bill of lading.

iv. be the sole original bill of lading or, if issued in more than one original, be the full set as indicated on the bill of lading.

v. contain terms and conditions of carriage or make reference to another source containing the terms and conditions of carriage (short form or blank back bill of lading). Contents of terms and conditions of carriage will not be examined.

vi. contain no indication that it is subject to a charter party.

b. For the purpose of this article, transhipment means unloading from one vessel and reloading to another vessel during the carriage from the port of loading to the port of discharge stated in the credit.

c. i. A bill of lading may indicate that the goods will or may be transhipped provided that the entire carriage is covered by one and the same bill of lading.

ii. A bill of lading indicating that transhipment will or may take place is acceptable, even if the credit prohibits transhipment, if the goods have been shipped in a container, trailer or LASH barge as evidenced by the bill of lading.

d. Clauses in a bill of lading stating that the carrier reserves the right to tranship will be disregarded.

Article 21: Non-Negotiable Sea Waybill

a. A non-negotiable sea waybill, however named, must appear to:

i. indicate the name of the carrier and be signed by:
- the carrier or a named agent for or on behalf of the carrier, or
- the master or a named agent for or on behalf of the master.

Any signature by the carrier, master or agent must be identified as that of the carrier, master or agent.

Any signature by an agent must indicate whether the agent has signed for or on behalf of the carrier or for or on behalf of the master.

ii. indicate that the goods have been shipped on board a named vessel at the port of loading stated in the credit by:
- pre-printed wording, or
- an on board notation indicating the date on which the goods have been shipped on board.

The date of issuance of the non-negotiable sea waybill will be deemed to be the date of shipment unless the non-negotiable sea waybill contains an on board notation indicating the date of shipment, in which case the date stated in the on board notation will be deemed to be the date of shipment.

If the non-negotiable sea waybill contains the indication "'intended vessel"' or similar qualification in relation to the name of the vessel, an on board notation indicating the date of shipment and the name of the actual vessel is required.

iii. indicate shipment from the port of loading to the port of discharge stated in the credit.

If the non-negotiable sea waybill does not indicate the port of loading stated in the credit as the port of loading, or if it contains the indication "intended" or similar qualification in relation to the port of loading, an on board notation indicating the port of loading as stated in the credit, the date of shipment and the name of the vessel is required. This provision applies even when loading on board or shipment on a named vessel is indicated by pre-printed wording on the non-negotiable sea waybill.

iv. be the sole original non-negotiable sea waybill or, if issued in more than one original, be the full set as indicated on the non-negotiable sea waybill.

v. contain terms and conditions of carriage or make reference to another source containing the terms and conditions of carriage (short form or blank back non-negotiable sea waybill). Contents of terms and conditions of carriage will not be examined.

vi. contain no indication that it is subject to a charter party.

b. For the purpose of this article, transhipment means unloading from one vessel and reloading to another vessel during the carriage from the port of loading to the port of discharge stated in the credit.

c. i. A non-negotiable sea waybill may indicate that the goods will or may be transhipped provided that the entire carriage is covered by one and the same non-negotiable sea waybill.

ii. A non-negotiable sea waybill indicating that transhipment will or may take place is acceptable, even if the credit prohibits transhipment, if the goods have been shipped in a container, trailer or LASH barge as evidenced by the non-negotiable sea waybill.

d. Clauses in a non-negotiable sea waybill stating that the carrier reserves the right to tranship will be disregarded.

Article 22: Charter Party Bill of Lading

a. A bill of lading, however named, containing an indication that it is subject to a charter party (charter party bill of lading), must appear to:

i. be signed by:
- the master or a named agent for or on behalf of the master, or
- the owner or a named agent for or on behalf of the owner, or
- the charterer or a named agent for or on behalf of the charterer.

Any signature by the master, owner, charterer or agent must be identified as that of the master, owner, charterer or agent.

Any signature by an agent must indicate whether the agent has signed for or on behalf of the master, owner or charterer.

An agent signing for or on behalf of the owner or charterer must indicate the name of the owner or charterer.

ii. indicate that the goods have been shipped on board a named vessel at the port of loading stated in the credit by:
    - pre-printed wording, or
    - an on board notation indicating the date on which the goods have been shipped on board.

    The date of issuance of the charter party bill of lading will be deemed to be the date of shipment unless the charter party bill of lading contains an on board notation indicating the date of shipment, in which case the date stated in the on board notation will be deemed to be the date of shipment.
iii. indicate shipment from the port of loading to the port of discharge stated in the credit. The port of discharge may also be shown as a range of ports or a geographical area, as stated in the credit.
iv. be the sole original charter party bill of lading or, if issued in more than one original, be the full set as indicated on the charter party bill of lading.

b. A bank will not examine charter party contracts, even if they are required to be presented by the terms of the credit.

Article 23: Air Transport Document

a. An air transport document, however named, must appear to:

i. indicate the name of the carrier and be signed by:
    - the carrier, or
    - a named agent for or on behalf of the carrier.

    Any signature by the carrier or agent must be identified as that of the carrier or agent.

    Any signature by an agent must indicate that the agent has signed for or on behalf of the carrier.
ii. indicate that the goods have been accepted for carriage.
iii. indicate the date of issuance. This date will be deemed to be the date of shipment unless the air transport document contains a specific notation of the actual date of shipment, in which case the date stated in the notation will be deemed to be the date of shipment.

    Any other information appearing on the air transport document relative to the flight number and date will not be considered in determining the date of shipment.
iv. indicate the airport of departure and the airport of destination stated in the credit.
v. be the original for consignor or shipper, even if the credit stipulates a full set of originals.
vi. contain terms and conditions of carriage or make reference to another source containing the terms and conditions of carriage. Contents of terms and conditions of carriage will not be examined.

b. For the purpose of this article, transhipment means unloading from one aircraft and reloading to another aircraft during the carriage from the airport of departure to the airport of destination stated in the credit.

c. i. An air transport document may indicate that the goods will or may be transhipped, provided that the entire carriage is covered by one and the same air transport document.
   ii. An air transport document indicating that transhipment will or may take place is acceptable, even if the credit prohibits transhipment.

PART II

Shipping and Logistics Law

5

Freight Forwarding Law

5.1 INTRODUCTION

In England, one of the earliest freight forwarders was Thomas Meadows and Company Limited of London, established in 1836.[1] The first international freight forwarders were innkeepers in London who held and re-forwarded the personal effects of their hotel guests.

Originally, a forwarder was an agent for a customer (usually a shipper) to arrange for carriage by contracting with various carriers. Forwarders provided advice on documentation and customs requirements in the country of destination. They would have a correspondent agent overseas who would look after the goods of their customers.

In modern times, forwarders provide a wide range of services, including the collection, packing and consolidation of cargo, warehousing, handling of shipping documents, and arranging for cargo insurance and customs clearance. They may be acting as a carrier (principal) or as an agent for the shipper or both. They operate either as a domestic carrier or otherwise with a corresponding agent overseas or with their own branch-office. Forwarders may own trucks to perform land carriage.

While a unimodal carrier, such as air or sea carrier, will operate one particular mode of carriage, a freight forwarder may offer door-to-door international multimodal carriage. Sometimes, a forwarder may act as a carrier to perform the whole carriage (unimodal or multimodal) or may also act as an

1. According to 'Understanding the Freight Business', written and published by the executive staff of Thomas Meadows and Company in 1972.

agent for a customer in forming separate contracts with different carriers. It is also possible for a forwarder to act as a carrier for one stage of the carriage and, as an agent of the customer, to form separate contracts with other carriers for the other stages.

These services are becoming important in today's logistics industry. They are also indispensable if Hong Kong is to become a logistics hub for international trade.

Different legal status

The law of freight forwarding cuts across several areas of law. The following is a brief discussion on the relevant law and the legal environment in which forwarders carry on their business.

Often, a forwarder acts as an agent for a customer for the arrangement of the carriage. Usually, there is an agency contract between the forwarder and the customer. The rights and obligations of the parties are basically determined by the terms of the contract. In performing the carriage, the forwarder has all the rights and liabilities of a carrier. If it is an international carriage, the forwarder is also subject to the international convention governing the particular mode of transport in issue.

In any case, the forwarder at fault is liable in the law of negligence to parties with whom there is no contract. Further, if the forwarder is in possession of the goods, there is liability as a bailee if the goods are damaged or lost.

In short, a firm grasp of the law of agency is essential for understanding the legal status of forwarders. Knowledge of the laws of contract, torts and personal property are necessary to determine forwarders' rights and obligations.

Standard trading terms

The rights and obligations of a forwarder to the customer are generally governed by the standard terms of the forwarder's form contract. Standard conditions drafted by leading organizations of the trade are often adopted. Examples are the Model Rules for Freight Forwarding Services of the International Federation of Forwarding Agents Association[2] (FIATA), the Standard Trading Conditions of the Hong Kong Association of Freight

2. The website is <www.fiata.com>.

Forwarding Agents[3] (HAFFA) and the 1989 Standard Trading Conditions of the British International Freight Association (BIFA).[4]

In Hong Kong, many forwarders are HAFFA members. The HAFFA STC are the most widely used conditions for freight forwarding in Hong Kong. As a FIATA member, HAFFA is the only body in Hong Kong that may issue FIATA documents, such as FIATA Bills of Lading and Forwarder's Cargo Receipts. Where appropriate, the HAFFA STC will be referred to in this chapter.

Incorporation issue

Although a forwarder would like to use the standard conditions of its form contract to govern the relationship with its customer, it has to prove that such a form contract has become the contract of the parties. This issue will arise when an oral contract is formed initially between a forwarder and its customer and the standard conditions are to be incorporated to the contract through an incorporation notice in the invoice from the forwarder.

Normally, the incorporation notice has to be given to another party on or before the formation of the contract. Moreover, the notice has to be reasonable in order to take effect.

If the incorporation notice has been given after the formation of the contract, it may take effect because of 'course of dealing'.

No standard trading terms are relevant, however, if they are not incorporated into the contract. All the relevant circumstances must be taken into account to determine whether a particular set of conditions applies to the contract in issue.

> *Poseidon Freight Forwarding Co Ltd v Davies Turner Southern Ltd*[5]
> D and P were freight forwarders and had a 'mutual agency relationship' between them. The effect of the arrangements was that one party would deliver cargoes shipped by the other. On one occasion, D misdelivered some goods shipped by P. D would be liable unless the BIFA Standard Trading Conditions applied. It was found that before the misdelivery, D had sent 7 documents and 69 faxes to P. The faxes stated that the BIFA Standard Trading Conditions were printed on the back. In fact, only the front page had been faxed and none of the faxes bore the terms. The terms were printed on the backs of the 7 documents. D argued that the conditions were incorporated into the contracts either by means of documents transmitted by D to P or by past dealing.

3. The website is <www.haffa.com.hk>.
4. The website is <www.bifa.org>.
5. [1996] 2 Lloyd's Rep 388.

> The court held that the BIFA Standard Trading Conditions were not incorporated into the relevant contracts since no reasonable notice of incorporation was given. From the facts, notice of incorporation had only been given for shipments of goods sent by D to P. For goods sent by P to D, the terms did not apply.

Even though a set of standard trading conditions has been incorporated into the contract, its validity now becomes the issue. In a commercial transaction, the shipper or the consignor is usually not a consumer. The Unconscionable Contracts Ordinance (Cap 458) that applies only to contracts formed by consumers is, therefore, not relevant. However, the Control of Exemption Clauses Ordinance (Cap 71) still applies.

> *Overland Shoes Ltd v Schenkers Ltd*[6]
> X was an international freight carrier. In this case, X acted as a forwarder for Y for the arrangement of the carriage of a consignment of shoes from China. The contract incorporated the British International Freight Association standard trading conditions. Similar to Clause 19(4) of the HAFFA STC, Clause 23A (Clause 21A of the 2000 edition) imposed upon Y a duty to make prompt payment 'without reduction or deferment on account of any claim, counterclaim or set off'. When X asked Y to pay the freight and other ancillary charges relating to the carriage, Y tried to set off the VAT (value added tax) that X owed to Y. Y contended that Clause 23A was invalidated by the Unfair Contract Terms Act 1977 (the UK equivalent of the Control of Exemption Clauses Ordinance) because it was unreasonable.
>
> The court held that Clause 23A was reasonable. The following facts were relevant. First, the British International Freight Association standard trading conditions were commonly used. Second, practitioners generally considered them fair and reasonable. Third, almost all interested parties had been consulted when they were drafted. Fourth, the clause did not limit liability but only provided a mechanism for resolving competing claims.

Freight forwarder as agent or principal

Traditionally, freight forwarders are professional intermediaries (or agents) between consignors or consignees and the carrier. The name 'freight forwarder' generally refers to

> any person which holds itself out to the general public to . . . provide and arrange transportation of property, for compensation, and which may assemble and consolidate shipments of such property, and

6. [1998] 1 Lloyd's Rep 498.

> performs or provides for the performance of break-bulk and distributing operations with respect to such consolidated shipments and assumes responsibility for the transportation of such property from point of receipt to point of destination and utilises for the whole or any part of the transportation of such shipments, the services of a carrier or carriers, by sea, land or air, or any combination thereof.[7]

Accordingly, a forwarder only agrees to arrange for the carriage of the goods for customers and does not contract to carry goods. As described in *Jones v General Express*,[8] forwarders are:

> willing to forward goods for you . . . to the uttermost ends of the earth. They do not undertake to carry for you, and they are not undertaking to do it either themselves or by their agent. They are simply undertaking to get somebody to do the work, and as long as they exercise reasonable care in choosing the person to do the work they have performed their contract.

In fact, many disputes between forwarders and their customers are concerned with the status of the forwarder in issue: whether they are an agent or a carrier. In other words, the point in dispute is whether they have, as a principal, formed a carriage of goods contract with the customer. This is particularly important in the context of international carriage of goods. Several international conventions regulate the rights and obligations of sea and air carriers. The frameworks provided by the conventions are different from the common law regime. Take the case of international carriage of goods by air. A forwarder acting as a carrier, is subject to the Warsaw Convention or the amended Convention and is presumed to be liable for the loss of or damage to the cargo, but the liability is generally limited. There is no similar international convention for forwarders. If not, the common law governs. To hold the forwarder liable, the client must prove the forwarder's fault caused the loss. But if that can be done, the forwarder is subject to unlimited liability.

Forwarder as agent

In his traditional role, a forwarder is an agent of the customer in the provision of services. Clause 4.1 of the HAFFA STC recognizes this role of forwarders. The effect of this clause is that the forwarder acts as an agent of the customer unless opting to act as a carrier. The fact that the forwarder issues its own transport document, charges an inclusive price or the goods

7. DJ Hill, *Freight Forwarders* (1972) 16.
8. (1920) 4 Ll L Rep 127.

are forwarded together with the goods of other shippers does not make the forwarder a carrier. In general, a forwarder is remunerated and liable as an agent. All the relevant circumstances must be considered to see whether a forwarder is an agent of a party for whom services were provided.

> *FH Bertling Ltd v Tube Developments Ltd*[9]
> By fax, T sent F a copy of a letter of credit financing a shipment of cargo to be sent to Saudi Arabia. T stated in the fax that T did not intend to have any contractual relationship with F. F took steps to ensure that the shipping documents were in compliance with the letter of credit. F claimed against T for the costs incurred in taking such steps.
>
> The court held that the mere despatch of the letter of credit did not necessarily make F the agent of T. Neither did the acts done by F to ensure compliance with the letter of credit. None of them could be taken as T's instructions to appoint F as T's agent.

The extent of the forwarder's authority as an agent is generally defined by the contract. Standard trading terms of forwarders often give wide discretion to forwarders to facilitate the arrangement of carriage. Clause 5 of the HAFFA STC provides an example of the wide powers and authorities given to forwarders. By virtue of clause 5, the forwarder may, among other things,

- enter into contract with other parties for the carriage or handling of the goods;
- depart from the instructions of the client if it's considered desirable or expedient;
- conclude contracts with third parties without seeking further instructions from the customer;
- inspect the goods or arrange for them to be inspected;
- consolidate the goods.

Further, clause 12 of the HAFFA STC gives the forwarder a general power to deviate. This liberty clause also states that 'the Customer or the Owner' bears the risk of deviation. The obligations of the owner under the HAFFA STC require further consideration. 'The Owner' is defined as

> the owner of the goods (including any packings, containers or equipment other than those provided by the Company or carriers) to which any business concluded under these Conditions relates and any other person who is or may become interested in them and including the consignee named on the front of the Shippers' Instructions and of the Company's form of transport document (including the Company's house air waybill or house bill of lading.)

9. [1999] 2 Lloyd's Rep 55.

This is a very wide definition, covering almost anyone having any interest in the goods. For example, an endorsee of the air waybill or bill of lading or a shipowner having a maritime lien on the goods is caught. Despite this wide definition, owners are not bound by the contract if they do not form the contract themselves or the customer has no authority to form the contract on their behalf.

Under clause 1.4, the customer guarantees any representations, warranties, undertakings and indemnities are jointly and severally made together with the owner. The effects of this clause are examined in greater detail.

The customer, ie, the shipper or consignor, may be an agent of the cargo owner. The cargo owner is, as the principal, bound by a statement made by the customer with proper authority. With respect to such a statement, clause 1.4 does not change the common law rule. Another common law rule is that an agent is generally not personally liable under the contract formed on a disclosed principal's behalf. But under clause 1.4, the customer is personally liable in such a case. If the customer is an agent of an undisclosed principal, the common law rule is that the forwarders can sue cargo owners who have revealed themselves. The shipper, as an agent, is still personally liable after the revelation. Clause 1.4 does not make the customer's liability more onerous in this respect.

The situation is different if the customer makes an undertaking without authority. The common law rule is that the principal who does not later ratify the undertaking is not liable. In such a case, owners are still liable under clause 1.4 because the undertaking is deemed to be made by them as well. Owners are not liable only if the customer has no authority to form the contract on their behalf in the first place or clause 1.4 is not binding on them.

At common law, an agent is not liable for the acts of his principal. Clause 1.4 does not change this rule because it does not hold the customer liable for the cargo owner's statements or acts.

Under clause 3 of the HAFFA STC, the customer warrants that, acting as an agent, there is authorization by the principal to act on their behalf for any business with the forwarder. A similar warranty of authority is implied at common law. However, clause 3 does something more. It imposes personal liability on the agent. In addition to suing the customer for want of authority, the forwarder may enforce the contract against the customer as the agent.

It is often the case that forwarders are acting for both the shipper and the carrier at the same time. In such a case, they are liable as an agent to both principals. If both principals trust and rely on the forwarder, a fiduciary duty is owed them. If there is a breach of duty by, say, making a secret profit, the forwarder may be ordered to account for the secret profit. An agent may, however, obtain remuneration or commission from third parties with the consent of the principal. By adopting the HAFFA STC, the customer agrees that the forwarder may receive remuneration from other parties. With consent of the carrier, the forwarder may charge the customer for service and also get commission from the carrier.

Forwarder acting as carrier

If a forwarder chooses to act as a carrier by undertaking to perform the carriage, there will be liability the same as a carrier. The HAFFA STC allow a forwarder to act as a carrier. Clause 7 of the HAFFA STC states forwarders may in their discretion issue a FIATA Air Waybill or a FIATA Multimodal Transport Bill of Lading naming themselves as the carrier. This is only an entitlement given to the forwarder. The clause does not impose an obligation on the forwarder to act as a carrier.

In the modern commercial world, the traditional distinction between forwarders and carriers blurs. Sometimes, it is not easy to make the distinction. In providing door-to-door service to customers, a forwarder often acts as a carrier with respect to some parts or the whole of the carriage.

A forwarder forming a carriage of goods contract with a customer in the way stipulated in clause 7 of the HAFFA STC acts as a carrier. The relevant transport document issued by the forwarder to the consignor, such as the FIATA Air Waybill or the FIATA Multimodal Transport Bill of Lading, governs the contract. For example, if a FIATA Multimodal Transport Bill of Lading is issued, the forwarder is liable for the goods under the conditions laid down in the bill for the period 'from the time the Freight Forwarder has taken the goods in his charge to the time of their delivery'.

The court will consider all the relevant circumstances to determine whether a forwarder acts as a carrier. Thus, even though a forwarder has no intention to act as a carrier, there may still be liability as a carrier.

> *Yuen Fung Metal Works Ltd v Negel (Hong Kong) Ltd*[10]
>
> A German customer ordered some goods from Y. Y contracted with N to arrange for carriage of the goods to Germany. N issued three original through bills of lading for the goods. The goods were carried to Germany by X. The goods were to be delivered only upon surrender of an original bill of lading. When the goods arrived at the border between Germany and Poland, they were delivered to the German

10. [1978] HKLR 588.

> customer without surrender of an original bill of lading. Y suffered loss when the customer refused to pay or return the goods. Y sued N for the loss. N claimed that it was merely a forwarder and should not be liable for the fault of X, the carrier.
>
> The court held that N was liable. N was in law a carrier when it issued the bills. X was therefore N's agent.

Clause 7 of the HAFFA STC mentions only the FIATA Air Waybill and the Multimodal Transport Bill of Lading. Thus, a forwarder chooses to act as a carrier only if one of the prescribed transport documents is issued. Other transport documents have not been mentioned in clause 7. A forwarder who issues, say, a bill of lading in relation to container carriage, is not a carrier under clause 7. But the court may still hold the forwarder liable as a carrier under the circumstances.

The forwarder may not call the contract a carriage of goods contract. However, the label is not material. The court will look at the contract as a whole to decide whether the forwarder acts as a carrier. For example, it was held in *Fyffes Group Ltd and Caribbean Gold Ltd v Reefer Express Lines Pty Ltd and Reefkrit Shipping Inc (The Kriti Rex)*[11] that 'a contract of affreightment' is a carriage of goods contract.

To determine whether a forwarder undertakes to carry the cargo, the main test is the parties' intention as expressed by the terms of the contract. As pointed out in *Yuen Fung Metal Works Ltd v Negel (Hong Kong) Ltd*,[12] the relationship between a forwarder and a customer 'must be ascertained from the contract itself'. In short, the question to ask is whether there is a contract to 'arrange the carriage' or 'carry'.

Where it is not clear from the agreement whether the forwarder undertakes to perform the carriage, the court would consider all other relevant circumstances. One factor is the method of charging. If an all-in rate is charged, the forwarder is more likely to be a carrier. Second, the obligation to provide information about transport arrangements to the customer is relevant. The forwarder is less likely to be a carrier if the obligation is not extensive. Third, if the forwarder often acted in the past as a carrier in similar situations, the court is more likely to hold that same view. Fourth, forwarders issuing a transport document in their own names are more likely to be liable as carriers. Fifth, if the customer receives a transport document that is enforceable against the party who actually performs the carriage, the forwarder is less likely to be a carrier. Finally, the assets of the forwarder are also relevant. Owning no ship, vehicle or other vessels for the carriage,

11. [1996] 2 Lloyd's Rep 171.
12. [1978] HKLR 588.

it is less likely the court would hold the forwarder formed the contract as a carrier.

All these are factors to consider only if it is unclear from the contract whether the forwarder agrees to carry the goods. Further, all the circumstances must be considered as a whole. As Cons J observed in *Yuen Fung Metal Works Ltd v Negel (Hong Kong) Ltd*,[13] the relationship between forwarders and their customers cannot solely be decided, for example, 'by the way they look at themselves or by whether they directly own the ships, trains or trucks which actually carried the goods'.

The risk of being held as a carrier is recognized by HAFFA. Clause 7.2 of the HAFFA STC states the forwarder is, if held to be liable as a carrier, entitled to all the rights, exceptions and limitations enjoyed by a carrier. With this clause, a forwarder cannot be said to have waived any rights a carrier may have if inadvertently acting as a carrier.

Subsequent assumption of liability as carrier

A forwarder who does not originally act as a carrier may assume the liability of carrier later by variation, waiver or estoppel.

Variation refers to alteration of the terms of contract by the parties. By agreement, a forwarder may subsequently take up the role of carrier. The rules in contract law apply generally.

Waiver means clear and unequivocal representation by words or conduct by which a party gives up contractual rights. In the case of undisclosed principal, the agent forms, on behalf of the principal, a contract with a third party who does not know there is a principal behind the agent. When the third party later knows of the undisclosed principal, they may elect to sue either the agent or the principal if the contract is breached. If the third party waives the right against the principal and chooses to sue only the agent, the agent will be liable as if being the principal. The agent may, of course, look to the principal for compensation.

Forwarders acting as an agent may be estopped from denying their liabilities as a principal by reason of their conduct. There are three types of estoppel: estoppel by representation, estoppel by convention and promissory estoppel.

Estoppel by representation may be pleaded as a defence where a party has made a representation. The defence is available to the representee if it is proved the representor intended the representee act upon the representation. The representee did act, relying on the representation, and in so doing,

13. [1978] HKLR 588.

acted detrimentally. Generally, inaction does not constitute a representation. But a person makes a representation by remaining silent in a situation where there is a duty to speak.[14] The following example shows how the doctrine operates. A forwarder agrees to act as the agent of a customer. After the agency agreement is formed, the forwarder tells the customers there is no need to take up certain insurance policies because the forwarder would be liable as a carrier if there is anything wrong. Relying on the forwarder's statement, the customer does not take up any insurance policies. Although the forwarder is in fact not a carrier in this case, the doctrine of estoppel by representation may operate. The forwarder is not allowed to deny liability as a carrier if the goods are later damaged in transit.

Where the parties have acted upon a common mistaken assumption, they cannot deny the truth of the assumption if the doctrine of estoppel by convention applies.

> *The Vistafjord*[15]
> P was the owner of a cruise-liner. D was a passenger sales agent. D agreed to sell tickets for P. In return, P promised to pay D 15% of the sales as commission. X needed a vessel for a promotional tour. D agreed to sub-charter P's cruise-liner to X. The negotiation with X was virtually done by P. Although D was in fact not entitled to commission in relation to the sub-charter, D and P assumed throughout that D would be entitled to commission. P later claimed that D was not entitled to any commission.
>
> The court found that D and P had acted under the assumption that D would get commission. Relying in faith on the assumption, D had not attempted to negotiate for other commission. It was held that the doctrine of estoppel by convention applied. D was entitled to commission.

Finally, the doctrine of promissory estoppel may operate and a party is not allowed to go back on their word. For the doctrine to apply, an unequivocal promise must have been made by the promisor to the promisee. In making the promise, the promisor intended the promisee would act upon the promise. Relying on the promise, the promisee acted to detrimentally by not enforcing certain legal rights. The following example shows how the doctrine operates. The HAFFA STC are incorporated into the contract formed by a forwarder and the customer. The forwarder then makes a promise to the customer that if the goods are damaged in transit, clause 7.5 will not be invoked (see below) so that the customer may hold the forwarder liable

14. *Greenwood v Martins Bank Ltd* [1933] AC 51.
15. [1988] 2 Lloyd's Rep 343.

as a common carrier. Relying on this promise, the customer does not take up insurance policies to cover the risk. The carrier charters a ship for the carriage. The goods are later damaged in a collision at sea caused solely by the negligence of the master of another ship. In such a case, the forwarder cannot invoke clause 7.5 to deny liability as a common carrier.

Forwarder as common carrier

If all the conditions are satisfied, a forwarder may be liable as a common carrier. In the old case of *Hellaby v Weaver*,[16] a forwarder who collected all the freight charge was held to be liable as a common carrier. Since the duties of a common carrier are onerous, forwarders who decide to act as a carrier usually reserve the right of refusal to retain the status of private carrier. Clause 7.5 of the HAFFA STC provides that:

> Notwithstanding any other provisions of these Conditions, the Company is never a common carrier and may in its sole discretion refuse to offer its services to any person.

By adopting the HAFFA STC, a forwarder who acts as a carrier is liable only as a private carrier.

Forwarder as bailee

Usually, a forwarder is in possession of the customer's goods when there is provision of services. In addition to other obligations, there is liability as a bailee. The extent of rights and obligations as a bailee is generally governed by the bailment agreement, ie, the contract formed with the customer.

It is not uncommon a forwarder may provide some services free of charge for the purpose of, say, maintaining a good relationship with the customer. A forwarder receiving goods in such a case becomes a gratuitous bailee. At common law, a gratuitous bailee is under a duty to take reasonable care of the goods. The standard of care required depends on the relevant circumstances of the case.[17] The court held in this case that the services are provided gratuitously is only one factor to be taken into account.

For services provided by the forwarder gratuitously, clause 2.2 of the HAFFA STC states they are provided on the basis that no liability of any

16. (1851) 17 LTOS 271.
17. *Houghland v R. R. Low (Luxury Coaches) Ltd* [1962] 1 QB 694.

kind, including that in bailment and tort, should be imposed on the forwarder. Two points should be noted here.

First, there may be bailment for reward in the absence of consideration specially given by the owner of the goods for which the services are provided.

> *Andrews v Home Flats*[18]
>
> X rented a flat from Y. Under the tenancy agreement, Y provided a room to X for storage. The rent was solely paid by X. X's wife made no payment. A trunk was deposited by X's wife in the storage room. It was lost. X's wife claimed against Y for damages. One of the defences put forward by Y was that Y was only a gratuitous bailee.
>
> The court held that there was a bailment for reward with respect to the storage of the trunk. Being paid for the storage of the trunk, Y was under the duty of a bailee for reward to X's wife, the owner of the trunk.

At common law, the forwarder is liable to the owner of the goods as a bailee for reward even though there is only remuneration from a third party. Thus, a forwarder may be liable as a bailee for reward even if obtaining no remuneration for the services provided for the carriage of the goods in issue. For example, a forwarder arranges for the carriage of some small consignment of goods for free. At the same time, the forwarder is paid for a large bulk of cargo which is to be shipped at the same time with the small consignment of goods. In such a case, the court may hold that the separate services provided for the two loads of goods are in fact parts of an overall transaction. The decision may be that there is a bailment for reward with respect to all the goods of the same shipment. In *Cohen v S.E. Rly Co*,[19] it was held that a railway or bus company which agreed to carry passengers' belongings for free was liable as a bailee for reward for their loss or damage.

Second, the validity of clause 2.2 should be considered. As an exemption clause, it is subject to the Control of Exemption Clauses Ordinance. It may be invalid if it is unreasonable in the circumstances. The court will consider all the circumstances to determine whether the clause is reasonable. Little weight may be given to the label of 'gratuitous service'. Clause 2.2 of the HAFFA STC may not be given its full effect in some cases.

18. (1945) 173 LT 408.
19. (1877) 2 Ex D 253.

5.2 UNIMODAL CARRIAGE OF FORWARDER

Carrier of goods by land

In Hong Kong, all the intentional carriages of goods are either by sea, air or multimodal transport. Carriage of goods by land is basically domestic in nature. The distinction between forwarder and carrier is not so material, since no international convention applies in this context. Often, a transport company agrees with its customers to handle, load, carry and unload the goods. The contract governs the legal relations of the parties.

In general, the HAFFA STC do not apply to domestic carriage of goods by land in Hong Kong. Under the HAFFA STC, a forwarder who does not choose to act as a carrier provides two kinds of services: 'Principal Services' and 'Ancillary Services'. 'Principal Services' are defined as 'the services of arranging for the transportation or carriage of goods by air and/or sea'. 'Ancillary Services' include 'services of arranging for the storage, warehousing, collection, delivery, local transportation, insurance, customs clearance, packing, unpacking and other handling of goods, and other services relating or ancillary to the Principal Services'. Unless the parties agree otherwise, the HAFFA STC apply only to carriage of goods by sea, air and multimodal transport with a sea leg or an air leg.

Carrier of goods by sea

In the context of carrier of goods by sea, the traditional role of forwarders is clear. They are often an independent intermediary. They may form a contract with the consignor or consignee. The relationship between a forwarder and the customer is governed by the contract.

A forwarder, when receiving the goods from the consignor and taking them to the carrier (or receives them from the carrier and sends them to the consignee), is often acting as the agent of the consignee or the consignor.

For carriage of goods by sea, it has been held that only in exceptional cases would forwarders put themselves into the position of a carrier.[20] This observation may not be correct in the light of the wide scope of modern forwarders' activities. As discussed above, forwarders may now provide other services that they did not do in the past. The names 'Non-Vessel Owning Carrier (NVOC)' and, in the USA, 'Non-Vessel Owning Common Carrier (NVOCC)' refer to

20. *Langley, Beldon and Gaunt v Morley* [1965] 1 Lloyd's Rep 297.

> one who holds himself out to provide transportation for hire by water in interstate commerce . . . who assumes or has liability for safe transport and who does not operate the vessel upon which the goods are transported. Thus an NVOCC (or NVOC) is comparable to a surface freight forwarder who undertakes to deliver the cargo to destination. An NVOCC (or NVOC) will issue a bill of lading to the shipper but does not undertake the actual transportation of the cargo. Instead, the NVOCC (or NVOC) delivers the shipment to an ocean carrier for transportation.
>
> Consequently the NVOCC (or NVOC) is a carrier with respect to its customers and a shipper in relation to the actual ocean . . . carrier who transports the cargo.[21]

The court has held that if forwarders issue a bill of lading in their own name, they will be regarded as a carrier but not an agent of their client.[22] Yet a forwarder who does not issue transport documents may be liable as a carrier. It is now clear that whether or not a forwarder acts as a carrier does not solely depend on the issue of bills of lading.[23] In many situations, particularly if forwarders hold themselves out as operating a groupage service, a forwarder who did not issue a bill of lading, sea waybill or consignment note may still be liable as a carrier. On the other hand, as an application of the doctrine of undisclosed principal, forwarders who, in their own name, charter a ship for the carriage of a client's goods may still act as an agent of the client. In such a case, there is no liability as a carrier. All the circumstances should be considered to determine the forwarder's liability.

> *Harlow & Jones Ltd v P. J. Walker Shipping & Transport Ltd*[24]
>
> P wanted to ship a load of steel plate from Hamburg to Manchester. P formed a contract with D for the arrangement of the carriage. D entered, in its own name, into a charterparty for the hire of a ship for the shipment of the goods. The ship was not able to arrive at the designated UK port as scheduled. The goods had to be discharged at another port. D did not co-operate with P. As a result, P suffered loss. P sued D for D's failure to procure shipment at the agreed rate. D argued that they were not liable because, as forwarders, they had merely acted as P's agent.
>
> The court held that D in fact acted as forwarders, even though D entered into the charterparty in its own name. As such, D was basically an agent of P. However, D in this case did not act merely as an agent. D had an obligation for the procurement of a seaworthy ship and was

21. Sorkin, *Goods in Transit*, Matthew Bender, New York, updated looseleaf, para 1.15[8].
22. *Troy v The Eastern Company of Warehouses* (1921) 8 Ll L Rep 17.
23. cf *Hanjin Shipping Co Ltd v Procter & Gamble (Philippines) Inc* [1997] 2 Lloyd's Rep 341.
24. [1986] 2 Lloyd's Rep 141.

responsible for the carriage and discharge of the goods. In particular, D had warranted the freight rate. D was therefore liable.

If a forwarder inadvertently acts as a carrier, it may have significant consequences. For example, if there is liability as a sea carrier, the forwarder may not be able to limit it under the Hague-Visby Rules because, say, the contract does not provide for the issue of a bill of lading.

Since the liability of a carrier could be onerous, standard terms often provide that the forwarder is not liable as a carrier in situations that may be so regarded at common law. For example, clause 4.3 of the HAFFA STC states that the forwarder acts as the agent of the customer, not a carrier, despite the fact that:

- the forwarder issues its own transport document, such as a house air waybill, air consignment note, house bill of lading or freight forwarder cargo receipt;
- an inclusive price is charged; and
- the goods are forwarded, carried, stored or consolidated with goods of other shippers.

Although the court will consider all the relevant circumstances to determine whether a forwarder acts as a carrier, it normally gives effect to the contract. Thus, it is less likely for a forwarder to be liable as a carrier if the HAFFA STC apply.

As the contract is only binding on the contracting parties, clause 4.3 has no effect on third parties. If the forwarder forms a contract with a third party, there may still be liability, for matters relating to that contract, as a principal but not an agent of the customer.

Carriage of goods by air

Many forwarders in HK are approved agents of the International Air Transport Association (IATA). They often act as consolidators, and some of them in fact issue their own air waybills. A forwarder who owns or charters no plane may still form a carriage of goods by air contract with his clients. Forwarders who issue an air waybill in their own name often assumes liability as a carrier: *Panalpina v Densil Underwear*.[25] Besides, the course of negotiation as a whole has a bearing on whether a forwarder acts as a carrier.

25. [1981] 1 Lloyd's Rep 187.

Salsi v Jetspeed Air Services Ltd[26]

S and J were airfreight brokers. They were negotiating for the carriage of some goods for a customer of S. Their legal status was not clear. S knew that J was not resourceful. It appeared that J, in negotiating with S, depended on parallel negotiation with an airline. There was equivocal reference to a commission. J used extensively the personal pronoun. Some examples were: 'We can do this for you . . .' and 'In fact we can take another 5/6 tons on these sizes . . . we can send a person to Rome . . . if you have a backload ex Lagos . . .' J issued invoices in its own name.

The court drew inference from the negotiations as a whole. It was held that S and J were contracting as principals. J's obligation 'was not to carry the goods, but equally it was not to procure a contract under which [S] would become entitled to require the goods to be carried (in the agency or brokerage situation). [J's] obligation was personally to procure that the goods were carried.'

Clause 4.3(a) of the HAFFA STC states explicitly that the issue of a house air waybill or air consignment note does not, by itself, prevent the forwarder from acting as an agent of the customer. It should, however, be noted that express terms in a contract may be overridden by oral statements made in the course of negotiation. The forwarder, giving an oral undertaking to act as a carrier, acts as a carrier in spite of clause 4.3. On the other hand, a forwarder is unlikely to be liable as a carrier if the customer receives an air waybill issued by a party who actually performs the carriage.

If the forwarder only contracts to arrange for the carriage, the Warsaw regime does not apply. Instead, the contract governs. The Warsaw regime is relevant only if the forwarder acts as a carrier. Yet, the legal regime of the Warsaw system regulates only liabilities of carriers 'during the carriage by air': Article 18(1). For acts done outside the carriage by air, the common law applies generally. It is thus important to know when the carriage begins and ends. Similar principles apply if the freight forwarder also acts as the agent of the carrier. Under the Warsaw Convention, 'carriage by air . . . comprises the period during which the . . . cargo is in charge of the carrier'. The carrier is in charge of the goods when they are delivered to the forwarder (agent of the carrier) in an airport. As such, once the cargo is delivered to the forwarder, the convention governs.

26. [1997] 2 Lloyd's Rep 57.

Forwarder: Multimodal transport

There is no international convention governing multimodal transport. Unless the relevant conventions for international carriage by sea or by air apply, the common law regime regulates the legal relations between a forwarder and the customers. Sometimes it may not be easy to determine whether a forwarder acts as a carrier.

> *Aqualon (UK) Ltd v Vallana Shipping Corp*[27]
>
> N formed a contract with P. Under the contract, N was under an obligation to arrange for carriage of all the goods of P from the Netherlands to the UK. N's remuneration was based on an all-in price. N hired carriers in the UK and the Netherlands for the carriage by road in the two countries respectively. For the sea leg, either N or the UK carrier would make the arrangements. The consignment note prepared by P named N as the carrier. Often, N would later delete N's name and insert the UK carrier's name. In the box marked 'sender's instructions', N stamped 'N (As Agents Only)'. On one occasion, the goods were damaged during the carriage by sea. P contended that N was a carrier. N claimed that it was liable only as an agent of P.
>
> The court held that N was liable as a carrier. The court would not hold a forwarder liable as a carrier in the absence of clear evidence. The terms of the contract, descriptions used, the course of dealings, the nature of the charge, the mechanism for calculating the charge, and the terms of the consignment note were relevant in deciding the issue. In this case, N never informed P of the regular alteration of the consignment note. P had no knowledge that N did not consent to act as a carrier. There was no inconsistency between the terms of contract and N's acting as a carrier. With the all-in price, N would be more likely to act as a carrier. The consignment note also indicated who the carrier was. In the circumstances, P was under the wrong but reasonable impression that N was the carrier. Since N failed to correct the impression, it was liable as a carrier.

If the HAFFA STC apply and forwarders choose to act as a carrier by issuing a FIATA Multimodal Transport Bill of Lading in their name, the bill and the HAFFA STC apply. In case of conflict between the two, the bill prevails.

Forwarder and other intermediaries

Nowadays, the roles of forwarders and other intermediaries may not be easy to distinguish. Forwarders generally undertake to provide all sorts of

27. [1994] 1 Lloyd's Rep 669.

services associated with the carriage of goods. It is therefore, from a legal point of view, not absolutely necessary to categorize an intermediary who provides certain services to a consignor as forwarder, shipping agent, loading broker, clearing agent or customs agent, provided that their legal status, such as an agent of the consignor or a carrier, is clear. To illustrate the point, it may be useful to give a brief overview of the traditional roles played by the loading broker.

Loading brokers are employed by sea or air carriers to handle the loading and discharge of cargo. The crucial point is that a loading broker acts as the agent of the carrier. Forwarders often, as described above, act as carriers, or as agents of consignors or consignees.

A general discussion of the traditional distinction between forwarder and loading broker is found in *Heskell v Continental Express Ltd.*[28]

> The shipper frequently employs a forwarding agent and the shipowner [or charterer] a loading broker. The forwarding agent's normal duties are to ascertain the date and place of sailing, obtain a space allocation . . . and prepare the bill of lading . . . and send [it] to the loading broker. [He arranges] for the goods to be brought alongside, making the customs entry and paying any dues on the cargo. After shipment he collects the completed bill of lading and sends it to the shipper. All the regular shipping lines . . . appear to entrust the business of arranging for cargo to a loading broker. He advertises the date of sailings in shipping papers . . . and prepares and circulates to his customers a sailing card. It is his business to supervise the arrangements for loading, though the actual stowage is decided upon by the cargo superintendent, who is in the direct service of the shipowner. It is the broker's business also to sign the bill of lading and issue it to the shipper or his agent in exchange for the freight. His remuneration is . . . by way of commission on freight, and that is doubtless an inducement to him to carry out his primary function . . . of securing enough cargo to fill the ship . . . [They] discharge well-defined and separate functions. But in practice the same firm is often both the loading broker and the forwarding agent, though two sets of dealings may be kept in separate compartments of the business. The firm generally acts as loading broker only for one line and does all that line's business, so that it is free in respect of other business to act as it will. But even in the case of the same transaction it appears to be customary for the firm to act both as loading broker and as forwarding agent.

Heskell v Continental Express Ltd was decided more than fifty years ago. This may not be an apt description of modern commercial practices.

28. (1950) 83 Ll L Rep 438.

Nowadays, an intermediary in the shipping or transport logistics industry may play the dual role of loading broker and forwarder (as described in *Heskell v Continental Express Ltd*) in the sense they act as the agent of both the consignor and the carrier. If the HAFFA STC apply, the functions of the loading broker are in fact within the scope of 'Ancillary Services'. In such a case, it may not be meaningful to distinguish the roles of forwarder and loading broker.

5.3 CHARGES, SET OFF AND LIEN

A forwarder is entitled to the agreed remuneration. If the parties have not agreed on the amount, as in the event that the services are provided without a contract, the forwarder may claim on a *quantum meruit* basis for a reasonable value of the services.

If the customer fails to pay the agreed charge, the forwarder may sue him under the contract. The customer may not set off any amount the forwarder owes against the charge. The rule of 'no set off against freight' applies equally to forwarders. The rationale is that it would be an anomaly if the law were that the rule applied to a carrier but not a forwarder who, having been contracted as a carrier,[29] would have been protected.

There is a common law lien for the charge due to the forwarder. Besides, standard trading terms usually give a contractual lien to the forwarder. He may exercise a right of lien to retain the property of the customer until the charge is paid. In addition to a contractual particular and general lien, clause 18.5 of the HAFFA STC also authorizes the forwarder to sell the retained properties if payment is not made within 14 days after the forwarder gives notice to the customer. The sale does not release the customer from the obligation to pay. The customer is still liable for the balance if the proceeds are not sufficient for the charge due to the forwarder.

Conversion

A forwarder who commits an act of conversion is liable accordingly and cannot refer to any terms in the contract to exclude liability. The intention or motive is generally irrelevant. Following the instructions of the customer does not reduce liability.

29. *Britannia Distribution Co Ltd v Factor Pace Ltd* [1998] 2 Lloyd's Rep 420; see also *Overland Shoes Ltd v Schenkers Ltd* [1998] 1 Lloyd's Rep 498.

Dangerous goods

At common law, a shipper gives an implied warranty to the carrier that the goods are fit for carriage. For dangerous goods, the shipper is under an obligation to inform the carrier of the precautions to be taken. In breach of the obligations, the shipper is liable for the loss of the carrier.

It is well settled that a forwarder who acts as an agent of a shipper is also subject to the same obligations.[30] The liabilities are strict. It is no defence for the forwarder to argue that there was no actual knowledge of the dangerous nature of the goods or that the forwarder was deceived by the principal. In such a case, the shipper is liable to indemnify the forwarder for the loss under clause 10 of the HAFFA STC. This clause also gives a wide discretion to forwarders themselves to deal with dangerous goods received from the customers.

Indemnity

At common law, forwarders who act as agents are entitled to indemnity. However, customers are not liable to indemnify forwarders for all the expenses. Generally, customers are required to indemnify forwarders for expenses, disbursement or liability reasonably incurred

- in the course of proper exercise of their authority;
- in the performance of their duties, eg, storage charge due to unforeseen delay in transport; or
- by reason of the principal's actions or fault (*Immediate Transportation Co Ltd v Speller, Willis & Co* (1920) 2 Ll L Rep 645), eg, failure to give notice of the dangerous nature of the goods.

The mere fact that the expense or liability arises in the course of carrying out the principal's instruction does not warrant a right to indemnity at common law. For example, a forwarder may commit the tort of conversion while delivering the goods to a third party on the instructions of the customer who is not the owner of the goods. If the owner sues for conversion, there may not be a common law right to indemnity against the principal who was innocent in causing the conversion.

Further, if a forwarder incurs personal liabilities that are imposed by legislation or trade custom, eg, to pay port charges or freight, there is a right to indemnity against the principal. Subject to the contract, the forwarder

30. *Brass v Maitland* (1856) 6 E & B 470.

cannot ask the principal for indemnity if voluntarily accepting other personal liabilities.

If forwarders do not have a common law right to indemnity, they could seek an express agreement of indemnity to protect their interests. In fact, the common law right to indemnity need not be invoked if the contract is formed on standard trading conditions. Without exception, standard trading conditions contain an extensive indemnity clause: see for example clause 9 of the HAFFA STC. The effect of clause 9 is that forwarders, as well as their servants and agents, are entitled to indemnity for almost all conceivable losses, expenses and liabilities. Such a catchall indemnity clause may still be valid. As indicated by *Overland Shoes Ltd v Schenkers Ltd*,[31] a set of standard terms which has been adopted after full consultation with interested parties and is widely used by practitioners should pass the reasonableness test laid down by the Control of Exemption Clauses Ordinance.

31. [1998] 1 Lloyd's Rep 498.

APPENDIX

Haffa Standard Trading Conditions
29 DECEMBER 1997 EDITION

1. DEFINITIONS AND INTERPRETATIONS

1.1 In these Conditions, the following words and expressions have the following meanings unless the context otherwise requires:

"Ancillary Services"
includes services of arranging for the storage, warehousing, collection, delivery, local transportation, insurance, customs clearance, packing, unpacking and other handling of goods and other services relating or ancillary to the Principal Services.

"Company"
means [(name of the company)], a member of the Hongkong Association of Freight Forwarding Agents Limited trading under these Conditions.

"Conditions"
means the entire undertakings, terms, conditions and clauses embodied herein and includes the Company's terms and conditions printed on the front of the Shippers' Instructions and of the Company's form of transport document (including the Company's house air waybill or house bill of lading).

"Customer"
means any person at whose request or on whose behalf the Company undertakes any business, or provides advice, information or services, and includes the party named as "shipper" or "consignor" on the front of the Shippers' Instructions and of the Company's form of transport document (including the Company's house air waybill or house bill of lading).

"FIATA"
means the International Federation of Freight Forwarders Associations.

"FIATA Air Waybill"
means the form of neutral air waybill together with the conditions governing such air waybill (1996) published by FIATA and recommended by FIATA for use by forwarders who choose to act in the capacity of a (contracting) carrier.

"FIATA Multimodal Transport Bill of Lading"
means the FIATA form of multimodal transport bill of lading together with the standard conditions governing such bill of lading (1992) for use by forwarders who choose to assume liabilities as a (contracting) carrier.

"Shippers' Instructions"
means any of the Company's form or forms of shipping instructions or orders containing the Customer's instructions to the Company.

"Hague Rules"
means The International Convention for the Unification of Certain Rules of Law relating to Bills of Lading signed at Brussels on 25th August 1924.

"Hague-Visby Rules"
means The International Convention for the Unification of Certain Rules of Law relating to Bills of Lading signed at Brussels on 25th August 1924 (as amended by the Protocol signed at Brussels on 23rd February 1968).

"Instructions"
means statements of the Customer's specific requirements and includes the instructions specified on the front of the Shippers' Instructions and of the Company's form of transport document (including the Company's house air waybill or house bill of lading).

"Principal Services"
means the services of arranging for the transportation or carriage of goods by air and/or sea.

"Services"
means the services to be provided by the Company and includes the Principal Services and the Ancillary Services.

"the Owner"
means the owner of the goods (including any packings, containers or equipment other than those provided by the Company or carriers) to which any business concluded under these Conditions relates and any other person who is or may become interested in them and including the consignee named on the front of the Shippers' Instructions and of the Company's form of transport document (including the Company's house air waybill or house bill of lading.)

"Warsaw Convention"
means The Convention for the Unification of Certain Rules relating to International Carriage by Air signed at Warsaw on 12th October 1929 or that Convention as amended at The Hague, 28th September 1955, whichever may be applicable.

1.2 References to statutory provisions shall be construed as references to those provisions as respectively amended or re-enacted or as their application is modified by other provisions from time to time and shall include any provisions of which they are re-enactments (whether with or without modification).

1.3 Unless the context requires otherwise, words importing the singular include the plural and vice versa, words importing a gender include every gender, references to persons include any body corporate or unincorporated, and references to Clauses are to Clauses of these Conditions. The headings are inserted for convenience only and shall not affect the construction of these Conditions.

1.4 All representations, warranties, undertakings, agreements, covenants, obligations, liabilities, guarantees and indemnities expressed in these Conditions or otherwise implied to be made given or assumed by the Customer shall be deemed to be made, given or assumed by the Customer and the Owner jointly and severally.

1.5 No omission or delay on the part of the Company in exercising its rights shall operate as a waiver thereof, nor shall any single or partial exercise by the Company of any such right preclude the further or other exercises thereof or the exercise of any other right which it has. The rights and remedies of the Company provided in these Conditions shall be cumulative and not exclusive of any rights or remedies otherwise provided by law.

1.6 Each of the provisions of these Conditions is severable and distinct from the others and if at any time one or more of such provisions is or becomes invalid, illegal or unenforceable, the validity legality and enforceability of the remaining provisions of these Conditions shall not in any way be affected or impaired thereby.

2. APPLICATION OF THE CONDITIONS/LEGISLATION COMPULSORILY APPLICABLE

2.1 All and any business undertaken by the Company is transacted subject to these Conditions and each of these Conditions shall be deemed to be incorporated in and to be a condition of any agreement between the Company and the Customer. All other terms and conditions are hereby excluded. Should any Customer wish to contract with the Company otherwise than subject to these Conditions, special arrangements can be made subject to revised charges having been agreed and having been paid in advance by the Customer to the Company and subject to such arrangements having been reduced into writing and signed by an authorised officer of the Customer and by an authorized officer of the Company. Save as aforesaid, no agent or employee of the Company has the Company's authority to waive or vary any of these Conditions.

2.2 All and any advice information or services provided by the Company gratuitously is provided on the basis that the Company will not accept any liability whatsoever therefor, whether in tort or bailment or otherwise.

2.3 If any legislation is compulsorily applicable to any business undertaken, these Conditions shall, as regards such business, be read as subject to such legislation and nothing in these Conditions shall be construed as a surrender by the Company of any of its rights or remedies or immunities or as an increase of any of its responsibilities or liabilities under such legislation, and, if any part of these Conditions be repugnant to such legislation to any extent, such part shall as regards such business be overridden to that extent and no further.

3. CONTRACTUAL STATUS OF CUSTOMER

The Customer entering into any transaction or business with the Company hereby expressly warrants to the Company that the Customer is either the Owner or the authorized agent of the Owner and that it is authorized to accept and is accepting these Conditions not only for itself but also for the Owner. Where the Customer acts as the agent of the Owner, the Customer also accepts personal liability to the Company (but without prejudice to any of the rights or remedies of the Company against the Owner) and so that in respect of such transaction or business the Company is entitled to enforce its rights or remedies (including without limitation the right to recover any sum payable to the Company) against the Customer and the Owner jointly and severally.

4. CONTRACTUAL STATUS OF THE COMPANY

4.1 Save as provided in Clause 7, Services are provided by the Company as agents on behalf of its Customers, except that, subject to Clause 16.2, the Company itself may provide (instead of arranging to provide) the Ancillary Services.

4.2 The Company shall be entitled to perform any of its Services or exercise any of its powers or discretions hereunder by itself or its parent, subsidiary or associated companies. In the absence of agreement to the contrary any contract to which these Conditions apply is made by the Company on its own behalf and also as agent for and on behalf of any such parent, subsidiary or associated company and any such company shall be entitled to the benefit of these Conditions.

4.3 It shall not be construed that any Services are provided by the Company other than as an agent of the Customer by reason only of any one or more of the following:

a. The Company issuing its own transport document including its house air waybill or air consignment note or house bill of lading or freight forwarder cargo receipt;
b. The Company charges an inclusive price;
c. The Customer's goods are forwarded, carried, transported, stored or otherwise handled together or in consolidation with other goods.

5. COMPANY'S AUTHORITY

5.1 The Company is hereby expressly authorised by the Customer as hereinafter provided.

5.2 The Company is authorized to act on behalf of the Customer to select, engage and enter into contract or arrangement (whether in the name of the

Customer or otherwise) with any carriers, truckmen, forwarders, receiving or delivery agents, warehousemen, packers and other persons (together "3rd Parties", and individually "3rd Party"):

a. for the carriage of the goods by any route or any carrier;
b. for the storage, packing, unpacking, (local) transportation, transhipment, loading, unloading or other handling of the goods by any person at any place or places and for any length of time

AND to do other acts or enter into other contracts or arrangements for any other purposes pursuant or relating or incidental to the Customer's instructions.

5.3 The Company is authorised (but is not obliged) to depart or deviate from the Customer's instructions in any respect if in the opinion of the Company such departure or deviation is necessary or desirable in the Customer's interests or is otherwise expedient.

5.4 The Company is authorised by the Customer to act or to enter into any contract or arrangement without prior consultation with or further authorization from the Customer, AND the Company is not required, unless specifically requested by the Customer in writing, to inform the Customer of the terms and conditions or details of the contracts or arrangements or acts entered into or taken by the Company.

5.5 Without prejudice to the generality of the foregoing, the Company is authorised to agree with any 3rd Party the charges payable to such 3rd Party without reference to or further authorization from the Customer, it being agreed that the difference between the charges payable by the Company to the 3rd Party(ies), and the charges payable by the Customer to the Company is the Company's commission or remuneration or profit. The Customer waives any and has no right of enquiry of the charges payable to the 3rd Party(ies) and the Company is not under any duty to account to the Customer for the Company's commissions, remunerations or profits.

5.6 The Company is authorised (but is not obliged) to inspect or arrange for the goods to be inspected.

5.7 The Company is not obliged to arrange for the Customers' goods to be carried, forwarded, packed, unpacked, stored or handled separately. The Company is authorised (but is not obliged) to consolidate or arrange to be consolidated the goods of the Customer with other goods.

5.8 The Customer expressly agrees to be bound in all respects by any act or contract or arrangement done or entered into by the Company pursuant to the aforesaid authorizations.

6. WHERE THE COMPANY CONTRACTS (ON BEHALF OF THE CUSTOMER) IN ITS OWN NAME

6.1 Where the Company enters into a contract on behalf of the Customer in its own name with any 3rd Party for any purposes, the Company is not itself a carrier for the purposes of the Carriage by Air Ordinance or the Carriage of Goods by Sea Ordinance or for any other purposes, nor does the Company make or purport to make any contract as a principal with the Customer for the carriage, storage, packing, unpacking, (local) transportation, transhipment, loading, unloading or other handling of the goods. The Company's sole obligation is to procure contracts for the carriage, storage, packing, unpacking, (local) transportation, transhipment, loading, unloading or other handling of goods by other persons.

6.2 In addition and without prejudice to the exceptions and limitations contained in these Conditions, the Company shall be entitled to the benefit of all exceptions and limitations in favour of any 3rd Party expressly contained or implied in the Company's contract with such 3rd Party. The Customer shall not seek to impose on such 3rd Party any liability greater than that accepted by such 3rd Party under such contract.

7. WHERE THE COMPANY CONTRACTS AS PRINCIPAL

7.1 The Company in its absolute discretion may, under certain circumstances notwithstanding the terms and conditions contained herein, issue a FIATA Air Waybill or a FIATA Multimodal Transport Bill of Lading naming the Company as the carrier and the principal. Where such a document is issued, the terms and conditions embodied in it shall be paramount in governing the relationship between the Customer and the Company in so far as those terms and conditions are inconsistent with or repugnant to these Conditions.

7.2 Where, in respect of a transaction, the Company is held by a court of competent jurisdiction to be a carrier, the Company shall be entitled to all the rights, immunities, exceptions and limitations conferred on the carrier by any applicable law or legislation, and these Conditions shall be overridden to the extent that they are inconsistent with such rights, immunities, exceptions and limitations, but without prejudice to the operation of Clause 2.3.

7.3 If the Company is or is deemed to be a carrier in respect of a carriage of goods by air, the following notices are hereby given:

> "If the carriage involves an ultimate destination or stop in a country other than the country of departure, the Warsaw Convention may be applicable and that the Convention governs and in most cases limits the liability of carriers in respect of loss of or damage or delay to cargo."
>
> The first carrier's name may be abbreviated on the face of the air waybill, the full name and its abbreviation being set forth in such carrier's tariffs,

conditions of carriage, regulations and timetables. The first carrier's address is the airport of departure shown on the face of the air waybill. The agreed stopping places (which may be altered by carrier in case of necessity) are those places, except the place of departure and the place of destination, set forth on the face of the air waybill or shown in carrier's timetables as scheduled stopping places for the route. Carriage to be performed under the air waybill by several successive carriers is regarded as a single operation.

7.4 If the Company itself performs (instead of arranging for the performance of) any of the Ancillary Services, the Company is entitled:

a. to perform any local transportation of the goods by any route or by any means;
b. to store, pack, unpack, load, unload or otherwise handle the goods at any place or places and for any length of time;

And to do all such other acts as may be necessary or incidental thereto in the absolute discretion of the Company. The Company may (but is not obliged to) depart or deviate from the Customer's instructions if in the opinion of the Company such departure or deviation is necessary or desirable in the Customer's interests or is otherwise expedient.

7.5 Notwithstanding any other provisions of these Conditions, the Company is never a common carrier and may in its sole discretion refuse to offer its services to any person.

8. CUSTOMER'S FURTHER WARRANTIES

The Customer further warrants and acknowledges that:

(a) Proper packing etc.
All the goods, the subject of any Service provided by the Company, have been properly and sufficiently packed and/or prepared, and that the Company has no liability for any loss of or damage to goods which are improperly or insufficiently packed or prepared, no matter how such loss or damage is caused.

(b) Transport Unit
Where the goods delivered by or on behalf of the Customer are already carried in or on containers, trailers, flats, tilts, railway wagons, tanks, igloos, or any other unit load device (each hereafter individually referred to as "transport unit") then,

The transport unit is in good condition, is suitable to carry the goods loaded therein or thereon, and is suitable for the intended carriage and other handling; and

The goods are suitable for carriage and other handling in or on the transport unit and has been properly and competently packed or loaded in or on the transport unit.

(c) Description of Goods
All descriptions, values and other particulars of the goods furnished to the Company for customs, consular and other purposes are true, complete and accurate, it being the duty of the Customer to provide such information to the Company and to ensure that such information is true, complete and accurate.

(d) Fitness of Goods
In addition and without prejudice to any provisions of Clauses 10 and 11, the goods are fit and suitable for the carriage (international as well as local), storage, packing, unpacking and other handling in accordance with, pursuant or related or incidental to the Customer's instructions.

(e) Delivery of Goods
The consignee or other person entitled to the delivery of the goods shall take delivery of the goods upon their arrival at destination and shall pay all necessary charges, taxes and duties and shall comply with all necessary formalities and procedures.

9. INDEMNITIES

9.1 The Customer shall save harmless and indemnify and keep indemnified the Company from and against all claims, liabilities, losses, damages, costs and expenses (including without limitation all duties, taxes, imposts, levies, deposits, fines and outlays of whatsoever nature levied by any authority) arising out of the Company acting in accordance with the Customer's instructions, or arising from a breach of warranty or obligation by the Customer, or arising from the Customer's inaccurate or incomplete or ambiguous information or instructions, or arising from the negligence of the Customer or Owner.

9.2 Advice and information, in whatever form as may be given by the Company, are provided by the Company for the Customer only and the Customer shall save harmless and indemnify and keep indemnified the Company from and against all claims, liabilities, losses, damages, costs and expenses arising out of any other person relying on such advice or information. Except under special arrangements previously made in writing, advice or information which is not related to specific instructions accepted by the Company is provided gratuitously and without liability and Clause 2.2 is applicable.

9.3 The Customer undertakes that no claim shall be made against any officer, servant, agent or sub-contractor of the Company which imposes or attempts to impose upon them any liability in connection with any services provided or to be provided by the Company. If any such claim should nevertheless be made the Customer shall indemnify the Company against all consequences thereof. Without prejudice to the foregoing every such officer, servant agent and sub-contractor shall have the benefit of all provisions herein benefiting the Company as if such provisions were expressly for his or its benefit. For the foregoing purposes, the Company contracts for itself as well as agents for all the aforesaid persons.

9.4 The Customer shall defend, indemnify and hold harmless the Company from and against all claims, costs and demands whatsoever and by whomsoever made or preferred in excess of the liability of the Company under the terms of these Conditions, and without prejudice to the generality of the foregoing this indemnity shall include (without limitation) all claims, costs and demands arising from or in connection with the negligence of the Company, its officers, servants, agents or sub-contractors.

9.5 The Customer shall defend, indemnify and hold harmless the Company in respect of any general average or any claims of a general average nature which may be made on the Company and the Customer shall provide such security as may be required by the Company in this connection.

10. DANGEROUS GOODS, ETC

Except under special arrangements previously made in writing, the Customer warrants that the goods are not goods (or consist of goods) included in the Dangerous Goods (Application and Exemption) Regulations of the Laws of Hong Kong Cap 295 or any modification thereof or the IATA Dangerous Goods Regulations prevailing at the time the Company confirms acceptance of the Customer's instructions, nor are goods (or consist of goods) of comparable hazard, nor are goods (or consist of goods) otherwise likely to cause damage. Should the Customer nevertheless deliver any such goods to the Company or cause the Company to accept or handle or deal with any such goods otherwise than under special arrangements previously made in writing, then whether or not the Company is aware of the nature of such goods, the Customer shall be liable for all expenses losses or damages whatsoever caused by or to or in connection with the goods howsoever arising, and shall indemnify the Company against all penalties, claims, damages, costs, expenses and any other liabilities whatsoever arising in connection therewith, and the goods may be destroyed or otherwise dealt with at the risk and expenses of the Customer or the Owner in the sole discretion of and without any liability to the Company or of any other person in whose custody or control the goods may be at the relevant time. The Company or such other person shall have the right to decide whether or when the goods are or become (or consist of goods which are or become) unfit for carriage (overseas or local), storage, packing, unpacking, handling, etc., or are or become goods (or consist of goods which are or become goods) of comparable hazard to the goods included in the IATA Dangerous Goods Regulations or the Dangerous Goods (Application and Exemption) Regulations or any modification thereof, or are or become goods (or consist of goods which are or become goods) which are otherwise likely to cause damage. A copy of the prevailing IATA Dangerous Goods Regulations is available for inspection by the Customer upon request. If such goods are accepted under arrangements previously made in writing, they may nevertheless be destroyed, or otherwise dealt with at the risk and expenses of the Customer or the Owner in the sole discretion of and without any liability to the Company or any other person in whose custody or control they may be at the relevant time on account of risk to other goods, property, life or health. The expression "goods likely to cause damage" includes but is not limited to goods likely to harbour or encourage vermin or other pests.

11. BULLION, ETC

Except under special arrangements previously made in writing the Company will not accept or deal with bullion, coins, precious stones, jewellery, valuables, antiques, pictures, livestock or plants. Should the Customer nevertheless deliver any such goods to the Company or cause the Company to handle or deal with any such goods otherwise than under special arrangements previously made in writing, the Company shall be under no liability whatsoever for or in connection with the goods or any part thereof (including without limitation any loss or damage or non-delivery or mis-delivery or delay) howsoever caused and notwithstanding that the value may be shown, declared or indicated on any documents accompanying the shipment.

12. DEVIATION

Subject to express instructions in writing given by the Customer and the acceptance of those instructions in writing by the Company, the Company reserves to itself absolute discretion as to the means, routes and procedures to be followed in the carriage, transportation, storage and other handling of goods. Further, if in the opinion of the Company it is at any stage necessary or desirable in the Customer's interests to depart from those instructions, the Company is hereby irrevocably authorised and shall be at liberty to do so, and any departure from the terms and conditions, or in the handling other than pursuant to the normal custom of handling the goods is done at the sole risk of the Customer or the Owner.

13. WAREHOUSING

Pending forwarding or delivery, goods may be warehoused or otherwise held at the risk of the Customer or the Owner at any place at the sole discretion of the Company and the cost therefor shall be for the account of the Customer.

14. DECLARATION OF VALUE, ETC

14.1 The Company shall not be obliged to make any declaration for the purpose of any statute or convention or contract as to the nature or value of any goods or as to any special interest in delivery, unless express instructions in writing were previously given to and accepted by the Company.

14.2 Without prejudice to the generality of Clause 14.1 where there is a choice of rates according to the extent or degree of the liability assumed by carriers, warehousemen or others, goods will be forwarded, dealt with, etc., at the Customer's or the Owner's risk and at such charges (including the lowest charges) as the Company may at its discretion decide, and no declaration of value (where optional) will be made, unless express instructions in writing to the contrary have previously been given by the Customer and accepted by the Company.

14.3 A mere statement or declaration of the value or nature of the goods for insurance or export or customs or other purposes is not and shall not be construed to be instructions to the Company to make any declaration for the purposes of Clause 14.1 and/or Clause 14.2 above.

15. DUTIES

The Customer shall be liable for any duties, taxes, levies, deposits or outlays of any kind levied by the authorities at any port or place for or in connection with the goods and for any payments, storage, demurrage, fines, expenses, loss or damage whatsoever incurred or sustained by the Company in connection therewith.

16. INSURANCE

16.1 No insurance will be arranged except upon express instructions given in writing by the Customer and accepted by the Company. All insurances arranged by the Company are subject to the usual exceptions and conditions of the policies of the insurance company or underwriters taking the risk. The Company shall not be under any obligation to arrange a separate insurance on each consignment but may declare it on any open or general policy. Should the insurers dispute their liability for any reason the insured shall have recourse against the insurers only and the Company shall not be under any responsibility or liability whatsoever in relation thereto notwithstanding that the premium upon the policy may not be at the same rate as that charged by the Company or paid to the Company by its Customer.

16.2 In so far as the Company agrees to arrange insurances, the Company acts solely as the agent of the Customer using reasonable effects to arrange such insurance. The Company does not warrant or undertake any such insurance will be accepted by the insurance company or underwriters.

17. NO DUTY TO PRESERVE RIGHTS

The Company shall not be under any duty or obligation to the Customer or the Owner to give any notice or otherwise take any action to preserve or protect the right of the Customer or the Owner in relation to any claim or remedy which the Customer or Owner may have against any third parties.

18. DISPOSAL OF GOODS/LIEN, ETC

18.1 Notice of arrival of the goods will be sent to the notify party or the consignee by ordinary methods. The Company is not liable for the non-receipt or delay in the receipt of such notices. Any charges including storages incurred pending collection will be for the account of the Customer.

18.2 Without prejudice to any other rights or remedies which the Company may have (including without limitation those under the other sub-Clauses of this Clause 18), if delivery of the goods or any part thereof is not taken by the consignee or other person entitled to the delivery of the same at the time and place when and where delivery should be taken, the Company shall be entitled (but is not obliged) to store or cause to be stored the goods or any part thereof at the sole risk of the Customer or the Owner, whereupon any liability which the Company may have in respect of the goods or that part thereof stored as aforesaid shall wholly cease and the cost of such storage shall upon demand be paid by the Customer to the Company.

18.3 Perishable goods which are not taken up immediately upon arrival or which are insufficiently addressed or marked or otherwise not readily identifiable, may be sold or otherwise disposed of without any notice to the Customer or the Owner and payment or tender of the net proceeds of any sale after deduction of charges and expenses shall be equivalent to delivery. All charges and expenses arising in connection with the sale or disposal of the goods shall be paid by the Customer.

18.4 The Company is entitled (but not obliged) to sell or dispose of (or cause to be sold or disposed) all non-perishable goods which in the opinion of the Company cannot be delivered either because they are insufficiently or incorrectly addressed or because they are not collected or accepted by the consignee or any other reason, upon giving 14 days' notice in writing to the Customer. All charges and expenses arising in connection with the storage and sale or disposal of the goods shall be paid by the Customer.

18.5 All goods (and documents relating to goods) shall be subject to a particular and general lien and right of detention for monies due either in respect of such goods, or for any particular or general balance or other monies due from the Customer or the Owner to the Company. If any such monies due to the Company are not paid within 14 days after notice has been given to the Customer that such goods are being detained, the goods and/or the documents may be sold by auction or otherwise at the sole discretion of the Company at the expense of the Customer, and the proceeds (net of the expenses in connection with such sale) applied in or towards satisfaction of such indebtedness, and the Company shall not be liable for any deficiencies or reduction in value received on the sale of the goods, nor shall the Customer be relieved from the liability merely because the goods have been sold.

18.6 The rights of the Company under this Clause 18 are independent and cumulative.

19. QUOTATIONS AND CHARGES

19.1 The Customer is primarily liable for the payment of all freight, fees, duties, charges and other expenses whether the same (or any of them) are to be pre-paid or to be collected.

19.2 The Customer shall pay to the Company all sums immediately when due without deduction or deferment on account of any claim, counterclaim or set-off. Payment to the Company is due as soon as an invoice is rendered. Payment shall be made in cash unless otherwise agreed by the Company.

19.3 The Company at its discretion may request an advance to cover fees, duties, charges, taxes and/or other expenses payable before the Company's invoice is rendered. Forthwith upon such request being made, the Customer shall make such advance to the Company.

19.4 Without prejudice to the foregoing provisions, when the Company is instructed to collect freight, duties, fees, charges or other expenses from any person other than the Customer, the Customer shall remain responsible for the payment of the same. The Customer shall forthwith upon demand pay the Company such freight, duties, fees, charges and other expenses or any balance thereof together with interest (if applicable) without deduction or deferment on account of any claim, counterclaim or set off (whether or not demand is made to such other person). Without prejudice to the generality of the foregoing, this provision shall apply if (*inter alia*) the goods are refused by the consignee or other person entitled to delivery or confiscated by the customs or other authorities or for any reason it is in the opinion of the Company not practicable or impossible to arrange for the delivery of the goods.

19.5 On all amounts overdue to the Company, the Company shall be entitled to interest calculated on a monthly basis from the date such accounts are overdue until payment thereof at 2% per month (compounded monthly) during the period that such amounts are overdue.

19.6 Quotations are given on the basis of immediate acceptance by the Customer and are subject to withdrawals or revisions by the Company. Further, unless otherwise agreed in writing by the Company, the Company, notwithstanding acceptance of the quotations by the Customer, shall be at liberty to revise quotations or charges with or without prior notice in the event of changes occurring in currency exchange risks, rates of freight, insurance premiums or any charges applicable to the goods.

19.7 Freight charges are usually quoted and charged on "chargeable weight" basis. Chargeable weight is the actual gross weight or volume weight, whichever is the higher. Volume weight is calculated by reference to the volume of the consignment (including packaging) divided by a certain factor. References to "per kilogramme" or "per ton" or "per pound" refer to the higher of the actual gross weight and the volume weight. Further details relating to the computation of freight charges will be provided to the Customer upon request. Customers are advised to obtain such details.

20. SUB-CONTRACTING

The Company shall be entitled to sub-contract on any terms the whole or any part of the Services and any and all duties whatsoever undertaken by the Company.

21. LIABILITY AND LIMITATION

21.1 Notwithstanding any negligence of the Company, its servants or agents or sub-contractors or other persons for whom the Company is responsible, the Company shall not be responsible or liable for any damage to or loss or non-delivery or mis-delivery of goods or for any delay or deviation in respect of the transportation or delivery or other handling of goods, unless it is proved that such damage, loss, non-delivery, mis-delivery, delay or deviation occurred whilst the goods were in the actual custody of the Company and under its actual control and that the damage, loss, non-delivery, mis-delivery, delay or deviation was due to the wilful neglect or wilful default of the Company or its own servants.

21.2 Notwithstanding any negligence of the Company, its servants or agents or sub-contractors or other persons for whom the Company is responsible, the Company shall not be liable for any non-compliance or mis-compliance with instructions given to it unless it is proved that such non-compliance or mis-compliance was caused by the wilful neglect or wilful default of the Company or its own servants.

21.3 Save as provided in Clause 21.1 or Clause 21.2, the Company shall be under no liability whatsoever and howsoever arising and whether in respect of or in connection with any goods or any instructions, business, advice, information or service or otherwise, and whether or not there is negligence on the part of Company, its servants or agents or sub-contractors or other persons for whom the Company is responsible.

21.4 Further and without prejudice to the generality of the preceding provisions of this Clause 21, the Company shall not in any event, whether under Clause 21.1 or Clause 21.2 or otherwise, be under any liability whatsoever for:

a. any special, incidental, indirect, consequential or economic loss or damage (including without limitation loss of market, profit, revenue, business or goodwill);
b. any loss or damage or expense arising from or in any way connected with fire or consequence of fire

in each case howsoever caused and whether or not resulting from any act or default or neglect of the Company or its servants or agents or sub-contractors or other persons for whom the Company is responsible.

21.5 Save where Clause 21.6 or Clause 21.7 is applicable, in no case whatsoever shall the liability of the Company howsoever arising and notwithstanding any lack of explanation exceed the value of the relevant goods or a sum of HK$200.00 per shipping package or unit or HK$10.00 per (weight) kilogram, whichever is the least.

21.6 If any one or more of the Hague Rules, the Hague-Visby Rules, the Hague-Visby Rules (as amended by the Protocol signed at Brussels on 21st December 1979), the Warsaw Convention and the Guadalajara Convention are compulsorily applicable, the relevant limitation amounts set out therein as applied by the applicable legislation will apply. In all other cases the limitation amounts detailed in Clause 21.5 will apply.

21.7 By special arrangement agreed in writing, the Company may accept liability in excess of the limit set out in Clause 21.5 if the Customer agrees to pay and has paid the Company's additional charges for accepting such increased liability. Details of the Company's additional charges will be provided upon request.

22. NOTICE OF CLAIM

22.1 Any claim against the Company must be in writing and delivered to the Company at its registered office or its principal place of business in Hong Kong within 14 days:

a. in the case of damage to goods, the date of delivery of the goods;
b. in the case of loss or non-delivery or mis-delivery or delay in delivery of goods, the date that the goods should have been delivered; and
c. in any other case, the date of the event giving rise to the claim.

22.2 No action shall lie against the Company if the claim is not made within the times and in the manner specified in Clause 22.1.

23. TIME BAR

Any right of action against the Company shall be extinguished if suit is not brought in the proper forum and written notice thereof received by the Company within 9 months from the date the goods arrived at the destination or the date the goods should have arrived at the destination (whichever date is the earlier).

24. COLLECT ON DELIVERY (C.O.D.) SHIPMENTS

Goods received with Customer's or other person's instruction to Collect on Delivery (C.O.D.) by bank drafts or otherwise, or to collect on any specified terms by time drafts or otherwise, are accepted by the Company only upon the express understanding that it will exercise reasonable care in the selection of a bank, correspondent, carrier or agent to whom it will send such item for collection, and the Company will not be responsible for any act, omission, default, suspension, insolvency or want of care, negligence, or fault of such bank, correspondent, carrier or agent, nor for any delay in remittance lost in exchange, or during transmission, or while in the course of collection.

25. GOVERNING LAW

These Conditions and any act or contract to which they apply shall be governed by and construed according to the laws of the Hong Kong Special Administrative Region. Any dispute arising out of these Conditions or any such act or contract shall be subject to the non-exclusive jurisdiction of the courts of the Hong Kong Special Administrative Region.

6

Sea Carriage Law

6.1 BILLS OF LADING

Introduction

A bill of lading is a document issued by or on behalf of the carrier of cargo by sea to the party (usually known as the 'shipper') with whom the carrier has contracted for carriage of goods. The bill of lading usually states the terms on which the cargo is consigned to and received by the ship. In the medieval world of shipping, a merchant shipper would simply contract with a vessel owner to carry his goods from one port to another. Frequently, the vessel owner and the shipper were the same, and when not, the merchant and shipowner might join in a 'partnership' or form a 'joint-venture' sharing the risks of losses, and the profits when the voyage and sale of cargo had been successful.

According to legal historians, bills of lading were commonly employed in the thirteenth century. They may have accompanied bills of exchange, which were used as instruments of trade finance. Freight charges were frequently stated in the bills of lading. This is an extract of a medieval bill of lading dated 1248:

> April twenty-fourth in the year of the Incarnation of the Lord 1248.
>
> We, Eustace Cazal and Peter Amiel, carriers, confess and acknowledge to you, Falcon of Acre and John Confortance of Acre, that we have had and received from you twelve full loads of brazil wood and nine of pepper and seventeen and a half of ginger for the purpose of taking them from Toulouse to Provence, to the fairs of Provence to be

> held in the coming May, at a price or charge of four pounds and fifteen solidi in Vienne currency for each of the said loads. And we confess we have had this from you in money, renouncing, etc. And we promise by this agreement to carry and look well after those said loads with our animals, without carts, and to return them to you at the beginning of those fairs and to wait upon you and do all the things which carriers are accustomed to do for merchants. Pledging all our goods, etc.; renouncing the protection of all laws, etc. Witnesses, etc.[1]

In modern practice, bills are usually issued and signed in sets of three duplicate originals: the shipper retains one copy, one is kept aboard the vessel, and one is to be delivered to the consignee at the port of discharge. The typical bill contains the names of the shipper and consignee, a description of the goods, including shipping marks used for identification purposes, stipulations for payment of freight, and details of the conditions of carriage. Many bills of lading contain the jurisdiction clauses stipulating the forum for litigation or arbitration arising out of claims, as well as the applicable law.

In a typical international sale of goods scenario, the shipper of goods is also the seller. The carrier will issue the bill of lading to a shipper/seller who owns the goods. The bill is initially in the possession of the person with property in the goods. The seller retains the bill of lading as security for the price. Under the contract of sale and purchase of goods, the bill will thereafter be delivered to the purchaser/consignee against payment of the purchase price. The property in the cargo is often intended to pass once payment has been effected or secured, unless the terms of the agreement stipulate otherwise.

Therefore, fundamental to all maritime transport is the performance of a carrier's obligation to deliver the cargo to the party entitled to its possession at the port of discharge. Traditionally, the carrier has honoured this obligation by ensuring delivery at the port of discharge on presentation of the bill of lading. The unique character of the bill of lading is that delivery of the goods has to be made against surrender of the document. On the one hand, it protects the holder of the bill of lading because it is a basic term of the contract of carriage that the carrier must only deliver goods against presentation of the bill of lading. On the other hand, such delivery serves to discharge the carrier from further obligations under the contract of carriage.

Essentially, a bill of lading has three related but separate elements that are separately examined in the latter part of this chapter. It functions as

1. Roy C Cave and Herbert H Coulson, *A Source Book for Medieval Economic History*, 159–60 (Biblo and Tannen, 1965).

- a receipt for the cargo shipped,
- evidence of the contract of carriage, and
- a document of title.

Express terms and terms implied at Common Law

In numerous shipping disputes, cargo owners and insurance companies have continually attempted to invalidate the express terms of bill of lading contracts. Shipowners and protection and indemnity clubs (P & I Clubs) also frequently incorporate every possible terms into the bill of lading to reserve for themselves privileges, rights and exemptions from liabilities. While there are different types of bills of lading performing different functions, contemporary practice and globalization have gradually given rise to a common format that has been used in most nations to facilitate the preparation of shipping documents. Annex 1 shows a sample bill of lading adopting the common format. The reverse side of the bill of lading contains the express terms of the bill.

Express terms

A typical bill of lading commonly contains the following terms, conditions and particulars:

- the signature by the master, or the agent for and on behalf of the carrier;
- the application of laws and jurisdiction clause;
- the parties to the carriage and the definitions of various terms;
- the scope of the voyage;
- frustration or governmental intervention, fire, delay;
- descriptions of goods;
- liability of shippers;
- rights and immunities of all servants and agents of the carrier (the 'Himalaya' clause);
- stowage on board and storage ashore;
- refrigeration;
- deck cargo;
- transshipment;
- custody, discharge, storage and liens;
- freight charges;
- package limitation;
- notice of loss or damage, time for filing suit, service of process;
- specified stowage, dock loading or discharge;

- statutory protection;
- surrender of bill of lading;
- dock receipts and mate's receipt; and
- lien.

Terms implied at common law

At common law, the carrier is under an obligation to provide aseaworthy vessel, to proceed on the voyage with reasonable dispatch, not to ship dangerous goods, and not to deviate from the agreed route without justification. These common law obligations are very similar to the obligations imposed on the carrier pursuant to the Hague-Visby Rules, which will be discussed in detail in the next section of this chapter.

Statutory implied terms: The Hague-Visby Rules

Under the Carriage of Goods by Sea Ordinance, the Hague-Visby Rules as set out in the schedule of the Ordinance shall have the force of law in Hong Kong. The full text of the Hague-Visby Rules can be found in Annex 2.

The Hague-Visby Rules shall apply to every bill of lading relating to the carriage of goods between ports in two different states if the bill of lading is issued in a contracting state, or the carriage is from a port in a contracting state, or the contract contained in or evidenced by the bill of lading provides that the Hague-Visby Rules or the legislation of any state giving effect to them are to govern the contract, whatever may be the nationality of the ship, the carrier, the shipper, the consignee, or any other interested person. Under the Carriage of Goods by Sea (Parties to Convention) Order 1985, the following states are certified by the Hong Kong Special Administrative Region as the contracting states:

- The United Kingdom of Great Britain and Northern Ireland (Great Britain and Northern Ireland, The Isle of Man, Bermuda, British Antarctic Territory, British Virgin Islands, Cayman Islands, Falkland Islands, Gibraltar, Montserrat, Turks and Caicos Islands)
- The Kingdom of Belgium
- Hong Kong Special Administrative Region, The People's Republic of China
- The Kingdom of Denmark
- The Republic of Ecuador
- The Arab Republic of Egypt
- The Republic of Finland

- The French Republic
- The German Democratic Republic
- The Republic of Lebanon
- The Kingdom of the Netherlands
- The Kingdom of Norway
- The Polish People's Republic
- The Republic of Singapore
- Spain
- The Democratic Socialist Republic of Sri Lanka
- The Kingdom of Sweden
- The Swiss Confederation
- The Syrian Arab Republic
- The Kingdom of Tonga

For the purposes of the application of the Hague-Visby Rules in Hong Kong, the contracting states specified in the Carriage of Goods by Sea (Parties to Convention) Order 1985 shall be conclusive evidence of the matters so certified. The Hague-Visby Rules shall also apply to the carriage of goods by sea in ships where the port of shipment is in Hong Kong, whether or not the carriage is between ports in two different states (s 3(2) of the Ordinance). However, the Hague-Visby Rules shall not apply to any contract for carriage of goods by sea, unless the contract expressly or by implication provides for the issue of a bill of lading or any similar document of title (s 3(3)). The Hague-Visby Rules shall also apply to any bill of lading if the contract contained in or evidenced by it expressly provides that the Hague-Visby Rules shall govern the contract, and any receipt which is a non-negotiable document marked as such if the contract contained in or evidenced by it is a contract for the carriage of goods by sea expressly providing the Hague-Visby Rules are to govern the contract as if the receipt were a bill of lading (s 3(4)).

In addition, the Hague-Visby Rules may govern the bill of lading by virtue of the 'clause paramount'. Some standard form bills of lading expressly incorporate the Hague-Visby Rules by reference, using a clause known as a 'clause paramount'. For example:

> General Clause Paramount—The Hague-Visby Rules contained in the International Brussels Convention 1924 as amended by the Protocol signed at Brussels on 23 February 1968 as enacted in the country of shipment shall apply to this contract. When no such enactment is in force in the country of shipment, the corresponding legislation of the country of destination shall apply, but in respect of shipments to which

> no such enactments are compulsorily applicable, the terms of the said Convention shall apply.

In *Pohang Iron & Steel Co Ltd v Norbulk Cargo Services Ltd*,[2] the court held that the time period laid down in the Hague Rules/Hague-Visby Rules included in a paramount clause was in reality part of the contract between the parties. The agreement was that an action would have to be brought within a year of the delivery of goods otherwise all liability would have been discharged. In consequence, the liability was extinguished after the year had passed and could not be revived. The court in *Seabridge Shipping AB v AC Orssleff's Eftf's A/S*[3] gave the following opinion concerning the 'clause paramount':

> The shipping trade commonly uses terms—'clause paramount', 'general clause paramount' . . . For over twenty years the meaning of 'clause paramount' has been certain. Persons in the shipping trade have been free to use the phrase 'general clause paramount' if they wished to incorporate the Hague-Visby Rules into trades where those rules are compulsorily applicable. Thus, on the evidence before me I see no warrant for departing from the views of shipping men which the court ascertained.

The scope of application of the Hague-Visby Rules

The scope of application of the Hague-Visby Rules can be found in the following illustrations from case law:

> *Pyrene Co Ltd v Scindia Navigation Co*[4]
> The plaintiff sold a piece of machinery to the Government of India for delivery from London. While the machine was being lifted on to the vessel by the ship's tackle, it was dropped and damaged through the fault of the ship. The carrier admitted liability, but sought to claim that the amount was limited under Art 4 r 5 of the Hague Rules (i.e., the earlier version of the Hague-Visby Rules). Art I (b) states that 'contracts of carriage' apply 'only to contracts of carriage covered by the a bill of lading or any similar document of title, in so far as such document relates to the carriage of goods by sea'. When the accident occurred, the bill of lading had not yet been issued.
>
> The court held that the carriage was still 'covered' by a bill of lading and therefore the Hague Rules applied.

2. [1996] 4 HKC 701.
3. [2000] 1 All ER (Comm) 415.
4. [1954] 2 QB 402.

Lord Devlin: 'The use of the words "covered" recognises the fact that the contract of carriage is always concluded before the bill of lading, which evidences its terms, is actually issued. When parties enter into a contract of carriage in the expectation that a bill of lading will be issued to cover it, they enter into it upon those terms which they know or expect the bill of lading to contain. Those terms must be in force from the inception of the contract; if it were otherwise the bill of lading would not evidence the contract but would be a variation of it. Moreover, it would be absurd to suppose that the parties intend the terms of the contract to be changed when the bill of lading is issued: for the issue of the bill of lading does not necessarily mark any stage in the development of the contract; often it is not issued till after the ship has sailed, and if there is pressure of office work on the ship's agent it may be delayed several days. In my judgment, whenever a contract of carriage is concluded, and it is contemplated that a bill of lading will, in due course, be issued in respect of it, that contract is from its creation "covered" by a bill of lading, and is therefore from its inception a contract of carriage within the meaning of the rules and to which the rules apply.'

The Morviken[5]

A piece of machinery was shipped at Leith, Scotland for carriage to Amsterdam. The bill of lading stated that Dutch law should apply and all actions should be brought before the Court of Amsterdam. When the machine was discharged, it was dropped and damaged. The owner of the machine claimed damages, but the Dutch carrier applied for the action to be stayed in view of the jurisdiction clause in the bill of lading.

By Dutch law the original Hague Rules applied, which entitled a carrier to limit his liability to £250, whereas under Art IV r 5 (a) of the Hague-Visby Rules as set out in the Carriage of Goods by Sea Act, the sum would be far higher, in the amount of £11,000.

The House of Lords held that the action would not be stayed, for the bill of lading was one to which the Hague-Visby Rules were expressly made applicable by Art X, and the clause was void under Art III r 8 because it lessened the liability of the carrier.

Lord Diplock: 'The bill of lading issued to the shippers by the carriers upon the shipment of the goods at the Scottish port of Leith was one to which the Hague-Visby Rules were expressly made applicable by Art X; it fell within both paragraph (a) and paragraph (b); it was issued in a Contracting State, the United Kingdom, and it covered a contract for carriage from a port in a Contracting State. For good measure, it also fell directly within s 1(3) of the Carriage of Goods by Sea Act itself. The first paragraph of condition 2 of the bill of lading,

5. [1983] 1 Lloyd's Rep 1.

prescribing as it does for a package maximum limit of liability on the part of the carriers for loss for damage arising from negligence or breach of contract instead of the higher per kilogramme maximum applicable under the Hague-Visby Rules, is ex facie, a clause in a contract of carriage which purports to lessen the liability of the carriers for such loss or damage otherwise than is provided in the Hague-Visby Rules. As such it is therefore rendered null and void and of no effect under Art III r 8 . . . [to reach any other conclusion] would leave it open to any shipowner to evade the provisions of Art III r 8 by the simple device of inserting in his bills of lading issued in, or for carriage from a port in, any contracting state a clause in standard form providing as the exclusive forum for resolution of disputes what might aptly be described as a court of convenience, viz. one situated in a country which did not apply the Hague-Visby Rules.'

Kum v Wah Tat Bank Ltd[6]

A carriage of goods between Sarawak and Singapore was covered by a mate's receipt marked 'non-negotiable'. The Privy Council held that the mate's receipts were not in law documents of title. Therefore, the carriage covered by the mate's receipt was beyond the scope of the Hague-Visby Rules.

Lord Devlin: 'If the mate's receipt had been a bill of lading, the legal position would be beyond dispute. Not only is the bill of lading a document of title, but delivery of it is symbolic delivery of the goods. But the mate's receipt is not ordinarily anything more than evidence that the goods have been received on board. This is so firmly settled by *Hathesing v Laing* (1873) LR 17 Eq 92 and *Nippin Yusen Kaisha v Ramjiban Serowgee* [1938] AC 429.'

Deviation

Art IV r 4 provides that 'any deviation in saving or attempting to save life or property at sea or any reasonable deviation shall not be deemed to be an infringement or breach of these Rules or of the contract of carriage, and the carrier shall not be liable for any loss or damage resulting therefrom'. Unjustified deviation is a breach of the shipowner's undertaking to proceed by a usual and reasonable route.

In *Renton v Palmyra*,[7] the vessel was chartered to carry a cargo of coal and to proceed from Swansea, where the coal was to be loaded with all possible despatch to Constantinople. The usual and customary route for the voyage was from Swansea, south of Lundy, thence in a straight line to

6. [1971] 1 Lloyd's Rep 439.
7. [1957] AC 149.

a point about five miles off Pendenn, on the north coast of Cornwall, and then with a slight alteration to the east to Finistere and so on. The ship had been fitted with heating apparatus designed to make use of the heat which might otherwise be wasted as steam and so to diminish the bill for fuel. This apparatus had not been working satisfactorily, and the owners arranged to send representatives of the engineers to make a test when the vessel started on her next voyage. Two engineers accordingly joined the boat, the intention being that they should leave the ship with the pilot somewhere off Lundy. The firemen on board the ship were not in possession of their full energies when the boat started, due to excessive drinking before they joined the ship. The result was that a proper head of steam necessary for making the test was not got up in time to enable the test to be made before the pilot was discharged. Accordingly they proceeded on the voyage until the ship was off St. Ives, when the ship was turned about five miles out of its course to enter at St. Ives Harbour in order that the engineers might be landed. After landing them, the ship did not go straight back to the recognized route that she ought to have pursued, but hugged too closely the dangerous coast of Cornwall, and ran on a rock called the Vyneck Rock, with the result that the vessel and cargo were totally lost. The accident took place at about 3.20 p.m. on 31 June 1929. There was a moderate wind from ENE, the weather was cloudy, but visibility was moderately good up to six miles. The cargo owners sought to recover damages for loss of their cargo. The carrier argued that under Art IV r 4, they were entitled to make the deviation which led to the disaster. The court held that the deviation was unreasonable. Lord Atkin stated:

> The true test seems to be what voyage at the time make and maintain, having in mind all the relevant circumstances existing at the time, including the terms of the contract and the interests of all parties concerned, but without obligation to consider the interests of any one as conclusive . . . The decision has to be that of the master or occasionally of the shipowner; and I conceive that a cargo owner might well be deemed not to be unreasonable if he attached much more weight to his own interests than a prudent master having regard to all the circumstances might think it wise to do . . . Was the deviation justifiable? . . . though the port of refuge was justifiably reached, the subsequent voyage might be so conducted as to amount to an unreasonable deviation . . . It is obvious that the small extra risk to ship and cargo caused by deviation to St. Ives, was vastly increased by the subsequent course. It seems to me not a mere error of navigation but a failure to pursue the true course from St. Ives to Constantinople which in itself made the deviation cease to be reasonable.

Due diligence to make the ship seaworthy

Under Art III r 1, the carrier shall be bound before and at the beginning of the voyage to exercise due diligence to make the ship seaworthy. In *Maxine Footwear Co Ltd*,[8] the court held that there was a failure to exercise due diligence to make the ship seaworthy, which caused the damage to and loss of the cargo. In this case, shortly before the vessel was due to sail, an attempt was made to thaw a frozen drainpipe with an acetylene torch. A fire was started in the cork insulation around the pipe. The fire eventually forced the master to scuttle the ship. The court was invited to interpret the meaning of the phrase 'before and at the beginning of the voyage'. Lord Somervell of Harrow averred:

> 'before and at the beginning of the voyage' means the period from at least the beginning of the loading until the vessel starts on her voyage . . . On that view the obligation to exercise due diligence to make the ship seaworthy continued over the whole of the period from the beginning of loading until the ship sank. There was a failure to exercise due diligence during that period. As a result the ship became unseaworthy and this unseaworthiness caused the damage to and loss of the [goods]. The [cargo owners] are therefore entitled to succeed.

A carrier cannot claim to have shed his obligation to exercise due diligence to make his ship seaworthy by selecting a firm of competent ship repairers or independent contractors to make the ship seaworthy. In *Riverstone Meat Co Pty Ltd v Lancashire Shipping Co Ltd* (The '*Muncaster Castle*'),[9] a consignment of ox tongue was carried by sea from Sydney to London under a bill of lading governed by the Australian Sea Carriage of Goods Act 1924, which embodied the Hague Rules. Shortly before the commencement of the voyage, the storm valve inspection covers had been removed during a survey of the ship. The covers had not been properly refitted by the independent firm of ship repairers, who had been instructed in connection with the survey. The court held that the carrier was liable for failing to exercise due diligence to make the ship seaworthy. Lord Keith of Avonholm stated:

> The obligation is a statutory obligation imposed in defined contracts between the carrier and the shipper. There is nothing novel in a statutory obligation being held to be incapable of delegation so as to free the person bound of liability for breach of the obligation . . .

8. [1959] AC 589.
9. [1961] AC 807.

> *Smith, Hogg & Co Ltd v Black Sea and Baltic General Insurance Co Ltd*[10] MacKinnon LJ said: 'The limitation and qualification of the implied warranty of seaworthiness by cutting down the duty of the shipowner to the obligation to use "due diligence to make the ship seaworthy" is a limitation or qualification more apparent than real, because the exercise of due diligence involves not merely that the shipowner personally shall exercise due diligence, but that all his servants and agents shall exercise due diligence . . . In most cases if the vessel is unseaworthy due diligence cannot have been used by the owner, his servants, or agents; if due diligence has been used the vessel in fact will be seaworthy. The circumstances in which the dilemma does not arise (e.g. a defect causing unseaworthiness, but so latent a nature that due diligence could not have discovered it) are not likely to occur often.'

Care of the cargo

In addition to his duties not to deviate from the reasonable route and to make the ship seaworthy before and at the beginning of the voyage, the carrier shall properly and carefully load, handle, stow, carry, keep, care for, and discharge the goods carried under Art III (2) of the Hague-Visby Rules. The interpretation and the meaning of the terms 'properly' and 'carefully' were debated in *Albacora SRL v Westcott & Laurance Line Ltd.*[11] At the port of discharge, the fish was found to have been contaminated by bacteria because the temperature in the cargo hold had been above the critical level of 5°C. The shipper and the carrier were not aware of the requirement of refrigeration, and the contract of carriage did not stipulate it. The cargo owner argued that the only proper way to carry the consignment on that voyage was in a refrigerated hold, and therefore the carrier was in breach of their obligation for not doing the same. The House of Lords ruled that the shipowner was not in breach of Art III r 2. Lord Reid:

> I agree with Viscount Kilmuir, LC, that here 'properly' means in accordance with a sound system (*GH Renton & Co Ltd v Palmyra Trading Corporation of Panama* [1957] AC 149 at p 166) and that may mean rather more than carrying the goods carefully. But the question remains by what criteria it is to be judged whether the system was sound. In my opinion, the obligation is to adopt a system which is sound in light of all the knowledge which the carrier has or ought to have about the nature of the goods. And if that is right, then the [carrier] did adopt a sound system. They had no reason to suppose that the goods required any different treatment from that which the goods in fact received.

10. [1939] 2 All ER 855.
11. [1966] 2 Lloyd's Rep 53.

Lord Pearce added: 'A sound system does not mean a system suited to all the weaknesses and idiosyncrasies of particular cargo, but a sound system under all the circumstances in relation to the general practice of carriage of goods by sea.'

Carrier's defences

Under Art IV r 2(a), neither the carrier nor the ship shall be responsible for loss or damage arising or resulting from an act, neglect or default in the 'navigation and management of the ship'. However, a clear distinction shall be drawn between the management of the ship and the management of the cargo. In the case of *Hourani v Harrison,*[12] the Court of Appeal had to consider the meaning to be attached to the words 'management of the ship' in Art IV r 2(a). In this case, loss of cargo was caused by the pilfering of the stevedore's men whilst the ship was being discharged. The court held that this did not fall within the expression 'management of ship', and therefore the carrier could not make use of this defence under the rules to exonerate his liability. The learned judges expressed the distinction as being between damage resulting from some act relating to the ship herself and only incidentally damaging the cargo, and an act dealing solely with the goods and not directly or indirectly with the ship herself. Atkin LJ said:

> There is a clear distinction drawn between goods and ship; and when they talk of the word 'ship', they mean the management of the ship, and they do not mean the general carrying on of the business of transporting goods by sea.

This test was applied with approval in *Goss Millerd Ltd v Canadian Government Merchant Marine Ltd.*[13] A ship carried cargo of tinplates. During the voyage, the ship sustained damage, and had to go into dock for repairs. While the repairs were being carried out, the hatches were left open so that the workmen could go in and out of the cargo hold more easily. When rain fell, the hatches were not properly closed. As a result, the tinplates were damaged by the rain. The House of Lords held that the carrier had failed properly and carefully to carry the goods as required by Art IV r 2(a). The term 'management of the ship' did not include lack of care of the cargo, or negligence in the management of the cargo stored in the hatches. Lord Hailsham LC:

12. 32 Com Cas 305 (1927).
13. [1929] AC 233.

> In my judgment [the carrier] have not even assumed that the persons who were negligent *were* their servants; but even if it can be assumed that the negligence in dealing with the tarpaulins was by members of the crew, such negligence was not negligence in the management of the ship, and therefore is not negligence with regard to which Art IV r 2(a) affords any protection.

The carrier is not liable for loss or damage arising from inherent defect, quality or vice of the goods under Art IV r 2(m). A concise statement in respect of this defence can be found in *The Barcore*.[14] Justice Gorell Barnes said:

> This cargo was not damaged by reason of the shipowner committing a breach of contract, or omitting to do something which ought to have been done, but it was deteriorated in condition by its own want of power to bear the ordinary transit in a ship.

The 'ordinary transit' refers to the kind of transit which the contract requires the carrier to afford. The defence is intended to give effect to the well-established principle that if an article is unfitted owing to some inherent defect or vice for the voyage which is provided for in the contract, then the carrier may escape liability when damage results from the activation of that inherent vice during the voyage.

In *Albacora SRL v Westcott & Laurance Line Ltd*,[15] a consignment of fish was damaged due to lack of refrigeration. Neither the shipper nor the carrier was aware that refrigeration was necessary, and the contract of carriage did not provide for it. The House of Lords held that there was an inherent vice in the fish in respect of that transit, and therefore the carrier was not liable for the damage. Lord Reid:

> The obligation under the Article is to carry goods properly and if that is not done there is a breach of contract. So it is argued that in the present case it is proved that the only proper way to carry this consignment on this voyage was in a refrigerated hold, and there the obligation of the [carrier] was to do that, even if the [cargo owner's] agents who were parties to the contract were aware that there was no refrigeration in this ship. This construction of the word 'properly' lead to such an unreasonable result that I would not adopt it if the word can properly be construed in any other sense . . . In my opinion, the obligation to adopt a system which is sound in light of all the knowledge which the carrier has or ought to have about the nature of the goods. And if that is right, then the respondents did adopt a sound system. They had no

14. [1896] P 294.
15. [1966] 2 Lloyd's Rep 53.

> reason to suppose that the goods required any different treatment from that which the goods in fact received.

According to Art VI (2)(c), the ship is not liable for loss and damage caused by 'perils of the sea'. At common law, the most authoritative judicial definition is that approved by Lord Bramwell in *Thames & Mersey Marine Insurance Co v Hamilton, Fraser & Co (The 'Inchmaree')*:[16]

> Every accidental circumstance not the result of ordinary wear and tear, delay, or of the act of the assured, happening in the course of the navigation of the ship, and incidental to the navigation, and causing loss to the subject-matter . . . In a seaworthy ship damage to goods caused by the action of the sea during transit not attributable to the fault of anybody, is a damage from a peril of the sea.

Lord Herschell in *The Xantho*[17] said:

> [the exception does not protect] against that natural and inevitable action of the wind and waves which results in what may be described as wear and tear.

The requirement of 'perils of the sea' will be met if the loss would only occur at sea and would not have occurred on land. In the same case, Lord Bramwell described the position in this way:

> The damage to the donkey-engine was not through its being in a ship or at sea. The same thing would have happened had the boilers and engines been on land, if the same mismanagement had taken place. The sea, waves and winds had nothing to do with it.

However, the defence of 'perils of the sea' will fail if evidence shows that it is the ship's unseaworthiness that constituted the dominant and operating cause of the loss. The decision in *The Miss Jay Jay*[18] clearly shows that for 'perils of the sea' defence to succeed is rather unusual, and the conclusions of the judges are nearly uniform in that the shipowner has a duty to provide a vessel that can 'deal adequately with adverse as well as favourable weather'. In *Sasson & Co v Western Assurance Co*,[19] the court pointed out that there was no weather, nor any other fortuitous circumstances, contributing to the incursion of the water. The water merely aggravated by its own weight through the opening of the decayed wood of the ship. The damage to the cargo was not a loss caused by a peril of the sea. In *Grant, Smith & Co*

16. (1887) 12 App Cas 484.
17. (1887) 12 App Cas 503.
18. [1985] 1 Lloyd's Rep 265.
19. [1912] AC 561.

v Seattle Construction and Dry Dock Co,[20] the court found that the loss of the dry dock that had capsized in the harbour was due to the ship's inherent unfitness for the work. Lord Buckmaster laid down the following principles:

> It is some condition of sea or weather or accident of navigation producing a result which, but for these conditions, would not have occurred.

Collision at sea is a peril of the sea. In *The Xantho*,[21] the House of Lords declared that a collision, whether caused by a sunken rock, or by an iceberg, or by another vessel, or whether that other vessel is or is not in fault, is a peril of the sea. According to *Smith v Scott*[22] and *Davidson v Burnard*,[23] no distinction is drawn between a loss caused by the negligence of the crew of the subject vessel and one caused by the negligence of the crew of another vessel.

Deck cargo, live animals and dangerous goods

Art I (c) states that 'goods' includes goods, wares, merchandise, and articles of every kind whatsoever except live animals and cargo which by the contract of carriage is stated as being carried on deck and is so carried. Hence, the Hague-Visby Rules are inapplicable to the carriage of live animals and cargo carried on deck in accordance with the bill of lading.

> *Svenska Traktor Skiebolaget v Maritime Agencies (Southampton) Ltd*[24]
> The bill of lading stated that the carrier 'has liberty to carry goods on deck and shipowners will not be responsible for any loss damage or claim arising therefrom'. It was held that the clause could not exclude the carrier's liability for damage to cargo carried on deck.
>
> Pilcher J: 'A mere general liberty to carry goods on deck is not in my view a statement in the contract of carriage that the goods are in fact being carried on deck. To hold otherwise would in my view do violence to the ordinary meaning of the words of Art I (c). I hold accordingly, that the plaintiff's [cargo] were being carried by the shipowners subject to the obligations . . . under Art III r 2 properly and carefully to load, handle, stow, carry, keep and care for the goods in question.'

According to Art IV (6), goods of an inflammable, explosive or dangerous nature to the shipment whereof the carrier, master or agent of the carrier has not consented with knowledge of their nature and character, may at

20. [1920] AC 162.
21. (1887) 12 App Cas 503.
22. (1811) 4 Taunt 126.
23. (1868) LR 4 CP 117.
24. [1953] 2 QB 295.

any time before discharge be landed at any place, or destroyed or rendered innocuous by the carrier without compensation. The shipper of such goods shall be liable for all damages and expenses directly or indirectly arising out of or resulting from such shipment. If any such goods shipped with such knowledge and consent shall become a danger to the ship or cargo, they may in like manner be landed at any place, or destroyed or rendered innocuous by the carrier without liability on the part of the carrier except to general average, if any.

> In *The 'Aconcagua'*[25]
>
> The vessel had been damaged following an on-board explosion of a cargo of chemical in a container which had been stowed in a position in the hold where it was surrounded on three sides by a bunker tank which was heated during the voyage. The cargo should have been stowed away from sources of heat. Having settled the shipowners' damages claim against them, the charterers claimed an indemnity from the shippers under Article IV Rule 6 of the Hague Rules. The High Court held that the cargo was of a dangerous nature of which the charterers neither had nor ought to have had knowledge. Further, a vessel was not unseaworthy because at the commencement of the voyage there was something which might need a correction which could readily be made. The operative fault lay not in the stowage of the container, but in the decision to use and heat the relevant bunker tank which was an 'act, neglect or default in the management of the ship' and therefore an excepted peril. Since the casualty was caused by the shipment of dangerous goods and by a cause for which charterers were not liable, the charterers could successfully claim an indemnity against the shippers.
>
> *Effort Shipping Co Ltd v Linden Management SA (The 'Giannis NK')*[26]
>
> A cargo of groundnut extractions was loaded onto the carrier's vessel to be shipped to the Dominican Republic. Unfortunately, unknown to both the shipper and the carrier, the cargo was infested with khapra beetles. In consequence, the vessel and its entire cargo, which included a separate cargo of wheat owned by another cargo owner, were excluded from the port of discharge. The vessel was arrested by the cargo consignees, and eventually the entire cargo had to be dumped at sea. The ship was given clearance to load her next charterparty after two and a half month's delay. The court held that pursuant to Art IV r 6, the shipper of the groundnut was liable to the carrier for the delay caused by the shipment of the dangerous goods.

25. [2010] 1 Lloyd's Rep 1.
26. [1998] 1 Lloyd's Rep 337.

> Lord Lloyd of Berwick: 'What are the consequences of the finding that the groundnut cargo was physically dangerous to the wheat cargo? Since the carrier did not consent to the shipment of the groundnut with knowledge of its dangerous character, the shippers are prima facie liable for all damages and expenses suffered by the carriers.'

The shipper of the groundnut then tried to invoke Art IV r 3. It provides that the shipper shall not be responsible for loss or damage sustained by the carrier or the ship arising or resulting from any cause without the act, fault or neglect of the shipper, his agent or his servants. The shipper argued that the infection of the cargo with khapra beetles was without the fault of the shipper. Nevertheless, the court rejected such argument and held that Art IV r 6 imposed strict liability on the shipper in association with the shipment of dangerous goods and they were not entitled to limit their liability by reference to Art IV r 3. Lord Steyn took the view that the best interpretation of the language of Art IV r 6 read in conjunction with Art IV r 3, seen against its contextual background, was that it created free-standing rights and obligations in respect of the shipment of dangerous cargo.

Time bar and package limitation

Art III r 6 provides that the carrier shall in any event be discharged from all liability whatsoever in respect of the goods unless suit is brought within one year of their delivery (or of the date when they should have been delivered).

In *Kenya Railways v Antares Co Pte Ltd*,[27] part of the machinery had been loaded on deck without authorization and had been seriously damaged in the course of the voyage. The cargo owner overlooked the 'demise clause' printed on the back of the bill of lading, which made it clear that MSC, the issuer of the bill, might not be the shipowner even though the bill of lading was issued on MSC's form. The cargo owner subsequently realized this and commenced action against the shipowner. Unfortunately, the suit was not brought within one year of the delivery of the machinery. The cargo owner put forward the argument that the unauthorized carriage of goods on deck constituted a 'fundamental breach' of the contract of carriage. By reason of the 'fundamental breach' (that is, the doctrine that a breach of contract may be so fundamental as to displace the exception clauses altogether), the carrier cannot rely on the time bar provision in Art III r 6. However, the court rejected the cargo owner's argument, holding that the action against the shipowner was time barred. Lloyd LJ:

27. [1987] 1 Lloyd's Rep 424.

> The sole question therefore is whether, on its true construction, Art III r 6 applies. It is clear that it does. It provides that the carrier shall in any event be discharged from all liability whatsoever unless suit is brought within one year . . . the word 'whatsoever' makes it clear that the time limit applies even where the carrier has committed . . . the unauthorised carriage of goods on deck . . . The [cargo owner] seeks a declaration that the [shipowner] is barred from relying on Art III r 6, by reason of their fundamental breach of contract. For the reasons I have given they are not entitled to that.

> *Pohang Iron & Steel Co Ltd v Norbulk Cargo Services Ltd*[28]
> Two bills of lading, under which cargo was shipped from Hamburg and discharged from the vessel in South Korea, contained a paramount clause whereby the Hague Rules were to stated to apply to the contract of carriage. The goods were found damaged on arrival. The court had to decide whether the cargo owner's claim was time barred pursuant to Art III r 6 of the Hague Rules in that the cargo owner's suit had not been brought within one year after delivery of the goods.
>
> Roger J: 'The period laid down in the Hague Rules is in reality part of the contract between the parties. The agreement was that action would have to be brought within a year of the delivery of goods otherwise all liability would be discharged. It is the liability which is extinguished after the year's period and it cannot be revived . . . There is no justification therefore in seeking to extend the meaning of Art III r 6 to allow proceedings which were a nullity to be treated as "bringing suit".'

The complaint of misdelivery is subject to the limitation provisions of Art III r 6, as decided by the court in *Cia Portorafti Commerciale SA v Ultramar Panama Inc (The 'Captain Gregos')*.[29] This is because Art II laid down the scope of operations and activities to which the Hague-Visby Rules apply. Art III r 2 sets out the carrier's obligation to properly and carefully load, handle, stow, carry, keep, care for and discharge the goods carried. The time limitation provisions applied because the carrier had been in breach of those obligations by wrongfully delivering the goods to the party that was not entitled to receive the cargo. The court rejected the argument that deliberate misappropriation of the cargo did not fall within the scope of that rule.

On some occasions, the bills of lading may prescribe shorter periods of limitation. In *Finagra (UK) Ltd v OT Africa Line Ltd*,[30] the bills of lading included a 'carrier's responsibility' clause providing that 'save as otherwise provided in the bill of lading' the Hague Rules would apply to the stage of carriage where the damage was caused was not known. Where that stage

28. [1996] 4 HKC 703.
29. [1990] 3 All ER 967.
30. [1998] 2 Lloyd's Rep 622.

was known the liability of the carrier would be determined by the relevant international convention. A time bar clause required suit to be commenced within nine months 'subject to any provisions in this clause to the contrary'. Legal proceedings were brought more than nine months but less than one year after delivery. The court held that the proviso to the time bar clause which provided that the clause applied 'subject to any provision of this clause to the contrary' meant that the one year time limit set out in Art III r 6 applied. Therefore, the claims in the action were not time barred. Rix J observed:

> Where the Hague Rules are incorporated by contract, the essential rule is to treat the rules as set out in the body of the contract in extenso, but rejecting provisions which are insensible, because inconsistent with the incorporating document . . . The presence of a repugnancy clause, whether the clause inherent in the rules in the form of Art III r 8, a fortiori a separate repugnancy clause in the contract itself, will always be relevant: but neither its presence nor its absence is necessarily decisive . . . Even if, however, I had felt myself forced to conclude that the two provisos were ineradicably in conflict with one another, I would consider myself obliged to give precedence to the second proviso. It is that which is related to the time bar clause itself. Thus, even if other provisions of the bill of lading were to have priority over the Hague Rules in case of conflict, nevertheless in the matter of time bar it should be the second proviso which prevails. The specific controls the general.

Under Art IV r 5, unless the nature and value of such goods have been declared by the shipper before shipment and inserted in the bill of lading, neither the carrier nor the ship shall in any event be liable for any loss or damage in connection with the goods in an amount exceeding 666.67 SDR[31] per package or unit or 2 units of account per kilogramme of gross weight of the goods lost or damaged, whichever is the higher. In *El Greco*,[32] the court treated the whole container as one unit, since the bill did not make clear on its face what number of packages or units were packed inside the container.

The bill of lading as a receipt for cargo shipped

A bill of lading serves as a receipt for the cargo shipped. Usually, a bill will contain statements as to the condition in which they were received by the

31. Special Drawing Right. Under section 7 of the Carriage of Goods by Sea Ordinance, the HK Monetary Authority may issue a certificate specifying in Hong Kong dollars the respective amounts which are to be taken as equivalent for a particular day to the sums expressed in SDR.
32. [2004] 2 Lloyd's rep 537 (Federal Court of Australia).

carrier, the quantity and description of the goods shipped, together with any identifying marks of the goods on shipment. A bill of lading is therefore a formal receipt and acknowledgement that goods of a certain kind, quantity, and condition have been delivered for shipment. Those statements and representations in the bill of lading are of significant evidential and commercial value. They form the important basis in an action for short delivery or damage to the cargo on discharge, where the burden of proof lies on the cargo owner to establish that the goods were damaged or lost while in the custody of the carrier. This can be illustrated by a simple example. Suppose a claimant now commences proceedings against the carrier for short delivery of part of a consignment of 30 cars. The claimant is able to discharge his evidential burden of proof of loss by establishing that shipment of 200 cars was acknowledged in the bill of lading, while only 170 cars were delivered from the vessel at the port of discharge.

Nonetheless, it is worth noting that some bills of lading may include none of these representations or acknowledgments. In order to minimize the evidential value of the bill in subsequent litigation, the carrier may simply include statements like 'weight and quantity unknown', 'condition unknown' or 'said by the shipper to contain' or words of similar effect in the bill of lading. In order to cope with this problem, Art III r 3 of the Hague-Visby Rules enables the shipper to demand the issue of a bill of lading containing certain specified particulars:

> After receiving the goods into his charge the carrier or the master or agent of the carrier shall, on demand of the shipper, issue to the shipper a bill of lading showing among other things:
>
> (a) The leading marks necessary for identification of the goods as the same are furnished in writing by the shipper before the loading of such goods starts, provided such marks are stamped or otherwise shown clearly upon the goods if uncovered, or on the cases or covering in which such goods are contained, in such a manner as should ordinarily remain legible until the end of the voyage.
>
> (b) Either the number of packages or pieces, or the quantity, or weight, as the case may be as furnished in writing by the shipper.
>
> (c) The apparent order and condition of the goods.

In *Monica Textile Corp v SS Tana*,[33] the American court held that when a bill of lading consisted of a number of containers followed by a description of the goods, statements like 'said to contain 25 packages' and 'particulars furnished by the shippers: not verified by the carrier' did not exonerate

33. 2nd Cir 1991 (23 December 1991).

the carrier from liability for short delivery. The carrier was held to have acknowledged having received these 25 packages. In contrast, the English courts adopted a different approach and held that such a qualification greatly reduced the evidentiary value of the statement so far as the shipper is concerned:

Noble Resources Limited v Cavalier Shipping Corporation (The 'Atlas')[34]
The court rejected the argument that Art III r 8 of the Hague-Visby Rules was to make any bill of lading containing clauses such as 'said to be' or 'weight unknown' prima facie evidence of quantity shipped.

Justice Longmore said: 'Do the Russian bills show the number of packages or weight 'as furnished in writing by the shipper'? In one sense it can be said they do, because the bills have figures which were in fact provided by the shipper in writing. But if the bills provide 'weight . . . number . . . quantity unknown' it cannot be said that the bills 'show' that number or weight. They 'show' nothing at all because the shipowner is not prepared to say what the number or weight is. He can, of course, be required to show it under Art III r 3, and unless and until he does so, the provisions of Art III r 4 as to *prima facie* evidence cannot come into effect. This seems to me to be right as a matter of language but there is authority to the same effect.' (Justice Longmore then referred to *Canadian and Dominion Sugar Co Ltd v Canadian National (West Indies) Steamship Ltd* [1947] AC 46 and *Attorney General of Ceylon v Scindia Steam Navigation Co Ltd* [1962] AC 60)

Attorney General of Ceylon v Scindia Steam Navigation Co Ltd[35]
Bills of lading acknowledged receipt of 100,652 bags of rice 'weight, contents and value when shipped unknown' for carriage from Burma to Ceylon. The clause did not qualify the statement as to the number of bags. On arrival, only 100,417 bags were delivered. The cargo owner claimed damages.

Lord Morris stated: 'Though by relying upon the bills of lading the plaintiff presented *prima facie* evidence that 100,652 bags . . . were shipped, the bills of lading were not even *prima facie* evidence of the weight or contents or value of such bags . . . It was for the plaintiff to prove what was in the missing bags . . . In this connection reference may again be made to the decision of Lush J in *Hogarth Shipping Co Ltd v Blyth, Greene, Jourdain & Co Ltd* [1917] 2 KB 534. In his judgment Lush J pointed out that if a certain number of bags had been lost, and if one had to ascertain what was in the bags that were lost, then as a matter of evidence one would almost necessarily infer that the lost bags were bags containing similar goods to those which were not lost.'

34. [1996] 1 Lloyd's Rep 642.
35. [1962] AC 60.

New Chinese Antimony Co Ltd v Ocean Steamship Co Ltd[36]

The bill of lading acknowledged shipment of 937 tonnes of antimony oxide ore with the clause 'weight, measurement, contents and value (for purpose of estimating freight) unknown'. On arrival the ore weighed only 861 tons. The cargo owner brought action to recover the difference in value between 861 and 937 tons. The Court of Appeal held that the shipowner was not responsible for any loss.

Viscount Reading CJ: 'I think that the true effect of this bill of lading is that the words "weight unknown" have the effect of a statement by the shipowners' agent that he has received a quantity of ore which the shippers' representative says weighs 937 tons but which he does not accept as being of that weight, the weight being unknown to him, and that he does not accept the weight of 937 tons except for the purpose of calculating freight and for that purpose only.'

Statements as to quantity or weight

At common law, a bill of lading is only prima facie evidence of the quantity or weight of the shipped goods.[37] It is still open to the shipowner to establish that the goods were never put on board the ship for carriage. For example, if the consignee of a bill of lading stating 'shipped . . . 100 cars' only received 90 cars at the port of discharge, then it is still possible for the shipowner to prove that the other 10 cars had never in fact been put on board. Nonetheless, there are legal authorities stating that the evidence provided must be 'clear, distinct and convincing' (see *Ace Imports Pty Ltd v Companhia De Navegacao Lloyd Brasileiro*).[38] The evidence must exclude beyond reasonable doubt the possibility of the goods having been received for shipment.[39]

Henry Smith & Co v Bedouin Steam Navigation Co Ltd[40]

A cargo of jute was shipped from Calcutta to Dundee, but on discharge 12 bales were missing from the 1,000 stated to have been shipped. The House of Lords held that the shipowner had not done enough to defeat the inference from the bill of lading.

Lord Shand said: 'The bills of lading—all evidence of acts by servants of the shipowners, form a strong and consistent body of proof that the shipment acknowledged under the captain's hand was actually made, and impose a heavy onus on the shipowner who alleges

36. [1917] 2 KB 664.
37. *Henry Smith & Co v Bedouin Steam Navigation Co Ltd* [1896] AC 70 at 80 per Lord Shand and Lord Davey.
38. (1987) 10 NSWLR 32 at 39.
39. *Rosenfield Hillas & Co Pty Ltd v The Ship Fort Laramie* (1923) 32 CLR 25 at 33 per Isaacs J.
40. [1896] AC 70 (HL).

> that nevertheless there was a deficiency, through non-shipment, in the quantity of goods shipped . . . the evidence must be sufficient to lead to the inference not merely that the goods may possibly not have been shipped, but that in point of fact they were not shipped.'

If the Hague-Visby Rules govern the bill of lading, then under Art III r 4 a representation of the quantity or weight of the goods is prima facie evidence of receipt by the carrier of the quantity or weight so described. However, proof to the contrary shall not be admissible when the bill of lading has been transferred to a third party acting in good faith. However, it must be noted that the operation of Art III r 4 is subject to the overriding provision in Art III r 3 in respect of the issuance of the bills of lading 'on demand of the shipper'. In other words, Art III r 4 only comes into operation if the relevant demand of the shipper to issue a bill of lading has been made within the meaning of Art III r 3:

> *Agrosin Pte Ltd v Highway Shipping Co Ltd (The 'Mata K')*[41]
> The bill of lading included the clause 'weight, measure . . . unknown'. On arrival it was discovered that there was a shortfall of about 2705 tonnes as compared with the total of the cargo stated in the three bills of lading, namely 24,024.7 tonnes. The bill of lading was governed by the Hague Rules. Evidence revealed that no relevant demand to issue the bill of lading was made by the shipper under Art III r 8. The court held that the carrier was not liable for short delivery, and the cargo owner could not rely on the representation regarding the quantity of the cargo stated on the bill of lading.
>
> Clarke J: 'In all the circumstances I have reached the conclusion that, on the allegations made by the [cargo owner] and on the facts set out in the material before the court, there was no demand such as should satisfy Art III r 3 of the Hague Rules. It follows that there is no basis on which the "weight . . . unknown" provision could be treated as null and void and of no effect under Art III r 8. It also follows that Art III r 4 has no application because the bill of lading is not "such a bill of lading" (that is a bill of lading of the kind referred to in Art III r 3) so that the bill is not prima facie evidence of the receipt of 11,000 tonnes of cargo under r 4. Finally (and crucially on the facts of this case) it follows that the bill of lading does not represent that 11,000 tonnes were shipped so as to be conclusive evidence against the [carrier] . . .'

Apparent good order and condition

If the bill of lading states that the goods have been shipped 'in apparent good order and condition', the shipowner will be estopped as against an

41. [1998] 2 Lloyd's Rep 642.

indorsee by such a statement as far as defects which ought to be apparent on reasonable examination. However, the words 'in apparent good order and condition' do not involve any representation associated with the inherent or internal quality of the goods carried. Such words only relate to the external condition of the consignment.

> *Silver v Ocean Steamship Co Ltd*[42]
> Cans of frozen eggs were shipped under a bill of lading stated to be 'in apparent order and condition'. When the cans arrived at the port of discharge, some were found in a damaged condition, being gashed or punctured. Some only had pinhole perforations.
>
> The Court of Appeal held that the shipowners were estopped as against the indorsees of the bill of lading from showing that the cans had gashes which would have been apparent when the goods were loaded, but they were entitled to show that they had perforations which would not have been apparent on a reasonable examination.
>
> Justice Scrutton: 'The ultimate damage was classed by the surveyors as (i) serious damage where the tins were gashed or punctured, damage easily discernible in handling each tin; (ii) minor damage, pinhole perforations, which on tins covered with rims were not easily discernible, but which were found when the tins were closely examined. I have considered the evidence and I find that the first class of damage was apparent to reasonable examination; the second, having regard to business conditions, was not apparent. The result of this is that the shipowner is estopped against certain persons from proving or suggesting that there was gash or serious damage when the goods were shipped.'

A 'clean' bill of lading is a bill of lading containing the words 'received in apparent good order and condition', or words of similar effect, without including any qualifying words. In *Canadian and Dominion Sugar Company Limited v Canadian National (West Indies) Steamships Limited*,[43] the bill of lading holder claimed against the carrier in respect of a quantity of sugar found to be damaged on arrival. If the statement at the head of the bill, 'received in apparent good order and condition', had stood by itself, the bill would have been a 'clean' bill of lading, which means no clause had qualified the statement as to the condition of the goods. But the bill did in fact on its face contain the qualifying words 'signed under guarantee to produce ship's clean receipt'. The Privy Council held that this qualifying clause reasonably conveyed to any businessman that if the ship's receipt was not clean, the

42. [1929] All ER Rep 611.
43. [1947] AC 46.

statement in the bill of lading as to apparent order and condition could not be taken to be unqualified. Lord Wright said:

> If the ship's receipt was not clean, the bill of lading would not be a clean bill of lading, with the result that the estoppel which could have been set up by the indorsee as against the shipowner . . . could not be relied on. That type of estoppel is of greatest importance in this common class of commercial transactions; it has been upheld in a long series of authoritative decisions . . . But if the statement is qualified, the estoppel fails.

Leading marks

Under Art III r 3 of the Hague-Visby Rules, after receiving the goods into his charge the carrier or the master or the agent of the carrier shall, on demand of the shipper, issue to the shipper a bill of lading showing the leading marks necessary for identification of the goods as the same are furnished in writing by the shipper before the loading of such goods starts, provided such marks are stamped or otherwise shown clearly upon the goods if uncovered, or on the cases of coverings in which such goods are contained, in such a manner as should ordinarily remain legible until the end of the voyage.

At common law, a bill of lading is evidence of the marks of identity of the goods. However, the shipowner will not be precluded from showing that the goods shipped were marked otherwise than as stated, unless the marks are material to the description of the goods:

> *Parsons v New Zealand Shipping Co*[44]
> Frozen carcasses of lamb were shipped, and the bills of lading, signed by the shipowner's agents, described the goods as '622 X, 608 carcases. 488 X, 226 carcases'. At the port of discharge some carcasses were found to be marked 522 X and others 388 X. The indorsees of the bill of lading claimed that the shipowner was estopped from denying the statements in the bill of lading, and was liable for failing to deliver the carcasses shipped.
>
> The Court of Appeal held that the shipowner was not liable because the description of the goods in the bills of lading did not affect or denote the nature, quality, or commercial value of the goods.
>
> Collins LJ: 'It is obvious that, where marks have no market meaning and indicate nothing whatever to a buyer as to the nature, quality, or quantity of the goods which he is buying, it is absolutely immaterial to him whether the goods bear one mark or another . . . the goods which the bill of lading represents as shipped continue to be the

44. [1901] 1 KB 548.

same goods, whichever out of any number of merely arbitrary marks are put on them, and will remain the same whether the marks were on them before shipment or are rubbed off or changed after shipment. In other words, they go to the identification only, and not the identity. The goods represented by the bill of lading to have been shipped have been shipped, and a mistaken statement as to marks of this class merely makes identification more difficult; it does not affect the existence or identity of the goods.'

The bill of lading as evidence of the contract of carriage

In common law, a bill of lading is not, in itself, the contract between the shipowner and the shipper of cargo, although it has often been regarded as a very convincingpiece of evidence of its terms. Since the bill is not the conclusive evidence of the terms and conditions on which the goods have been received, it is still possible for the shipper to demonstrate that there are other terms of the contract not contained in the bill. This is because the bill of lading is signed after the contract of carriage has come into existence. The bill of lading is signed by the carrier only and handed by the carrier to the shipper, usually after the cargo has been put on the vessel for carriage. Without doubt, if the shipper finds that the bill of lading includes terms with which he is not satisfied, or that it fails to include some terms for which he previously expressed, the shipper may, if the ship has not yet departed, demand his goods back. However, the shipper is not thereby precluded from adducing evidence that there was a contract made before the bill of lading was signed, and that it was different from that which is found in the bill of lading document. The shipper has not participated in the preparation of the bill of lading, nor does he sign it. Therefore, since the bill of lading is not the contract itself, evidence as to the true terms of the contract is admissible. This common law position can be found in the following authorities:

> *The Ardennes* [1951] 1 KB 55
>
> A bill of lading gave the shipowner liberty to proceed by any route and to carry the goods beyond their port of destination. Instead of proceeding direct to London from Cartagena, the ship went first to Antwerp. When she arrived in London, the shipper had to pay a higher import duty since the goods had arrived after the proper time. Before the bill of lading had been issued the shipowner had promised the shipper that the ship would go direct to London.
>
> The King's Bench Division held that the bill of lading was not in itself the contract between the shipowner and the shipper, and that the shipper was entitled to show that the prior promise constituted the contract on which he could make the shipowner liable for the payment of higher duty.

Cho Yang Shipping Co Ltd v Coral (UK) Ltd[45]

The Korean shipowner sought to recover freight from the Coral (UK) Ltd under three bills of lading related to the carriage of 20 containers. The bill of lading named Coral as the shipper and consigned the goods to their order. The bill contained the words 'Freight Prepaid as Arranged'. The court took the view that the words 'freight prepaid' in the bill of lading was not conclusive. In the absence of some other consideration, the shipper was normally contractually liable to the carrier for the freight. The personal liability was that of the person with whom the performing carrier had contracted to carry the goods. This person was normally the shipper. However, the court held that the inference that the shipper was agreeing to pay the freight was only the usual inference, but it was not a necessary inference, and a different inference might in a particular case be appropriate. The court drew the inference that there was no agreement by Coral to pay freight to the plaintiff in the absence of evidence that Coral's agent had authority to reach agreement concerning freight on behalf of Coral.

Hobhouse LJ: 'In English law the bill of lading is not the contract between the original parties but is simply evidence of it . . . Therefore, as between the shipper and carrier, it may be necessary to inquire what the actual contract between them was; merely to look at the bill of lading may not in all cases suffice. It remains necessary to look at and take into account evidence bearing upon the relationship between the shipper and the carrier and the terms of the contract between them.'

However, if the bill of lading has been indorsed for value in favour of a bona fide third party, that third party will not be affected by any additional terms or alleged agreement between the carrier and the shipper. The bill of lading becomes the conclusive evidence of the terms of the contract of carriage. In *Leduc v Ward*,[46] it was held that the indorsee of the bill of lading was not affected by the alleged agreement between the shippers and the carriers that the ship would proceed to Glasgow.

Generally speaking, only the parties to a bill of lading have any rights under it. In *Scruttons Ltd v Midland Silicones Ltd*,[47] the bill of lading contained the terms by which the shipowner was able to limit its liability if the goods were damaged through its negligence. The cargo was damaged by the stevedores during the loading operation. The House of Lords held that the stevedores were not entitled to claim to be able to limit their liability in accordance with the terms of the bill of lading. The stevedores were not parties to the bill of lading, and so were liable to pay in full for the damage

45. [1997] 2 Lloyd's Rep 641.
46. (1888) 20 QBD 475.
47. [1962] 1 All ER 1.

which had been caused. Lord Reid in this case, however, indicated how it might prove successful in a future case:

> I can see a possibility of success of the agency argument if (first) the bill of lading makes it clear that the stevedore is intended to be protected by the provisions in it which limit liability, (secondly) the bill of lading makes it clear that the carrier, in addition to contracting for these provisions on his own behalf, is also contracting as agent for the stevedore that these provisions should apply to the stevedore, (thirdly) the carrier has authority from the stevedore to do that, or perhaps later ratification by the stevedore to do that, or perhaps later ratification by the stevedore would suffice, and (fourthly) that any difficulties about consideration moving from the stevedore were overcome.

It was mainly on this passage that the 'Himalaya clause' (named after the vessel involved in *Adler v Dickson*)[48] was subsequently to be found. In *Adler*, a female passenger suffered injury during a voyage on the *Himalaya* as a result of the negligence of a member of the ship's crew. Her ticket contained an exclusion clause, freeing the shipowner from liability for death or injury howsoever caused. The court held that the exclusion clause would not prevent the passenger from bringing a valid claim against the negligent crew member direct in tort.

> *New Zealand Shipping Co Ltd v AM Satterthwaite & Co Ltd (The 'Eurymedon')*[49]
>
> A bill of lading in respect of a drilling machine stated that the carrier acted as agent 'for all persons who are or might be his servants or agents from time to time', and that the limitation of liability provisions in the bill of lading were available to such servants or agents. The machine was damaged while being discharged by the stevedores.
>
> The Privy Council held that the stevedores were entitled to limit their liability under the 'Himalaya' clause in the bill of lading. The discharge of the goods for the benefit of the shipper was the consideration for the agreement by the shipper that the stevedores should have the benefit of the limitation provisions in the bill of lading.
>
> Lord Wilberforce: '. . . to give the appellant the benefit of the exemptions and limitations contained in the bill of lading is to give effect to the clear intentions of a commercial document, and can be given within existing principles. They see no reason to strain the law of the facts in order to defeat these intentions. It should not be overlooked that the effect of denying validity to the clause would be to encourage actions against servants, agents and independent contractors

48. [1955] 1 QB 158.
49. [1974] 1 Lloyd's Rep 534.

in order to get around exemptions (which are almost invariable and often compulsory) accepted by shippers against carriers.'

New Zealand Shipping establishes that if a carrier acts as an agent for a third party, such as a stevedore, the third party can enforce the terms of the bill of lading against the shipper if the bill of lading makes it clear that the stevedore is protected by the provisions, and if any difficulties about consideration moving from the third party are overcome. On the other hand, an exclusive jurisdiction clause may not be classified as an 'exception, limitation, condition or liberty' or a provision for the 'benefit' of the carrier as provided for in the 'Himalaya' clause:

> *The 'Mahkutai'*[50]
> The shipowner appealed against a decision by the Hong Kong court setting aside an order granting a stay of proceedings upon an exclusive jurisdiction clause. The exclusive jurisdiction clause was contained in a bill of lading between the charterer and the cargo owner. The shipowner was not a party to the bill of lading. The shipowner intended to rely on a 'Himalaya' clause which stated that 'all exceptions, limitations, provisions, conditions and liberties herein benefiting the carrier' should expressly accrue to the benefit of agents and sub-contractors. The shipowner argued that the Himalaya clause should allow the shipowner to take advantage of the exclusive jurisdiction clause.
>
> The Privy Council dismissed the shipowner's appeal. One of the reasons was that the exclusive jurisdiction clause was not covered by the terms of the 'Himalaya' clause as the jurisdiction clause did not benefit one party over another. It constituted a mutual agreement on the relevant jurisdiction for dispute resolution.

One must draw a distinction between *The 'Mahkutai'* and another case, *Pioneer Container*.[51] As the Privy Council stated: '[*Pioneer Container*] was however concerned with a different situation, where a carrier of goods subcontracted part of the carriage to a shipowner under a "feeder" bill of lading, and that shipowner sought to enforce an exclusive jurisdiction clause contained in that bill of lading against the owners of the goods. The Judicial Committee held that the shipowner was entitled to do so, because the goods owner had authorized the carrier so to subcontract "on any terms", with the effect that the shipowner as sub-bailee was entitled to rely on the clause against the goods owner as head bailor. [*The 'Mahkutai'*] is however concerned not with a question of enforceability of a term in a *sub-bailment* by the sub-bailee against the head bailor, but with the question whether a

50. [1996] 2 AC 650.
51. [1994] 2 AC 324.

subcontractor is entitled to take the benefit of a term in the *head contract*. The former depends on the scope of the *authority* of the intermediate bailor to act on behalf of the head bailor in agreeing on his behalf to the relevant term in the *sub-bailment*; whereas the latter depends on the scope of the *agreement* between the head contractor and the subcontractor, entered into by the intermediate contractor as agent for the subcontractor, under which the benefit of a term in the *head contract* may be made available by the head contractor to the subcontractor. It does not follow that a decision in the former type of case provides any useful guidance in a case of the latter type.'

The bill of lading as a document of title

By virtue of the 'custom of merchants', a bill of lading is a document that represents title to the goods.

> *Lickbarrow v Mason*[52]
>
> The jury on a special verdict found that: 'by the custom of merchants, bills of lading, expressing goods or merchandises to have been shipped by any person or persons to be delivered to order to assigns, have been, and are, at any time after such goods have been shipped, and before the voyage performed, for which they have been or are shipped, negotiable and transferable by the shipper or shippers of such goods to any other person or persons by such shipper or shippers endorsing such bills of lading with his, her or their name or names, and delivering or transmitting the same so indorsed, or causing the same to be so delivered or transmitted to such other person or persons. And that, by the custom of merchants, indorsements of bills of lading in blank, that is to say, by the shipper or shippers with their names only, have been, and are, and may be, filled up by the person or persons to whom they are so delivered or transmitted as aforesaid, with words ordering the delivery of the goods or contents of such bills of lading to be made to such person or persons; and, according to the practice of merchants the same, when filled up, have the same operation and effect, as if the same had been made or done by such shipper or shippers when he, she, or they indorsed the same bills of lading with their names as aforesaid.'

In *Enichem Anic SpA v Ampelos Shipping Co Ltd (The 'Delfini')*,[53] Mustill LJ described the function of a bill of lading in the following manner:

- It is a symbol of constructive possession of the goods which (unlike many such symbols) can transfer constructive possession by indorsement and transfer: it is a transferable 'key to the warehouse'.

52. (1794) 5 TR 683.
53. [1990] 1 Lloyd's Rep 252 at 268.

- It is a document which, although not itself capable of directly transferring property in the goods which it represents, merely by indorsement and delivery, nevertheless is capable of being part of the mechanism by which property is passed.

> *E. Clemens Horst Co v Biddell Bros*[54]
> A contract was made for the sale of hops to be shipped from San Franscisco to London. The buyer refused to pay for the goods until they were actually delivered at the port of discharge.
> The House of Lords held, among other issues, that possession of the bill of lading was in law equivalent to possession of the goods.
> Earl Loreburn LC: '. . . that delivery of the bill of lading when the goods are at sea can be treated as delivery of the goods themselves, this law being so old that I think it is quite unnecessary to refer to authority for it.'

The passing of possession of the goods is a concept different from the passing of property or title to the goods. In *The Future Express*,[55] the English Court of Appeal held that endorsement and delivery of the bill of lading does not pass property to the goods if the transferor and transferee do not intend that it should do so. In *Sewell v Burdick (The 'Zoe')*,[56] the House of Lords held that the indorsement and delivery of a bill of lading did not necessarily pass an absolute or general interest in the property in the goods to the indorsee, but only such an interest in the property as the parties intended to transfer. Lord Selborne said:

> In principle, the custom of merchants as found in *Lickbarrow v Mason* seems to be as much applicable and available to pass a special property at law by the indorsement (when that is the intent of the transaction) as to pass the general property when the transaction is, e.g. one of sale . . . so long at all events as the goods are in transitu, there seems to be no reason why the shipper's title should be displaced any further than the nature and intent of the transaction requires.

A further illustration of this principle is found in *Sanders Brothers v Maclean & Co.*[57] Bills of lading were issued in a set of three in relation to a cargo of old iron flange carried from the Black Sea to Philadelphia. The buyer of the cargo refused to pay for it because only two out of three bills of lading were tendered to him. The court held that the indorsement of only one bill out of a set is sufficient to pass the property in the goods, and

54. [1912] AC 18.
55. [1993] 2 Lloyd's Rep 542.
56. (1884) 10 App Cas 74.
57. (1883) 11 QBD 327.

the buyer had no right to refuse payment. A classic statement was made by Bowen LJ:

> A cargo at sea while in the hands of the carrier is necessarily incapable of physical delivery. During this period of transit and voyage, the bill of lading by the law merchant is universally recognised as its symbol, and the indorsement and delivery of the bill of lading operates as a symbolical delivery of the cargo. Property in the goods passes by such indorsement and delivery of the bill of lading, whenever it is the intention of the parties that the property should pass, just as under similar circumstances the property would pass by an actual delivery of the goods . . . And it is plain that the purpose and idea of drawing bills of lading in sets—whatever the present advantage or disadvantage of the plan—is that the whole set should not remain always in the same hands . . . the shipper or his vendees may prefer to retain one of the originals for their own protection against loss, or to transfer it to their correspondents. In such cases they are in the habit of treating the remainder of the set as the effective documents and as sufficient for all purposes of negotiating the goods comprised in the bill of lading.

According to the law of personal property and the Hong Kong Sale of Goods Ordinance, the passing of property depends on the intention of the party as expressed in the terms of the contract, the conduct of the parties, and the circumstances of the case:

> *S 19 Sale of Goods Ordinance*
> (1) Where there is a contract for the sale of specific or ascertained goods the property in them is transferred to the buyer at such time as the parties to the contract intend it to be transferred.
> (2) For the purpose of ascertaining the intention of the parties regard shall be had to the terms of the contract, the conduct of the parties, and the circumstances of the case.

Thus, in the absence of contrary provisions stipulated in the contract of sale of goods, the property will normally pass on shipment of the goods if the bill of lading is in the buyer's name. If the bill of lading names the seller as the consignee, or is issued to the order of the seller, then property in the cargo will ordinarily only pass when the bill of lading is transferred to the buyer and the price is paid:

> *S 21 Sale of Goods Ordinance*
> (1) Where there is a contract for the sale of specific goods . . . the property in goods does not pass to the buyer until the conditions imposed by the seller are fulfilled.

(2) Where the goods are shipped, and by the bill of lading the goods are deliverable to the order of the seller or his agent, the seller is prima facie to be taken to reserve the right of disposal.

Title to cargo and title to sue under the bill of lading

The Bill of Lading and Analogous Shipping Documents Ordinance

The Bill of Lading and Analogous Shipping Documents Ordinance came into force on 1 March 1994 (see Annex 3). It covers not only bills of lading but also sea waybills and ship's delivery orders. The Ordinance envisages two significant departures from the previous legislation. First, title to sue will no longer be linked to property in the goods. Secondly, the transfer of rights under a contract of carriage will be effected independently of any transfer of liabilities. Title to sue is now vested in the lawful holder of a bill of lading, the consignee identified in a sea waybill or the person entitled to delivery under a ship's delivery order, irrespective of whether or not they are owners of the goods covered by the document (s 4 of the Ordinance). The 'lawful holder' of a bill of lading is defined in s 2(2) as a person in possession of the bill in good faith who is either:

(a) identified in the bill as consignee, or
(b) an indorsee of the bill, or
(c) a person who would have fallen with categories (a) or (b) if he had come into possession of the bill before it ceased to be a document of title.

There is a clear separation of rights from liabilities. Liabilities under the contract of carriage are no longer transferred with title to sue. They will only attach to persons in whom rights of suit are vested when they either take or demand delivery of the goods, or make a claim under the contract of carriage, or take or demand delivery of the goods before the rights of suits are vested in them (s 5).

It is necessary to divorce rights from liabilities as it is soon realized that extending rights of suit to a wide class of persons might be problematic if such persons were also to be subject to the liabilities of the shipper of the goods. If the shipper's rights and liabilities were transferred to all holders, including those holding the bill merely as security, it would mean that such people, including banks who take up shipping documents in the normal course of financing international transactions, would suddenly find themselves liable for freight, demurrage and other charges.

The new law effectively protects a bank which is holding the bill as security for a credit from incurring liabilities until it seeks to enforce its security by claiming delivery of the goods or commencing proceedings against the carrier. It should be noted, however, that no distinction is drawn between pre-shipment and post-shipment liabilities in the Ordinance. It would be very unfair for the final holder of a bill of lading to be liable in respect of such matters as the shipper's breach of warranty in shipping dangerous goods, demurrage incurred at the port of loading and dead freight. The consignee often stands in no relation to the goods at the moment of shipment and to make him liable in respect of such pre-shipment liabilities is to make him subject to retrospective liability for acts with which he had nothing to do. Legislative reform is obviously needed to remedy this deficiency. For instance, a line should be drawn between pre-shipment and post-shipment liabilities and that special provision should be made for liability in respect of the carriage of dangerous cargo.

Common problems associated with bills of lading

Bills of lading issued under charterparties: Who is the carrier?

When a shipper of cargo is issued a bill of lading with respect to goods carried aboard a vessel under a charterparty, the question arises whether the contract is with the charterer or the shipowner or both. In practice, the bill of lading may be issued in the name of the shipowner, the charterer, a sub-charter or the agent of any one of them. These complicated situations can be illustrated in the following diagrams:

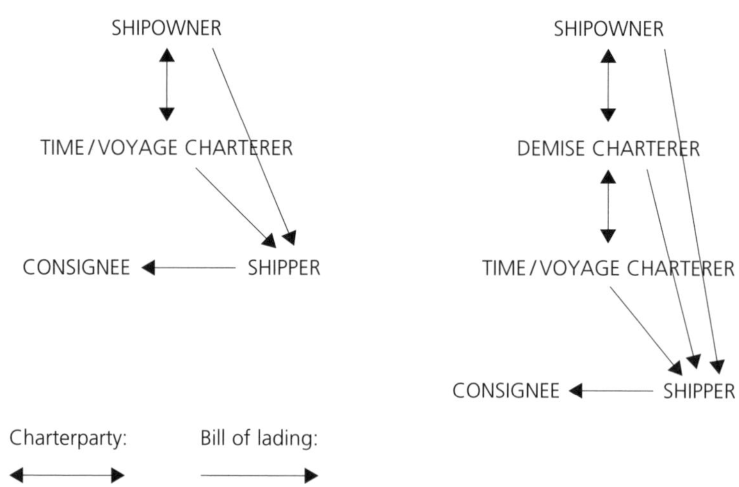

Art I of the Hague-Visby Rules states that a 'carrier' includes 'the owner or the charterer who enters into a contract of carriage with a shipper'. Some courts look to principles of agency law, including implied and apparent authority, to determine whether the bill is issued on behalf of the owner or the charterer. Relevant factors may include the type of charterparty, who signed the bill of lading, whose form was used, and under whose authority the bill of lading was issued. However, the authority of the charterer to issue bills of lading still chiefly depends on the express terms of the charterparty. For instance, NYPE charter clause 30 states, 'The master shall sign the bills . . . for the cargo as presented in conformity with . . . mates receipts. However, the charterers may sign bills . . . on behalf of the master, with the shipowner's prior written authority, always in conformity with . . . mates receipts.'

Whether the bill of lading is a shipowner's bill of lading or the charterer's bill of lading is an issue of construction depending on the terms of the bill of lading and all the surrounding circumstances. For example, the bill of lading may include a 'demise clause' stipulating:

> if the ship is not owned or chartered by demise to the company by whom this bill of lading is issued (as may be the case notwithstanding anything which appears to the contrary) the Bill of Lading shall take effect as a contract with the Owner or demise charterer, as the case may be, as principal made through the agency of the said company or line who act as agents only and shall be under no personal liability whatsoever in respect thereof.

If the court gives recognition to the 'demise clause', that means the shipper or the holder of the bill of lading has a direct contractual relationship with the shipowner. As a result, the shipper or the holder of the bill could directly sue the shipowner as the carrier for breach of the contract of carriage. In *The Berkshire*,[58] the court upheld the validity of the 'demise clause' contained in a bill of lading issued by the sub-charterer. Brandon J stated:

> I see no reason not to give effect to the demise clause in accordance with its terms . . . it follows that the bill of lading is, by its express terms, intended to take effect as a contract between the shippers and the shipowners made on behalf of the shipowners . . . as agents only. All the demise clause does is to spell out in unequivocal terms that the bill of lading is intended to be a shipowner's bill of lading . . . In my view, so far from being an extraordinary clause, it is an entirely usual and ordinary one.

58. [1974] 1 Lloyd's Rep 185.

In contrast, the 'demise clause' was not given effect in *Homburg Houtimport BV v Agrosin Private Ltd, The Starsin*.[59] The shipowner, not the charterer, was identified as the carrier under the 'Identity of Carrier' and the 'Demise Clause' on the reverse of the bill of lading. However, on the face of the bill, the charterer's agent had signed the signature box, which was described as 'the carrier'.The master had not signed the bill. The court held that on the proper interpretation of the bill, the shipowner was not the carrier. The bill wasthe charterer'sbill, essentially because of the signature of the charterer's agent 'as the carrier' on the front page of the bill. Lord Bingham averred:

> a very cursory glance at the face of the bill is enough to show that the master has not signed the bill. It has instead been signed by agents for [the charterer] which is described as 'The Carrier'. I question whether anyone engaged in maritime trade could doubt the meaning of 'carrier' . . . I have great difficulty in accepting that a shipper or transferee of a bill of lading would expect to have to resort to the detailed conditions on the reverse of the bill . . . in order to discover who he was contracting with. And I have even greater difficulty in accepting that he would expect to do so when the bill of lading contains, on its face, an apparently clear and unambiguous statement of who the carrier is.

Bills of lading issued to charterers

Where the charterer is also the shipper of the goods, prima facie the bill of lading will merely act as a receipt for and a document of title to the goods. It will not be the evidence of the contract of carriage, because the charterparty is the contract of carriage between the shipowner and the charterer (*The 'Al Battani'*).[60] If a time-charterer enters into a voyage sub-charterparty, and if the shipowner issues a bill of lading to the voyage sub-charterer on shipment of the goods, the bill of lading does act as evidence of a contract of carriage between the shipowner and the voyage sub-charterer (*The 'Al Battani'*).

If a bill of lading held by the charterer for goods on a chartered ship is issued or indorsed to a bona fide third party for value, then the bill of lading will become the conclusive evidence of the contract of carriage so far as the relationship between the indorsee and the shipowner is concerned. As between the shipowner and the indorsee, the bill of lading must be considered to contain the contract, because the former has given it for the purpose of enabling the charterer to pass it on as the contract of carriage in

59. [2003] 1 Lloyd's Rep 571.
60. [1993] 2 Lloyd's Rep 219.

respect of the goods. Upon indorsement of the bill by the charterer in favour of the indorsee, a new contract comes into operation between the shipowner and the consignee on the terms of the bill of lading (*Hain Steamship Co Ltd v Tate and Lyle Ltd*).[61]

Master's authority's to sign bills of lading

In *Grant v Norway*,[62] twelve bales of silk were arranged for shipment in an ocean carrier and the ship's master signed the bill of lading evidencing the receipt of these twelve bales on board. The original bill of lading was indorsed in favour of a third party for value. Subsequently, it was discovered that the twelve bales of cargo had never been physically placed on board the ship named in the bill of lading. The court held that the master's act of signing the bill of lading for the cargo which was never put on board was an act outside the scope of his implied authority. As a result, the innocent holder of the bill of lading could not claim against the shipowner, because the bill of lading did not bind the shipowner. This grossly unfair decision, fortunately, has been overridden by the Hague-Visby Rules Art III r 4 which provides that a bill of lading shall be prima facie evidence of the receipt by the carrier of the goods as therein described. Proof to the contrary shall not be admissible when the bill of lading has been transferred to the third party acting in good faith.

Presentation of non-negotiable bills of lading

Non-negotiable bills of lading (also known as 'straight' bills of lading) are frequently used in maritime transport. In a straight bill of lading, limited number of parties are involved in the transaction, typically a shipper (the party sending the consignment), a carrier (the party transporting the goods), and a particular named consignee. This is different from a negotiable bill of lading expressed to be 'to order' without naming a particular consignee. It is generally accepted that a straight bill of lading is not a negotiable document. In other words, the holder of a straight bill of lading cannot transfer title to the goods during transit. In recent years, serious legal and practical problems have occurred concerning whether delivery of goods against presentation of straight bills of lading is required. The following authorities from England, Singapore and Hong Kong indicated that delivery of the goods has to be made against surrender of the straight bills of lading. If the carrier

61. (1936) 55 Ll L Rep 159.
62. (1851) 20 LJPC 93.

delivers the goods to a party without presentation of the straight bill of lading, it does so entirely at its own peril:

The Rafaela S[63]

Four containers of printing machinery were damaged in the course of their carriage by sea from Felixstowe, England to Boston, USA. The bill in question was a straight bill, which provided that: 'One of the Bills of Lading must be surrendered duly endorsed in exchange for the goods.'

The House of Lords held that the contract for the carriage was covered by 'a bill of lading or any similar document of title' within the meaning of Art 1(b) of the Hague-Visby Rules. The court had no difficulty in regarding the straight bill as a document of tile, since on its express terms it must be presented to obtain delivery of the cargo. The court even went further by commenting that even where there was no express provision to that effect, production of the straight bill was still a necessary pre-condition of requiring delivery.

Voss Peer v APL Co Pte Ltd[64]

The carrier released cargo to a party without the production of the straight bill of lading. The carrier argued that since the delivery was made to a named consignee, production of the straight bill was not required.

The Singapore Court of Appeal rejected the carrier's argument and held that whether it was a negotiable bill or a straight bill, the carrier should never deliver the cargo without the production of the straight bill.

Carewins v Bright Fortune Shipping Ltd & Anor[65]

The HK Court of Final Appeal held: 'It is the law of Hong Kong that a carrier of goods shipped under a straight bill of lading is potentially liable for breach of contract or in conversion if it releases those goods without production of the original bill of lading . . . The shipper's ability to withhold the bill of lading – the metaphorical key to the warehouse – pending payment by the consignee is a highly important feature of the recognized mercantile arrangement. This applies just as much to the relationship between shipper and consignee under a straight bill as between the parties to an order bill. It is true that a carrier is able to see who is the intended consignee on the face of the bill, but that does not mean that he is justified in assuming that such person is entitled, as against the shipper, to possession of the goods. If the named consignee is unable to produce the bill of lading it may

63. [2005] 2 AC 423.
64. [2002] 2 Lloyd's Rep 707 (Singapore Court of Appeal).
65. [2009] 5 AC 160.

very well be because he has not paid for the goods and is not entitled to possession, as numerous decided cases show.'

The difference between a straight bill of lading and a sea waybill must be noted. A seaway bill often contains a clause requiring the delivery of the cargo to the named consignee on production of evidence of identity at the place of discharge. Hence, it is not necessary for the named consignee to produce the original sea waybill in order to obtain delivery of the goods.

Electronic bills of lading and the Electronic Transactions Ordinance

The advent of advanced communications technology and the internet makes paper bills of lading extremely outmoded. The aim of electronic bills of lading is to completely remove the paper element of international trade transactions. The need to present the paper bill of lading at the port of discharge means the bill must be sent to the consignee physically by air before or at the same time as the cargo. With speedier vessels and quicker turn-around times, this is no longer guaranteed. Under the electronic data interchange procedure, the shipper (eg, Company A) can supply the information to be contained in the bill of lading to the carrier online. Having received the cargo on board the vessel, the carrier would digitally sign a data message that would become the electronic bill of lading. The message of the carrier decrypted by a public key can only be encrypted by someone having the private key of the same key pair. The shipper is provided with a private key to control the cargo during carriage. The shipper can indorse the electronic bill to a third party (eg, Company B) by digital signing and transmission to the carrier. On each transfer the existing private key is cancelled and replaced by a new key issued by the carrier to the transferee. At the port of discharge, only the party holding the current private key is entitled to delivery of the cargo.

A digital signature is truly secure only when certified by a trustworthy certification authority. For instance, a paper document can be copied and presented to outside parties without the knowledge of the sender. A paper document can also be forged and successfully passed on as an original. The same things can happen to electronic data interchange and internet-delivered messages. BOLERO launched its commercial operation www.bolero.net as a joint venture between the TT Club (Through Transport Club) and SWIFT (the Society for Worldwide Interbank Financial Telecommunication). Under the BOLERO system, encryption prevents the viewing of a document by any party other than the intended receiver holding the current private key.

Digital signatures ensure that signed documents cannot be altered. BOLERO is a certification authority that certifies digital signatures. It could issue a certificate which:

- lists its issuer (ie, BOLERO) by name;
- lists a public key;
- lists by name, or otherwise indicates, a user (ie, a person who is enrolled as a user of the BOLERO system) holding the private key corresponding to the listed public key;
- is digitally signed by its issuer.

BOLERO is not simply a certification authority. It is also a title registry. The title registry maintains an endorsement chain for each BOLERO bill of lading, reflecting the transfer of rights and obligations between parties. Similar to a paper bill of lading, each BOLERO bill can be created, transferred, amended and surrendered. In each case, only the authorized party holding the current private key can instruct the title registry to complete the transaction.

The legal implications

With the potential widespread use of electronic bills of lading in Hong Kong, questions like whether the electronic bills of lading will be recognized by Hong Kong laws, and whether the BOLERO certificates will be admissible as evidence in the Hong Kong courts, will be increasingly faced by legal advisers, carriers, bankers and merchants.

The United Nations Commission on International Trade Law (UNCITRAL) is attempting to develop uniform international rules that would validate the use of Electronic Data Interchange (EDI). In 1995, the Commission adopted the Model Law on Legal Aspects of Electronic Data Interchange and Related Means of Communication. The Model Law is intended to serve as a model to countries in order to create uniform law and practice involving the use of computerized systems in international trade.

Article 4 of the UNCITRAL Model Law provides that information should not be denied effectiveness, validity or enforceability solely on the grounds that it is in the form of a data message. Article 8 also provides that nothing in the application of the rules of evidence shall apply so as to prevent the admission of a data message in evidence on the ground that it is a data message, or if it is the best evidence that the person adducing it could reasonably be expected to obtain, on the grounds that it is not in its original form. In Hong Kong, the Electronic Transactions Ordinance is based on the Model Law. Section 5 of the Ordinance provides:

> (1) If a rule of law requires information to be or given in writing or provides for certain consequences if it is not, an electronic record satisfies the requirement if the information contained in the electronic record is accessible so as to be usable for subsequent reference.
> (2) If a rule of law permits information to be or given in writing, an electronic record satisfies that rule of law if the information contained in the electronic record is accessible so as to be usable for subsequent reference.

Section 9 of the Ordinance provides:

> Without prejudice to any rules of evidence, an electronic record shall not be denied admissibility in evidence in any legal proceedings on the sole ground that it is an electronic record.

According to s 6 of the Ordinance, if a rule of law requires the signature of a person or provides for certain consequences if a document is not signed by a person, a digital signature of the person satisfies the requirement but only if the digital signature is supported by a recognized certificate. Through the use of public and private key pairs and recognized certificates issued by recognized certification authorities, individuals and businesses can establish the identity of the opposite party in electronic transactions, and ensure the integrity of the electronic messages received. Section 34 of the Ordinance states that the Hong Kong Postmaster General is one of the recognized certification authorities. Although BOLERO is both a central registry of bills of lading and a certification authority providing security controls and procedures to protect the integrity and prove the authenticity of electronic messages, it is not yet a recognized certification authority in Hong Kong under the Electronic Transactions Ordinance.

In order to use the BOLERO service, all users (such as carriers, banks, shippers or consignees) must enter into a standard form contract with the BOLERO association, becoming members of the association and binding themselves to the conditions of use of the service contained in the BOLERO Rulebook. For instance, under the BOLERO Rulebook, no user shall contest the validity of any transaction, statement or communication made by means of a signed message on the grounds that it was made in electronic form instead of by paper and/or signed or sealed. Each user agrees that a signed message will be admissible before any court or tribunal as evidence of the message or portion thereof. In the event that a written record of any message is required, a copy produced by a user, which BOLERO has authenticated, shall be accepted by that user and any other user as primary evidence of the message. The BOLERO Rulebook defines a 'signed message' as a document bearing a digital signature which can be verified by using the public key listed in a certificate issued by BOLERO and which was a valid certificate when the digital signature was created.

The existing shipping legislation in Hong Kong

As examined earlier in this chapter, the Hague-Visby Rules (The Hague Rules as amended by the Brussels Protocols 1968 and 1979) are scheduled to the Carriage of Goods by Sea Ordinance. Since Hong Kong is one of the contracting states to this international convention, the Hague-Visby Rules apply to every bill of lading issued in Hong Kong or relating to a carriage from Hong Kong. The object of these rules is to protect cargo owners from widespread exclusion of liability by sea carriers. The rules state the carrier's obligations and set out a list of 'excepted perils' exemptions from liability or immunities of the carrier. In addition, the rules provide a system for the package limitation of the cargo claims by reference to the amount or value of the cargo. Article I of the Hague-Visby Rules states that the rules apply only to 'contracts of carriage covered by a bill of lading or any similar document'. Unfortunately, it is not entirely clear whether a BOLERO electronic bill of lading is within the definition.

On the other hand, the Bills of Lading Analogous Shipping Documents Ordinance (Cap 440) governs the transfer of rights and liabilities under bills of lading, and representations in bills of lading. Under the Ordinance, any 'lawful holder' of a bill of lading has a right of action under the contract of carriage against the carrier. Although the Ordinance provides that the Secretary for Trade and Industry may by regulation make provision for the application of the Ordinance to cases where a telecommunication system or any other information technology is used for effecting transactions relating to bills of lading, no regulations have been made yet.

Need for legislative reform?

Because the current legislation in Hong Kong has failed to bring electronic bills of lading into practice, the participants in BOLERO can only rely on the terms of the private contract contained in the BOLERO Rulebook to give effect to any international convention such as the Hague-Visby Rules or any national law giving effect to such international convention. The need for legislative reform in Hong Kong is pressing. In fact, Hong Kong is lagging behind when compared with the other major jurisdictions. In Australia, the Sea-Carriage Document Act brings electronic and computerized sea-carriage documents into practice. Section 6 of the Act stipulates that the statute applies in relation to a sea-carriage document in the form of a data message in the same way as it applies to a written sea-carriage document. It also applies in relation to communication of a sea-carriage document by means of a data message in the same way as it applies to the communication of a sea-carriage document by other means. The Hong Kong legislature should, as a matter of urgency, seriously consider making regulations or amending the existing legislation to cope with the issues arising from the forthcoming widespread use of electronic bills of lading in Hong Kong.

The Rotterdam Rules: An overview

In December 2008, the United Nations General Assembly adopted the 'United Nations Convention on Contracts for the International Carriage of Goods Wholly or Partly by Sea'. A signing ceremony was held in September 2009 in Rotterdam, the Netherlands, and the rules embodied in the convention became known as 'The Rotterdam Rules'. The Rotterdam Rules contain many innovative provisions, including some that allow for electronic transport records and others that fill the perceived loopholes in current transport regimes.

The goal of the Rotterdam Rules, according to the United Nations, is to create a modern and uniform law that will cover door-to-door container transport that includes an international sea leg but is not limited to port-to-port carriage of cargo. The Working Group on Transport Law, under the United Nations Commission on International Trade Law (UNCITRAL), has been working on the draft since 2002, involving thirteen rounds of consultation and negotiation with global inter-governmental and non-governmental organizations. The key provisions of the Rotterdam Rules are summarized below.

Door-to-door multimodal transportation

The rules intend to cover door-to-door rather than tackle to tackle (or port-to-port) basis. Art 1.1 states that 'contract of carriage means a contract in which a carrier, against the payment of freight, undertakes to carry goods from one place to another. The contract shall provide for carriage by sea and may provide for carriage by other modes of transport in addition to the sea carriage'.

Carrier's liabilities and shippers' obligations

A carrier is 'a person that enters into a contract of carriage with a shipper'.[66]

The carrier is obliged to exercise due diligence to make the ship seaworthy before and at the beginning of the voyage, and to keep the ship seaworthy during the voyage by sea.[67] The defence of navigation fault is removed from the list of defences,[68] so that a carrier can no longer be exonerated from liability arising from fault or error in navigation.

A shipper is obliged to furnish information to the carrier about the dangerous nature of the goods in a timely manner. 'Dangerous goods' are defined as goods which 'by their nature or character are, or reasonably appear likely to become, a danger to persons, property or the environment'.[69]

Performing parties

A 'performing party' is defined as 'a person other than the carrier that performs or undertakes to perform any of the carrier's obligation under a contract of carriage'.[70] A 'maritime performing party' is defined as a person who 'performs or undertakes to perform any of the carrier's obligations during the period between the arrival of the goods at the port of loading of a ship and their departure from the port of discharge of a ship.[71] Obviously, performing parties include the carrier's agents, employees and independent contractors. A carrier is liable for the acts or omissions of any performing party.[72] A maritime performing party is subject to the obligations and liabilities imposed on the carrier, and even without the Himalaya clause, a maritime performing party can enjoy the carrier's defences, provided that it 'received the goods for carriage in a Contracting State, or delivered them in a Contracting State, or performed its activities with respect to the goods in a port in a Contracting State.'[73]

66. Art 1.5.
67. Art 14.
68. Art 17.
69. Art 32.
70. Art 1.6.
71. Art 1.7.
72. Art 18.
73. Art 19.1(a). A list of the Contracting States is shown at <www.rotterdamrules.com/>.

Right of control

A 'controlling party' can exercise the 'right of control'. The right of control is limited to the right to give or modify instructions concerning the cargo, the right to take delivery of the cargo at the agreed destination, and the right to replace the consignee.[74] In general, the shipper is the controlling party, unless the consignee (or a third party) is named as the controlling party when the contract of carriage was entered into. The controlling party can transfer the right of contract to another party.[75]

Volume contact

A 'volume contract' is 'a contract of carriage that provides for the carriage of a specified quantity of goods in a series of shipments during an agreed period of time. The specification of quantity may include a minimum, a maximum or a certain range.'[76] The carrier and shipper, under a volume contract, may reduce or increase their rights and liabilities imposed by the rules.[77]

Electronic transport record

The rules give full recognition to the use of electronic transport record, including 'messages issued by electronic communication under a contract of carriage'.[78] An electronic transport record shall include the carrier's electronic signature, which shall identify the signatory and indicate the carrier's consent of the issuance of the record.[79]

Package and time limitations

The package limitation is 835 special drawing rights (SDRs) per package or other shipping unit[80] and weight limitation to 3 SDRs per kilogramme[81] of the gross weight of the cargo, whichever gives the higher amount.[82] A cargo claim must be instituted against the carrier within 2 years.[83]

74. Art 50.1.
75. Art 51.
76. Art 1.2.
77. Art 80.
78. Art 1.18.
79. Art 38.
80. About US$1,260.
81. About US$4.44.
82. Art 59.
83. Art 62.

6.2 VOYAGE CHARTERPARTIES

Overview

A voyage charterparty is a contract to carry specified goods on a defined voyage or series of voyages. The shipowner is remunerated by the payment of freight, which is usually calculated by reference to the quantity of cargo shipped, but may be fixed as a lump sum for the complete voyage. Under a voyage charterparty, the shipowner agrees to provide the ship in a seaworthy condition. The vessel will proceed to the port of loading with reasonable dispatch, as well as load the cargo and carry the cargo to the agreed destination without unjustifiable deviation.

The principal obligations of the charterer are to: (1) furnish full cargo complying with the charterparty, in reasonable time to enable it to be loaded within the permitted laytime; (2) nominate safe loading or discharging ports or berths if the charterparty requires the charterer to do so; (3) perform the charterer's part in the loading and discharging operations; and (4) pay the freight punctually in the agreed manner.

The approach voyage and the loading operation

The voyage charterparty usually contain an express term that the vessel is to proceed to the port of loading with a reasonable dispatch. Where the charterparty specifies a date by which the ship is to arrive at the port of lading, such clause may be construed as a condition precedent of the contract, the breach of which amounts to a repudiation of the charterparty.[84] The charterparty may also contain a cancelling clause under which the charterer is given the option of cancelling the charterparty if the vessel is not ready to load by a specified deadline.

When the ship has arrived at the port of loading and is ready to load, and the charterer has been duly notified, the charterer's duty is to provide the full cargo as agreed in the charterparty within the time specified in the contract. An implied undertaking at common law is that the ship on which goods are to be loaded must be seaworthy at the time when the loading begins. The vessel must be reasonably fit for the loading of the agreed cargo and for encountering ordinary perils which are likely to arise during the loading.[85]

84. *Shadforth v Higgin* (1813) 3 Camp 385.
85. *McFadden v Blue Star Line* [1905] 1 KB 697.

The voyage and the unloading

Once loading is complete, the vessel must then be seaworthy for carrying the cargo on the voyage. The vessel must be seaworthy at the commencement of the voyage and fit to encounter ordinary perils of the voyage:

> *Stanton v Richardson*[86]
> Under a voyage charterparty, the ship was to take a cargo of wet sugar. After the loading of the sugar, it was discovered that the pumps were not of adequate capacity to remove the drainage from the sugar. As a result, the cargo had to be discharged.
>
> It was held that the ship was unseaworthy for carrying the agreed cargo, as evidence showed that adequate pumping device could not have been provided within a reasonable time.

It is the shipowner's duty to take reasonable care of the cargo. If the goods, from their nature, require airing or ventilating, the master must adopt the proper methods for the purpose. If the cargo is composed of live animals, the master must provide an adequate amount of drinking water.[87] In addition, the vessel must proceed to the port of discharge without delay or deviation from the proper course of navigation. As Lord Atkin said: 'the departure from the voyage contracted to be made is a breach by the shipowner of his contract.'[88]Deviation is only justified where it is for the purpose of saving life, or the ship cannot safely keep its course owing to stress of weather, or the vessel is attempting to avoid imminent danger.[89]

The discharge of the cargo from the ship is the joint act of the shipowner and the consignee. It is the shipowner's duty to discharge the cargo out of the holds and to deliver it to the consignee. It is the consignee's duty to take delivery of the goods within the period fixed by the terms of the contract. If the consignee fails to take delivery of the goods, the master may warehouse them at the consignee's risk and expense under the terms of the contract.

Laytime and demurrage

Commencement of laytime

A charterparty usually contains a provision as to the time to be occupied in the loading and discharging of the cargo. This is usually called the 'laytime'

86. (1874) LR 9 CP 390.
87. *Vallee v Bucknall Nephews* (1900) 16 TLR 362.
88. *Hain SS CO Ltd v Tate and Lyle Ltd* [1936] 2 ALL ER 597 at 601, HL.
89. *Scaramango v Stamp* (1840) s Asp MCL 295; *Kish v Taylor* [1912] AC 604.

or 'lay days'. In *Nielsen v Wait*,[90] Lord Esher MR said, 'There must be a stipulation as to the time to be occupied in the loading and in the unloading of the cargo. There must be a time, either expressly stipulated, or implied. If it is not expressly stipulated, then it is a reasonable time which is implied by the law; but either the law or the parties fix a time. Now, when they do fix a time how do they fix it? Why, they allow a certain number of days, during which, although the ship is at the disposal of the charterer to load or to unload the cargo, he does not pay for the use of the ship. That is the meaning of "lay days".'

Unless there are terms in the charterparty changing the standard requirements as to when laytime begins, laytime commences when:

- the ship reaches the agreed destination and therefore becomes an 'arrived ship'; and
- the ship is ready to load or discharge; and
- the notice of readiness to *load* has been tendered. Apart from special contract or custom or course of dealing, the shipowner is not bound to give notice of readiness to *discharge* at the discharging port to the charterer or to the consignee under the bill of lading.

The destination may be expressly stipulated in the charterparty. For example, the agreed destination may be a port, dock or berth. If a berth is the specified destination, the ship becomes an 'arrived ship' only when that particular berth is available.

> *EL Oldendorff & Co GmbH v Tradex Export SA ('The Johanna Oldendorff')*[91]
> By a voyage charterparty the carrier undertook that the vessel should load a bulk grain cargo and 'therewith proceed to London or Avonmouth or Glasgow or Belfast or Liverpool/Birkenhead or Hull'. The charterer gave instructions to proceed to the port of Liverpool. The anchorage was 17 miles from the usual discharging berth, but was the usual place where grain vessels lay while awaiting a berth.
>
> The House of Lords held that under a port charterparty, the vessel became an 'arrived ship' when, if she could not proceed immediately to a berth, she had reached a position within the port where she was at the immediate and effective disposition of the charterer.
>
> Lord Reid: 'Before a ship can be said to have arrived at a port she must, if she cannot proceed immediately to a berth, have reached a position within the port where she is at the immediate and effective disposition of the charterer. If she is at a place where waiting ships

90. (1885) 16 QBD 67.
91. [1973] 2 Lloyd's Rep 285.

usually lie, she will be in such a position unless in some extraordinary circumstances proof of which would lie in the charterer.'

Stag Line Ltd v Board of Trade[92]

The charterparty gave the charter an express right to nominate 'one or two safe East Canada or Newfoundland, place or places as ordered by charterers and/or shippers'. The charterer at first requested the ship to proceed to the port of Miramichi, and on arrival the charterer required the ship to load at Millbank, a specific place within the port. As no berths were available at Millbank, the ship had to wait for six days, and the shipowner claimed demurrage.

The English Court of Appeal held that the charterparty expressly allowed the charterer to nominate a 'place'. The 'place' that the charterer was entitled to nominate would include a berth a Millbank. The ship, therefore, did not become an 'arrived ship' until she arrived at the berth, and demurrage was not payable.

Lord Oaksey: 'I think that it is important to remember the settled principle of law with reference to the result of having an express power given to the charterers to nominate the berths at which the ship should lie . . . what the parties were endeavouring to say was that they might proceed to one or two safe ports in East Canada or Newfoundland, and then, within those ports, the charterers would order the place or places to which the vessel was to go.'

Metallgesellschaft Hong Kong Ltd v Chinapart Ltd[93]

The charterparty stipulated, among other terms, that the discharge rate of the cargo should be at 1,000 metric tons per working day and demurrage per day at US$5,000 or pro-rata. The shipowner claimed demurrage amounting to US$132,964.75. It was argued that the shipowner was under an obligation to tender a notice of readiness to discharge to the consignee.

The Hong Kong Court of Appeal held that there was no implied obligation or duty of care to notify the consignee of the time of the ship's arrival. Laytime should have commenced to run when the ship arrived at the port and was ready to discharge.

The court approved the following statement in *the Halsbury's Law of England* (4th ed) vol 43 para 656: 'Apart from special contract or custom or course of dealing, it is the consignee's duty to use due and reasonable diligence to discover when the ship arrives with the cargo on board; and the master is, therefore, under no obligation, in the absence of special contract or custom or course of dealing, to give notice of his arrival or readiness to unload, whether the ship is a general ship or whether she is working under a charterparty. In either case time begins

92. [1950] 1 All ER 1105.
93. [1990] 1 HKC 114.

> to run against the consignee as soon as the ship is ready to unload, and it is immaterial that he was in fact ignorant of her arrival.'

The tendering of notice of readiness to load is normally a pre-condition to the running of laytime. Absence of notice may prevent laytime from commencing. The shipowner must give notice to the charterer because readiness to load is a state of affairs that is with the peculiar knowledge of the shipowner and not within the knowledge of the charterer. However, a notice of readiness is not valid *unless* when the notice is tendered, the ship is physically and legally capable of receiving or discharging the cargo. Broadly speaking, a ship is physically ready to load if the charterer is given complete control of all the holds and every portion of the ship available for the cargo. In addition, the cargo holds must be free from contamination. A ship is not legally ready to load if, under the regulations of the port, the charterer is prevented from obtaining access to the ship owing to a quarantine restriction.

> *Shipping Developments Corporation SA v V/O Sojuzneftexport ('The Delian Spirit')*[94]
>
> Notice of readiness was given at 0100 hours on 19 February while the ship was at anchorage at the port of Tuapse. She arrived at the berth at 1320 hours on 24 February. However, free pratique was not given until 1600 hours on 24 February. The charterer argued that the notice of readiness was not valid.
>
> The court held that the mere fact that free pratique had not been granted did not mean that the ship was not ready to load.
>
> Lord Denning: 'I can understand that, if a ship is known to be infected by a disease such as to prevent her from getting her pratique, she would not be ready to load or discharge. But if she has apparently a clean bill of health, such that there is no reason to fear delay, then even though she has not been given her pratique, she is entitled to give notice of readiness, and laytime will begin to run.'
>
> *Christensen v Hindustan Steel Ltd*[95]
>
> At 0900 hours on 28 October 1967, the master tendered a notice that the vessel would be ready to load on 29 October.
>
> The court held that the notice of readiness was not valid unless it indicated that the vessel was presently ready to load or discharge. It was insufficient if the notice merely indicated that she would be ready at a future time.
>
> Donaldson J: 'The whole purpose of a notice of readiness is to inform the shippers or consignees that the vessel is presently ready to

94. [1971] 1 Lloyd's Rep 506.
95. [1971] 1 Lloyd's Rep 395.

> load or discharge and the period of time within which they have agreed to load or discharge the vessel is measured from that moment, whether or not the counting of laytime is postponed . . . In the present case the notice was on its face one of anticipated readiness and impliedly reported to the charterers that the vessel was not ready at the time at which it was given. Accordingly it cannot be relied upon as a notice of actual readiness, even if in fact the vessel was ready.'

Duration and interruption of laytime

If no definite period of laytime is fixed, the loading or discharging must be completed within a reasonable period of time. In *Carlton SS Co v Castle Mail*,[96] Lord Herschell was of the opinion that there is no such thing as reasonable time in the abstract. The question is whether, having regard to all the obligations of the contract, to its condition, to its restrictions, and to its limitations, more than a reasonable time has been taken in the performance of any one of these obligations in respect of which the parties have not, by their contract, expressed any limit of time for its performance. In some occasions, the charterparty may state that the charterer must load or discharge the goods 'with all dispatch according to the custom of the port' or 'in the usual and customary time'. At common law, what constitutes a reasonable time for loading and unloading depends on the circumstances such as strikes, action of the harbour authorities and engagements of the charterers and consignees. The charterer must employ reasonable diligence and must not act negligently or unreasonably. On the other hand, the shipowner must accept the impediments arising from practice or custom, and the actual circumstances of the particular voyage.

When laytime is fixed, it can be provided expressly by reference to days or by reference to hours. The parties have complete freedom and discretion in deciding the laytime provisions. For example:

- The cargo shall be loaded in [ ] days and discharged in [ ] days.
- The cargo shall be loaded and discharged in [ ] working days.
- The cargo shall be loaded and discharged in [ ] weather working days.
- The cargo shall be loaded in [ ] running hours.
- The cargo shall be loaded at the rate of [ ] tons per weather working days.
- The cargo shall be discharged at the rate of [ ] tons per workable hatch per weather working day.

96. [1898] AC 486.

Hicks v Raymond and Reid[97]

A ship carried grain from Taganrog to London. The bills of lading were silent as to the time for unloading. The unloading was interrupted because of a dock strike. The consignees failed to find any other person to perform the labour in order to get the cargo unloaded.

The House of Lords decided that the shipowner was not entitled to damages for the delay. The consignees had done their best and made all reasonable efforts in trying to discharge the cargo by means of other labour.

Lord Watson: 'When the language of a contract does not expressly, or by necessary implication, fix any time for the performance of a contractual obligation, the law implies that it shall be performed within a reasonable time. The rule is of general application, and is not confined to contracts for the carriage of goods by sea. In the case of other contracts, the condition of reasonable time has been frequently interpreted, and has invariably been held to mean that the party upon whom it is incumbent duly fulfils his obligation, notwithstanding protracted delay, so long as such delay is attributable to causes beyond his control, and he has neither acted negligently nor unreasonably.'

Lord Herschell: '. . . the only sound principle is that "reasonable time" should depend on the circumstances which actually exist. If the cargo has been taken with all reasonable despatch under these circumstances I think the obligation of the consignee has been fulfilled.'

Hulthen v Steward & Co[98]

A charterparty provided that the vessel was to be unloaded with 'customary steamship dispatch'. There was a delay in the discharging operation because of the crowded state of the dock at which the vessel lay.

The House of Lords ruled against the shipowner's claim for demurrage.

Lord Halsbury: 'There are two forms in which charterparties of this character can be made. In one case a specific number of days are given within which the discharge is to be taken, and if those days are exceeded, quite apart from the circumstances, the demurrage is due. If, on the other hand, the parties choose to agree not to a definite number of days, but to a charterparty such as is to be found here, they necessarily import into it the circumstances under which the discharge takes place . . . Without going through the cases it appears to me that every . . . attempt to impose an absolute unconditional burden upon the charterers has always failed, because in this, as in every other contract where no specific time is mentioned, it is to be measured by the legal test, namely, what is reasonable under the circumstances of the case.'

97. [1893] AC 22.
98. (1903) 88 LT 702.

The laytime clause may stipulate that laytime is to be suspended in the event of, say, a strike or bad weather. The clause may also exclude Sundays and holidays from the calculation of laytime. At common law, bad weather includes rain, ice or surf that prevents loading or unloading. Strike, according to the judicial decisions, means a general concerted refusal by workmen to work in consequence of an alleged grievance. The general principle is that if the charterer has agreed to load or unload within an agreed period of laytime, the charterer is answerable for the non-performance of that engagement, whatever the nature of the impediments, unless the delay is covered by the laytime exceptions in the charterparty, or arises through the fault of the shipowner. In contrast, if the delay can be avoided by the charterer but the charterer does not take any reasonable action to keep away from the delay, the charterer is responsible for the time lost and the laytime continues to run despite the laytime exceptions.

> *Damoskibsselskabet Svendborg v Love & Stewart Ltd*[99]
>
> The charterer had failed to make adequate arrangements for the supply of a consignment of coal. Had the charterer done so, the delay caused by a strike would have been avoided.
>
> The court held that the charterer was liable for the delay. The charterer was not exonerated from its liability despite the presence of the exceptions clause in the charterparty in relation to strike. It was the neglect or inaction on the part of the charterer that caused the delay.

> *William Alexander & Sons v Aktieselskabet Dampskibet Hansa*[100]
>
> A delay of seven days occurred during the discharging of the cargo because of a shortage of stevedores. However, this event was not covered by the laytime exceptions in the charterparty.
>
> The House of Lords held that the charterer was liable for the seven days' demurrage. The delay was not covered by the exceptions clause, and the shortage of labour at the port was not attributed to negligence or fault of the shipowner.
>
> Lord Shaw of Dunfermline: 'The law is perfectly well settled . . . The person who hires a vessel detains her, if at the end of the stipulated time he does not restore her to the owner, he is responsible for all the various vicissitudes which may prevent him from doing so . . . This is the prescription of the general law. To avoid its application either, (1) the contract of parties must be absolutely clear; or (2) it must be established that the failure of the charterer's duty arose from the fault of the shipowners or those for whom they are responsible.'

99. (1913) SC 1043.
100. [1920] AC 88.

Duration, continuance and interruption of demurrage period

Delay to the ship beyond the laytime is a breach of contract. Demurrage is the amount agreed contractually in advance by the parties payable as liquidated damages for delay beyond laytime. On the other hand, if laytime has expired, but there is no demurrage stipulation in the contract, damages for detention become payable (see Figure 1). Damages for detention are also due if the demurrage clause provides that demurrage shall run for a certain period of time and that particular period of time has been exhausted (see Figure 2).

Figure 1

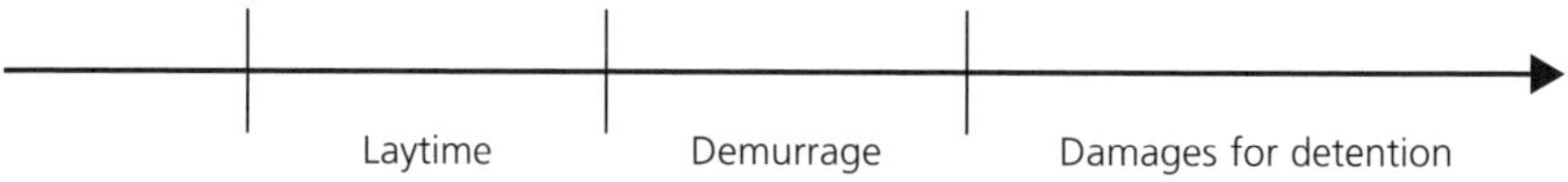

Figure 2

The well-established principle is that '*once on demurrage, always on demurrage*'. Demurrage continues to run and the exceptions that prevent laytime from running do not generally interrupt demurrage. Once a ship is on demurrage, no exceptions will operate to prevent demurrage continuing to be payable unless the exceptions clause is clearly worded so as to have that effect.

> *Margoronis Navigation Agency Ltd v Henry W Peabody & Co of London Ltd*[101]
>
> The charterer had loaded 12,588 tons of a cargo by 29 December. Only 11 tons remained to be loaded, which should normally only have taken another 40 minutes. The charterer decided to slow down the loading operation. The loading was only completed on 2 January, which was still within the period of laytime.
>
> The court held that so long as loading had not been completed, and the laytime had not expired, the charterer was entitled to keep the vessel in port for the whole of the laytime without incurring any liability for demurrage or damages for detention.
>
> Roskill J: '. . . where a charterparty prescribes that a charterer is to have a fixed time to load . . . a charterer is entitled to have that time

101. [1964] 1 Lloyd's Rep 173.

for loading . . . A charterer is entitled to have that time to load, but, once he has loaded, he must not use that time for some other purpose. But, so long as he has not completed loading, that time is his, and he is under no obligation to accelerate that rate of loading so as to shorten the time to which he is otherwise entitled.'

Compania Naviera Aeolus SA v Union of India[102]

After the laytime had expired, a strike took place which further interrupted the unloading works.

The House of Lords held that once the vessel was on demurrage, no exceptions would operate to prevent demurrage continuing to be payable unless the exceptions clause was clearly worded so as to have that effect. Since the strike clause was not clearly worded to have the effect of relieving the charterer from the payment of demurrage, the charterer was held liable to pay demurrage for the whole period after the expiration of the laytime.

Lord Reid: 'If a strike occurs before the end of the laytime, neither party can be blamed in any way. But if it occurs after demurrage has begun to accrue, the owner might well say: true, your breach of contract in detaining my ship after the end of the laytime did not cause a strike, but if you had fulfilled your contract, the strike would have caused no loss because my ship would have been on the high seas before it began: so it is more reasonable that you should bear the loss than I should do. So it seems to me right that if the [charterers] are to escape from paying demurrage during this strike, they must be able to point to an exceptions clause which clearly covers this case.'

Fina Supply Ltd v Shell UK Ltd ('The Poitou')[103]

The demurrage provision stated that 'time would not count for demurrage when spent or lost in berthing the vessel during bad weather'. When the shipowner claimed demurrage, a dispute arose as to the construction of this provision.

The court held that according to the evidence, the vessel was prevented from berthing from 1645 hours on 15 February to 1301 hours on 16 February. It was established from the terminal log books that during that period the winds were gusting in excess of 30 m.p.h. The court formed the opinion that time was lost by the vessel in berthing due to bad weather. The charterer was not liable for demurrage for that period of time.

Waller J: 'If one there poses the question in a common-sense way, "was time lost by Poitou in berthing due to bad weather?", it seems to me that there is only one answer which is that it was . . . on this issue, I am in favour of the [charterer].'

102. [1962] 3 All ER 670.
103. [1991] 1 Lloyd's Rep 452.

Dispatch

The charterparty may contain a dispatch provision stating that dispatch money is payable by the shipowner if loading or discharging is achieved before the expiration of the laytime. At common law, dispatch is presumed to be payable for all time saved including Sunday, unless the charterparty expressly provides otherwise. Also, the laytime exceptions are not taken into consideration in the calculation of dispatch money, unless the laytime and time saved by dispatch are dealt with in the same clause and demurrage in another clause.

> *Mawson Shipping Co Ltd v Beyer*[104]
>
> The charterparty stated that the cargo was to be loaded at the average rate of 500 units per running day of 24 consecutive hours (Saturdays and non-working holidays excepted). The charterparty also provided that 'owners agree to pay charterers £10 per day for all time saved in loading'.
>
> The court held that Sunday and non-working holidays were to be counted among dispatch days. The laytime exceptions were not applicable.
>
> Bailhache J: 'Prima facie the presumption is that the object and intention of these dispatch clauses is that the shipowners shall pay to the charterers for all time saved to the ship, calculated in the way in which, in the converse case, demurrage would be calculated; that is, taking no account of the lay day exceptions.'

6.3 TIME CHARTERPARTIES

The time charter is a common method of shipowners to charter a vessel to charterers having sufficient cargo for the vessel. A time charter is a contract for services to be rendered to the charterers by the shipowner through the use of the vessel by the shipowner's own servants, the master and the crew, acting in accordance with such directions as to the cargo to be loaded and the voyages to be undertaken as by the terms of the charterparty (*Scandinavian Trading Tanker Co A.B. v Flota Petrolera Ecuatoriana (The 'Scaptrade')* [1983] 2 Lloyd's Rep 253, HL). The period of hire is usually stated: during this period, the vessel is let and the shipowners place the service of the vessel with those of her master and crew at the charterer's disposal upon delivery. However, the shipowner remains in possession of the vessel.

---

104. [1914] 1 KB 304.

Time charter trip charterparty

A time charter trip is not a voyage charter nor a time charter, but a time charter for a period measured by the duration of a particular voyage.

Melvin International SA v Poseidon Schiffahrt GmbH (The 'Kalma') [1999] 2 Lloyd's Rep 374
By a time charter-party the claimant agreed to let its vessel to the charterers for one time charter trip via safe port(s) safe berth(s) Black Sea to the Far East always afloat within Institute Warranty limits. The charterers guaranteed a minimum duration of 57.5 days within given trading limits. The court held that it had become a frequent practice to time charter for a period measured by the duration of a certain voyage instead of a stated number of months or days. The voyage was then not merely the measure of the charter but became the subject matter of the contract, so that the charterers must send the ship on that particular voyage.

Time and voyage charter

In a time charter, 'hire' is consideration paid by the charterers to the shipowners for the use of the vessel for a period of time, eg, 6 or 12 months. In a voyage (or trip) charter, 'freight' is consideration paid by the charterers to the shipowners for the carriage of goods from one place to another, ie, a voyage.

Itex Itagrani Export SA v Care Shipping Corporation and Others (The 'Cebu') (No. 2) [1990] 2 Lloyd's Rep 316
It was held that the word 'hire' was used for sums payable under time charters and the word 'freight' was restricted to voyage charterparties and bills of lading.

The amount of freight payable can be agreed in a lump sum, but it usually depends on the quantity of cargo loaded in the loading port or discharged in the unloading port. The time which is required to complete a voyage would not be a determining factor in drafting a voyage charterparty. The fundamental difference between voyage and time charter is that in a time charterparty it is the charterer who bears the risk of delay, whilst in a voyage charterparty it is the shipowner who bears the risk of delay.

Agent and principal relationship in a bill of lading of time charter

Chartering of a vessel may involve more than two parties which may complicate the agent and principal relationship. Withdrawal of a vessel due to non-payment of hire by one party along the chain of parties can impact them all.

> *Dry Bulk Handy Holding Inc and Another v Fayette International Holdings Ltd and Another (The 'Bulk Chile')*[105]
>
> The first claimant (DBHH) was the owner of the vessel and chartered the vessel to CSAV on the NYPE 1946 form. DBHH chartered the vessel to KLC, then to Fayette in a trip charterparty both on the NYPE 1946 form. Fayette entered into a voyage charterparty with Metinvest on the Gencon form. KLC failed to pay the first two hire invoices. DBHH sent to Fayette and Metinvest a 'Notice of Lien'. DBHH withdrew the vessel from KLC's service.
>
> The Court held that the bill of lading contract between the shipowner and the shipper was not a contract by which the shipowner contracted to provide a service to the shipper in consideration of the shipper promising to confer a benefit (in the form of payment of freight) on an independent third party, but rather that the nominated recipient was, as between the shipowner and the shipper, to be regarded as the shipowner's agent. On that basis there was no reason why the shipowner's contract with the shipper should be taken to preclude the shipowner from cancelling his nominated agent's authority to act on his behalf in receiving the freight, before such payment had been made, and requiring it to be made to himself.

Preliminary voyage

Before the current charter or charterparty comes to an end, the shipowner would fix another charterparty for the vessel regardless of whether it is a voyage or time charter. If a charterparty is fixed for the next employment of the vessel, the master will discharge the remaining cargo in the last discharge port of the current charterparty and proceed to the first load port agreed on the next charterparty usually in ballast. The voyage between the last discharge port of the current charterparty and the first load port of the next charterparty is considered as the preliminary voyage.

In *Société Franco Tunisienne D'Armement v Sidermar S.p.A.* [1961] 2 QB 278, the shipowners chartered their vessel to the charterers from Masulipatan to Genoa. The terms of the charterparty were, *inter alia*, that the vessel should proceed with all convenient speed to Masulipatan and there load. Pearson J at p. 306 in the above case stated: 'They had to bring the vessel by what is called the preliminary voyage to Masulipatan, and there they would have to assist in the loading of the cargo on board'. In *Monroe Brothers,*

105. [2013] 2 Lloyd's Rep 38, CA.

Limited v Ryan [1935] 2 KB 28, it was stated at p. 32 that 'the nearest case to the present one is *Hudson v Hill* (1874) 43 L J (C P) 273. In that case the vessel was chartered for a homeward voyage from Barbados, but was allowed to take a cargo to Rio on her outward voyage, and it was held that the exceptions in the charter for the voyage from Barbados applied during the outward voyage to Rio. That case is cited in *Scrutton on Charterparties*, 13th edition, 124, as authority for the proposition that the excepted perils in the charter apply to the preliminary voyage to the port of loading. It was held in *Barker v M'Andrew* (1865) 18 C B (N S) 759, where the ship was lying at her port of loading but had to move from the place where she was lying to a loading berth, that the voyage to which the exceptions related commenced as soon as she broke ground to go to that berth. It was also held in *Bruce v Nicolopulo* (1855) 11 Ex 129; 24 L J (Ex) 321 that the preliminary voyage to the port of loading was to be considered part of the voyage within the contract. In *Crow v Falk* (1846) 8 Q B 467 the vessel was at Liverpool and was to load and sail thence with a cargo, and it was held that the voyage did not commence, and the exception did not apply, until she departed Liverpool. That case, however, was doubted in *Barker v M'Andrew* (1865) 18 C B (N S) 759 and also in *The Carron Park* (1890) 15 P D 203.'

Trading limit

Charterers are entitled to order the vessel to proceed anywhere within trading limits which are not inconsistent with the charter-party.

> *The 'Aragon'*[106]
> The charterers chartered the vessel from the owners under a charter-party on the NYPE Time Charter form. The charter-party provided that the charterers were to hire the vessel: . . . for the period necessary to perform one time charter trip via safe port(s) East Coast Canada within . . . trading limits. The owners protested but agreed that the vessel would sail for the Gulf and load a cargo without prejudice to the rights of the parties.
>
> The Court held that 'U.S.A. East of Panama Canal' meant what it said and excluded the U.S. Gulf. Award remitted to arbitrator for further consideration. The phrase 'U.S.A. East of Panama Canal' meant that part of the U.S.A. which could be reached from Europe westbound without passing through the Panama Canal or going round Cape Horn. The question stated in the award would be answered 'yes'.Judgment for the charterers.

106. *Segovia Compagnia Naviera SA v R. Pagnan & Fratelli (The 'Aragon')* [1975] 2 Lloyd's Rep 216.

Period of hire and delivery

A charterparty usually shows when and how the charter commences and ends. Shipowners and charterers may agree that the period of hire is, for instance, six months. The charterers then take delivery of the vessel at the commencement of the time charter and re-deliver it to the shipowners upon the expiry of the period of hire, ie, after six months. In situations where the charterers want to extend the period of hire beyond the sixth month, the shipowners may not agree to any extension.

The express extension of period of hire, which is usually provided in the charterparty, has to be clear and unambiguous.

> *Partenreederei M.S. Karen Oltmann v Scarsdale Shipping Co Ltd (The 'Karen Oltmann')* [1976] 2 Lloyd's Rep 708
>
> The vessel was chartered 'for a period of 2 years 14 days more or less' by clause 1 of a Baltime charter, and further clause 26 provided that 'charterers to have the option to redeliver the vessel after 12 months' trading subject to giving 3 months' notice'. The charterers exercised the option after 22 months.
>
> The court held that the words 'after 12 months' trading' were used in the agreed sense of 'on the expiry of' and not 'at any time after the expiry of'. If charterers have an option to vary the duration of the charter they have to exercise their right in time or may lose it.

In *The Avisfaith* [1969] 2 Lloyd's Rep 257, by a time charterparty, the defendants' vessel was chartered to the plaintiffs for a period. Clause 1 stated 'of 24/30 calendar months in Charterers' option declarable at the end of the 22nd month with a margin of 15 days more or less at Charterers' option from the time the Vessel is delivered'. Clause 28 provided that '[f]or time lost on account of unseaworthiness of ship and for any time of the vessel being unable to render services provided for in the Time charter Party, through circumstances beyond Charterers' control, time for hire to be accordingly prolonged at Charterers' discretion'.

> *Empresa Cubana de Fletes v Aviation & Shipping Company, Ltd (The 'Avisfaith')* [1969] 2 Lloyd's Rep 257
>
> The charterers purported to declare their option to hire the vessel for a period of 30 months. The shipowners contended that the charterers had lost their right to the option by failure to exercise it within the 22 months. The charterers contended that they did not lose their right to exercise the option by failing to exercise it in timely fashion. The time lost under Clause 28 (which amounted to at least 18 days) could be added to the 22 months under Clause 1.

The court held that Clause 1 created an option in favour of the charterers on condition that the charterers declared their intention to exercise the option not later than the end of the 22nd month. Accordingly, the charterers had failed to exercise their rights and had lost them. Time under Clause 28 could not affect the period of 22 months under Clause 1. Judgment was given for the defendant shipowners.

The word 'month' used in a charterparty means 'calendar month' unless the terms express the contrary (*Jolly v Young* (1794) 1 Esp 186; *Hart v Middleton* (1845) 2 C&K 9). The word 'day' usually means a calendar day, which begins and ends at midnight (*The Katy* [1895] P 56, CA). A part of a day is counted as one whole day where nothing contrary is expressed (*Glassington v Rawlins* (1803) 3 East 407; *R v St Mary, Warwick* (1853) 1 E & B 816; *Angier v Steward* (1884) 1 Cab & Ell 357).

Period of hire after detention, confiscation and sale of a vessel

Payment of hire is closely related to the period of hire under a charterparty. The issue becomes complicated where a vessel is detained, confiscated and being sold by an authority in a shipment of unlawful cargo. Subject to the issue of mitigation, charterers are liable to indemnify owners. Clause 4 of the Shelltime 4 form constitutes an absolute warranty as to the lawfulness of the cargo, not merely an undertaking that the cargo is lawful to the best of the charterer's belief.

The 'Greek Fighter'[107]

The vessel and her cargo were detained in the UAE. The vessel was subsequently confiscated by the UAE authorities and sold at public auction. At the time of the detention, the vessel was on time charter.

The Court held that as to the owners' claim for hire up to the date of confiscation of the vessel, if Fal were not acting unlawfully in causing Iraqi oil to be loaded onboard the vessel, (a) the vessel would have been off-hire throughout the period of detention up to the first period of hire covered by the First Hire Agreement, after which hire would have fallen due in accordance with the hire agreements, (b) the vessel would have been effectively redelivered on 25 November 2002 at which point hire would have ceased to accrue, (c) the owners would have been entitled to be compensated in damages for breach of the redelivery clause, and (d) hire paid in respect of the period from

107. *Ullises Shipping Corporation v Fal Shipping Co Ltd (The 'Greek Fighter')* [2006] 1 Lloyd's Rep 99.

the commencement of the detention to 23 February 2002 would not have been due because the vessel would be off-hire, and would therefore be recoverable by Fal by way of restitution.

Period of hire and arrest of charterers

The authorities established the following exception to the general rule that the innocent party had an option whether or not to accept a repudiation: (i) the burden was on the contract-breaker to show that the innocent party had no legitimate interest in performing the contract rather than claiming damages; (ii) that burden was not discharged merely by showing that the benefit to the other party was small in comparison to the loss to the contract-breaker; (iii) the exception to the general rule applied only in extreme cases, ie, where damages would be an adequate remedy and where an election to keep the contract alive would be unreasonable.

> *Ocean Marine Navigation Limited v Koch Carbon Inc. (The 'Dynamic')*[108]
> By a charter-party on an amended NYPE form the claimant owners chartered their vessel to the defendant charterers for a time charter trip. The vessel was delayed while being repaired in Singapore. From Singapore the vessel sailed for Myrtle Grove. Discharge was completed. The charterers arrested the vessel to obtain security for various performance claims.
>
> The Court held that cl 60 was never intended to deal with an arrest by the charterer during the currency of the charter-party; the charterers were liable for hire under cl 60 if there was an arrest during the currency of the charter-party and the arrest was the consequence of a deliberate act or omission by the charterers or their agents.

Charterers' option for period in time charter

The correct approach to the construction issue is that one should have regard to the factual matrix so as to consider the contract as expressed by the parties in the same context as the parties do at the time they make it.

> *Nippon Yusen Kubishiki Kaisha v Golden Strait Corporation (The 'Golden Victory')*[109]
> By a charter-party in the amended Shelltime form, the defendant owners let their vessel to the claimant charterers NYK. The memorandum of agreement (MOA) provided inter alia: Charterers' Option For

108. [2003] 2 Lloyd's Rep 693.
109. [2003] 2 Lloyd's Rep 592.

> Period: – the Charterers shall have their option . . . but such option shall be declared . . .
>
> The Court held that there was an option to charter back on precisely the same terms as the existing charter to GOL at three and five years; a charter back was not a redelivery, which was an option not acceptable to the Golden Ocean Group; the appeal would be dismissed.

Whether owners had no legitimate interest in insisting the charter remained alive

The earning of hire after purported redelivery is not dependent on any performance by the charterers of their obligations. With only 94 days left of a five-year time charter in a difficult market where a substitute time charter is impossible, and trading on the spot market very difficult, it would be impossible to characterize the owners' stance in wishing to maintain the charter as unreasonable, let alone wholly unreasonable.

> *The 'Aquafaith'*[110]
>
> The vessel was chartered on an amended NYPE form. The charter included a warranty that the vessel would not be re-delivered before the minimum period of 59 months. The charterers stated that they would redeliver the vessel on dropping the last outward sea pilot after discharge in China under the then current voyage. The owners did not accept the repudiation and sought to affirm the charterparty.
>
> The Court held that the appeal would be allowed. The arbitration award would be varied to declare that: (i) the owners were entitled to refuse the purported redelivery by the charterers on 9 August 2011 and were entitled to hire in accordance with the terms of the charterparty; and (ii) the charterers were not entitled to insist on redelivery on 9 August 2011.

Legitimate last voyage

The time charter market fluctuates and is interdependent with other freight markets, eg, the spot market. The profit margins of shipowners and charterers may differ when they fix a vessel in a different cycle of the freight market. Charterers may try to employ a vessel as long as the charterparty permits them to do so in a rising market, while shipowners would think otherwise. This is one of the reasons why disputes regularly arise in connection with the issue of legitimate last voyage.

110. *Isabella Shipowner SA v Shagang Shipping Co Ltd (The 'Aquafaith')* [2012] 2 Lloyd's Rep 61.

In *The World Renown* [1992] 2 Lloyd's Rep 115, Lord Donaldson in the Court of Appeal stated as follows at p. 118:

1. A charter for a fixed period will have a small implied tolerance or margin in its duration.
2. A charter for a fixed period with an expressed tolerance or margin—in this case '15 days more or less'—will have no further implied tolerance or margin.
3. In either of these cases, in the absence of a 'last voyage' clause, the charterers will be in breach of contract if the vessel is redelivered after the expiry of the fixed period extended by the implied or expressed tolerance or margin, unless the late delivery arises out of a cause for which the shipowners are responsible.
4. A 'last voyage' clause is needed and will protect the charterer if he orders the vessel to undertake a last voyage which can reasonably be expected to enable the vessel to be redelivered punctually, but without fault on his part in the event that such redelivery proves impossible.
5. If a 'last voyage' clause is to protect a charterer from being in breach by late redelivery in circumstances in which he has ordered a voyage which is likely to or must have this result, the intention to provide this protection must be clearly expressed.

An illegitimate last voyage is where charterers give orders for the employment of the vessel which cannot reasonably be expected to be performed by the final terminal date. In this case, the charterers will seek to avail themselves of the services of the vessel at a time when the shipowner had never agreed to render such services.

> *Hyundai Merchant Marine Co Ltd v Gesuri Chartering Co Ltd (The 'Peonia')* [1991] 1 Lloyd's Rep 100, CA
>
> By a charterparty in the NYPE form the disponent shipowners let their vessel to the charterers for a period. Hire was to continue until the hour of the day of the vessel's redelivery to the shipowners within specified areas. If a sub-charter to Singapore and Butterworth had been performed the vessel would have been redelivered to the shipowners no earlier than 19 July 1988. The time charter 12 month period expired on 11 June 1988.
>
> The shipowners, alleging that the proposed voyage was illegitimate, called for voyage orders which would enable the vessel to be redelivered within the charter period or alternatively for payment of hire at an enhanced rate for the duration of the voyage outside the charter period. The charterers accepted neither condition and the vessel was withdrawn.

The court held that it was an order that the charterer was not entitled to give and in giving it the charterer committed a breach of contract. If the shipowner complied with the order he would be entitled to payment of hire at the charter-party rate until redelivery of the vessel and, provided he did not waive the charterers' breach, to damages (being the difference between the market rate and the charter rate) for the period between the final terminal date and redelivery.

There is a presumption that a charter is intended to continue in operation until the end of a legitimate last voyage, ie, one which is expected at its commencement to end within the charter period, unless one of the parties is responsible for the delay (*Timber Shipping Co SA v London & Overseas Freighters Ltd (The 'London Explorer')* [1972] AC 1, HL as per *Lord Reid and Lord Cross of Chelsea*).

In *The 'London Explorer'*, the shipowners claimed to be paid at the charterparty rate for the excess period during which the charterers had kept the ship. The charterers contended that they were liable to pay only damages for breach of contract in failing to redeliver the ship on time. The market rate had fallen since the date of the charterparty.

Timber Shipping Co SA v London & Overseas Freighters Ltd (The 'London Explorer') [1972] AC 1, HL

By a time charterparty the charterers chartered the vessel from the shipowners for '12 months 15 days more or less in charterers' option' from the time of delivery. Clause 4 of the charterparty provided for 'hire to continue until the hour of the day of her redelivery'. If all had gone well, the vessel would have been redelivered to the shipowners at Houston well before 29 December 1968 (or, alternatively, 13 January 1969). As a result of successive dock strikes at New Orleans and Houston, she was not redelivered to the owners until 24 April 1969. The off-hire clause in the charterparty did not cover strikes.

The court held that (1) clause 4 of the charterparty obliged the charterers to pay at the contractual rate until actual redelivery. There was a breach by the charterers in failing to redeliver when they should have observations of time being of the essence in relation to charterparties.

Lord Reid in *The 'London Explorer'* stated at p. 14 that '*Gray & Co v Christie & Co* (1889) 5 TLR 577 was the earliest case cited and has come to be regarded as the leading case. . . . This case has been regarded as authority for some very wide propositions. . . . there is a presumption that a definite date for the termination of a time charter should be regarded as an approximate date only.'

> *Gray & Co v Christie & Co* (1889) 5 TLR 577
>
> By a charterparty, Grays let their ship to Christie for a period which expired on September 26. The charter rates had risen and the shipowners objected to the vessel being sent on a voyage which would not end by that date. Nevertheless, Christies sent her on a voyage which was expected to end on September 30. There was some unexplained delay and the ship was not redelivered until October 13. Thus, the shipowners claimed payment for the last 17 days at the higher current market rate. Their claim failed.

An order which is originally permissible may become illegitimate if circumstances change. The original order having become ineffectual, the charterers are obliged to replace it with one which they are entitled to give. There is an anticipatory breach on the part of the charterers if they refuse to issue another valid order, and the shipowners may treat the charterparty as at an end.

In *The Gregos* [1995] 1 Lloyd's Rep 1, the shipowners let their vessel by a time charter to the charterers for a period and the charter provided *inter alia*: 'Clause 4. Hire . . . to continue until the hour of the day of her redelivery.'

> *Torvald Klaveness A/S v Arni Maritime Corporation (The 'Gregos')* [1995] 1 Lloyd's Rep 1, HL
>
> Redelivery of the vessel became due on 18 March with no further tolerance or margin. The schedule became unworkable because another vessel had grounded in the river. The shipowners were unwilling to proceed with the laden voyage to Fos. A without prejudice agreement was entered into and the vessel performed her voyage. She was redelivered 8 days late, partly due to the time spent in negotiations.
>
> The court held that by then an order originally permissible had become illegitimate. The charterers' persistence in the original order, which had been rendered invalid by the changed circumstances, showed that they did not intend to perform their obligations under the charter. They evinced an intention no longer to be bound by the charter and this was an anticipatory breach which entitled the shipowners to treat the contract as ended. The appeal would be allowed.

Termination of a time charter

There is no reason to accord to the word 'terminated' anything other than its ordinary meaning of coming or being brought to an end, however that result might have occurred. In the event the circumstances permitting termination are to be confined to those where there have been a legitimate and consensual contractual cancellation, the clause would have said so.

Aktieselskabet Dampskibsselskabet Svendborg and Another v Mobil North Sea Ltd and Others[111]

The defendants Mobil entered into an agreement with R&B for the provision of a drilling unit. An invitation to tender in respect of anchor-handling tug supply vessels was distributed on behalf of Mobile. Maersk let their vessel to Mobil. In late 1997 and early 1998 there were discussions between Mobil and Maersk relating to the provision by Maersk in substitution for existing vessel of a larger and more sophisticated anchor handling vessel. Clause 4.3 of the charter was deleted and amended.

The Court held that the underlying purpose of the charter-party was expressly to support Mobil's operation with the rig; and it was entirely consistent with common sense that Mobil should be able to dispense with the services of the tug in any circumstances where the rig contract ceased. On the facts and the evidence the claimants failed to establish that Mobil had elected not to exercise its rights of termination under cl 4.3 and had affirmed the contract.

Reasonable contemplation and foreseeable consequence

In the absence of special knowledge, a type of loss is within the reasonable contemplation of the parties if having regard to the terms of the contract against its commercial background the parties reasonably assume liability for it; or, if having regard to the nature and object of the contract it is sufficiently likely to result from the breach to make it proper to hold that it flows naturally from the breach; or, if it would reasonably have been contemplated by the contracting parties as being likely to happen in the ordinary course of things as a result of the breach.

Transfield Shipping Inc v Mercator Shipping Inc (The 'Achilleas')[112]

By a charterparty on the NYPE form as amended the vessel was chartered for a period. By an addendum the vessel was fixed in direct continuation for a further period. The maximum duration of the extended period expired on 2 May 2004. By April 2004 market rates had more than doubled. The charterers gave notice of redelivery between 30 April and 2 May. On 21 April the owners fixed a period charter of about four to six months with Cargill. The charterers fixed the vessel under a subcharter. The owners did not object to the fixture. By 5 May it had become clear that the vessel would not be available to Cargill before the cancelling date. By that time rates had fallen sharply.

The Court held that the appeal would be allowed. It was not sufficient that the type of loss complained of was a foreseeable consequence

111. [2001] 2 Lloyd's Rep 127.
112. [2008] 2 Lloyd's Rep 275, HL.

of the defendant's breach of contract. The type of loss had to be within the reasonable contemplation of the parties at the time the contract was made.

Late redelivery

A clause providing that charterers should pay market rate of hire from the 30th day prior to the maximum period date until actual redelivery in late redelivery of a time charter is a penalty.

Lansat Shipping Co Ltd v Glencore Grain BV (The 'Paragon')[113]

The owners chartered their vessel on the NYPE form. In the event, the vessel was redelivered six days late. The charterers paid hire at the market rate for the six days the vessel was overdue, but the owners contended that they were entitled to hire based on the market rate for 30 days before the latest date for redelivery.

The Court held that the appeal would be dismissed. Clause 101 was a penalty. Its primary purpose was to deter the charterers from breaching their obligation to redeliver the vessel in time, and was not a genuine pre-estimate of damage resulting from a breach of contract.

IMT Shipping and Chartering GMBH v Chansung Shipping Co Ltd (The 'Zenovia')[114]

The vessel was chartered by her owners on an amended NYPE form for a period. The market was rising, and it later became clear that an extra voyage could be squeezed in before the contractual redelivery date. The owners did not accept that the charterers were entitled to change the expected redelivery date. The owners withdrew the vessel from the chartered service.

The Court held that the words 'without prejudice' indicated that what was said could not usually be relied upon unless it led to a binding agreement. Whether or not the charter contained a relevant implied term, a notice of approximate redelivery expressly given 'without prejudice' could not generate an obligation on the charterer not deliberately to do anything which might prevent the approximate date given therein being met. The 30-day notice so qualified could not give rise to a promissory estoppel.

Charterer's option to extend the last voyage

The effect of cl 18 on Shelltime 3 form is to expose the owners to a final round voyage of no fixed length, which it is clear from the outset would

113. [2009] 1 Lloyd's Rep 658.
114. [2009] 2 Lloyd's Rep 139.

extend very considerably beyond the final terminal date. But that is not a problem which derives from including any margin period of 'days more in charterers' option' in the charter period during which cl 18 might be operated.

> *Petroleo Brasiliero SA v Kriti Akti Shipping CoSA (The 'Kriti Akti')*[115]
> The charterers chartered the vessel from the owners under a time charter on the Shelltime 3 form. During the charter the vessel was off-hire on various occasions. The charterers told the owners that they were exercising their option to extend the final date of the charter to June 14. While the vessel was discharging, the charterers ordered her to carry out another voyage. The owners took the view that the charter had already expired and refused to comply with the charterers' orders unless the hire was increased.
>
> The Court held that the arbitrators and the Judge were right to conclude that any period which the charterers elected to take as an extension of the basic period under cl 50 counted as part of 'the period of this charter' for all relevant purposes, including specifically those of cl 18. The appeal would be dismissed.

Payment of hire

It is important charterers comply with the clause of payment of hire as it may allow the shipowners to withdraw the vessel from the service of the charterers. Where charterers are in default on payment of hire, the shipowners may withdraw the vessel or sue for the hire due.

In *The Brimnes* [1974] 2 Lloyd's Rep 241, charterers instructed their bankers by telex to credit the shipowners' account one day after the hire was due, and on the same day the shipowners issued notice of withdrawal.

> *Tenax Steamship Co Ltd v The 'Brimnes' (Shipowners) (The 'Brimnes')* [1974] 2 Lloyd's Rep 241, CA
> A time charterparty provided that 'otherwise failing the punctual and regular payment of the hire', the shipowners were to be at liberty to withdraw the vessel from the charterers' service. The payment of hire was due on 1 April 1970 and was not received by the shipowners. On 2 April the shipowners withdrew the vessel. On the same day the charterers' agents sent an order by telex to the bank instructing it to transfer the hire.
>
> The court held that the shipowners had effectively withdrawn the vessel, for the withdrawal had preceded the payment of the hire.

115. [2004] 1 Lloyd's Rep 712, CA.

In *The Brimnes* [1972] 2 Lloyd's Rep 465, Justice Brandon held that the shipowners were entitled to withdraw the vessel because the receipt of the transfer order by the bank was not analogous to the receipt of a cheque, and thus, the withdrawal had taken place before payment. Late payment of hire was not of itself repudiatory entitling the shipowners, in the absence of a withdrawal provision, to terminate the charter. Lord Salmon in *The Laconia* [1977] AC 850 stated that '[c]ertainty is of primary importance in all commercial transactions. I am afraid that ever since 1971 when *The Georgios C* [1971] 1 QB 488 was decided a great deal of doubt has been generated about the effect of clauses conferring the right upon shipowners to withdraw their vessels when charterers fail to pay hire in accordance with the terms of the charterparties in the well-known New York Produce Exchange, Baltime and Shelltime forms. No such doubt existed between 1949 (when the *Tankexpress* case [1949] AC 76 was decided) and 1971. My Lords, I hope that the doubts which have troubled the waters since 1971 will now be finally dispelled by this decision of your Lordships' House.'

> *Mardorf Peach & Co Ltd v Attica Sea Carriers Corporation of Liberia (The 'Laconia')* [1977] AC 850, HL
>
> Shipowners let their vessel to charterers on a time charter, which by clause 5 required payment of hire 'in cash' semi-monthly and gave the shipowners liberty to withdraw the vessel "failing the punctual and regular payment of the hire'. The seventh instalment fell due on a Sunday when London banks were closed. At that date the market rate was rising. On Monday the shipowners' agents notified the charterers that they were contemplating withdrawal of the vessel for breach of the obligation to pay hire. At about 3 p.m., the charterers' London bank delivered a payment order for the hire instalment to the shipowners' bank.
>
> At 6.55 p.m. on Monday the shipowners' agents gave notice to the charterers that the vessel was withdrawn from the charter. The charterers, in order to complete a voyage, agreed to pay $8 per ton pending a reference to arbitration of the question whether the shipowners were entitled to withdraw the vessel.
>
> The court held that on the proper construction of clause 5 of the charterparty, once a punctual payment of any instalment had not been made, a right of withdrawal accrued to the shipowners, who must give notice of its exercise within a reasonable time. Unless the default was waived, the charterers could not avoid the consequences by tendering an unpunctual payment.

In *The Chikuma* [1981] 1 WLR 314, where payment of hire was to be made 'in cash', the charterers were not considered to have paid unless what the shipowners had received was the equivalent of or as good as cash.

A/S Awilco v Fulvia S.p.A. di Navigazione (The 'Chikuma') [1981] 1 Lloyd's Rep 371, HL; [1981] 1 WLR 314, HL

By the terms of a time charterparty the payment of hire was to be made in cash monthly in advance, otherwise the shipowner could withdraw the vessel from the charterers. One instalment fell due on 22 January 1976. On that date the sum due was credited to the shipowner's bank account, but interest could not run on it until 26 January. On 24 January the shipowners withdrew the vessel.

The court held that there would be judgment for the shipowners, for the payment which had been made was not payment in cash or its equivalent because the money could not immediately earn interest.

Withdrawal of vessel

The withdrawal of a vessel under a time charter may seriously disrupt the shipment of cargo scheduled by the charterers but it is a right given to the shipowners by the terms of the charterparty. It is important for the shipowners to be paid promptly in accordance with the terms of the charterparty as they are required to pay for the mortgage of the vessel, pay the wages of the master and crew, stock up on stores and provisions, and cover other expenses of the vessel, eg, insurance.

Scandinavian Trading Tanker Co A.B. v Flota Petrolera Ecuatoriana (The 'Scaptrade') [1983] 2 Lloyd's Rep 253, HL

Under a time charterparty, clause 8 provided for payment of the hire monthly in advance. The clause further provided that 'in default of such payment shipowners may withdraw the vessel from the service of the charterers, without prejudice to any claim shipowners may have on charterers under this charter'. When the charterers failed to pay on time the hire instalment that fell due on 8 July 1979, the shipowners sent a telex withdrawing the vessel on 12 July.

The court held that where the withdrawal clause so provided, the shipowner was entitled to withdraw the services of the vessel from the charterer if the latter failed to pay an instalment of hire in precise compliance with the provisions of the charter. The shipowner committed no breach of contract if he did so and the charterer had no remedy in damages against him. Lord Diplock stated that: '[t]he freight market is notoriously volatile. It rises during the period of a time charter, the charterer is the beneficiary of the windfall which he can realise if he wants to by sub-chartering at the then market rate. What withdrawal of the vessel does is to transfer the benefit of the windfall from charterer to shipowner.'

When the shipowners know there has been an underpayment of hire for the last month of the charter, they can elect between two courses of action:

either to withdraw the vessel and refund any unearned hire, or to retain the money as paid for the last month and allow the charterers to continue to use the vessel for the rest of the period for which they have paid hire, relying on a final settlement of accounts:

> *China National Foreign Trade Transportation Corporation v Evlogia Shipping Co SA of Panama (The 'Mihalios Xilas')* [1978] 1 WLR 1257, CA
>
> A time charterparty provided inter alia: Clause 6 – 'in default of payment' of hire on the stipulated terms the shipowners had the right to withdraw the vessel. By Clause 39 payment of hire was to be every calendar month in advance, 'except for the last month's hire' when items of the shipowners' liability could be deducted up to the expected redelivery time.
>
> When the ninth month's hire was due, the charterers estimated that she would be redelivered by the end of the ninth month. However, the shipowners' view was that the charterparty period must extend into the tenth month. The charterers deducted from the hire some $31,000 for disbursements made on the shipowners' behalf, and the shipowners accepted the underpaid hire and asked for details of deductions made, but withdrew the vessel under clause 6.
>
> The court held that by retaining the hire paid while demanding further details of the deductions over a period of days the shipowners indicated that they had elected to treat the charterparty as continuing for the rest of the period paid for. Accordingly their withdrawal of the vessel while retaining the money was wrongful.

Lord Denning MR in *The Mihalios Xilas* [1978] 1 WLR 1257 stated that '[w]here a withdrawal clause carries such rigorous penalties, the courts should construe it sensibly as applying only to a clear and substantial default by the charterers. As a matter of construction, the deductions made by the charterers, if honestly made on reasonable grounds, would not put them 'in default of payment' within clause 6. There is no term in the charterparty entitling the charterers to be repaid the unearned hire. They can claim it only as damages for wrongful withdrawal.'

The 'Brimnes'[116] is not followed in a recent case, *The 'Astra'*,[117] in payment of hire. After the charterparty was concluded, market rates of hire fell. The charterers sought reductions in the rate of hire, and threatened to declare bankruptcy unless the owners agreed. The parties entered into an agreement set out in Addendum No. 1. The charterers did not pay the reduced hire. The owners issued an anti-technicality notice. The charterers asked

116. [1972] 2 Lloyd's Rep 465.
117. [2013] 2 Lloyd's Rep 69.

for more time to pay, and the parties entered into Addendum No. 2, which extended the reduced rate of hire. The charterers failed to pay the reduced rate of hire due and also failed to pay the full rate of hire due. The owners served an anti-technicality notice. The owners withdrew the vessel from the charterers' service and terminated the charterparty.

Kuwait Rocks Co v Amn Bulkcarriers Inc (The 'Astra')[118]

The appellant charterers chartered the vessel from the respondent owners for a period on the NYPE 1946 form as amended. Clause 5 gave the owners a contractual right to withdraw the vessel, failing the punctual and regular payment of hire. Clause 31 was an anti-technicality clause.

The Court held that the obligation to make punctual payment of hire was a condition of the contract, breach of which entitled the owners to withdraw the vessel and claim damages for loss of bargain.

Dry Bulk Handy Holding Inc and Another v Fayette International Holdings Ltd and Another(The 'BulkChile')[119]

The first claimant (DBHH) was the owner of the vessel and chartered the vessel to the second claimant (CSAV). DBHH, chartered the vessel to KLC. KLC entered into a trip charterparty with Fayette and all charters were on the NYPE 1946 form. Fayette entered into a voyage charterparty with Metinvest on the Gencon form. KLC failed to pay the first two hire invoices. DBHH sent to Fayette and Metinvest a 'Notice of Lien'. DBHH withdrew the vessel from KLC's service.

The Court held that the message of 5 March to the master was even more explicit. Both messages had to be read in the context of the owners having already made it clear on 19 February that whilst they would carry the cargo to destination and there discharge it in the usual way in the event of withdrawal from KLC, they would expect the sub-charter hire to be paid to them. It was no impediment that the owners were in any event under a contractual obligation to the shippers to carry the cargo to destination and there discharge it. There was no reason why a request, giving rise to a right to remuneration, could not be made for performance of services some, or even all, of which the recipient of the request was obliged to perform under a contract with a third party.

Withdrawal must be final

Where the charterers for some reason do not remit the hire to the shipowners as mutually agreed on the charterparty, some shipowners may try to

118. ibid.
119. [2013] 2 Lloyd's Rep 38, CA.

put pressure on the charterers by withdrawing the service of the vessel, eg, refusing to continue the cargo operation.

The charter in *The Agios Giorgis* [1976] 2 Lloyd's Rep 192 provided, *inter alia*, clause: 'the shipowners were to have a lien upon all cargoes and all sub-freights for any amounts due under the charter including general average contributions and the charterers were to have a lien on the ship for all monies paid in advance and not earned'.

> *Steelwood Carriers Inc of Monrovia, Liberia v Evimeria Compania Naviera SA of Panama (The 'Agios Giorgis')* [1976] 2 Lloyd's Rep 192
>
> When a monthly payment of a time charter in the New York Produce Exchange form was due, the charterers deducted a sum which subsequently proved to be excessive in respect of the speed warranty. The shipowners rejected the claim and instructed the master not to allow discharge at Norfolk. The cargo was detained for two days.
>
> The court held that the shipowners were entitled by reason of the withdrawal provisions in Clause 5 of the charter to withdraw the vessel. However, the shipowners were not entitled to effect a partial or temporary suspension or withdrawal. Apart from Clause 18, the shipowners were not entitled to refuse to allow discharge to begin at Norfolk as and when they did.

In *The Agios Giorgis* the natural interpretation of the word 'withdrawal' was held to be equivalent to 'cancellation'. There was a breach by the master of the obligations undertaken by clause 8 of the charter.

A charterer in a time charter may cancel a charterparty if an owner fails to obtain the approval of some oil majors agreed in the charter.

> *BS & N Ltd (BVI) v Micado Shipping Ltd (Malta) (The 'Seaflower')*[120]
>
> By a time charter-party the defendant owners let their vessel to the claimant charterers. The charter contained a majors approval clause. Owners guarantee to obtain within 60 days EXXON approval in addition to present approvals. The vessel was delivered. Owners advised charterers that the vessel would be ready for Exxon inspection by the end of January/early February. The charterers' response by telex dated Dec. 30, 1997 was to terminate the charter and redeliver the vessel.
>
> The Court held that under cl 46 if during the period of hire 'even' one major approval was lost and the owners failed to obtain the reinstatement to the same within 30 days the charterers had the right to cancel (or continue with the hire discounted by U.S. $250); that provision clearly applied to the loss of Exxon if Exxon approval had ever been obtained; and clearly pointed to the importance to the charterers

120. [2000] 1 Lloyd's Rep 341, CA.

of the majors' approval being maintained and seemed to treat a failure to reinstate within 30 days as a breach of condition.

Notice of withdrawal

Some charterparties do not provide expressly whether the shipowners should send a notice of withdrawal to the charterers before exercising the right of withdrawal.

> *Tankexpress A/S v Compagnie Financière Belge des Petroles SA* [1949] AC 76, HL
>
> A clause in a time charterparty provided that payment of hire was to be made 'in cash monthly in advance in London'. In default the shipowners had the right to withdraw the vessel. In accordance with the practice which had always been followed, the charterers had sent a cheque every month to the shipowners instead of paying cash. A cheque was delayed in transmission, and the shipowners claimed that they were entitled to withdraw because payment had not been made in cash.
>
> The court held that the claim failed, for if they wanted to insist on payment in cash, the shipowners must give reasonable notice of such intention.

In *Tankexpress A/S v Compagnie Financière Belge des Petroles SA* [1949] AC 76, there was no express notice in the charterparty to be given to the charterers in default of payment of hire, and the court might have considered the fact that the charterers had sent a cheque every month to the shipowners instead of paying cash.

In *The Brimnes* [1974] 2 Lloyd's Rep 241, the court held that the charterers' regular late payment did not clearly evince an intention not to be bound by the contract.

Withdrawal and bunkers consumed

Although the scope of the indemnity in clause 13 (employment of the vessel and indemnity) of Shelltime 3 form is very wide, it only protects the owners against losses arising from risks or costs which they have not expressly or implicitly agreed in the charterparty to bear.

> *Ene Kos 1 Ltd v Petroleo Brasileiro SA (The 'Kos')*[121]
>
> The vessel was chartered on the Shelltime 3 form. The charterparty contained a standard form of withdrawal clause providing that if hire

121. [2012] 2 Lloyd's Rep 292, SC.

was not paid when due the owners should have the right to withdraw the vessel. The owners withdrew the vessel for non-payment of hire.

The Court held that the owners were entitled to recover the sums claimed under clause 13 of the charterparty. The loss claimed by the owners was the consequence of the charterers' order to load the parcel of cargo which was on board the vessel when it was withdrawn.

Bunkers

The position under the NYPE and similar forms of charter is that property in the bunkers passes to the charterer at the time of delivery of the vessel. Delivery occurs when the owner places the ship at the charterer's disposal.

Daebo Shipping Co Ltd v The Ship 'Go Star'[122]

The vessel was chartered under a string of time charters on the 1981 NYPE. Daebo chartered the vessel to Nanyuan. Daebo issued an invoice to Nanyuan for the first hire payment and for the value of the bunkers. Nanyuan purported to cancel or withdraw from the Nanyuan sub-charter. It did not pay Daebo's invoice, and arranged an alternative carrier for its cargo.

The Court held that delivery marked the time of commencement of the charterer's obligation to pay hire. It was not necessary that the charterer performed some act so as to 'take over' the bunkers. The expression 'take over' in the NYPE form of charter signified a change of ownership consequent upon and concurrent with delivery of the ship to the charterer. The charterer could not assert that it had not taken over the ship and its bunkers once a ship was made available or placed at its disposal, by delivery.

Proceedings against shipowners for price of bunkers supplied to time charterers

Owners are entitled to be put in the same position as if they are parties to the bunker supply contract containing clause 19.1 of Balttime form notwithstanding their averment that they are not a party.

Jewel Owner Ltd and Another v Sagaan Developments Trading Ltd (The 'MD Gemini')[123]

The owners chartered out their passenger vessel on the Baltime form which provided for New York arbitration. The defendant was the unpaid supplier of bunkers to the time-charterers. Sagaan brought

122. [2013] 1 Lloyd's Rep 18, Federal Court of Australia.
123. [2012] 2 Lloyd's Rep 672.

proceedings in the US against the owners, ISP and the charterers. The claim was for the price of the bunkers.

The Court held that a contractual exclusive jurisdiction clause ought to be enforced unless there were strong reasons not to do so. In the absence of such a clause, the court would generally only restrain the claimant from pursuing proceedings in the foreign court if such pursuit would be vexatious or oppressive on grounds of forum non conveniens on the basis that: (a) England was clearly the more appropriate forum (the natural forum); and (b) justice required that the claimant in the foreign court should be restrained from proceeding there.

Bunkers on redelivery

The words 'price actually paid' means the price paid when the bunkers are stemmed. The conclusion reached by the arbitrators was correct principally because of the specific requirement in the clause for 'first-in-first-out' assessment.

Eitzen Bulk A/S v TTMI Sarl (The 'Bonnie Smithwick')[124]

TTMI chartered their vessel to Eitzen on the Shelltime 4 (2003) form. Clause 15 of the head charter provided: '15. Charterers shall accept and pay for all bunkers on board at the time of delivery, and owners shall on redelivery accept and pay for all bunkers remaining on board, at the price actually paid, on a first-in-first-out basis.' The market in charter rates and bunker prices had dropped considerably.

The Court held that the arbitrators had specifically addressed Eitzen's argument that TTMI's construction of clause 15 was unworkable, and had found that although a charterer *might* have difficulty in obtaining copies of relevant invoices, it was not impossible. The court could not go behind that conclusion.

Bunkers and shipowners' duty as bailees

Shipowners are not in breach of their duties as bailees whether before or after redelivery of vessel. The bunkers are held by the charterers as bailees, and by the shipowners as sub-bailees on terms.

Angara Maritime Ltd v Oceanconnect UK Ltd and Another (The 'Fesco Angara')[125]

The vessel was time-chartered. The charterers entered into a contract with bunker suppliers for the provision of bunkers. The bunker

124. [2012] 1 Lloyd's Rep 407.
125. [2011] 1 Lloyd's Rep 61.

suppliers were never paid, and the charterers subsequently went into administration. The bunker supply contract contained a retention of title clause whereby the suppliers retained title in the bunkers.

The Court held that the shipowners were entitled to the protection of section 25. On the facts, there had been a voluntary act of delivery of the bunkers by the charterers to the shipowners, and there was nothing to put the shipowners on enquiry as to the bunker suppliers' rights.

Arresting ship to recover price of bunkers

The statutory requirement is that the person be in control of the ship, not that the person be in control of the company which is the charterer of the ship.

Chimbusco Pan Nation Petro-Chemical Co Ltd v The Owners and/or Demise Charterers of the Ship or Vessel 'Decurion'[126]

The vessel was owned by Maruba. The plaintiffs supplied bunkers to *Decurion* and to 10 other vessels but did not receive payment. Maruba accepted that the plaintiffs had an in rem claim against *Decurion* for bunkers supplied to that ship, but not for bunkers supplied to the 10 other vessels.

The Court held that the judge did not err in finding that Maruba was not in control of the other 10 vessels. On the evidence as a whole, the plaintiffs fell short of proving that the defendants were 'in control of' the 10 other vessels.

Bunker suppliers supplying bunkers to time-charterers and asserting maritime lien over vessel

For Article 17 of the Brussels Convention to apply, the agreement as to exclusive jurisdiction had to be clearly and precisely demonstrated not just by looking at the words of the contract but at all the circumstances.

Andromeda Marine SA v O W Bunker & Trading A/S (The 'Mana')[127]

The defendant was a Danish company which, pursuant to a request from the American time-charterers, supplied bunkers to the vessel, owned by the claimants, a Panamanian company. The bunker supply contract contained an exclusive English jurisdiction clause. The time-charterers went into chapter 7 bankruptcy before making payment for the bunkers.

The Court held that the court did not have jurisdiction to hear Andromeda's claim. In any event, on the facts, OW Bunker had never

126. [2013] 2 Lloyd's Rep 407.
127. [2006] 2 Lloyd's Rep 319.

actually asserted that Andromeda was party to the bunker supply contract. OW Bunker had relied on a maritime lien.

Bunkers 'wasted' and connection with cargo

Clause 27(c)(ii) of Shelltime 4 form is concerned with what would be regarded in the marine market as 'cargo claims'. It refers only to claims of the sort which are normally brought by bill of lading holders claiming loss or damage arising in relation to the cargo and measured by reference to the cargo.

> *Borgship Tankers Inc. v Product Transport Corporation Ltd (The 'Casco')*[128]
> The claimants chartered the vessel from the defendant owners on the Shelltime 4 form, and sub-chartered her for a voyage. The sub-charterers subsequently cancelled the sub-charter on the ground that the vessel's cargo tanks were in a very poor condition.
>
> The Court held that in the present case the claim was not a cargo claim in the sense understood in the marine market. The charterers claimed damages for loss of use of the vessel.

Shipowner's lien on unpaid hire

Head charterers may sub-charter a vessel to sub-charterers who then may sub-charter her further to sub-sub-charterers. The shipowners only have a contractual relationship with the head charterers and would not have control over the sub-charterers or sub-sub-charterers despite the fact that the shipowners may have the right to a lien on the cargo and sub-freight of the head charterparty.

In *The Cebu* (No. 2) [1993] QB 1, clause of all three charterparties, ie, a head charterparty, a sub-charterparty and a sub-sub-charterparty provided that 'the shipowners shall have a lien on all cargoes and sub-freights for any amounts due under this charter . . .'. The issue for decision in the case was whether the shipowners effectively exercised rights under their contractual lien on sub-freight under clause of the head charterparty in respect of hire due under a sub-sub-time charter.

> *Care Shipping Corporation v Itex Itagrani Export SA (The 'Cebu') (No. 2)* [1993] QB 1
> By three time charters, the Cebu was chartered, sub-chartered and sub-sub-chartered. The head charterer owed the shipowners at least one month's hire. The sub-sub-charterers paid the hire instalment due

128. [2005] 1 Lloyd's Rep 565.

> under their charter to the sub-charterers. The shipowners claimed that there had been an equitable assignment and that the sub-sub-charterers had paid the moneys to the sub-charterers at their peril.
>
> The court held that the contractual lien on sub-freights conferred by Clause 18 was an equitable assignment enforceable by the shipowners, but that in the shipping trade the contemporary use of 'freight' was restricted to a voyage charter and bills of lading, and 'hire' to a time charter.

When the shipowners in *The 'Cebu' (No. 2)* chartered the vessel on the New York Produce Exchange form, it was distinguished between freight and hire and the lien on freight in clause was restricted to a voyage charter. The shipowners had no lien enforceable against the sub-sub-charterers for hire payments due to the sub-charterers. Wording which may create ambiguity on a charterparty should be avoided in the early stages of negotiation.

Owner has right to suspend services pending non-payment of hire

It cannot be said that a charterparty cannot properly work without a five banking days' notice being given since the charterers are already on notice on the charterparty that non-payment entitles the owners to suspend performance there and then.

> *Greatship (India) Ltd v Oceanografia SA De CV (The 'Greatship Dhriti')*[129]
> By a charterparty on an amended Supply time 1989 form the owners agreed to charter their vessel to the charterers for two years. The owners purported to suspend the provision of the services of the vessel for non-payment of hire.
>
> The Court held that the owners were not required to give the charterers five banking days' notice before exercising their right of suspension.

On-off hire survey

NYPE 93, Clause 3 'On-Off Hire Survey' states that '[p]rior to delivery and redelivery the parties shall, unless otherwise agreed, each appoint surveyors, for their respective accounts . . . [and] conduct joint on-hire/off-hire surveys'.

129. [2013] 2 Lloyd's Rep 359.

Daebo Shipping Co Ltd v The Ship 'Go Star'[130]

The vessel was chartered under a string of time charters on the 1981 NYPE. Daebo chartered the vessel to Nanyuan sub-charter. Daebo issued an invoice to Nanyuan for the first hire payment and for the value of the bunkers. Nanyuan purported to cancel or withdraw from the Nanyuan sub-charter.

The Court held that delivery also marked the time of commencement of the charterer's obligation to pay hire. It was not necessary that the charterer performed some act so as to 'take over' the bunkers. The expression 'take over' in the NYPE form of charter signified a change of ownership consequent upon and concurrent with delivery of the ship to the charterer.

Title to claim, shipowners or time charterers?

The owners of the vessel 'Ibaraki Maru' in *The Ibaraki Maru* ([1986] AC 1) were also the time charterers and they chose to pursue the claim in the name of the time charterers. Under the bareboat charter, the bareboat charterer was liable for the cost of repairs occasioned by a collision, and under the time charter the daily hire payable by the time charterer to the bareboat charterer was reduced while the vessel was undergoing repairs.

Candlewood Navigation Corporation Ltd v Mitsui O.S.K. Lines Ltd (The 'Ibaraki Maru') [1986] AC 1, PC

The shipowner let the vessel to the bareboat charterer and by a time charter of the same date the bareboat charterer let it back to the shipowner. Another vessel owned by the defendant negligently collided with it, causing damage. Temporary repairs to the vessel were delayed by a union ban. The final repairs were effected in Japan.

The court held that the first plaintiff, suing as time charterer and not as shipowner of the damaged vessel, was not entitled to recover from the defendant the hire paid to the bareboat charterer and loss of profits while the vessel was not operational.

In *The Ibaraki Maru*, it was held to be a principle of common law that if a wrong was done to a chattel, a person who merely had a contractual right in relation to the chattel and not a proprietary or possessory right could not bring an action against the wrongdoer for injury to their contractual right. The principle had been applied so that it had become well established that a time charterer could not recover damages for pecuniary loss caused by damage to the chartered vessel by a third party.

130. [2013] 1 Lloyd's Rep 18, Federal Court of Australia.

Seaworthiness

It is not easy to find a short and precise statement reflecting the exact meaning of seaworthiness. One definition which was offered in *Dixon v Sadler* (1839) 5 M & W 405, 151 ER 172 is also used in the Marine Insurance Ordinance. The Marine Insurance Ordinance s 39(4) states: 'A ship is deemed to be seaworthy when she is reasonably fit in all respects to encounter the ordinary perils of the seas of the adventure insured'.

There is an objective test sometimes called the 'Prudent Shipowner's Test' in connection to seaworthiness. In *McFadden v Blue Star Line* [1905] 1 KB 697, it was stated on p 706 that 'to be seaworthy, a vessel must have that degree of fitness which an ordinary, careful and prudent shipowner would require his vessel to have at the commencement of the voyage, having regard to all the probable circumstances of it'.

> *McFadden v Blue Star Line* [1905] 1 KB 697
> In the course of loading of a cargo of cotton, a ballast tank was filled with seawater, the defective sea-cock allowed water to flow in, the water pressure in the ballast tank forced a defective valve chest, and water flowed through the valve chest and then a sluice door and into the cargo hold, and the cotton was thereby damaged.
>
> The court held that the defective packing of the valve chest, being an existing defect at the time of the loading of the goods, was a breach of warranty.

Undertaking of seaworthiness

In *The Hongkong Fir* [1961] 2 Lloyd's Rep 478, the delay which had occurred owing to the incompetence of the engine-room staff and the effort of the shipowners to remedy matters were not such as to deprive the charterers of substantially the whole of the benefit of the use of the vessel under the charterparty.

> *Hongkong Fir Shipping Company, Ltd v Kawasaki Kisen Kaisha, Ltd (The 'Hongkong Fir')* [1961] 2 Lloyd's Rep 478, CA
> The charterers' time chartered the vessel on the Baltime form for 24 months. While on her voyage from Liverpool to Osaka, the vessel was at sea for 8½ weeks and off-hired due to repairs which took 5 weeks. At Osaka, it took another 15 weeks for repairs to make her seaworthy. Subsequently, repudiation of the charter was made by the charterers.
>
> The court held that in the circumstances of this case the breach of the unseaworthiness obligation was not such as entitled charterers to accept it as a repudiation and to withdraw from the charter.

Seaworthy at the beginning of a voyage

Shipowners are normally required to agree on a time charterparty that 'the vessel [is] tight, staunch and strong and in every way fit for the voyage', and 'to exercise due diligence to make the ship seaworthy before and at the beginning of the voyage'. In *The Fjord Wind* [2000] 2 Lloyd's Rep 191, the charter provided *inter alia*: 'It is this day mutually agreed . . . that the said vessel being tight, staunch and strong and in every way fit for the voyage, shall with all convenient speed proceed to [the river Plate] . . . Shipowners shall be bound before and at the beginning of the voyage to exercise due diligence to make the ship seaworthy'.

> *Eridania S.p.A. v Rudolf A Oetker (The 'Fjord Wind')* [2000] 2 Lloyd's Rep 191, CA
>
> The vessel owners let her to the first defendant disponent owners under a time charter on the New York Produce Exchange form. The disponent owners entered into a voyage charter with the first plaintiff. The vessel had problems in the main engine and the estimate for the repairs rose to 95 days. The cargo was transhipped into a substitute vessel.
>
> The court held that the vessel was unseaworthy when she left Rosario because there was present a defect which meant that it could not operate on an ordinary voyage and since it failed on the voyage down river. The consequences were extremely foreseeable. If the owners had known that such a defect was present the bearing would fail in those circumstances, they would have rectified the defect.

In *The Fjord Wind*, clause 35 expressly applied 'before and at the beginning of the voyage', which must include the loading process. The shipowners had to exercise due diligence to make her seaworthy for the loading process and thereafter they were to exercise due diligence to make her seaworthy for the cargo-carrying voyage itself. The expression 'before and at the beginning of the voyage' was apt to include the whole period before the beginning of the voyage. The disponent owners' obligations as to seaworthiness at each stage was the same, namely to exercise due diligence to make the vessel seaworthy.

Seaworthiness and deck cargo

Cargo is stowed on deck in some particular trades, eg, timber products. Bills of lading which incorporate the Hague Rules or the Hague Visby Rules are free to exclude the carriers' liability, including unseaworthiness of the vessel, under the rules in respect of cargo carried on deck.

In *The Imvros* [1999] 1 Lloyd's Rep 848, by a charter-party on the New York Produce Exchange form (as amended), the shipowners let their vessel to the charterers for a one-time charter trip, ie, minimum duration 70 days. The charter provided, *inter alia*, clause 8: 'Charterers are to load, stow . . . lash . . . the cargo at their expense under the supervision of the Captain . . . ' Additional clause 48: 'Lashing . . . although done by crew, crew to be considered as Charterers' servants'. Additional clause 62: Bills of Lading: '. . . (c) In the event that cargo is shipped on deck . . . Charterers are to ensure that . . . Bills of Lading are claused as follows: Carried on deck at Shippers' risk without responsibility for loss or damage however caused.' Additional clause 91: 'Deck cargo . . . Charterers are permitted to load cargo on the vessel's deck . . . provided always that the permissible loads . . . are not exceeded . . . The vessel is not to be held responsible for any loss of or damage to the cargo carried on deck whatsoever and howsoever caused.'

> *Transocean Liners Reederei GmbH v Euxine Shipping Co Ltd (The 'Imvros')* [1999] 1 Lloyd's Rep 848
>
> As instructed by the charterers, the vessel proceeded to Brazil and loaded a cargo of sawn timber in bundles both above and below deck. The loading and stowage of the cargo was carried out under the directions of the charterers' supercargo and the lashing of the cargo was carried out by the crew. The bills of lading were not claused as required. The lashings for deck cargoes were in contravention of the IMO Code of Practice for Ships Carrying Timber Deck Cargoes. In the voyage to Durban, part of the deck cargo was lost and the vessel damaged.
>
> The court held that for cargo carried on deck the shipowners had no responsibility for loss or damage however caused. The exclusion covered any cause and there was no justification for excluding unseaworthiness as a cause.

As to the express absolute obligations of seaworthiness, in *The Imvros*, the obligation to load and lash was expressly placed on the charterers by clauses 48, 87 and 91 and both the latter clauses referred in terms to that obligation extending to the seaworthiness of the vessel. The references to loading and lashing under the supervision of the captain and to the satisfaction of the master and to the entire satisfaction of the master in the clauses were not expressed as qualifications on the obligations of the charterers but in the language of a right to be satisfied or to supervise its performance. A right to intervene did not normally carry with it a liability for failure to do so, let alone relieve the actor from liability.

On its true construction, clause 13(b) of the Hague-Visby Rules only covers loss, damage or liability effectively caused by the carriage of deck cargo and does not cover loss, damage or liability caused by negligence or unseaworthiness.

> *Onego Shipping & Chartering BV v JSC Arcadia Shipping (The 'Socol 3')*[131]
> The vessel was chartered on the NYPE form as amended. The charterparty incorporated the Hague-Visby Rules (the Rules). Whilst en route to Alexandria the deck cargo shifted in high seas and some of it was lost overboard. The vessel proceeded to Halmstad as a port of refuge.
>
> The Court held that the legal consequence of the tribunal's findings was that the casualty was effectively caused by the owners' negligence and by unseaworthiness for which the owners were contractually responsible.

Seaworthiness and dangerous cargo

An operative fault does not lie in the stowage of a container which contains dangerous cargo, but may be in the negligence of crew in using and heating the relevant bunker tank.

> *Compania Sud Americana De Vapores SA v Sinochem Tianjin Import and Export Corporation (The 'Aconcagua')*[132]
> A cargo of calcium hypochlorite stowed in a container was loaded. While in the course to San Antonio, the calcium hypochlorite self-ignited and exploded. The bunker tank in No 3 hold had been heated during the voyage in order to allow a transfer of bunkers for fuel oil.
>
> The Court held that the heating of the bunker tank did not constitute or result from unseaworthiness. The vessel could not be treated as unseaworthy at the commencement of the voyage unless the heating of the bunker tank was bound to occur which, on the facts of the present case, it was not. Whether or not that particular bunker tank was used would depend on an operational decision made during the voyage.

There is a clear distinction between the owners' entitlement to supervise, on the one hand, and a duty to do so—owed to the charterers—on the other. An owner may be liable under a bill of lading to a cargo owner if the stowage is such as to render the vessel unseaworthy and the owners are guilty of a lack of due diligence in looking after the vessel and the goods.

131. [2010] 2 Lloyd's Rep 221.
132. [2010] 1 Lloyd's Rep 1.

Compania Sud American Vapores v MS ER Hamburg Schiffahrtsgesellschaft MBH & Co KG[133]

Clause 8 of the charter-party on amended NYPE form provided that the charterers were 'to load, stow, lash, secure, unlash, trim and discharge and tally the cargo at their expense under the supervision of the Captain' and clause 24 expressly incorporated the Hague-Visby Rules. Following an explosion on board the vessel, the owners brought a claim against the charterers.

The Court held that the question was not whether the owners were under a duty to intervene in the loading process, but rather whether they owed that duty to the charterers. The effect of the unamended clause 8 of the NYPE form is to transfer that responsibility to the charterers from whom the owners are entitled to an indemnity. There is no authority which supports a charterers' argument that the owners have a duty to the charterers to intervene to avoid unseaworthiness. The appeal was dismissed.

Seaworthiness and Inter-Club Agreement

The Inter-Club Agreement provides that claims for loss of or damage to cargo due to unseaworthiness should be apportioned 100 per cent to owners, and that short delivery claims are to be apportioned 50 per cent to owners and 50 per cent to charterers, save where there is clear and irrefutable evidence that the shortage is due to act, neglect or default on the part of owners' or charterers' servants or agents, in which case the party whose servants or agents are at fault should bear the claim in full.

Kamilla Hans-Peter Eckhoff KG v A C Oersslef's Eftf A/B (The 'Kamilla')[134]

The vessel was chartered on an amended NYPE form which incorporated the NYPE Inter-Club Agreement. The vessel was unseaworthy in that the No 2 hold hatch covers were not completely watertight, and a small amount of seawater entered the hold and wetted the cargo.

The Court held that the appeal would be dismissed. The test for causation was whether the act or default complained of was a proximate cause of the alleged damage. The arbitrators found that the admitted unseaworthiness and the decision of the DCP to prohibit the import of the cargo were not mere coincidences, and that the unseaworthiness was an effective cause of the whole loss. The arbitrators were correct in holding that the damages claimed were 'due to unseaworthiness' within the meaning of the Inter-Club Agreement.

133. [2006] 2 Lloyd's Rep 66.
134. [2006] 2 Lloyd's Rep 238.

Seaworthiness and due diligence

The exercise of due diligence is equivalent to the exercise of reasonable care and skill. Lack of due diligence is negligence.

> *Papera Traders Co Ltd and Others v Hyundai Merchant Marine Co Ltd and Another (The 'Eurasian Dream')*[135]
> The Court held that on the facts and the evidence the claimants had proved that the loss and damage was caused by the unseaworthiness of the vessel; the fire would not have broken out if the master and crew had been properly instructed and trained.
>
> The defendants failed to prove that they and those for whom they were responsible exercised due diligence to make the ship seaworthy, the defendants as bill of lading carriers were liable for want of due diligence by the owners/manager and for want of due diligence of the master in so far as the defendants or the owners or managers delegated to him their duties as to seaworthiness. There would be judgment for the claimants.

Seaworthiness and immunity in carriage of goods by sea

The Carriage of Goods by Sea Ordinance is usually incorporated into the bill of lading and bill of lading holders may claim against the carrier for loss of or damage to cargo contending that the carrier could not rely on the immunity of the Carriage of Goods by Sea Ordinance on the ground of unseaworthiness.

> *Leesh River Tea Co Ltd v British India Steam Navigation Co Ltd* [1967] 2 QB 250, CA
> The plaintiffs consigned chests of tea, which were loaded on board the ship. The defendants employed stevedores for the cargo work at Port Sudan. One or more of the stevedores removed and stole a small brass plate which was the cover plate of a storm valve in the area of the No. 2 hold. The vessel was unseaworthy from the time it left Port Sudan and remained unseaworthy for the remainder of the voyage. The vessel encountered heavy weather, seawater entered the hold through the storm valve, and the tea was damaged. Neither the ship's officers nor the crew could reasonably have detected the absence of the cover plate.
>
> The court held that the defendants were not entitled to rely on the immunity in Article IV (1) of the Schedule to the Carriage of Goods by Sea Act 1924, exempting them from loss or damage arising from unseaworthiness, because that immunity was confined to unseaworthiness

135. [2002] 1 Lloyd's Rep 719.

> arising before or at the beginning of the voyage and did not cover unseaworthiness arising, as here, in the course of the voyage.

In *Leesh River Tea Co Ltd v British India Steam Navigation Co Ltd*, it was held that even if the stevedores could be regarded as the shipowners' servants for the purposes of Article IV (2) (a), the theft of the cover plate was not an act, neglect or default in the management of the ship. The stevedore or stevedores who stole the plate were not agents or servants of the shipowners when they stole the plate, since the theft of the plate (part of the ship) was in no way incidental to or a hazard of the process of discharge and loading of the cargo which they were employed to handle, but was an act quite unconnected with the cargo and outside the course of their employment. Therefore, the defendants were not liable for the damage to the tea.

Safe port

Time charterers are required by the terms of the charterparty to nominate only safe ports for the vessel under the charter in respect of the operations of the vessel, eg, loading, unloading, bunkering, change of crew, waiting for instructions, etc. Lord Justice Sellers in *The Eastern City* [1958] 2 Lloyd's Rep 127 stated that '[i]f it were said that a port will not be safe unless, in the relevant period of time, the particular ship can reach it, use it and return from it without, in the absence of some abnormal occurrence, being exposed to danger which cannot be avoided by good navigation and seamanship, it would probably meet all circumstances as a broad statement of the law'.

> *Leeds Shipping Company, Ltd v Société Française Bunge (The 'Eastern City')* [1958] 2 Lloyd's Rep 127
>
> In a voyage charterparty the charterer directed a vessel to Mogador. Dragging of anchor was suspected due to a deterioration in the weather conditions. The vessel was grounded in squall while attempting to leave port. There was a lack of shelter and a liability to the onset of a high wind which could not be predicted. The port was very near some rocks and the anchorage was very restricted.
>
> The court held that the grounding could not be attributed to any negligence in the navigation or handling of the Eastern City. The stranding was the result of the breach of charterparty in nominating an unsafe port.

In *The Eastern City*, the vital factors of unsafety at Mogador were the lack of reliable holding ground in the anchorage area, the lack of shelter, the liability to the sudden onset of high wind which could not be predicted and which might quickly cause an anchor to drag, and, in the circumstances, the restricted area of the anchorage and its close proximity to the rocks and shallows with a high wind or gale from the south or somewhere to the west of south.

There is no inherent inconsistency between a safe port warranty and a named loading or discharging port. Effect has to be given to all the terms of the charter which are not inconsistent. The identification of a named port or anchorage, thereby limiting the charterer's choice as to the location of performance, is not inconsistent with a warranty that it is safe.

> *STX Pan Ocean Co Ltd v Ugland Bulk Transport AS (The 'Livanita')*[136]
> The vessel was chartered on a modified NYPE form for one time charter trip. She was assisted by ice breakers to the outbound convoy area. During the outbound convoy her hull was damaged by ice.
>
> The Court held that there was no evidence that either party knew or ought reasonably to have anticipated anything about the likely conditions at St Petersburg more than the other, nor that either knew or should reasonably have known that St Petersburg was unsafe at the time the charter was entered into, nor that it was unsafe as sought to be alleged.

Safe port and delay

In *SS. Knutsford, Limited v Tillmanns & Co*, it was held that 'inaccessible' and 'unsafe' in the bills of lading must be read reasonably and with a view to all the circumstances. The port must be inaccessible in a commercial sense so that a ship cannot enter without inordinate delay. There is no ground for saying that a delay of three days on a long journey could be regarded as inordinate delay.

> *SS. Knutsford, Limited v Tillmanns & Co* [1908] AC 406, HL
> The vessel went from Middlesbrough to Japan and then Vladivostock. When she arrived within forty miles of Vladivostock she found she could not get into the port by reason of ice. The vessel tried for three days in vain to get through the ice and then went back to Nagasaki and discharged her cargo. The day after she turned back the ice cleared off, the access was safe, and other vessels entered.
>
> The court held that the master was not justified under all the circumstances in this case in failing to deliver the goods at the port for which they were shipped merely because that port was at the moment of their arrival inaccessible on account of ice for three days only.

136. [2008] 1 Lloyd's Rep 86.

In *The Hermine*, it was held that a shipowner cannot rescind a charter-party merely because there has been a commercially unacceptable delay, ie, delay exceeding a reasonable time. Before the shipowner can seek to treat the charterer's conduct as a repudiation of the charterer's obligation to load, the delay in such a case has to be such as would frustrate the adventure.

> *Unitramp v Garnac Grain Co Inc (The 'Hermine')* [1979] 1 Lloyd's Rep 212, CA
>
> A vessel was delayed as she was leaving Destrehan, which was about 140 miles from the open sea on the Mississippi. The factors causing the delay were the lowering of the river draft because of the continuing accretion of silt from floods, and the congestion due to severe fog in the port and the grounding of two vessels off the channel and at the entrance to the pass respectively. The vessel was delayed for 37 days but sustained no physical damage.
>
> The court held that the port was still a safe port. An obstruction which merely caused delay did not render a port unsafe unless the delay was sufficient to frustrate the adventure.

A port is not safe if the vessel is exposed to a danger which cannot be avoided by good navigation and seamanship.

> *Gard Marine & Energy Ltd v China National Chartering Co Ltd (The 'Ocean Victory')*[137]
>
> OLH time-chartered the vessel to Sinochart on an amended NYPE form. Sinochart in turn sub-chartered the vessel to Daiichi for a time charter trip on an amended NYPE form. The vessel arrived off Kashima. At that time the weather was recorded as no more than force 3 or 4. Thereafter, the weather began to deteriorate. Long waves tended to cause greater ranging or surging of a vessel moored alongside a berth than swell waves.
>
> The Court held that the casualty was caused by the unsafety of the port in breach of the safe port warranty.

When the charterers order the vessel under charter to proceed to a port, the port should be prospectively safe in respect of the vessel. In *The Evia* [1983] 1 AC 736, clause 2 of the charterparty provided as follows: 'the vessel to be employed . . . between good and safe ports'.

> *Kodros Shipping Corporation of Monrovia v Empresa Cubana de Fletes (The 'Evia')* [1983] 1 AC 736, HL
>
> By a time charterparty in the Baltime form, the shipowners chartered the Evia to the charterers for a period. In March 1980 the charterers

137. [2014] 1 Lloyd's Rep 59.

> ordered the Evia to load a cargo for Basrah. She arrived on 1 July, berthed on 20 August and completed discharge on 22 September. By that date, large-scale hostilities had broken out between Iran and Iraq. From September 22 onwards no ship was able to leave the waterway.
>
> The court held that since Basrah had been prospectively safe at the time of nomination by the charterers and the unsafety had arisen after her arrival due to an unexpected and abnormal event, there had been no breach by the charterers of Clause 2 at the time of nomination.

In *The Evia*, the charterers' contractual promise to the charterparty was that at the time when the order was given for the vessel to go to a particular port or place, that port or place was prospectively safe for the vessel to get to, stay at so far as necessary, and in due course leave. If while the vessel was in such a port or place some unexpected or abnormal event suddenly occurred that made it unsafe, the charterers' contractual promise did not extend to making the charterers liable for any resulting loss or damage.

Safe port and war zone

The situation becomes complex if a vessel is ordered by the charterers to proceed to a port which is close to an area of hostility.

In *The Chemical Venture*, the approach to the port necessarily involved passage through a narrow and unsafe channel. The charterers did not exercise due diligence and were in breach of their obligation under clause 3 of the charterparty.

> *Pearl Carriers Inc v Japan Line Ltd (The 'Chemical Venture')* [1993] 1 Lloyd's Rep 508
>
> Under a time charter, the vessel was sub-chartered for a single voyage loading at one or two safe port(s) in the Arabian Gulf excluding Iran/Iraq. The charterers advised the master to proceed to the Persian Gulf for loading. While in the channel the vessel was struck by a missile fired from an Iranian jet and severely damaged.
>
> The court held that by construing the telex exchanges objectively, it was clear that the shipowners had made an unequivocal representation. They either waived their right to contend that the charterers were in breach or they were estopped from so contending.

In *The Product Star (No. 2)* [1993] 1 Lloyd's Rep 397, several factors were relevant in the appraisal of the validity of the shipowners' refusal to proceed to Ruwais. The shipowners made no attempt to consult the master, the refusal was made at short notice, no attack had been made at any material time on a ship trading to or from the UAE and no explanation was offered why the managers of the *Product Star* traded another vessel, the *East Star*, in the Gulf between September 15 and 25, 1987. These factors undermined if not invalidated the shipowners' reasons for declining to proceed and called in question the shipowners' good faith and in any event strongly suggested that their refusal was arbitrary.

> *Abu Dhabi National Tanker Co v Product Star Shipping Ltd (The 'Product Star') (No. 2)* [1993] 1 Lloyd's Rep 397, CA
>
> The time charterers instructed the master to proceed to Ruwais to load. Without previous warning the shipowners by a telex asked the charterers to nominate another load port. The master advised the charterers that the vessel was drifting until further instructions. The charterers purported to accept the shipowners' conduct as a repudiation of the charter and indicated an intention to claim damages. On September 29 the shipowners purported to accept the charterers' conduct as a repudiation of the charter.
>
> The court held that there was no material on which a reasonable shipowner could reasonably have considered that the risk of proceeding to Ruwais in September 1987 was a different risk from that which already existed at the date of the charter.

As to the telex message sent by the shipowners, in which they indicated their refusal to proceed, the court held that it exhibited an intention no longer to be bound by the charter and constituted a plain and deliberate abrogation of it.

Ice-bound port

Only some ships are built to ice classes, with additional scantling enabling them to navigate through different concentrations of ice at seas. The structure of a ship which is not built to ice class may be severely damaged if she forces her way through ice. Shipowners, therefore, include an Ice Clause into the charterparty, stating what the shipowners and master can do in situations of ice.

Where the shipowners want to exercise the discretion of refusing to allow the vessel to force ice as mentioned on the charterparty, they have to be aware of the situation of the port nominated by the charterers. The master has to report the weather conditions including the ice situation to the shipowners before the shipowners can decide whether to follow the instructions of the charterers or refuse them.

In *Limerick Steamship Company, Limited v W. H. Stott and Company, Limited* [1921] 2 KB 613, by clause 16 of the charterparty the ship was not to be ordered to any ice-bound port nor was she obliged to force ice, and the charterers were to be liable for the detention of the vessel from any of the causes set out in that clause.

> *Limerick Steamship Company, Limited v W. H. Stott and Company, Limited* [1921] 2 KB 613, CA
>
> A steamer was chartered for one Baltic round voyage in the wintertime. The charterers ordered her to Abo, a port in Finland. That port would normally be frozen up so that vessels could not enter or leave, but it is kept open all the year by means of icebreakers. The vessel on her voyage to and from Abo encountered ice; sometimes the ice was thin enough for the vessel to force her way through without the help of icebreakers, but at others she failed to force her way through the ice. Eventually with the help of an icebreaker she reached Abo. She did however sustain considerable damage from the ice.
>
> The court held that Abo was not an ice-bound port within the meaning of the charterparty as it was a port which, owing to the use of icebreakers, vessels could enter and leave at any time of the year. Therefore the charterers had not committed a breach of their charterparty in ordering the vessel to that port.

Where a clause of the charterparty allows the vessel not to force ice, it is at her own risk if she does force ice.

Loading and discharge of cargo

A charterparty may provide that the captain, although appointed by the shipowners, must be under the orders and direction of the charterers as regards to loading, stowing and trimming the cargo at their expense, under the supervision of the captain.

> *Canadian Transport Company, Limited v Court Line, Limited (The 'Ovington Court')* [1940] AC 934, HL
>
> Under the time charter the vessel was chartered for a voyage from Rotterdam to the North Pacific. The ship apparently proceeded to Vancouver and there loaded a cargo. At the port of discharge a claim was made by the holders of bills of lading of wheat in bulk against the shipowners for damage to the goods. The damage was due to improper stowage. By the terms of the charterparty the charterers were to 'load, stow and trim the cargo at their expense under the supervision of the captain'.
>
> The court held that the reservation of the captain's right to supervise the stowage had not had the effect of relieving the charterers of their primary duty to stow safely.

Following *The Ovington Court*, where the master exercises supervision and limits the charterers' control of the stowage, the charterers' liability will be limited to a corresponding degree.

Instruction of cargo receiver

In *The Goodpal* [2000] 1 Lloyd's Rep 638, the receivers at the first port of discharge stood in the charterers' position only to the extent that they were responsible for carrying out the discharging operation at the first port in respect of the specified amount of cargo and no more. Once that had been done they were no longer the delegates of the charterers and there were no further orders they could properly give to the master.

> *Merit Shipping Co Inc v T.K. Boesen A/S (The 'Goodpal')* [2000] 1 Lloyd's Rep 638
>
> By a charterparty the shipowners let their vessel to the charterers, who sublet the vessel for a time charter trip. At the first port of discharge the receivers requested the master to discharge some 450 tonnes in excess of the charterers' instructions. The master agreed to this against a letter of indemnity of the receivers.
>
> When the vessel arrived at Fangcheng she could only discharge 13,451.50 tonnes as against 13,993.90 tonnes on the bill of lading. The consignees duly arrested the vessel.
>
> The court held that the receivers at the first port had instructed the master to allocate to them part of the cargo which the shipowners were liable to deliver to other receivers at the second port. The charterers did not at any time vary their original instructions at the second port and they could not have been entitled to do so.

Loading and stevedore damage

Charterers' obligation to load, stow and trim the cargo and discharge, requires them to do so with due care and the primary responsibility for stowage is, however, imposed on the master.

> *MSC Mediterranean Shipping CoSA v Alianca Bay Shipping Co Ltd (The 'Argonaut')*[138]
>
> By a charter-party dated the owners let their vessel to the charterers for one time charter trip in the NYPE form. In course of unloading several places on board were damaged and/or pierced by falling granite blocks.

138. [1985] 2 Lloyd's Rep 216.

The Court held that the effect of cl 8 was to confer the primary duty on the owners and in the absence of actual intervention by the charterers, as distinct from stevedores employed by them, the owners' liability would not be avoided; the award in relation to the damage done at Marina di Carrara would be upheld but in relation to the damage caused at Sete there was no warrant for exonerating the master from 'responsibility' for the damage caused merely because the stevedores seemed to be competent and the master had not yet realized that steel plates should have been placed in the holds as a precautionary measure; there was no such officious intervention by the charterers themselves as would relieve the owners from liability; the charterers' appeal would be allowed and the owners' appeal dismissed.

Dangerous cargo

Injurious or dangerous cargo clauses are common in time charterparties and charterers agree that no cargo injurious to the vessel will be loaded on board the vessel under charter. In *The Athanasia Comninos*, Mustill J said at p. 282: 'It has been established for more than a century that a shipper, party to a contract of carriage, is under certain contractual obligations as to the suitability for carriage of the goods which he ships, and as to the giving of warnings concerning any dangerous characteristics of the goods: *Williams v East India Company* (1802) 3 East 192; *Brass v Maitland* (1856) 6 E & B 470. These obligations are not confined to cases where the goods are tendered to a common carrier, but are capable of applying, in appropriate circumstances, to all contracts for the carriage of goods by sea.'

The Athanasia Comninos [1990] 1 Lloyd's Rep 277

The Athanasia Comninos was under a time charter to the Atlantic and Great Lakes. Explosions occurred on board the vessels while she was in the course of carrying coal from Sydney, Nova Scotia to Birkenhead. The explosion arose from the ignition of a mixture formed between air and a quantity of methane gas which had been emitted from the coal after loading. Very substantial damage was caused to the forward part of the vessel and four seamen who were present in the forecastle at the time suffered serious personal injuries.

The court held that on the evidence given at the trial the cause of the explosion was unknown. The shipowners' claim failed as against all three defendants and was dismissed.

In *The Athanasia Comninos*, it was held impossible to say in the abstract that coal in general, or Devco coal in particular, is either dangerous or safe. The carriage of coal involves hazards greater than those associated with inert goods. However, they are hazards which could be overcome if the shipowner had the necessary knowledge, skill and equipment. Mustill J stipulated that 'it is not correct to start with an implied warranty as to the shipment of dangerous goods and try to force the facts within it; but rather to read the contract and the facts together, and ask whether, on the true construction of the contract, the risks involved in this particular shipment were risks which the plaintiffs contracted to bear'.

In seeking to determine the nature and character of goods declared as dangerous goods, it is necessary to take into account: (a) the hazard history of the goods and (b) the significance of the description of UN number in the IMDG Code; and (c) any other information of which a prudent carrier ought to have been aware.

> *Compania Sud Americana De Vapores SA v Sinochem Tianjin Import and Export Corporation (The 'Aconcagua')*[139]
>
> A cargo of calcium hypochlorite stowed in a container was loaded on board the vessel. While in the course to San Antonio, the calcium hypochlorite self-ignited and exploded. The bunker tank in No 3 hold had been heated during the voyage in order to allow a transfer of bunkers for fuel oil.
>
> The Court held that a prudent carrier was likely to have less knowledge about a product than a specialist manufacturer or distributor, although the owners of vessels specially constructed for the purpose of carrying particular cargoes, e.g., LPG, might have particular specialist knowledge.

Strict liability on dangerous cargo

In *The Giannis NK*, the word 'dangerous' had to be given a broad meaning. Dangerous goods were not confined to goods of an inflammable or explosive nature or their like. Goods might be dangerous within the meaning of Art IV, r 6 if they were dangerous to other goods even though they were not dangerous to the vessel itself.

> *Effort Shipping Co Ltd v Linden Management SA (The 'Giannis NK')* [1998] 1 Lloyd's Rep 337, HL
>
> The vessel loaded a cargo of groundnut extraction meal pellets at Dakar. Cargoes of bulk wheat pellets had been loaded into other holds at previous loading ports. She arrived in the Dominican Republic and live insects and shed skins were found in the cargo. The vessel was ordered to leave the port after the vessel was fumigated twice.

139. [2010] 1 Lloyd's Rep 1.

> The vessel sailed back to San Juan. The United States Department of Agriculture identified the specimens as Khapra beetle. The vessel proceeded out to sea and jettisoned both the groundnuts and the balance of the wheat still on board.

The court held that in the absence of informed consent on behalf of the carriers (the shipowners) to the dangerous shipment, the shippers were prima facie liable for all damages and expenses directly or indirectly arising out of or resulting from the shipment of dangerous goods. In *The Giannis NK*, it was held that Art IV, r 6 imposed strict liability on shippers in relation to the shipment of dangerous goods, irrespective of fault or neglect on their part.

Packaging of dangerous cargo

The charterparty in *The Marie H* [1998] 2 Lloyd's Rep 71 permitted the loading of dangerous cargo provided it was packed in accordance with IMO regulations; it also permitted stowage of such cargo on deck. The charter contained the usual terms that the '. . . Captain shall be under the orders and directions of the charterer as regards employment and agency'. This necessarily carried with it the term implied by law that the shipowners were entitled to an indemnity against loss or damage arising in consequence of complying with the charterers' orders and directions.

> *Deutsche Ost-Afrika-Linie GmbH v Legent Maritime Co Ltd (The 'Marie H')* [1998] 2 Lloyd's Rep 71
> The shipowners let their vessel to the charterers under a charterparty in the New York Produce Exchange form for a time charter trip. The vessel loaded one crate of IMO class 1.1 explosive cargo. In the Bay of Biscay the vessel ran into a severe tropical storm. Both the vessel and the deck cargo suffered severe and extensive damage and the vessel had to put into Lisbon as a port of refuge for repairs and inspection and repacking of the cargo. Further repairs were carried out at Durban. All this involved delay to the time charter trip.
>
> The court held that on reading the award as a whole, the arbitrator was proceeding correctly on the basis that the time and expenses that he allowed to the shipowners were caused by the nature of the cargo, namely explosives, rather than the heavy weather.

Employment and navigation of vessel

The 'employment' referred to in the employment and indemnity clause of a time charterparty is 'employment of the ship'.

Larrinaga Steamship Company, Ltd v The Crown (The 'Ramon de Larrinaga') [1945] 78 Ll L Rep 167, HL

The vessel was chartered by the Ministry of Shipping. While loading at Newport notice was given that after the next voyage, which was to be to St. Nazaire, the vessel would return to Cardiff for a joint survey. The weather was bad and the vessel grounded and was damaged.

Lord Wright at p. 173 said that these sailing orders were merely dealing with matters of navigation in regard to carrying out the orders to proceed to Cardiff. It was the duty of the master to exercise his judgment in such matters of navigation.

In *The Hill Harmony,* a desire to avoid heavy weather was not justification in respect of a vessel which was fit to sail the ordered route. The choice of route was in the absence of some overriding factor a matter of the employment of the vessel rather than navigation.

Whistler International Ltd v Kawasaki Kisen Kaisha Ltd (The 'Hill Harmony') [2001] 1 AC 638, HL

By a time charterparty the disponent shipowners chartered their vessel to the charterers for a period. The master, who had experienced very bad weather which caused some damage to the ship on a previous voyage, refused to take the shorter northern great circle route recommended by a weather routing service. He instead took the longer more southerly rhumb line route. The voyages took an extra 6.556 and 3.341 days respectively. The charterers deducted hire in respect of the additional days at sea and the cost of the extra bunkers consumed.

The court held that the shipowners were in breach of their obligation to proceed with utmost dispatch since the master's refusal to comply with the charterers' order that the vessel proceed by the shortest route had been without good reason.

It was stated in *The Hill Harmony* that the contractual obligation of a shipowner under a time charter to proceed with the utmost dispatch would, both as a matter of mercantile policy and of the use of English, not have been met if the master unnecessarily chose a longer route.

Employment and agency clause

The only relevant instructions were received from the receivers at the first port of discharge who could not reasonably be thought to be standing in the position of the charterers in relation to the balance of the cargo to be discharged at the second port. See *The 'Goodpal'* below.

Merit Shipping Co Inc. v T.K. Boesen A/S (The 'Goodpal')[140]

By a charterparty in the NYPE form the owners let their vessel to the charterers who sub-chartered to Finagrain for a time charter trip. At the first port of discharge the receivers requested the master to discharge a total of 19,445 tonnes, some 450 tonnes in excess of the charterers' instructions. The master agreed to this against a letter of indemnity of the receivers.

The Court held that the owners' decision to accede to the requests of the receivers at the first port of discharge which caused the under-delivery at the second port of discharge; that decision was an act, neglect or default of the owners for the purposes of the Inter-Club agreement. The appeal would be dismissed.

Owners' or charterers' bill of lading in a time charter

A time charterparty gives charterers the right to determine the form of the bills of lading and, if necessary to sign them on behalf of the owners. It contemplates that the time charterers bring bills of lading contracts into existence which bind the owners of the vessel and that is reflected in turn in the terms of the bills of lading; and the attestation clause contemplates that the master would sign on behalf of the owners.

Fetim B.V. and Others v Oceanspeed Shipping Ltd (The 'Flecha')[141]

The owners of the vessel, entered into a time charter with the charterers for a time charter trip. The vessel loaded a cargo of forest products at various Far Eastern ports. The bills of lading were all issued on the form of Continental Pacific Shipping.

The Court held that the bill of lading was on the form of Continental Pacific Shipping; it plainly contemplated that the contract would be a contract between the owner of the goods and the owner of the vessel; the forms of signature taken by themselves might well suggest that the time charterers were contracting as carriers; but if the contract was looked at as a whole and in its wider context it was plain that this was a case in which the contract as set out in the printed form was intended to be a contract between the owners of the vessel and the owners of the goods.

Since 1994, the practice of the market has been adopted by the ICC Uniform Customs and Practice for Documentary Credits. Art 23 indicates that banks do not in practice examine the contents of the terms

140. [2000] 1 Lloyd's Rep 638.
141. [1999] 1 Lloyd's Rep 612.

and conditions of carriage on the reverse of a bill of lading. Against such a commercial background it would create an unacceptable trap to allow the detailed conditions on the back of a bill of lading to prevail over an unequivocal statement of the identity of the carrier on the face of the bill.

> *Homburg Houtimport B.V. v Agrosin Private Ltd and Others (The 'Starsin')*[142]
> The defendants' vessel was on time charter to CPS. The bills of lading were all liner bills of lading on CPS form. The definition clause, cl 1 defined 'carrier' as the party on whose behalf the bill of lading was signed. All the bills were signed by a signing party 'As agents' for carrier CPS.
>
> The Court held that the bill contained or evidenced a contract made with CPS as carrier; CPS was the sole carrier under the bill of lading; the owners were not parties to the contract of carriage and were not liable under the bills of lading; these bills were charterer's bills.

Off hire

An off-hire clause gives charterers the right to stop the payment of hire when the vessel is not at their disposal. In *The Mareva A.S.* [1977] 1 Lloyd's Rep 368, Kerr J stated that '[t]he owners provide the ship and the crew to work her. So long as these are fully efficient and able to render to the charterers the service then required, hire is payable continuously. But if the ship is for any reason not in full working order to render the service then required from her, and the charterers suffer loss of time in consequence, then hire is not payable for the time so lost.'

> *Royal Greek Government v Minister of Transport (The 'Ilissos')* [1949] 82 Ll L Rep 196, CA
> The Minister of Transport hired the vessel on a time charter. The Royal Greek Government, the disponent owners, claimed hire while the ship was unable to proceed owing to the officers and crew refusing to sail except in convoy. The Minister of Transport counterclaimed for the return of sums paid on the ground that the vessel was off hire.
>
> The court held that the provision 'deficiency of men' meant 'numerical insufficiency' and was directed to putting the ship off hire when an adequate complement of officers and crew for working the ship was not available. There was no such 'deficiency of men' or

142. [2003] 1 Lloyd's Rep 571, HL.

'inability to get or to complete a crew'. The crew's wilful refusal to work was not an 'accident' and was liable for hire during the period of the delay.

In *The Ilissos*, Bucknill LJ said that '. . . in interpreting such a charter-party as this . . . the charterer will pay hire for the use of the ship unless he can bring himself within the exceptions. I think he must bring himself clearly within the exceptions. If there is a doubt as to what the words mean, then I think those words must be read in favour of the owners because the charterers are attempting to cut down the owners' right to hire.'

Off hire and order of port authority

In *The Jalagouri*, clause 72 covered detention or threatened detention for any reason and required the shipowners to provide security to prevent such detention under any circumstances. Until security was provided the vessel was not permitted to stay at her berth. Service required the vessel to be at her discharging berth to enable the charterers to discharge the whole of her cargo. Accordingly, clause 72 required the shipowners to provide security and the vessel was off hire under clause 53 until this was done.

Nippon Yusen Kaisha Ltd v Scindia Steam Navigation Co Ltd (The 'Jalagouri') [2000] 1 Lloyd's Rep 515

By a time charter, the shipowners let their vessel to the charterers. The vessel collided with a breakwater and there was an ingress of water into hold 3. At discharge at India the cargo in this hold was found to be damaged. Following discharge of the sound cargo the port authorities ordered the vessel off her berth and would not allow her to complete her discharge without a guarantee for the costs of storing the damaged parts or of clearing them from the port area. The vessel was allowed to reberth when a guarantee was provided.

The court held that the vessel was detained within the meaning of Clause 72 and/or delayed and so was off hire for the disputed period under the terms of Clause 53. The charterers were not in breach of Clause 8.

Off hire and inspection of US Coast Guard

The main part of clause 85 of NYPE form is intended to apply to matters which lie on the owners' side of responsibility, essentially the vessel and crew, whereas the proviso is intended to apply to matters relating to the charterers' employment of the vessel and crew for their trading purposes, which lies on the other side of the line.

Hyundai Merchant Marine Co Ltd v Furnace Withy (Australia) Pty (The 'Doric Pride')[143]

The vessel was hired by the claimant charterers from the defendant disponent owners under a charter-party on the NYPE form as amended. The charterers instructed the master to proceed to New Orleans. The master was notified by telex from the United States Coast Guard (USCG) that the vessel had been targeted as a 'High Interest Vessel', and as such was prohibited from entering the lower Mississippi and was directed to a waiting position until inspected by a USCG boarding team.

The Court held that the judge had been correct to conclude that, in the circumstances of the present case, the vessel's detention on the approach to a port of calling was a matter for which the owners, and not the charterers, were responsible. The fact that the vessel was a first time caller at the United States was a matter which went to the status of the vessel.

Off hire and tank cleaning

In *The Berge Sund*, the charterers' orders were, in part expressly and at all relevant times by implication to carry out further cleaning. That was the service required and the vessel was fully fit to carry it out. Cleaning was in the ordinary way an activity required by a time charterer and it was the charterer's choice what cargoes were loaded and consequently what cleaning was required.

Sig. Bergesen D.Y. & Co and Others v Mobil Shipping and Transportation Co (The 'Berge Sund') [1993] 2 Lloyd's Rep 453

By a charter-party the shipowners let their vessel to the charterers for a period. The surveyors in Ras Tanura appointed conducted tests on the ship's tanks. It was found that the No. 3 and No. 2 tanks were contaminated, and further cleaning was ordered. Eventually the master decided to wash the tanks with a caustic soda solution. On 1 January 1983 the condition of the tanks was found to be satisfactory but there were problems with obtaining further coolant and a cargo nomination; the port was also closed, so the vessel was only able to load on 9 January 1983.

The court held that (1) despite the wide words of Clause 8(a)(i) of the charter the vessel would not be off hire during the time occupied in 'ordinary' or 'normal' cleaning and to attribute to the parties the intention that the vessel should be off hire during all the cleaning time was untenable.

143. [2006] 2 Lloyd's Rep 175, CA.

An issue between the parties is whether the plaintiffs are in breach of cls 10 and 53 of the time charter by reason of the failure of the master and crew to carry out the tank cleaning process with reasonable skill and speed after the previous cargo is discharged; or whether in one particular respect, the vessel's machinery was defective.

> *Century Textiles & Industry Ltd (Tia Century Shipping) v Tomoe Shipping Co (Singapore) Pte Ltd (The 'Aditya Vaibhav')*[144]
> The plaintiffs let their vessel to the defendants under a time charter. On each occasion of tank inspection the inspectors found that the tanks were not sufficiently clean for loading the cargo of lube oil. The voyage sub-charterers cancelled the sub-charter.
>
> The Court held that the state of the tanks were not sufficient to raise any inference either that the crew had failed to work properly or that the Butterworth machines were not working efficiently. On the whole of the evidence the presence of traces of soft residues was not sufficient to give rise to an inference that the tanks had not been properly cleaned before Jan. 13. The plaintiffs were entitled to recover hire for the whole of the period for which it was claimed.

Off hire and de-fouling hull

Charterers are entitled to give an order as to employment of a vessel that involves going to a warm water port to discharge her cargo.

> *Action Navigation Inc. v Bottigliere Di Navigazione S.p.A. (The 'Kitsa')*[145]
> The vessel was chartered on an amended NYPE form. Discharge of the cargo was delayed, and the vessel remained at the port for over three weeks, during which time her hull became seriously fouled by barnacles.
>
> The Court held that that type of risk was one that was foreseeable and foreseen by both parties at the time the charter-party was concluded, they were entitled to conclude that the relevant type of risk was one that the owners agreed to accept at the time the charter-party was made.

Off hire and period of drydocking

Charterers cancel a charterparty on the basis that the vessel has been off-hire for a period of more than 30 days. Under the charterparty a dry docking is going to take about 15 days.

144. [1993] 1 Lloyd's Rep 63.
145. [2005] 1 Lloyd's Rep 432.

HBC Hamburg Bulk Carriers GmbH & Co KG v Tangshan Haixing Shipping Co Ltd (The 'Fu Ning Hai')[146]

The vessel was owned by the head owners who chartered her to the owners who, by a charterparty, time-chartered her to the charterers. CLAUSE 56 – Off-hire: If the vessel has been off-hire for a period of more than 30 days, the charterers are at liberty to cancel the balance of this charterparty . . .

The Court held that the charterers' consent was being asked for because the charterers were entitled to have the use of the vessel themselves prior to the dry docking. The charterers' reply was not an acceptance of the owners' offer but a counterproposal. What the charterers wanted to ensure was that as a condition of their agreement to the proposal the owners were prepared to modify clause 70 and not require the vessel to be placed off hire for dry docking 'DSLOP one safe port Singapore/Japan range' since Gresik was not within that range. It was held that the appeal and the challenge for serious irregularity would both be dismissed.

Off hire and defects

In *Poseidon Schiffahrt GmbH v Nomadic Navigation Co Ltd (The 'Trade Nomad')* [1998] 1 Lloyd's Rep 57, the second berth was in a position no less favourable to the charterers than the first berth because the charterers altered their discharging schedule sending the vessel to the second berth instead of waiting for her to re-enter the first berth after she had been repaired. It is a perfectly tenable result of the application of the words of the off-hire clause. The meaning of restoring a vessel from her defects is discussed in the case below.

Poseidon Schiffahrt GmbH v Nomadic Navigation Co Ltd (The 'Trade Nomad')[147]

By a charter-party in the Shelltime 4 form the owners let their vessel to the charterers for the carriage of oil and liquid gas cargoes. While the vessel was discharging in Singapore, a boiler breakdown interrupted discharging; the ship had been ordered off the berth by the port authorities; and the charterers ordered her to shift to a second berth. The charterers contended that the vessel was off-hire from June 30 to July 9.

The Court held that the arbitrator had not erred when he rejected the charterers' arguments and held that cl 3(i) and (iii) applied only to defects occurring after delivery of the vessel under the time charter; the phraseology of cl 3(i)—'exercise due diligence so as to maintain

146. [2007] 2 Lloyd's Rep 223.
147. [1998] 1 Lloyd's Rep 57.

or restore the vessel'—strongly suggested that the vessel had become defective after the commencement of that service; and the concept of 'restoring' the vessel suggested that its condition was to be returned to what it once was, i.e. on delivery, not to what it ought then to have been. The appeal would be dismissed.

Off hire and drifting

An off-hire clause is concerned with the service immediately required of the vessel, and not with 'the chartered service' as a whole or the entire maritime adventure or adventures which may be undertaken in the course of the chartered service.

Minerva Navigation Inc v Oceana Shipping AG (The 'Athena')[148]

The vessel was chartered by head owners to charterers, who in turn sub-chartered her to Transatlantica. Both charterparties were on amended NYPE forms. The master, in accordance with the owners' instruction, stopped in international waters outside Libya. The charterers sent a message to the master saying that the vessel was not complying with their instructions to proceed to the roads off Benghazi.

The Court held that the service immediately required of the vessel whilst drifting in international waters was to proceed to the roads at Benghazi.

Vetting procedure required by charterers

A tribunal is asked to determine whether owners are obliged to provide a vessel with vetting procedures required by charterers.

Seagate Shipping Ltd v Glencore International AG (The 'Silver Constellation')[149]

The vessel was chartered and sub-chartered on back-to-back terms on the NYPE form. Disputes arose between the parties relating to the RightShip approval system. The system had been set up in 2001 with the aim of identifying vessels that were suitable and safe for the carriage of iron ore or coal cargoes.

The Court held that the appeals would be allowed on the first issue, but dismissed on the second issue:(1) On the true construction of the charters the owners were not required to obtain and retain RightShip approval; (2) The arbitrators were correct in concluding that the words 'under orders and directions of the Charterers as regards

148. [2013] 2 Lloyd's Rep 673.
149. [2008] 2 Lloyd's Rep 440.

employment' in clause 8 of the charters obliged the owners to permit a RightShip inspection for the purpose of RightShip approval.

Deviation

Charterers have ordered the vessel to load and carry a cargo to China, but where their order as to route is an order, the charterers are not entitled to give that order. The vessel is not without orders and the decision to proceed to China via the Cape is in fulfilment of the owners' duty to prosecute that voyage with due dispatch.

> *Pacific Basin IHX Ltd v Bulkhandling Handymax AS (The 'Triton Lark')*[150]
> The vessel was chartered by the charterers from the owners on the NYPE form incorporating the BIMCO Standard War Risk Clause for Time Charters CONWARTIME 1993. The charterers instructed the vessel to carry a cargo of potash from Hamburg to China via Suez and the Gulf of Aden. The owners refused to proceed via Suez and the Gulf of Aden on account of a risk from pirates and instead proceeded via the Cape of Good Hope.
>
> The Court held that the arbitrators were entitled to conclude that there was no deviation. Their decision was explicable by reference to sub-clause 8 of CONWARTIME 1993 and by reference to the owners' duty to prosecute the voyage to China with due despatch.

Frustration and detention

A charterer is liable under the issue of mitigation to indemnify the owners for breach of Shelltime 4, clauses 4 and 28 of a time charter. Clause 4 constitutes an absolute warranty as to the lawfulness of the cargo, not merely an undertaking that the cargo is lawful to the best of the charterer's belief.

> *Ullises Shipping Corporation v Fal Shipping Co Ltd (The 'Greek Fighter')*[151]
> The vessel and her cargo were detained in UAE. The vessel was subsequently towed to Abu Dhabi, confiscated by the UAE authorities and sold at public auction. At the time of the detention, the vessel was on time charter to Fal. Fal contended that the charter was frustrated by the detention of the vessel.
>
> The Court held that Clause 4 constituted an absolute warranty as to the lawfulness of the cargo, not merely an undertaking that the cargo was lawful to the best of the charterer's belief.

150. [2012] 1 Lloyd's Rep 151.
151. [2006] 1 Lloyd's Rep 99.

Frustration and refusal of port authorities to issue a document

An application of the doctrine of frustration required a multi-factorial approach, including consideration of the terms of the contract, its matrix or context, the parties' knowledge, expectations, assumptions and contemplations, in particular as to risk, as at the time of contract, the nature of the supervening event, and the parties' reasonable and objectively ascertainable calculations as to the possibilities of future performance in the new circumstances.

> *Edwinton Commercial Corporation v Tsavliris Russ (Worldwide Salvage & Towage) Ltd (The 'Sea Angel')*[152]
>
> The tanker *Tasman Spirit* (the casualty) grounded near the approaches to the port of Karachi. The defendant salvors (Tsavliris) entered into a Lloyd's Standard Form of Salvage Agreement with the owners of the casualty. Tsavliris engaged a number of shuttle tankers, including Sea Angel (the vessel), to lighten the casualty by transhipping the cargo to a larger tanker *Endeavour II*. The vessel completed the discharge of cargo but was unable to depart Karachi because the Karachi Port Trust (KPT) refused to issue a 'No Demand Certificate' (NDC), a prerequisite to port clearance.
>
> The Court of Appeal held that in general terms the contractual risk of such delay was on Tsavliris. The risk of such detention was in general terms foreseeable, that the particular risk was within the provisions of SCOPIC, and that it was now common ground that there was no frustration until the strategy of commercial negotiation had initially failed by 13 or 17 October 2003.The judge's conclusion that the charter had not been frustrated by 13 or 17 October 2003 showed the doctrine of frustration working justly, reasonably and fairly.

Arbitration clause

It is common to incorporate a clause of arbitration in a charterparty. There are disputes to the process or award in arbitration.

> *Poseidon Schiffahrt GmbH v Nomadic Navigation Co Ltd (The 'Trade Nomad')*[153]
>
> While the vessel was discharging in Singapore, a boiler breakdown interrupted discharging and the ship had been ordered off the berth by the port authorities; and the charterers ordered her to shift to a second berth. The charterers contended that the vessel was off-hire from June 30 to July 9.

152. [2007] 2 Lloyd's Rep 517, CA.
153. [1998] 1 Lloyd's Rep 57.

The Court held that when a dispute was referred to arbitration, the parties to that reference would therefore already have consented to an appeal on questions of law from such award as may be made; and when the award was made and the losing party wished to appeal, all the other parties to the reference would have consented to the appeal being heard on a question of law affecting the rights of the parties since the other party to the arbitration agreement continued to be bound by the consent agreement.

Challenge of jurisdiction of arbitration tribunal

It is common ground that the jurisdiction challenge under section 67 of the 1996 Act involves a full rehearing.

Hyundai Merchant Marine Co Ltd v Americas Bulk Transport Ltd (The 'Pacific Champ')[154]

The claimant (HMM) was the bareboat charterer of the vessel under a charter. Negotiations commenced on the telephone between Mr B, for HMM, and Mr S, for the defendant (ABT), for the possible sub-charter of the vessel to ABT.

The Court held that there was no consensus and therefore no binding contract before or at the time of the second recap, and therefore no arbitration agreement. It followed that HMM's challenge concerning the tribunal's jurisdiction under section 67 of the 1996 Act succeeded. The award would be set aside. HMM was entitled to a declaration that there was no valid binding contract.

Lawyer or commercial/professional person as arbitrator

In interpreting the findings of a tribunal consisting of experienced commercial and professional men, as opposed to lawyers, the court should look at the substance of such findings, rather than their form, and should approach a reading of the award in a fair, and not in an unduly literal way.

Bottiglieri Di Navigazione SpA v COSCO Qingdao Ocean Shipping Company (The 'Bunga Saga Lima')[155]

The vessel was chartered on an amended NYPE form. When the vessel was delivered both the charterers and the owners knew that the holds were dirty with coal residues. However, the charterers made no objection. The vessel failed cargo hold cleanliness inspections for the second cargo at Rostock.

The Court held that the tribunal was correct that Clause 13 of the fixture note made it clear that the warranty was only that the vessel

154. [2013] 2 Lloyd's Rep 320.
155. [2005] 2 Lloyd's Rep 1.

would have clean holds upon delivery or arrival at first loadport. The only entitlement to place the vessel off-hire provided by clause 13 was to do so in the event that the holds were not clean upon delivery or arrival at the first loadport. No other right was given to charterers to place off-hire at a subsequent loadport.

Whether bill of lading incorporating 'Law and Arbitration Clause' of charterparty had effect to incorporate English law and court jurisdiction clause

There is a good arguable case that general words of incorporation are sufficient to incorporate a proper law clause. In addition, whatever the effect of the words 'and arbitration' in the bill of lading clauses, the express references to the governing law of the charterparty amounts to an irrefutable case that the parties to the bill of lading intend their contract to be governed by the same law as is applicable to the charterparty, provided that the law so chosen was usual and proper for the trade.

> *Caresse Navigation Ltd v Office National De L'electricite and Others (The 'Channel Ranger')*[156]
>
> The vessel was chartered on an amended NYPE form. The coal cargo was found to have self-heated at the discharge port. The shipowners brought proceedings against the receivers and the various cargo insurers claiming a declaration of non-liability. The defendants issued an application challenging the jurisdiction of the English court.
>
> The Court held that the defendants' jurisdiction challenge would be dismissed. As to whether the jurisdiction clause in the charterparty was incorporated into the bill of lading, the question was essentially one of construction rather than incorporation. The real question was what the parties should reasonably be understood to have meant by the words 'law and arbitration clause' which plainly contemplated the incorporation of at least one kind of ancillary clause. Accordingly the bill of lading contained a term requiring any dispute to be submitted to the exclusive jurisdiction of the English court.

Which charter is referred to in bill of lading: time or voyage charterparty

As to the issue of which Charter is referred to in B/L, ie Time or Voyage charterparty, the courts will be inclined to favour the incorporation of terms of that charter which are the more (or the most) appropriate to regulate the legal relations of the parties to the bill of lading contract. Where each

156. [2014] 1 Lloyd's Rep 337.

(or more than one) of the charterparties is equally appropriate for this purpose, the courts may determine the issue by holding the relevant charterparty to be that one which governed the contractual relations between the original parties to the bill of lading and in pursuance of which the bill was issued.

> *Re Yaoki*[157]
>
> There are two Charterparties, the head charter which is a Time Charter on Shelltime 4 Form ('Time CP') with S.H. Marine Ltd as the charterer and a Voyage Charter ('Voyage CP') with S.H. Marine Ltd as owner and BP Singapore as the voyage charterer. The Time CP contains a London Arbitration Clause but the Voyage CP contains no arbitration clause, only an Exclusive Jurisdiction Clause for London High Court.
>
> The Court held that in the issue of which Charter referred to in B/L.-Time or Voyage CP, the defendants have satisfied that the Time CP is the relevant Charter Party referred to in the B/L. The Arbitration Clause at Clause 3 on the back of the B/L and the Arbitration Clause on the Front of the B/L are binding on the Plaintiff and that therefore a mandatory stay of Hong Kong proceedings must be ordered.

6.4 FREIGHT AND LIEN

Freight

Freight is the consideration payable for the safe carriage and delivery of the cargo in a merchantable condition to the agreed destination. Generally, freight is payable in accordance with the express provisions of the charterparty or the bill of lading. For instance, freight for the hire of a ship under a voyage charterparty will be calculated for a voyage. When there is a time charterparty, the freight will be calculated for a specified time. Bill of lading freight, where no provision exists to the contrary, is calculated on the amount delivered according to weight, quantity or measurement. In the absence of special contractual provisions, freight is earned only upon the arrival of the goods at the port of discharge ready to be delivered to the consignee.

> *Hunter v Prinsep*[158]
>
> Freight was payable 'on a right and true delivery of the homeward bound cargo' for a voyage from Honduras to London. Due to bad weather, the vessel was driven ashore. The wreck and cargo were sold without the consent of the cargo owner.

157. [2006] HKEC 848.
158. (1808) 10 East 378.

The court held that although the ship was prevented by excepted perils from completing the voyage, no freight was payable. This was because where freight was payable on delivery, it would not be earned unless the goods were actually delivered.

Lord Ellenborough: 'The shipowners undertake that they will carry the goods to the place of destination, unless prevented by the dangers of the seas, or other unavoidable casualties, and the freighter undertakes that if the goods be delivered at the place of their destination, he will pay the stipulated freight . . . If the ship be disabled from completing her voyage, the shipowner may still entitle himself to the whole freight, by forwarding the goods by some other means to the place of destination; but he has no right to any freight if they be not so forwarded . . .'

Asfar & Co v Blundell[159]

While a vessel carrying dates was sailing up the Thames, she was hit by another vessel and was sunk. The vessel was filled with water. Although the dates were subsequently recovered, they were in a state that was unfit for human food. They were sold for distillation into spirit.

The court held that no freight was payable because the goods delivered were not in merchantable condition. In other words, the goods delivered were, for commercial purposes, something different from those shipped.

Lord Esher: 'There is a perfectly well-known test which has for many years been applied to such cases as the present—that test is whether, as a matter of business, the nature of the thing has been altered . . . But if the nature of the thing is altered, and it becomes for business purposes something else, so that it is not dealt with by business people as the thing which it originally was, the question for determination is whether the . . . original article of commerce has become a total loss.'

The following authorities demonstrate that even if the goods are delivered in a damaged condition, or even though there has been a short delivery of the cargo, the cargo owner or the charterer is liable to pay the freight without deduction. For instance, wheat may be damaged, but may still be the thing dealt with as wheat in the ordinary course of business. Rice may be short delivered, but the remaining portion of the rice on board the vessel ready to be discharged cannot be considered as a total loss. After paying the freight to the carrier, the cargo owner or the charterer may take a separate court action to claim for damage to, or short delivery of, the cargo.

159. [1896] 1 QB 123.

Dakin v Oxley[160]

Owing to the negligence of the master, coal carried under a charterparty was damaged. The charterer refused to pay for the full freight, claiming that the damaged cargo was not worth its freight.

The court held that the charterer was liable for the freight, although he had a separate right to claim for the damage to the coal by cross-action.

Willes CJ: 'If [the cargo] has arrived, though damaged, the freight is payable by the ordinary terms of the charterparty . . . our law does not allow deduction in that form, and as at present administered, for the sake, perhaps, of speedy settlement of freight and other liquidated demands, it affords the injured party a remedy by cross-action only.'

Aries Tanker Corporation v Total Transport Ltd[161]

A cargo of petroleum was carried from the Arabian Gulf to Rotterdam under a voyage charterparty. Freight was to be calculated on the basis of the quantity of cargo. On arrival at Rotterdam, the charterer discovered that there was a short delivery, and withheld part of the freight.

The House of Lords held that the charterer was not entitled to withhold the freight.

Lord Wilberforce: 'I am therefore firmly of the opinion that the rule against deduction has to be applied to this charterparty so that the charterers' claim for short delivery cannot be relied on by way of defence. On any view, therefore, of the time bar, and even assuming the latter to be only procedural, it must defeat the claim.'

Lord Salmon: 'A rule of law, particularly a rule of commercial law which has stood so long and upon the faith of which many thousands of contracts of carriage have been made containing a provision that the contract shall be governed by the law of England, cannot now be successfully challenged in our courts . . . I am, however, by no means satisfied that no reason or justification exists. It may well be, for example, that the whole incidence of insurance cover in respect of freight is based upon the rule. However this may be, it is a rule so well recognised and accepted as the law of England that if it is to be altered, it can, in my view be altered by Parliament alone. If it were to be altered by statute, it would presumably not be altered retrospectively and the alteration would therefore apply only to contracts made after the statute came into operation.'

RAF Forwarding (HK) Ltd v Wong Angela t/a JMT Co[162]

In a carriage of goods by air dispute in Hong Kong, the Court of First Instance held that the common law principle that a claim in respect

160. (1864) 10 LT 268.
161. [1977] 1 Lloyd's Rep 334.
162. [1999] 2 HKC 135.

of cargo could not be pursued by way of deduction from a claim for freight was not confined to contracts of carriage by sea and by road. It was equally applicable to air freight. The air carrier's entitlement to freight was not subject to any claim by way of set-off by the cargo owner.

Le Pichon J: 'In my judgment, the common law rule for freight also applies to a contract of freight by air. That is the basis upon which the parties contracted and the [air carrier's] entitlement to freight is not subject to any claim by way of set-off or counterclaim by the [cargo owner].'

One must carefully learn the lessons from the above cases. It is unwise for the charterer to rely on the alleged short delivery or cargo damage in order to raise a set-off against the claim for freight. The time bar for most cargo claims is one year, while the time bar for claims for freight is usually six years. If the charterer relies on the set-off, the danger is that the carrier may sue for the withheld freight after one year of the discharge of the goods. By that time, the charterer's claim for short delivery or cargo damage may have already been time-barred. In consequence, it can neither be pleaded as a defence nor a counterclaim against the carrier's claim for freight.

Types of freight

Lump-sum freight refers to the freight of an agreed amount to be paid for the use of the whole or part of a vessel and is not calculated in accordance with the quantity of cargo carried. If the ship carrying the cargo becomes incapable of completing the voyage owing to excepted perils, the carrier is still entitled to claim the full lump freight even if only part of the cargo is delivered.

William Thomas & Sons v Harrowing Steamship Co[163]

The ship was driven ashore near the port of discharge. The charterparty contained a clause exonerating the shipowner from liability for loss through perils of the sea. Part of the cargo was washed ashore and was afterwards collected on the beach by the master's directions, the residue being lost by perils of the sea.

The House of Lords decided that even though the charterparty provided that freight should be payable on 'unloading and right delivery of the cargo', the shipowner was entitled to receive the whole lump sum freight. The shipowner had already performed the contract of carriage by delivering the cargo so far as the shipowner was not prevented by perils of the sea.

163. [1915] AC 58.

> Lord Shaw of Dunfermline: '. . . the shipowner under the circumstances collected all the available cargo, and . . . delivery was made of all the props . . . except those which had been lost by perils of the sea. In my opinion the freight contracted for has been earned—there has been right and true delivery, taking into account that exception of perils of the sea which the contract itself contains.'

Pro-rata freight is payable for short delivery at the agreed destination, or for delivery short of destination, only if the cargo owner has expressly or implicitly agreed to pay pro rata freight. For example, in *St. Enoch Shipping Co Ltd v Phosphate Mining Co*,[164] a war broke out during the carriage, rendering the ship unable to go to Hamburg, the agreed port of discharge. The ship therefore discharged the cargo at Runcorn. The court ruled against the shipowner's claim for pro rata freight, holding that pro rata freight is only payable where there is an express agreement or one can be implied from the circumstances of the case. Since the voyage was not completed, the court decided that no freight was payable. On the other hand, under the principle of restitution, it is possible that the carrier can recover freight to the extent that the merchant has been incontrovertibly benefited by the carriage of goods. In *Procter & Gamble Philippine Manufacturing Corporation v Peter Cremer GmbH & Co*,[165] the court cited the following passage on the law of restitution:

> The general principle should be that restitution should always be granted when, as a result of the plaintiff's services, the defendant has gained a financial benefit readily realisable without detriment to himself or has been saved expense which he inevitably must have incurred. (Goff and Jones, *The Law of Restitution* [3rd edn, 1986, 148].)

> *Hain Steamship Co v Tate & Lyle*[166]
>
> Due to the failure of a messenger to deliver the instructions for loading from the telegraph office, the vessel deviated from her route without any justification. Nevertheless, the shipowner carried the cargo safely to the destination and claimed for freight stated in the bill of lading against the holder of the bill.
>
> The House of Lords held that where the deviation had not been waived by the bill of lading holder, the shipowner was not entitled to rely on any terms of the contract, including the 'cesser clause'. A 'cesser clause' is a provision in the charterparty stating that the charterer's liability under the charterparty will cease once the cargo has been put on board the ship. For liabilities arising after loading, the shipowner

164. [1916] 2 KB 624.
165. [1988] 3 All ER 843.
166. [1936] 2 All ER 597.

must proceed against the bill of lading holder. In this case, since the unjustified deviation had not been waived by the bill of lading holder, the shipowner could not claim for freight under the 'cesser clause' against the holder of the bill. The court further decided that the holder had not waived the deviation merely by obtaining delivery of the cargo on presentation of the bill of lading.

Lord Atkin: 'My Lords, the effect of a deviation upon a contract of carriage by sea has been stated in a variety of cases but not in uniform language. Everyone agreed that it is a serious matter . . . The party who is affected by the [deviation] has the right to say, I am not now bound by the contract, whether it is expressed in charterparty, bill of lading or otherwise. He can, of course, claim his goods from the ship; whether and to what extent he will become liable to pay some remuneration for carriage I do not think arises in this case . . . once he elects to treat the contract as at an end, he is not bound by the promise to pay the agreed freight any more than by his other promises. But on the other hand, as he can elect to treat the contract as ended, so he can elect to treat the contract as subsisting: and if he does this with knowledge of his rights, he must in accordance with the general law of contract be held bound . . . There must be acts which plainly show that the shipper intends to treat the contract as still binding.'

Advance freight is the whole or the portion of freight to be paid before or shortly after the cargo is loaded in advance of the carriage. At common law, there is a strong presumption that freight is only payable on delivery of the cargo. Hence, very clear wording must be included in the contract of carriage stipulating the sum is intended to be advance freight. Furthermore, once the freight has been paid in advance, it is not recoverable if the cargo is lost by a peril covered by an exception in the contract. However, if the goods are lost by a peril not excepted, the freight paid in advance is still recoverable.

De Silvale v Kendall[167]

The shipper paid an advance freight of £192 for a voyage from Liverpool to Maranham and back to Liverpool. The vessel arrived at Maranham, but she was lost by capture on her homeward voyage.

The court held that the shipper was not entitled to recover the advance freight in the absence of express stipulation that the shipper shall be entitled to recover it back.

Bayley J: 'It is suggested as a ground, that the freight has failed by the non-performance of the voyage, and thus the plaintiff has derived no benefit from it; but what benefit has the defendant derived? He also has lost as well as the plaintiff, and the question is, whether

167. (1815) 4 M & S 37.

> he is to bear a further loss. Now in order to maintain money had and received, it is in general incumbent upon the plaintiff to show that the defendant has money of the plaintiff which in equity and good conscience he ought not to detain from him. But here the question raised is not whether the defendant has money which he ought not to detain, but whether out of his own money he shall be bound to make good that which the plaintiff has lost. It seems to me that the defendant shall not be so bound.'

> *Bank of Boston Connecticut v European Grain & Shipping Ltd (The 'Dominique')*[168]
>
> The shipowner had assigned their freight earnings in favour of a third party, BBC. The charterparty stated that 'freight shall be prepaid within five days of signing and surrender of final bills of lading, full freight to be earned on signing of bills of lading'. The bills of lading were signed, but unfortunately the ship was arrested and failed to complete the voyage. BBC claimed the freight due and owing.
>
> The House of Lords held that the shipowner's right to advance freight accrued on completion of the signing of the bills of lading. The shipowner's right survived the arrest of the ship and the termination of the charterparty. BBC was entitled to the freight, and the charterer was not permitted to set off damages flowing from the breach of the charterparty against the freight claim.

Back freight will be payable in relation to expenses incurred where the master has acted for the benefit of the cargo owners. Back freight refers to the expense recoverable where, in the cargo owner's interest, cargo is warehoused, further shipped or forwarded in another vessel to such destination as may be most convenient for the cargo owner. This will usually happen when delivery of the goods at the agreed port of discharge is impossible. Dead freight (ie, charges for unoccupied carriage capacity), on the other hand, is the term for a claim arising from a breach of contract by a charterer to provide a full and complete cargo to a vessel in accordance with the terms of the charterparty.

> *Cargo ex 'Argos'*[169]
>
> A shipper shipped petroleum on board a ship bound from London to Harve. At Harve the port authorities refused to allow the petroleum to be landed because there were large quantities of munitions in the port. The master tried to discharge the cargo in the outer ports for delivery into lighters, but eventually decided that it would be best to go back to London.

168. [1989] 2 WLR 440.
169. (1837) LR 5 PC 134.

The Judicial Committee of the Privy Council held that the vessel had done all that was required on her part. The master had dealt with the cargo for the benefit of the shipper as seemed best to him. The shipowner was entitled to full freight.

Sir Montague E Smith: 'It seems to be a reasonable inference from the facts, that after the four days during which the petroleum had been lying in the harbour had expired, the authorities would not have allowed it to remain there. It was still in the master's possession, and the question is, whether he should have destroyed or saved it. If he was justified in trying to save it, their Lordships think he did the best for the interest of the [cargo owner] in bringing it back to England . . . Their Lordships have no doubt that bringing the goods back to England was, in fact, the best and cheapest way of making them available to the [cargo owner] . . . If the goods had been of a nature which ought to have led the master to know that on their arrival they would not have been worth the expenses incurred in bringing them back, a different question would arise.

Hunter v Fry[170]

A charterparty stipulated that the charterer was to load a 'full and complete cargo'. The vessel might have carried 400 tons of the kind of goods loaded. However, the charterer only loaded 336 tons.

The court held that the charterer had to load a full and complete cargo. Since the charterer had failed to do so, the shipowner was entitled to damages for dead freight.

Abbott CJ: 'Here, [the charterer] has covenanted to load and put on board a full and complete cargo, and to pay so much per ton for every ton loaded on board . . . Upon the whole, I am of opinion that the owner was bound to take on board such a number of tons of goods as the ship was capable of containing without injury; and, therefore, that the [shipowner] is entitled to have a verdict for £918 which is the difference between the sum actually paid for freight, and that which would have been payable if the shipper had loaded on board a full and complete cargo.'

The Archimidis[171]

The charterer was held liable for dead freight even though it had furnished a full cargo to the vessel. Both the shipowner and the charterer were aware that it was impossible for the vessel to proceed on its voyage after loading, as it could not leave the port safely due to certain prevailing conditions rendering the port unsafe.

170. (1819) 2 B & Ald 421.
171. [2008] 1 Lloyd's Rep 597.

Payment of freight

Generally speaking, freight is payable according to the terms of the contract of carriage. A recent illustration can be found in *Tradigrain SA v King Diamond Shipping SA ('The Sprios C')*.[172] The shipowner entered into a time charterparty with the charterer, and the charterer entered into a voyage sub-charterparty with the sub-charterer. The sub-charterparty stated that freight would be payable not to the shipowner, but to a third party. The bill of lading issued by the master, however, provided that the freight was payable 'as per charterparty'. The shipowner argued that the voyage sub-charterer was under an obligation to pay the bill of lading freight and the discharge port demurrage to the shipowner directly. The Court of Appeal held that the common intention was that the way in which freight would be collected would be entrusted to the time charterer. If the time charterer and the voyage sub-charterer agreed between themselves to vary the arrangements for the payment of freight, such a variation would remain within the scope of authority delegated to the time charterer. Hence, the voyage sub-charterer was entitled to pay the freight as they had done in compliance with the sub-charterparty.

Lien

Generally, a lien is the right of a creditor to retain the properties of his debtor until the debt is paid. A shipowner may retain the goods as security for the payment of freight or other charges. This right is called the shipowner's lien. A shipowner's lien is a possessory lien. In other words, it can only attach to goods in the possession of the shipowner. In fact, this right is solely based on the possession of the goods. Thus, a shipowner who has transferred the possession and control of his ship under a charter by demise has no lien on goods consigned on board the ship. Two types of shipowner's lien are: the common law lien and the contractual lien.

At common law, a shipowner may retain goods for freight or some other charges due to him. The common law lien is independent of any contract between the parties. It may be taken as the application of the generally accepted rule that if a man delivers goods to another and asks for work to be done on them, the person who has done the work is entitled to retain the goods until paid: see *Scarfe v Morgan*.[173]

172. [2000] 2 Lloyd's Rep 319.
173. (1838) 4 M & W 270.

Nowadays, all contracts for the carriage of goods by sea give shipowners a right to retain goods for freight or other charges due to them. A shipowner may, of course, sue a shipper who fails to pay the freight agreed under contract. At the same time, the shipowner may also exercise his contractual right to retain the goods until the freight is paid. The extent of the right is generally governed by the contract and usually wider that the common law lien. Originally a means to force the shipper to pay, the shipowner's lien is often exercised to obtain other forms of security, such as a bank guarantee, for the amount due: see *The Thorscan*.[174]

Shipowner's lien and maritime lien distinguished

Although both the shipowner's lien and the maritime lien may entitle a shipowner to retain the properties of another party, they are, to a great extent, different. Basically, the shipowner's lien is exercised against the properties of a party who has not paid the freight or some other money due to a shipowner. The shipowner may only retain the properties of that party. The maritime lien, on the other hand, is about the rights to retain and sell ships or other maritime properties. It is associated with an action in rem and gives a right against the maritime property in issue notwithstanding any changes of ownership. Referring to the contractual lien for general average contribution, Lord Diplock commented in *Castle Insurance Co Ltd v Hong Kong Islands Shipping Co Ltd*[175] that:

> The lien attaches to the preserved cargo at the time when the sacrifice is made or the liability to the expenditure incurred. The lien is a possessory lien and it is the duty of the vessel to exercise the lien at the time of discharge of the preserved cargo in such a way as will provide equivalent security for contributions towards general average sacrifices made or expenditure incurred not only by those concerned in the ship but also by those concerned in cargo in respect of which a net general average loss has been sustained. The lien, being a possessory one and not a maritime lien, is exercisable only against the consignee, but it is exercisable whether or not the consignee was owner of the consignment at the time of the general average sacrifice or expenditure that gave rise to the lien.

It should be noted that, under American law, a contractual shipowner's lien is treated as a maritime lien exercisable against goods owned by a third party. The lien is, however, valid only to the extent of the freight that

174. [1998] 4 HKC 536.
175. [1983] 1 HKC 32 at 36.

party owed under his carriage of goods contract: *American Steel Barge Co v Chesapeake & Ohio Coal Agency*.[176]

Common law lien

A shipowner has a common law lien for freight, general average contribution or expense incurred for the protection or preservation of the goods. However, there is no lien for dead freight, which is in fact not freight in the strict sense. There is also no shipowner's lien for demurrage, damages for detention, pilotage or port charges.

A shipowner's lien at common law is good against all claims to take away the goods. If the shipper is not the owner of the goods, a shipowner's lien is nevertheless good against the owner who may not be liable to pay the freight. Since the common law lien for freight is created independent of the contract, it is generally not affected by the contract of carriage of goods. For example, where the ship was on charter and a bill of lading has been issued to the charterer who is also the consignee, the lien is good against the charterer for freight due under the charterparty, unless the terms of the bill of lading are inconsistent with it. For example, a contract providing that the goods are to be shipped to Hong Kong but freight is payable in advance in Moscow is inconsistent with the common law lien.

At common law, no shipowner's lien exists for freight unless the obligations to pay the freight and deliver the goods are contemporaneous. A shipowner cannot retain the goods for advance freight or freight payable after the delivery of the goods. If only part of the whole freight is payable on delivery, there is no lien for other parts payable at a later time. Generally, there is no common law lien for freight payable independent of the delivery of goods.

> *Nelson v Association for Protection of Wrecked Property*[177]
> The bills of lading provided that the freight was payable at Liverpool, the port of discharge, 'ship lost or not lost'. The ship was wrecked. The shipowners, who took no effort to save the ship and cargo, abandoned the voyage. They allowed the agent of the underwriters on cargo to save and forward the cargo. The cargo was saved and forwarded to Liverpool. The shipowners claimed a lien on the cargo for the freight.
>
> The court held that 'the right to lien does not arise unless the payment of the freight is to be on the delivery of the cargo; if the freight is payable without delivery of the cargo, the lien does not accrue'. The shipowners had no lien at all in this case, since the shippers were

176. 115 F2d 669 (1st Circuit 1902).
177. (1874) 43 LJCP 218.

under an obligation to pay the freight even if the ship and cargo did not arrive.

Further, if the delivery is made by another party, the shipowner has no common law lien on the cargo: *Nelson v Association for Protection of Wrecked Property*.[178]

The shipowner's common law lien is a particular lien. It can be exercised on only the goods of the consignment for which the freight in issue is payable. It does not cover goods of the same shipper which are in the shipowner's possession under other contracts. This is so even if such goods are on board the same ship. Generally, all goods covered by the same contract may be retained for only part of the freight, provided they are all delivered to the same consignee. In such cases, it makes no difference that the goods are covered by different bills of lading. If the goods under the same contract are to be delivered or the bills have been indorsed to different consignees, only the goods consigned to the same consignee may be retained.

Contractual lien

If the shipowner loses his common law lien because, for example, he does not have the possession of the ship under a charter by demise or bareboat charter, he may still rely on any contractual liens he may have. The scope of a contractual lien is a matter for the shipowner and the shipper. The parties may, subject to the law, agree on any terms they prefer. In practice, there are liens for advance freight, dead freight, demurrage, damages for detention, or other costs or expenses. The lien clause may be widely worded. Besides, a party who has no personal obligation to pay the amount due may also be subject to the lien for it.

> *Miramar Maritime Corp v Holborn Oil Trading Co Ltd*[179]
> The bill of lading provided that 'all the terms whatsoever of the said charter . . . apply to and govern the rights of the parties concerned in this shipment . . .' The terms of the charter made the charterers exclusively liable for demurrage. The shipowners exercised their lien on cargo against the bill of lading holders for demurrage. The lien was later lifted when the receivers gave a guarantee for any sums on which a shipowner's lien might be exercised. The shipowners then sued the receivers for the amount in issue.
>
> The court held that (1) because of the demurrage clause in the charter, the receivers were not liable to pay demurrage; and (2) the

178. ibid.
179. [1983] 2 Lloyd's Rep 319.

shipowners were nevertheless entitled to the amount under the guarantee since they had a lien on the cargo.

However, the court construes lien clauses rather strictly. A lien must be expressly given and subject to other express terms of the contract. For example, a lien for 'all freights, primages and charges' does not include a lien for interest on freight: *Clemens Horst v Norfolk*;[180] and a lien for dead freight is not covered by a lien for 'all charges whatsoever': *Red 'Superior' v Dewar & Webb*.[181]

If the contract expressly provides for a lien but also contains terms contradictory to it, the court, applying the normal rules of construction, will give effect to the intention of the parties.

> *Foster v Colby*[182]
> The charterparty gave shipowners an 'absolute lien on the cargo for all freight'. The freight was stated to be a lump sum, part of which was payable on departure, another part on delivery, and the 'remainder in cash two months from the vessel's report inwards at London or Liverpool, and after right delivery of the cargo, or under [a certain discount] per annum, at freighter's option'.
>
> The court held that the part of freight became payable after the return of the ship.

A contractual lien has its root in the contract and is thus binding upon the parties. It is not good against any claims made by third parties. This is so even if the third party has all along had notice of the existence of the lien. If the owner of the goods, who is not a party to the contract, requests the return of the goods, the shipowner commits an act of conversion if he refuses to do so.

In practice, the master usually issues bills of lading to shippers. If the charter is not a charter by demise, the master is the agent of the shipowner although the charterer is entitled to the freight. The bill of lading freight is, in normal cases, higher than the chartered freight. The difference between them represents the profit of the charterer. The practice is that the charterer's agents collect the freight and then transfer the money to him. If the shipowner gives notice of claim to the freight to the agents, they are under obligation to receive the freight on behalf of the shipowner and to forward any collected freight they hold.

180. (1906) 11 Com Cas 141.
181. [1909] 2 KB 998.
182. (1858) 3 H & N 705.

Wehner v Dene[183]

A ship was on a time charter and a sub-time charter. The master signed bills of lading issued under the sub-charter. The sub-freight was collected by the agents of the sub-charterers. The shipowners gave notice to the agents of their claims for the collected freight. At the time of the notice, a certain amount of the head charter freight had already been due. Further hire became due after the notice.

The court held that the shipowners were in the capacity of a party to the bill of lading contract. The agents appointed by the sub-charterers were, in the absence of evidence indicating otherwise, presumed to be the agents of the shipowners and the sub-charterers. The shipowners were therefore entitled to the freight collected by the agents. They were however bound to account to the sub-charterers for the amount after deducting only the head charter freight already due at the date of receipt. No deduction was allowed for the hire due after the notice.

Where the shipowner is not a party to the bill of lading contract, he cannot exercise his lien against the shipper, unless the contract contained in the bill of lading incorporates the lien clause in the charter. Usually, terms such as 'paying freight and all other conditions as per charter' are sufficient to incorporate the lien clause: *Gray v Carr*.[184] The situation is more complicated if the hire is higher than the bill of lading freight.

Gardner v Trechmann[185]

The bill of lading stated that 'freight for the said goods' at a certain rate was payable on delivery. It also stated that 'all extra expenses in discharging to be borne by receivers, and other conditions as per charterparty'. The ship was chartered at a rate higher than the bill of lading rate. The charter gave the shipowner a lien for freight.

The court held that there was no shipowner's lien for the difference between the rates. The express terms of the bill of lading could not be altered by the clauses of the incorporated charter. A general incorporation of the charter could not bring in a lien for the full chartered rate.

A contractual lien is a general lien attaches to all the properties of the charterer or shipper in the shipowner's possession for money due to him. The following are the lien clauses in some common forms of charter:

'Barecon 2001' Standard Bareboat Charter—Clause 18

The Owners to have a lien upon all charges, sub-hires and sub-freights belonging or due to the Charterers or any sub-charterers and any Bill

183. [1905] 2 KB 92.
184. (1871) LR 6 QB 522.
185. (1885) 15 QBD 154.

> of Lading freight for all claims under this Charter, and the Charterers to have a lien on the Vessel for all moneys paid in advance and not earned.

'Gencon' Standard Voyage Charter—Clause 8

> The Owners shall have a lien on the cargo and all sub-freights payable in respect of the cargo, for freight, dead freight, demurrage, claims for damages and for all other amounts due under this Charterparty including costs of recovering the same.

'NYPE 93' Standard Time Charter—Clause 23

> The Owners shall have a lien upon all cargoes and all sub-freights and/or sub-hire for any amounts due under this Charter Party, including general average contributions, and the Charterers shall have a lien on the Vessel for all monies paid in advance and not earned, and any overpaid hire or excess deposit to be returned at once.

Both 'Barecon' and 'NYPE' give shipowners a general lien on 'all cargoes and sub-freights'. The validity and extent of the lien clauses require detailed discussion. As stated above, a lien under a charterparty is, unless incorporated into the sub-charter or bill of lading, binding only upon parties to the charterparty. So far as the cargoes of the charterers are concerned, the shipowner's lien is valid. It is however not exercisable on cargoes of third parties, such as a shipper who makes a carriage of goods contract with the charterer which does not incorporate the lien clause of the charterparty. In fact, 'Barecon 89' provides a general lien only 'upon all cargo and sub-freights belonging to the Charterers'. In contrast, the lien under 'NYPE' is 'upon all cargoes . . . for any amounts due under this Charter Party'. It was held in *The Agios Giorgis*[186] that, if the lien clause had not been included in the contract, the shipowner had no right to retain cargo not belonging to the charterers and on which no freight was owing to the shipowner. Such a clause, however, imposes an obligation on the charterer to procure a contractual lien upon cargoes of third parties in favour of the shipowner: *The Aegnoussiotis*.[187] If the charterer fails to do so and the shipowner is not able to exercise a lien on the goods of third parties, the charterer may not take advantage of his own fault. The practical result is that the 'lien' on third parties' goods, though not exercisable against third parties, is good against the charterer.

The lien clauses of the 'Barecon' and 'NYPE' also provide for liens on sub-freights or bill of lading freights. As a lien is based on possession, there

186. [1976] 2 Lloyd's Rep 192.
187. [1977] 1 Lloyd's Rep 268.

can be no shipowner's lien against the charterer on sub-freights due to a time or voyage charterer in the strict sense, since the shipowner is not in possession of the freights.

> *The Lancaster*[188]
>
> *The Lancaster* was time-chartered. The charter provided that the shipowners had 'a lien upon all cargoes and all sub-freights for any amounts due under this charter . . .' The ship became a constructive total loss after a collision. The brokers began to pay the insurance money to the mortgagees of the ship. The charterers argued that they were entitled to a certain amount in respect of advance hire which had not been earned at the time of collision. In order to decide the issue, the court needed to consider, among other things, the nature of a lien on sub-freights.
>
> The court held that a time or voyage charter was simply a contract of services. The obligation of the shipowners under such contracts was to make available to the charterers the services of the ship and crew. The charterers never gave the possession of sub-freight to the shipowners. As a result, a lien on sub-freight could not be a possessory lien.

A lien on sub-freights operates as an equitable charge on freights due from shippers to the charterer: *The Nanfri*.[189] In practice, such a 'lien' gives the shipowner a right to intercept before sub-freights are paid to the charterer. Once the charterers receive them, the right to intercept ends.

> *Tagart, Beaton & Co v Fisher*[190]
>
> The time charter gave the shipowner a lien on all cargoes and sub-freights any amount due under the charter. Agents were appointed by the charterers to collect freights from consignees. Before the agents transferred the freights paid by consignees to the charterers, the shipowner purported to exercise a lien on the freights.
>
> The court held that since the freights had been paid to the agents of the charterers, the shipowner lost the lien.

It is not clear whether the term 'sub-freight' includes 'sub-hire'. In the shipping industry, the term 'freight' customarily refers to the remuneration payable for the carriage of goods under a voyage charter or bill of lading, whilst the term 'hire' is used for the amount payable for the use of a ship under a time charter. It was thus held in *Itex Itagrani Export SA v Cape Shipping Corp (The 'Cebu' (No. 2))*[191] that a shipowner's lien 'on all cargoes and sub-freight' did not apply to sub-time charter hire. In the earlier case

188. [1980] 2 Lloyd's Rep 497.
189. [1979] 1 Lloyd's Rep 201 at 210.
190. [1903] 1 KB 391.
191. [1990] 2 Lloyd's Rep 316.

of *The 'Cebu'*,[192] a wider meaning was given to the term 'freight'. The court concluded that it should refer to any monies obtained by the time charterer as a result of the employment of the chartered ship. The better view seems to be that the law should give effect to trade customs and *The Cebu (No. 2)* should be followed.

A charterer who sub-charters a ship would not usually be involved in the actual carriage of goods. For obvious reasons, he wants to be relieved from liabilities under the head charterparty as soon as possible. Modern forms of charter often provide that the charterer is relieved from liabilities when the goods are shipped. A clause in the charter party to this effect is called a 'cesser clause'. For example, clause 31 of the Baltimore Form C Berth Grain Charter Party provides that, 'Charterers' liability under this Charter to cease on cargo being shipped'. This clause is widely worded. It purports to relieve charterers from all liabilities after the cargo has been shipped. On the other hand, clause 8 of the Gencon Charter states that the charterer remains responsible for freight and demurrage to the extent that the shipowner has not been able to obtain payment by exercising the lien on the cargo. If the shipowner is able to exercise a lien, the charterer's liability ceases, actual exercise of the lien being irrelevant. For the charterer to be released from liabilities, it is usually crucial that the shipowner is able to exercise the lien.

> *The Sinoe*[193]
>
> The relevant charterparty included a clause which stated that 'Charterers' liability shall cease as soon as the cargo is on board, where Owners having an absolute lien on the cargo for freight, deadfreight demurrage and average'. The terms of the charterparty was incorporated into the bill of lading. Liability to pay demurrage arose at the port of discharge. Since the cargo had been effectively discharged and the enforcement of any lien was in fact prohibited by the local government, no lien was exercisable. The charterers argued that their liabilities had ceased, since a right of lien was created and whether or not the shipowners were able to exercise it was irrelevant.
>
> The court held that the charterers were liable for demurrage. For their liability to cease, the lien must be exercisable at the time the cargo was discharged.

Generally, a 'cesser clause' is effective in releasing the charterer from liabilities only if, and to the extent that, the shipowner has obtained a right of lien as an alternative remedy. If the shipowner can prove it was impossible or illegal to exercise the lien, he does not lose his right: *The Sinoe*.[194]

192. [1983] 1 Lloyd's Rep 302.
193. [1972] 1 Lloyd's Rep 201.
194. ibid.

Whether difficulty or inconvenience is sufficient depends on the facts of the case.

> *The Tropwave*[195]
> The charter provided that (1) the receivers of the cargo were liable to pay the discharge port demurrage and (2) if they did not pay within a certain period, the charterers should pay. The discharge port demurrage was not paid. The shipowner did not try to exercise any lien. It was later found that it would have been impracticable and ineffective, since the ship would have been sent to the back of the queue for departure.
> The court held that 'if the owner can show that the right of lien was either legally or practically an ineffective right the failure to attempt to exercise it will not debar his claim'. On the facts, the shipowner's lien amounted to a practically ineffective right in law.

As the rights and obligations of the parties are governed by the charterparty, it is still possible, by using special wording, to release the charterer from liabilities under the charterparty upon the shipment of the cargo.

Rights and obligations of shipowner

The general rule is that, as mentioned above, a shipowner's lien only entitles the shipowner to retain the goods until he is paid. The shipowner is generally not entitled to sell the goods subject to the lien. He may sell the goods only if they have been abandoned by all parties concerned and thus become his property: *Enimont Overseas AG v Ro Jugstanrev Zadar (The 'Olib')*.[196] The rights and obligations of shipowners exercising common law liens and contractual liens are not the same and are better considered separately.

A shipowner exercising a common law lien is entitled to take all reasonable steps to maintain the lien. He may, for example, bring them back to the port of departure or other reasonable place, or land the goods at the port of destination and store them in a warehouse. The right to compensation for expenses incurred in maintaining the lien requires some consideration. The general rule is that one exercising his right of lien is not entitled to claim against the owner of the goods expenses incurred in keeping the goods: *Somes v British Empire Shipping Co.*[197] A shipowner may, seemingly in his capacity as a bailee, claim compensation for the costs of warehousing the goods: *Anglo-Polish SS Line v Vickers.*[198] If the exercise of the lien causes

195. [1981] 2 Lloyd's Rep 159.
196. [1991] 2 Lloyd's Rep 108.
197. (1860) 8 HLC 338.
198. (1924) 19 Ll LR 121 at 125.

delay in the discharge of the goods, the shipowner may also claim demurrage for the delay: *Lyle v Cardiff Corp.*[199]

The rights and obligations of a shipowner exercising a contractual lien are mainly governed by the contract. His right to do what is reasonable to maintain the lien is, subject to the contract, similar to that in relation to the common law lien.

Exercise of lien

The two conditions for the exercise of a lien are:

(1) possession of the goods, and
(2) a demand of the money for which the lien is exercised.

The possession element has already been discussed. For the demand element, the shipper is relieved from tendering by a demand for an excessive amount only if he cannot calculate the correct amount to be paid. He is relieved from tendering if the demand is made for a wrong cause or amount and the claimant makes it clear that the retained goods will not be released until full payment of the amount claimed: *Albemarle Supply Co Ltd v Hind & Co.*[200]

Usually, the shipowner exercises his lien by refusing to deliver the goods. The exercise of the shipowner's lien often causes delay in delivery of the cargo subject to it, as well as in the voyage of the ship. Whether the shipowner or the charterer is responsible for the losses sometimes depends on the time at which the shipowner begins to exercise the lien. The time is also relevant because, as mentioned above, the shipowner obtains certain rights only after its exercise.

There is no rule that the lien must be exercised at the port of discharge. The shipowner may exercise his lien by anchoring the ship off the port of discharge, if exercising the lien at the port of discharge would involve unnecessary expense or cause congestion in the port: *Santiren Shipping Ltd v Unimarine (The 'Chrysovalanaou-Dyo').*[201] On the other hand, refusing to carry the cargo further is generally not a proper way of exercising a lien.

> *The Mihalios Xilas*[202]
>
> The *Mihalios Xilas* was time-chartered under a charterparty which provided for payment of hire monthly in advance. At the port of departure,

199. (1899) 5 Com Cas 87 at 94.
200. [1928] 1 KB 307.
201. [1981] 1 Lloyd's Rep 159.
202. [1978] 2 Lloyd's Rep 186.

the charterers paid the advance freight and other charges only after the shipowners had instructed the master not to load until payment. At a bunkering port en route to the port of discharge, the charterers again failed to pay the hire due and other charges. The shipowners gave instructions to the master not to sail until further instructions. In the ensuing dispute on the hire, the charterers argued, among other things, that the shipowners had exercised their lien by halting the ship at the bunkering port.

The court held that a shipowner did not exercise a lien on cargo simply by refusing to carry it further, unless it was impossible to exercise it at the port of discharge and the shipowner would lose possession of the cargo by carrying it further. This is because 'the essence of the exercise of a lien is the denial of possession of the cargo to someone who wants it', and at that point in time no one wants to take delivery of the goods. On the facts, no lien had been exercised.

Usually, it is in the interests of the shipper or consignee to have the goods released as soon as possible. For this purpose, he often provides a bank guarantee or agrees to provide other forms of security to the shipowner. The guarantee may be voidable if its provision is made under economic duress, ie, the financial consequences resulting from the continuous retention of the goods are so serious that no other practical alternative is open to the shipper.

The Alev[203]

The charterers failed to pay the hire to the shipowners. The shipowners claimed the hire and other charges against the holder of the bills of lading, who was not liable to pay any freight. The shipowners threatened to, among other things, to exercise their lien on the cargo, abandon or sell the cargo. Eventually, the holder agreed to pay the amount due to the shipowners and refrain from arresting the ship.

The court held that there was economic duress in this case. The shipowner had knowingly made an illegitimate threat. They were under an obligation to deliver the cargo. By threatening to abandon or sell the cargo, they purported to assert inconsistent rights over the goods. In short, the consent of the holder was overborne. The agreement was avoided because the holder did not enter into it voluntarily.

Loss of lien

If the shipowner parts with the goods, he puts an end to the lien on them. A shipowner who has returned the goods to the shipper for some particular purpose does not lose his lien if the goods are to be afterward returned to

203. [1989] 1 Lloyd's Rep 138.

the shipowner. A shipowner's lien is also not affected if the shipowner loses the possession of the goods by operation of law, such as temporary seizure under legal process: *Wilson v Kymer*.[204]

Usually, a shipowner who has parted with the goods with the intention of making delivery to the owner loses possession of them and thus his lien. However, if the departure is made by storing the goods in a warehouse pending delivery to the consignee, he regains the lien by retaking the goods: *Levy v Barnard*.[205] To protect the innocent shipowner, his lien is not lost if he is induced to part with the goods by fraud: *Wallace v Woodgate*.[206] Delivery of the goods and the collection of freight under bills of lading do not end the lien under the charterparty: *Christie v Lewis*[207] and *Molthes Rederi v Ellerman's Line*.[208]

Conduct of the shipowner which is inconsistent with a lien also puts an end to it. Examples of such conduct are agreeing to make the freight payable after delivery of the goods (*Foster v Colby*)[209]and delivering the goods against a bill of exchange (*Tamvaco v Simpson*).[210]

204. (1813) 1 M & S 157 at 163.
205. (1818) 8 Taun 149.
206. (1824) R & M 193.
207. (1821) 2 B & B 410.
208. [1927] 1 KB 710.
209. (1858) 3 H & N 705.
210. (1866) LR 1 CP 363.

ANNEX 1

A Sample Bill of Lading

Any Container Line **BILL OF LADING**

SHIPPER/EXPORTER	BOOKING NUMBER	BILL OF LADING NUMBER
Export-Import Trade Software, Ltd 201 Arnold Ave. Suite J Pt. Pleasant, NJ 06611 UNITED STATES OF AMERICA	123WEST EXPORT REFERENCES Exporter File Number - N-China123456 Transaction Numner - 45678TN Letter of Credit Number - BB1234566 Forwarder Reference Number - 40/123NN	

CONSIGNEE	FORWARDING AGENT	FMC NO. CHB NO.
EXits China Inc. Friendship Hotel Software Road West Bejing 100001 Mao Sector CHINA [MAINLAND]		

NOTIFY PARTY	ALSO NOTIFY - ROUTING & INSTRUCTIONS
	Keep cargo dry and away from heat.

VESSEL	VOYAGE	FLAG	PLC OF RECEIPT BY PRECARRIER	RELAY POINT	POINT AND COUNTRY OF ORIGIN OF GOODS
SS Neversink	001	US	PORT OF LOADING: NEW YORK, NY	LOADING PIER: Shed 1 Pt Newark	TYPE OF MOVE: Breakbulk
PORT OF DISCHARGE: SHANGHAI, CHINA			PLACE OF DELIVERY BY ON CARRIER	ORIGINALS TO BE RELEASED AT	

PARTICULARS FURNISHED BY SHIPPER

MARKS & NO'S/CONTAINER NO'S	NO.OF PKGS.	DESCRIPTION OF GOODS	WEIGHT	MEASUREMENTS
Cartons 1/15 Freight PrePaid	1	Skid containing 15 cartons 10 ea Quick Assistant Software-6 Diskettes "Smart Software for Exporters" 5 ea Software Kit, For PC's Interlink These commodities, technology or software were exported from the United States in accordance with the Export Administration Regulations. Diversion contrary to U.S. Law prohibited. This is a sample B/L generated by the Quick Assistant for Export Documentation Please call EXits, Inc. @732/899/9030 to order your software.	2450.00 lbs	64.00 CF

FREIGHT CHARGES	RATED AS	PER	RATE	TO BE PREPAID IN U.S. DOLLARS	TO BE COLLECTED IN U.S. DOLLARS	FOREIGN CURRENCY
SUBJECT TO SECTION 7 OF CONDITIONS, IF SHIPMENT IS TO BE DELIVERED TO THE CONSIGNEE WITHOUT RECOURSE ON THE CONSIGNOR, THE CONSIGNOR SHALL SIGN THE FOLLOWING STATEMENT; 'THE CARRIER SHALL NOT MAKE DELILVERY OF THIS SHIPMENT WITHOUT PAYMENT OF FREIGHT AND OTHERLAWFUL CHARGES.'			TOTALS			

IN WITNESS WHEREOF THE CARRIER BY ITS AGENT HAS SIGNED

SIGNATURE OF CONSIGNOR

RECEIVED THE GOODS OR PACKAGES SHIPPER'S LOAD AND COUNT GOODS HEREINAFTER MENTIONED IN APPARENT GOOD ORDER AND CONDITION UNLESS OTHERWISE INDICATED TO BE RELAYED AS HEREIN PROVIDED, THE RECEIPT, CUSTODY, CARRIAGE, DELIVERY, AND TRANSSHIPPING OF THE GOODS ARE SUBJECT TO THE TERMS APPEARING ON THE FACE AND BACK HEREOF, AND CARRIER'S TARIFFS ON FILE WITH THE INTERSTATE COMMERCE COMMISSION AND/OR THE FEDERAL MARITIME COMMISSION, WASHINGTON, D.C

LIABILITY LIMITED TO AMOUNT SPCIFIED IN SEC 16 UNLESS INCREASED VALUE DECLARED BY SHIPPER AS SPECIFIED BELOW:

ORIGINAL BILLS OF LADING ALL OF THE SAME TENOR AND DATE ONE OF WHICH BEING ACCOMPLISHED THE OTHERS TO STAND VOID.

DECLARED VALUE

BY CARRIER

*APPLICABLE ONLY WHEN USED AS A THROUGH BILL OF LADING AFTER MENTIONED IN APPARENT GOOD ORDER AND CONDITION UNLESS

**INDICATE WHETHER ANY OF THE CARGO IS HAZARDOUS MATERIAL UNDER DOT, IMCO,OR OTHER REGULATIONS AND INDICATE THE CORRECT COMMODITY NUMBER IN DESCRIPTION OF PACKAGES AND GOODS ABOVE.

BY FOR SHIPPER

DATE

ANNEX 2

CARRIAGE OF GOODS BY SEA ORDINANCE
THE HAGUE-VISBY RULES
(THE HAGUE RULES AS AMENDED BY THE BRUSSELS PROTOCOLS 1968 AND 1979)

ARTICLE I

In these Rules the following words are employed, with the meanings set out below:

(a) "Carrier" includes the owner or the charterer who enters into a contract of carriage with a shipper.
(b) "Contract of carriage" applies only to contracts of carriage covered by a bill of lading or any similar document of title, in so far as such document relates to the carriage of goods by sea, including any bill of lading or any similar document as aforesaid issued under or pursuant to a charter party from the moment at which such bill of lading or similar document of title regulates the relations between a carrier and a holder of the same.
(c) "Goods" includes goods, wares, merchandise, and articles of every kind whatsoever except live animals and cargo which by the contract of carriage is stated as being carried on deck and is so carried.
(d) "Carriage of goods" covers the period from the time when the goods are loaded on to the time they are discharged from the ship.

ARTICLE II

Subject to the provisions of Article VI, under every contract of carriage of goods by sea, the carrier, in relation to the loading, handling, stowage, carriage, custody, care and discharge of such goods, shall be subject to the responsibilities and liabilities, and entitled to the rights and immunities hereinafter set forth.

ARTICLE III

1. The carrier shall be bound before and at the beginning of the voyage to exercise due diligence to:

    (a) Make the ship seaworthy.
    (b) Properly man, equip and supply the ship.
    (c) Make the holds, refrigerating and cool chambers, and all other parts of the ship in which goods are carried, fit and safe for their reception, carriage and preservation.

2. Subject to the provisions of Article IV, the carrier shall properly and carefully load, handle, stow, carry, keep, care for, and discharge the goods carried.

3. After receiving the goods into his charge the carrier or the master or agent of the carrier shall, on demand of the shipper, issue to the shipper a bill of lading showing among other things:

   (a) The leading marks necessary for identification of the goods as the same are furnished in writing by the shipper before the loading of such goods starts, provided such marks are stamped or otherwise shown clearly upon the goods if uncovered, or on the cases or coverings in which such goods are contained, in such a manner as should ordinarily remain legible until the end of the voyage.
   (b) Either the number of packages or pieces, or the quantity, or weight, as the case may be, as furnished in writing by the shipper.
   (c) The apparent order and condition of the goods.

   Provided that no carrier, master or agent of the carrier shall be bound to state or show in the bill of lading any marks, number, quantity, or weight which he has reasonable ground for suspecting not accurately to represent the goods actually received, or which he has had no reasonable means of checking.
4. Such a bill of lading shall be prima facie evidence of the receipt by the carrier of the goods as therein described in accordance with paragraph 3(a), (b) and (c). However, proof to the contrary shall not be admissible when the bill of lading has been transferred to a third party acting in good faith.
5. The shipper shall be deemed to have guaranteed to the carrier the accuracy at the time of shipment of the marks, number, quantity and weight, as furnished by him, and the shipper shall indemnify the carrier against all loss, damages and expenses arising or resulting from inaccuracies in such particulars. The right of the carrier to such indemnity shall in no way limit his responsibility and liability under the contract of carriage to any person other than the shipper.
6. Unless notice of loss or damage and the general nature of such loss or damage be given in writing to the carrier or his agent at the port of discharge before or at the time of the removal of the goods into the custody of the person entitled to delivery thereof under the contract of carriage, or, if the loss or damage be not apparent, within three days, such removal shall be prima facie evidence of the delivery by the carrier of the goods as described in the bill of lading.

   The notice in writing need not be given if the state of the goods has, at the time of their receipt, been the subject of joint survey or inspection.

   Subject to paragraph 6b is the carrier and the ship shall in any event be discharged from all liability whatsoever in respect of the goods, unless suit is brought within one year of their delivery or of the date when they should have been delivered. This period may, however, be extended if the parties so agree after the cause of action has arisen.

   In the case of any actual or apprehended loss or damage the carrier and the receiver shall give all reasonable facilities to each other for inspecting and tallying the goods.

7. An action for indemnity against a third person may be brought even after the expiration of the year provided for in the preceding paragraph if brought within the time allowed by the law of the Court seized of the case. However, the time allowed shall be not less than three months, commencing from the day when the person bringing such action for indemnity has settled the claim or has been served with process in the action against himself.
8. After the goods are loaded the bill of lading to be issued by the carrier, master, or agent of the carrier, to the shipper shall, if the shipper so demands, be a "shipped" bill of lading, provided that if the shipper shall have previously taken up any document of title to such goods, he shall surrender the same as against the issue of the "shipped" bill of lading, but at the option of the carrier such document of title may be noted at the port of shipment by the carrier, master, or agent with the name or names of the ship or ships upon which the goods have been shipped and the date or dates of shipment, and when so noted, if it shows the particulars mentioned in paragraph 3, shall for the purpose of this article be deemed to constitute a "shipped" bill of lading.
9. Any clause, covenant, or agreement in a contract of carriage relieving the carrier or the ship from liability for loss or damage to, or in connection with, goods arising from negligence, fault, or failure in the duties and obligations provided in this article or lessening such liability otherwise than as provided in these Rules, shall be null and void and of no effect. A benefit of insurance in favour of the carrier, or similar clause shall be deemed to be a clause relieving the carrier from liability.

ARTICLE IV

1. Neither the carrier nor the ship shall be liable for loss or damage arising or resulting from unseaworthiness unless caused by want of due diligence on the part of the carrier to make the ship seaworthy, and to secure that the ship is properly manned, equipped and supplied, and to make the holds, refrigerating and cool chambers and all other parts of the ship in which goods are carried fit and safe for their reception, carriage and preservation in accordance with the provisions of paragraph 1 of Article III.

   Whenever loss or damage has resulted from unseaworthiness the burden of proving the exercise of due diligence shall be on the carrier or other person claiming exemption under this article.
2. Neither the carrier nor the ship shall be responsible for loss or damage arising or resulting from:

   (a) Act, neglect, or default of the master, mariner, pilot, or the servants of the carrier in the navigation or in the management of the ship.
   (b) Fire, unless caused by the actual fault or privity of the carrier.
   (c) Perils, dangers and accidents of the sea or other navigable waters.
   (d) Act of God.
   (e) Act of war.

(f) Act of public enemies.
(g) Arrest or restraint of princes, rulers or people, or seizure under legal process.
(h) Quarantine restrictions.
(i) Act or omission of the shipper or owner of the goods, his agent or representative.
(j) Strikes or lockouts or stoppage or restraint of labour from whatever cause, whether partial or general.
(k) Riots and civil commotions.
(l) Saving or attempting to save life or property at sea.
(m) Wastage in bulk or weight or any other loss or damage arising from inherent defect, quality or vice of the goods.
(n) Insufficiency of packing.
(o) Insufficiency or inadequacy of marks.
(p) Latent defects not discoverable by due diligence.
(q) Any other cause arising without the actual fault or privity of the carrier, or without the fault or neglect of the agents or servants of the carrier, but the burden of proof shall be on the person claiming the benefit of this exception to show that neither the actual fault or privity of the carrier nor the fault or neglect of the agents or servants of the carrier contributed to the loss or damage.

3. The shipper shall not be responsible for loss or damage by the carrier or the ship arising or resulting from any cause without the act, fault or neglect of the shipper, his agents or his servants.
4. Any deviation in saving or attempting to save life or property at sea or any reasonable deviation shall not be deemed to be an infringement or breach of these Rules or of the contract of carriage, and the carrier shall not be liable for any loss or damage resulting therefrom.

5. (a) Unless the nature and value of such goods have been declared by the shipper before shipment and inserted in the bill of lading, neither the carrier nor the ship shall in any event be or become liable for any loss or damage to or in connection with the goods in an amount exceeding 666.67 units of account per package or unit or 2 units of account per kilogramme of gross weight of the goods lost or damaged, whichever is the higher.
(b) The total amount recoverable shall be calculated by reference to the value of such goods at the place and time at which the goods are discharged from the ship in accordance with the contract or should have been so discharged.

The value of the goods shall be fixed according to the commodity exchange price, or, if there be no such price, according to the current market price, or, if there be no commodity exchange price or current market price, by reference to the normal value of goods of the same kind and quality.

(c) Where a container, pallet or similar article of transport is used to consolidate goods, the number of packages or units enumerated in the bill of lading as packed in such article of transport shall be deemed the number of packages or units for the purpose of this paragraph as far as these packages or units are concerned. Except as aforesaid such article of transport shall be considered the package or unit.

(d) The unit of account mentioned in this article is the special drawing right as defined by the International Monetary Fund. The amounts mentioned in subparagraph (a) of this paragraph shall be converted into national currency on the basis of the value of that currency on a date to be determined by the law of the Court seized of the case.

(e) Neither the carrier nor the ship shall be entitled to the benefit of the limitation of liability provided for in this paragraph if it is proved that the damage resulted from an act or omission of the carrier done with intent to cause damage, or recklessly and with knowledge that damage would probably result.

(f) The declaration mentioned in subparagraph (a) of this paragraph, if embodied in the bill of lading, shall be prima facie evidence, but shall not be binding or conclusive on the carrier.

(g) By agreement between the carrier, master or agent of the carrier and the shipper other maximum amounts than those mentioned in subparagraph (a) of this paragraph may be fixed, provided that no maximum amount so fixed shall be less than the appropriate maximum mentioned in that subparagraph.

(h) Neither the carrier nor the ship shall be responsible in any event for loss or damage to, or in connection with, goods if the nature or value thereof has been knowingly mis-stated by the shipper in the bill of lading.

6. Goods of an inflammable, explosive or dangerous nature to the shipment whereof the carrier, master or agent of the carrier has not consented with knowledge of their nature and character, may at any time before discharge be landed at any place, or destroyed or rendered innocuous by the carrier without compensation and the shipper of such goods shall be liable for all damages and expenses directly or indirectly arising out of or resulting from such shipment. If any such goods shipped with such knowledge and consent shall become a danger to the ship or cargo, they may in like manner be landed at any place, or destroyed or rendered innocuous by the carrier without liability on the part of the carrier except to general average, if any.

ARTICLE IV BIS

1. The defences and limits of liability provided for in these Rules shall apply in any action against the carrier in respect of loss or damage to goods covered

by a contract of carriage whether the action be founded in contract or in tort.

2. If such an action is brought against a servant or agent of the carrier (such servant or agent not being an independent contractor), such servant or agent shall be entitled to avail himself of the defences and limits of liability which the carrier is entitled to invoke under these Rules.
3. The aggregate of the amounts recoverable from the carrier, and such servants and agents, shall in no case exceed the limit provided for in these Rules.
4. Nevertheless, a servant or agent of the carrier shall not be entitled to avail himself of the provisions of this article, if it is proved that the damage resulted from an act or omission of the servant or agent done with intent to cause damage or recklessly and with knowledge that damage would probably result.

ARTICLE V

A carrier shall be at liberty to surrender in whole or in part all or any of his rights and immunities or to increase any of his responsibilities and obligations under these Rules, provided such surrender or increase shall be embodied in the bill of lading issued to the shipper. The provisions of these Rules shall not be applicable to charter parties, but if bills of lading are issued in the case of a ship under a charter party they shall comply with the terms of these Rules. Nothing in these Rules shall be held to prevent the insertion in a bill of lading of any lawful provision regarding general average.

ARTICLE VI

Notwithstanding the provisions of the preceding articles, a carrier, master or agent of the carrier and a shipper shall in regard to any particular goods be at liberty to enter into any agreement in any terms as to the responsibility and liability of the carrier for such goods, and as to the rights and immunities of the carrier in respect of such goods, or his obligation as to seaworthiness, so far as this stipulation is not contrary to public policy, or the care or diligence of his servants or agents in regard to the loading, handling, stowage, carriage, custody, care and discharge of the goods carried by sea, provided that in this case no bill of lading has been or shall be issued and that the terms agreed shall be embodied in a receipt which shall be a non-negotiable document and shall be marked as such.

Any agreement so entered into shall have full legal effect.

Provided that this article shall not apply to ordinary commercial shipments made in the ordinary course of trade, but only to other shipments where the character or condition of the property to be carried or the circumstances, terms and conditions under which the carriage is to be performed are such as reasonably to justify a special agreement.

ARTICLE VII

Nothing herein contained shall prevent a carrier or a shipper from entering into any agreement, stipulation, condition, reservation or exemption as to the responsibility and liability of the carrier or the ship for the loss or damage to, or in connection with, the custody and care and handling of goods prior to the loading on, and subsequent to the discharge from, the ship on which the goods are carried by sea.

ARTICLE VIII

The provisions of these Rules shall not affect the rights and obligations of the carrier under any statute for the time being in force relating to the limitation of the liability of owners of sea-going vessels.

ARTICLE IX

These Rules shall not affect the provisions of any international Convention or national law governing liability for nuclear damage.

ARTICLE X

The provisions of these Rules shall apply to every bill of lading relating to the carriage of goods between ports in two different States if:

(a) the bill of lading is issued in a contracting State, or
(b) the carriage is from a port in a contracting State, or
(c) the contract contained in or evidenced by the bill of lading provides that these Rules or legislation of any State giving effect to them are to govern the contract, whatever may be the nationality of the ship, the carrier, the shipper, the consignee, or any other interested person.

ANNEX 3

BILLS OF LADING AND ANALOGOUS SHIPPING DOCUMENT ORDINANCE (Cap 440)

Section 1 Short Title

(1) This Ordinance may be cited as the Bills of Lading and Analogous Shipping Documents Ordinance.
(2) (Omitted as spent)
(3) Nothing in this Ordinance shall have effect in relation to any document issued before this Ordinance comes into operation.

Section 2 Interpretation

(1) In this Ordinance:
'bill of lading', 'sea waybill' and 'ship's delivery order' shall be construed in accordance with section 3;

'contract of carriage':

(a) in relation to a bill of lading or sea waybill, means the contract contained in or evidenced by that bill or waybill; and
(b) in relation to a ship's delivery order, means the contract under or for the purposes of which the undertaking contained in the order is given;

'holder', in relation to a bill of lading, shall be construed in accordance with subsection (2);
'information technology' includes any computer or other technology by means of which information or other matter may be recorded or communicated without being reduced to documentary form;
'telecommunications system' means a system for the conveyance, through the agency of electric, magnetic, electro-magnetic, electro-chemical or electro-mechanical energy, of:

(a) speech, music and other sounds;
(b) visual images;
(c) signals serving for the impartation (whether as between persons and persons, things and things or persons and things) of any matter otherwise than in the form of sounds or visual images; or
(d) signals serving for the actuation or control of machinery or apparatus.

(2) References in this Ordinance to the holder of a bill of lading are references to any of the following persons:

(a) a person with possession of the bill who, by virtue of being the person identified in the bill, is the consignee of the goods to which the bill relates;

(b) a person with possession of the bill as a result of the completion, by delivery of the bill, of any endorsement of the bill or, in the case of a bearer bill, of any other transfer of the bill;

(c) a person with possession of the bill as a result of any transaction by virtue of which he would have become a holder falling within paragraph (a) or (b) had not the transaction been effected at a time when possession of the bill no longer gave a right (as against the carrier) to possession of the goods to which the bill relates,

and a person shall be regarded for the purposes of this Ordinance as having become the lawful holder of a bill of lading wherever he has become the holder of the bill in good faith.

(3) References in this Ordinance to a person's being identified in a document include references to his being identified by a description which allows for the identity of the person in question to be varied, in accordance with the terms of the document, after its issue; and the reference in section 3(3)(b) to a document's identifying a person shall be construed accordingly.

(4) Without prejudice to sections 4(2) and 6, nothing in this Ordinance shall preclude its operation in relation to a case where the goods to which a document relates:

(a) cease to exist after the issue of the document; or

(b) cannot be identified (whether because they are mixed with other goods or for any other reason),

and references in this Ordinance to the goods to which a document relates shall be construed accordingly.

Section 3 Shipping documents etc to which Ordinance applies

(1) This Ordinance applies to the following documents–

(a) bills of lading;

(b) sea waybills; and

(c) ship's delivery orders.

(2) References in this Ordinance to a bill of lading do not include references to a document which is incapable of transfer either by endorsement or, as a bearer bill, by delivery without endorsement; but subject to that, do include references to a received for shipment bill of lading.

(3) References in this Ordinance to a sea waybill are references to any document which is not a bill of lading but:

(a) is such a receipt for goods as contains or evidences a contract for the carriage of goods by sea; and

(b) identifies the person to whom delivery of the goods is to be made by the carrier in accordance with that contract.

(4) References in this Ordinance to a ship's delivery order are references to a document which is neither a bill of lading nor a sea waybill but which contains an undertaking:

(a) that is given under or for the purposes of a contract for the carriage by sea of the goods to which the document relates, or of goods which include those goods; and
(b) by the carrier to a person identified in the document that he will deliver the goods to which the document relates to that person.

Section 4 Rights under shipping documents

(1) Subject to this section, a person who:

(a) becomes the lawful holder of a bill of lading;
(b) becomes (without being an original party to the contract of carriage) the person to whom delivery of goods to which a sea waybill relates is to be made by the carrier in accordance with that contract; or
(c) becomes the person to whom delivery of goods to which a ship's delivery order relates is to be made in accordance with the undertaking contained in the order,

shall (by virtue of becoming the holder of the bill or, as the case may be, the person to whom delivery is to be made) have transferred to and vested in him all rights of suit under the contract of carriage as if he had been a party to that contract.

(2) Where, when a person becomes the lawful holder of a bill of lading, possession of the bill no longer gives a right (as against the carrier) to possession of the goods to which the bill relates, that person shall not have any rights transferred to him by virtue of subsection (1) unless he becomes the holder of the bill:

(a) by virtue of a transaction effected in pursuance of any contractual or other arrangements made before the time when such a right to possession ceased to attach to possession of the bill; or
(b) as a result of the rejection to that person by another person of goods or documents delivered to the other person in pursuance of any such arrangements.

(3) The rights vested in any person by virtue of the operation of subsection (1) in relation to a ship's delivery order:

(a) shall be so vested subject to the terms of the order; and
(b) where the goods to which the order relates form a part only of the goods to which the contract of carriage relates, shall be confined to rights in respect of the goods to which the order relates.

(4) Where, in the case of any document to which this Ordinance applies:

(a) a person with any interest or right in or in relation to goods to which the document relates sustains loss or damage in consequence of a breach of the contract of carriage; and
(b) subsection (1) operates in relation to that document so that rights of suit in respect of that breach are vested in another person, the other person shall be entitled to exercise those rights for the benefit of the person who sustained the loss or damage to the same extent as they could have been exercised if they had been vested in the person for whose benefit they are exercised.

(5) Where rights are transferred by virtue of the operation of subsection (1) in relation to any document, the transfer for which that subsection provides shall extinguish any entitlement to those rights which derives:

(a) where that document is a bill of lading, from a person's having been an original party to the contract of carriage; or
(b) in the case of any document to which this Ordinance applies, from the previous operation of that subsection in relation to that document,

but the operation of that subsection shall be without prejudice to any rights which derive from a person's having been an original party to the contract contained in, or evidenced by, a sea waybill and, in relation to a ship's delivery order, shall be without prejudice to any rights deriving otherwise than from the previous operation of that subsection in relation to that order.

Section 5 Liabilities under shipping documents

(1) Where section 4(1) operates in relation to any document to which this Ordinance applies and the person in whom rights are vested by virtue of that subsection:

(a) takes or demands delivery from the carrier of any of the goods to which the document relates;
(b) makes a claim under the contract of carriage against the carrier in respect of any of those goods; or
(c) is a person who, at a time before those rights were vested in him, took or demanded delivery from the carrier of any of those goods,

that person shall (by virtue of taking or demanding delivery or making the claim or, in a case falling within paragraph (c), of having the rights vested in him) become subject to the same liabilities under that contract as if he had been a party to that contract.

(2) Where the goods to which a ship's delivery order relates form a part only of the goods to which the contract of carriage relates, the liabilities to which any person is subject by virtue of the operation of this section in relation to that order shall exclude liabilities in respect of any goods to which the order does not relate.

(3) This section, so far as it imposes liabilities under any contract on any person, shall be without prejudice to the liabilities under the contract of any person as an original party to the contract.

Section 6 Representations in bills of lading

A bill of lading which:

(a) represents goods to have been shipped on board a vessel or to have been received for shipment on board a vessel; and
(b) has been signed by the master of the vessel or by a person who was not the master but had the express, implied or apparent authority of the carrier to sign bills of lading,

shall, in favour of a person who has become the lawful holder of the bill, be conclusive evidence against the carrier of the shipment of the goods or, as the case may be, of their receipt for shipment.

Section 7 Regulations

(1) The Secretary for Trade and Industry may by regulation make provision for the application of this Ordinance to cases where a telecommunications system or any other information technology is used for effecting transactions corresponding to:

(a) the issue of a document to which this Ordinance applies;
(b) the endorsement, delivery or other transfer of such a document; or
(c) the doing of anything else in relation to such a document.

(2) Regulations under subsection (1) may:

(a) make such modifications as the Secretary for Trade and Industry considers appropriate in connection with the application of this Ordinance to any case mentioned in that subsection; and
(b) contain supplemental, incidental, consequential and transitional provisions.

Section 8 Hague-Visby rules

The provisions of this Ordinance shall have effect without prejudice to the application, in relation to any case, of the provisions of the International Convention for the unification of certain rules of law relating to bills of lading signed at Brussels on 25 August 1924 as amended by the Protocol signed at Brussels on 23 February 1968 (known as the Hague-Visby Rules), so far as those Rules have the force of law in Hong Kong.

7

Air Carriage Law

7.1 INTRODUCTION

In Hong Kong, legislation is what mainly governs the legal relations between air carriers and consignors or consignees. The application of the common law is very limited.[1] The most important piece of legislation is the Carriage by Air Ordinance (Cap 500) ('the CAO'). The main purposes of the CAO are to provide for a unified regime for air carriage and to incorporate the main international conventions into Hong Kong law.

Under the CAO, there are two types of carriage of goods by air: international and non-international. For international carriage of goods by air, the CAO gives effect to the Convention for the Unification of Certain Rules Relating to International Carriage by Air, signed at Warsaw on 12 October 1929 ('the Warsaw Convention') and the Warsaw Convention as amended by the Hague Protocol signed on 28 September 1955 ('the amended Convention'). The Warsaw Convention and the amended Convention are respectively set out in schedule 4 and schedule 1 to the CAO. If the carrier who actually performs the carriage is not the carrier concluding the carriage of goods by air contract with the consignor, the rights and obligations of the carriers are governed by the Guadalajara Convention 1961. The Guadalajara Convention is set out in schedule 2 to the CAO. A modified form of the amended Convention set out in schedule 3 ('the modified

1. See *Antwerp Diamond Bank NV v Brink's Inc* [2013] 1 HKLRD 396, as to right to sue at common law the Court held that at common law, a pledge could be created only by delivery or constructive delivery of the item pledged to give the bank the possessory right to found an action for conversion.

amended Convention') applies to non-international carriage, and carriage of mail and postal packages.

Many of the provisions of the Warsaw Convention, the amended Convention and the modified amended Convention are identical.[2] This chapter considers the amended Convention in detail. Relevant provisions in the Warsaw Convention are also mentioned where there are substantial changes between the two conventions. Topics to be discussed are application of convention, air waybill, rights and obligations of carriers, consignors and consignees, limits of carriers' liabilities and the procedures of bringing claims.

Two other conventions should also be mentioned. Hong Kong has not ratified Additional Protocol No 4 to amend the amended Convention signed at Montreal on 25 September 1975 ('Montreal Additional Protocol No 4'). Although Montreal Additional Protocol No 4 has come into force, it is for the moment not part of Hong Kong law. The Convention for the Unification of Certain Rules for International Carriage by Air, signed at Montreal on 28 May 1999 ('the Montreal Convention'), will enter into force after ratification by 30 countries.

The Montreal Convention entered into force on 4 November 2003 and it entered into force in the PRC on 31 July 2005. In consultation with the government of the Hong Kong Special Administrative Region, the government of the PRC decided to apply the Convention in the Hong Kong Special Administrative Region of the PRC from 15 December 2006. The Montreal Convention is set out in schedule 1A to the CAO.

7.2 THE AMENDED CONVENTION

Application

In general, only one convention governs one carriage of goods by air. The first issue in an air cargo claim is to decide which convention applies. Generally speaking, either the Warsaw Convention or the amended Convention applies to an international carriage of goods performed by an air transport undertaking for reward, while the modified amended Convention applies to non-international carriage by air. However, the Warsaw Convention and the amended Convention do not apply to carriage performed in extraordinary

2. See *Nantong Angang Garments Co Ltd v Hellmann International Forwarders Ltd* [2012] HKEC 782 (CA), the carriage 'would have been subject to the Warsaw Convention'. It is unnecessary to distinguish between the Warsaw Convention and the Amended Convention, the articles relevant to this appeal (Articles 5 to 14) are not in any material respect different. The Court contents to work on the basis of the Amended Convention.

Table 7.1 The Warsaw-System Conventions

1. Warsaw Convention 1929	6. Warsaw Convention 1929 supplemented by Guadalajara Convention 1961
2. Warsaw-Hague Convention 1955	7. Warsaw-Hague Convention 1955 supplemented by Guadalajara Convention 1961
3. Warsaw-MAP 1 Convention 1975	8. Warsaw-MAP 1 Convention 1975 supplemented by Guadalajara Convention 1961
4. Warsaw-Hague-MAP 2 Convention 1975	9. Warsaw-Hague-MAP 2 Convention 1975 supplemented by Guadalajara Convention 1961
5. Warsaw-Hague-MAP 4 Convention 1975	10. Warsaw-Hague-MAP 4 Convention 1975 supplemented by Guadalajara Convention 1961

circumstances outside the normal scope of an air carrier's business. Further, the Warsaw Convention does not apply to carriage performed by way of experimental trial.

The conventions also apply to carriage performed by the state or legally constituted public bodies.

Reward

The amended Convention applies to all international carriage performed by aircraft for reward. The term 'reward' should have the same meaning as 'remuneration' but not 'profit'.[3] Thus, a carriage performed by a non-profit organization is also governed by the amended Convention.

Gratuitous carriage

The amended Convention also applies to gratuitous carriage performed by an air transport undertaking. The term 'air transport undertaking' is not defined. It seems that a party who uses an aircraft for any commercial purposes is an air transport undertaking.[4] In this case, a company providing air

3. See *Corner v Clayton* [1976] 2 Lloyd's Rep 295.
4. See *Gurtner v Beaton* [1993] 2 Lloyd's Rep 369.

transport services to employees with its own aircraft is taken to be an air transport undertaking.

Only one type of carriage is not covered by the CAO. This is gratuitous carriage not performed by an air transport undertaking. For this type of carriage, the common law applies. For example, the carriage of a notebook computer by a pilot of a private plane who does this for free for a friend is not governed by the CAO. If the computer is damaged during the carriage, the rights and obligations between the pilot and the owner of the notebook are determined by the common law. In Hong Kong, this kind of carriage is rare. It should, however, be noted that, by virtue of section 19 of the CAO, the amended Convention also applies to gratuitous carriage performed by the Hong Kong government.

International carriage

The amended Convention only applies to 'international carriage'. The term does not refer to international carriage in the ordinary sense. It is defined as:

> any carriage in which, according to the agreement between the parties, the place of departure and the place of destination, whether or not there be a break in the carriage or a transshipment, are situated either within the territories of two High Contracting Parties or within the territory of a single High Contracting Party if there is an agreed stopping place within the territory of another State, even if that State is not a High Contracting Party.

For a carriage between two countries, the test is whether the places of departure and destination are, 'according to the agreement', situated in two High Contracting Parties. The carriage of goods contract is, therefore, crucial to determine whether it is an 'international carriage'. What actually happens is irrelevant. For example, if only the place of departure stated in the contract is situated in a High Contracting Party, the carriage is not an international carriage. It does not become an 'international carriage' even if it in fact ends, due to an emergency, in another High Contracting Party.

If the place of departure and the place of destination are within a High Contracting Party, the issue is whether there is an 'agreed stopping place' in another country. The carriage of goods contract is again relevant.

> *Grein v Imperial Airways* [1936] All ER 1258
> The freight in issue was a journey from London to Antwerp and back. The negligence of the air carriers caused an accident upon the return journey. As a result, a passenger died. The air carriers could limit their liability if the journey was an 'international carriage' as defined by the

Warsaw Convention. The question was whether the journey was one from London to London with Antwerp as a stopping place or one from London to Antwerp with an undertaking given by the carrier to carry the passenger back from Antwerp to London.

The court held that the phrase 'the place of departure and the place of destination' had the *prima facie* meaning of 'the place at which the contractual carriage begins and the place at which the contractual carriage ends.' The phrase 'agreed stopping place' referred to 'a place where, according to the contract the machine by which the contract is to be performed will stop in the course of performing the contractual carriage, whatever the purpose of the descent may be, and whatever rights the passenger may have to break his journey at that place.' It included, in the case of a return journey, the farthest place out to or back from which the passenger was, under the contract, to be carried. The journey in issue was therefore an 'international carriage'.

High contracting parties

It is likely that Hong Kong is named as both the place of departure and the place of destination in a carriage of goods by air contract. The local market of commercial carriage of goods by air for reward by helicopter in Hong Kong is relatively small. It is practically impossible for an 'agreed stopping place' within another state to be inserted into the contract. If the amended Convention applies, the place of departure and the place of destination are normally situated in Hong Kong and another state. It is, thus, more important to know which countries are High Contracting Parties to the Warsaw Convention and the amended Convention.

The Chief Executive may, by order in the Gazette, certify which countries are High Contracting Parties to the Warsaw Convention or the amended Convention. It should be noted that countries which are High Contracting Parties to the amended Convention are, normally, also High Contracting Parties to the Warsaw Convention. An order given by the Chief Executive is conclusive evidence of the status of the countries certified. In other words, countries not so certified are, for all purposes, not High Contracting Parties to the relevant convention.[5] A country which has denounced the convention is still taken to be a High Contracting Party unless and until it is removed from a new order superseding the existing one.

Hong Kong trading partners that are High Contracting Parties to both the Warsaw Convention and the amended Convention include Argentina, Australia, Austria, Belgium, Brazil, Bulgaria, Canada, Chile, China, Congo,

5. *Philippson v Imperial Airways Ltd* [1938] 1 All ER 759 (Court of Appeal); [1939] AC 332 (House of Lords).

Cuba, Cyprus, Denmark, Egypt, France, Germany, Greece, Hungary, Iceland, India, Iran, Iraq, Ireland, Israel, Italy, Japan, Kuwait, Laos, Lebanon, Luxembourg, Malaysia, Mexico, Nepal, the Netherlands, New Zealand, Norway, Pakistan, the Philippines, Poland, Portugal, Saudi Arabia, Singapore, South Africa, Spain, Swaziland, Sweden, Syria, Tunisia, Turkey, Russia, the United Kingdom, Vietnam and Zambia.

Some High Contracting Parties to the Warsaw Convention are Barbados, Botswana, Burma, Cambodia, Ethiopia, Gambia, Indonesia, Jamaica, Liberia, Malta, Mongolia, Sri Lanka, Tanzania, Uganda and, most importantly, the United States of America.

If the carriage is not within one country, the countries of departure and destination must both be High Contracting Parties to the same convention for it to apply. If the two countries are High Contracting Parties to the amended Convention, the amended Convention applies. As Hong Kong is signatory to both the Warsaw Convention 1929 and the Hague Protocol 1955, whether the Warsaw Convention or the amended Convention applies to a carriage that departs from or arrives in Hong Kong depends on the status of the other country. For example, the Warsaw Convention governs a carriage between Hong Kong and the USA, since the USA is a High Contracting Party to the Warsaw Convention but has not ratified the amended Convention. In the controversial case of *Re Korean Airlines Disaster*, 1 September 1983 (District Court, District of Columbia), a court in the USA held that a carriage from the USA to South Korea was governed by the Warsaw Convention, despite the fact that South Korea had ratified only the amended Convention. Since Hong Kong has joined both conventions, this ruling should not be a concern to Hong Kong carriers. On the other hand, a Hong Kong–New Zealand carriage is subject to the amended Convention, since New Zealand is a High Contracting Party to it.

In deciding which convention applies, the nationality of the aircraft is irrelevant.

Extraordinary and experimental carriage

The amended Convention and the Warsaw Convention do not apply to international carriage performed in extraordinary circumstances outside the scope of an air carrier's business. The following Canadian case provides an example of carriage in extraordinary circumstances.

> *Johnson Estates v Pischke* [1989] 3 WWR 207 (Sask)
>
> An aircraft with three passengers crashed in the mountains. The pilot (D) was a licensed instructor. D had no licence to carry passengers in

international flight. By performing the carriage, D wanted to gain more flying hours. The passengers were family members. Two were under D's instruction and the other wanted to pick up a dog.

The court held that the carriage was performed in extraordinary circumstances outside the normal scope of an air carrier's business.

Article 34 of the Warsaw Convention further states that the Convention does not apply to carriage performed by way of experimental trial by air navigation undertakings with the view to the establishment of a regular line of air navigation. At the time the Warsaw convention was prepared, aviation was still in its infancy, the inclusion of this exception was reasonable. Now that air transport is common, it is generally not relevant to an air carrier.

Successive carriage

It is possible for one consignment of goods to be covered by many contracts performed by different carriers. For example, the cargo may be carried by Carrier A from Hong Kong to New York under one contract and then by Carrier B from New York to Boston under another. If the carriage as a whole, ie, Hong Kong–New York–Boston, is one single operation, it is an 'international carriage' governed by the Warsaw Convention. If it is not a single operation, there are two separate carriages, one from Hong Kong to New York and the other from New York to Boston. If so, only the carriage performed by Carrier A is an 'international carriage'. The part performed by Carrier B is domestic in nature since it is within the same country. The Warsaw Convention governs only the part performed by Carrier A.

In such a case, the intention of the parties determines whether the carriage as a whole is an 'international carriage' or there are several carriages covering the consignment in segments. It is taken to be 'one undivided carriage' if the parties have regarded it as a single operation. It was held in *Parke, Davis & Company v BOAC* (New York City Court, 30 January 1958) that such a carriage might still be an undivided one, despite the fact that different carriers issued separate air waybills covering different parts of the carriage. Besides, the international character is not lost by the fact that some of the contracts are performed entirely within one country.

7.3 AIR WAYBILL

As discussed above, the agreement between the parties has an important bearing on the issue of 'international carriage'. The air waybill is, therefore, an important document to determine the applicability of the amended

Convention. If the amended Convention applies, the air waybill still has important roles to play.

The amended Convention does not touch upon every aspect of the legal relation among the consignor, the carrier and the consignee. For matters not governed by it, the terms of the contract of carriage regulate the parties' rights and obligations. In the case of combined carriage, Article 31 allows terms relating to other modes of carriage to be incorporated into the contract. On the other hand, Article 32 provides that the amended Convention does not restrict the carrier's right of refusal. The parties may also incorporate other terms into the contract. However, such terms cannot contravene the amended Convention. For example, the amended Convention sets the limits of the carrier's liability for damage of cargo during the carriage by air. A lower limit laid down in the contract is of no effect.

To fully understand the legal relation between the carrier and the consignor, the terms of the contract of carriage must be referred to. For example, the amended Convention does not govern cargo damaged or lost outside the carriage by air. Instead, the terms of the contract of carriage determine the extent of carrier's liabilities. Usually, the contract is contained in or evidenced by an air waybill. In practice, the IATA air waybill is widely used. Condition 4 of the IATA air waybill reads as follows:

> Except as otherwise provided in carrier's tariffs or Conditions of Carriage, in carriage to which the Warsaw Convention [and the amended Convention] does not apply, carrier's liability shall not exceed U.S.D. 20.00 or the equivalent per kilogramme of goods lost, damaged or delayed, unless a higher value is declared by the shipper and a supplementary charge paid.

The term 'carriage' is defined in Article 1 of the IATA air waybill as 'Carriage of cargo by air or by another means of transport, whether gratuitously or for reward.' For loss of or damage to cargo caused by an event occurring outside the 'carriage by air' defined by the amended Convention, Condition 4, therefore, applies.

Under the amended Convention, the carrier may require the consignor to make out an air waybill. On the other hand, the consignor may require the carrier to accept the air waybill. The original of an air waybill is divided into three parts. One is marked 'for the carrier' and signed by the consignor. Another is marked 'for the consignee' and signed by both the consignor and the carrier. The third is signed by the carrier and should be returned to the consignor after the cargo has been accepted.

Separate waybills

Where there are several packages, the carrier may ask for separate waybills covering different packages. The carrier must sign the air waybill before the goods are loaded on board the aircraft. Although it is the responsibility of the consignor to make out the air waybill, the carrier may do so upon the request of the consignor. In doing so, the carrier is prima facie deemed to act on behalf of the consignor. In other words, the consignor is still liable to third parties for the accuracy of statements inserted by the carrier, unless the consignor can prove the carrier was not his agent in making out the air waybill.

Content of air waybill

Article 8 of the amended Convention specifies what must be contained in the air waybill:

(1) It must indicate the places of departure and destination.
(2) If the two places are within a High Contracting Party, at least one agreed stopping place must be indicated.
(3) There must be a notice that the carriage is governed by the amended Convention and the carrier's liability is limited in respect of loss of or damage to the goods in most cases. The notice must also mention the possibility that, if the place of destination is not within the country of departure, the Warsaw Convention may apply.

If the Warsaw Convention applies, the air waybill must contain the following particulars:

(a) place and date of its execution;
(b) place of departure and of destination;
(c) agreed stopping places; provided that the carrier may reserve the right to alter the stopping places in case of necessity, and that if the carrier exercises that right the alteration shall not have the effect of depriving the carriage of its international character;
(d) name and address of the consignor;
(e) name and address of the first carrier;
(f) name and address of the consignee, if the case so requires;
(g) nature of the cargo;
(h) the number of packages, the method of packing and the particular marks or numbers upon them;

(i) weight, quantity and volume or dimensions of the cargo;
(j) apparent condition of the cargo and its packing;
(k) freight, if it has been agreed upon, the date and place of payment, and the person who is to pay it;
(l) if the cargo is sent for payment on delivery, the price of the cargo, and, if the case so requires, the amount of the expenses incurred;
(m) amount of the value declared in accordance with paragraph (2) of Article 22;
(n) number of parts of the air waybill;
(o) documents handed to the carrier to accompany the air waybill;
(p) time fixed for the completion of the carriage and a brief note of the route to be followed, if these matters have been agreed upon;
(q) a statement that the carriage is subject to the rules relating to liability established by the Warsaw Convention.

The parties may also insert other statements or information in the air waybill.

In general, the validity of the contract for carriage and the application of the amended Convention are not affected by the absence, irregularity or loss of the air waybill.

Air waybill and unlimited liability

The air waybill is crucial if the carrier wants to limit his liability. Under the amended Convention, the carrier cannot limit liability if:

(a) the cargo is, with the consent of the carrier, loaded on board the aircraft without an air waybill having been made out; or
(b) the air waybill does not contain the notice required by Article 8.

If the Warsaw Convention applies, the carrier is subject to unlimited liability if

(a) the carrier accepts goods without an air waybill having been made out; or
(b) the air waybill does not contain all the items set out in Article 8 (a) to (i) and (q).

Evidential value of air waybill

The air waybill is prima facie evidence of the contract of carriage, receipt of the cargo and conditions of carriage. The statements in it are also evidence of the stated weight, dimensions, packing of the cargo or number of packages. Other evidence may, however, be adduced to rebut these statements.

Statements in the air waybill relating to the quantity, volume and condition of the cargo are generally not evidence against the carrier. The exceptions to this rule are:

- The statements have been checked by the carrier in the presence of the consignor and are so stated in the air waybill.

Or

- The statements relate to the apparent condition of the cargo.

Negotiable air waybill

An air waybill may be negotiable or non-negotiable. Negotiable in this context means that the air waybill is transferable. A negotiable air waybill is not a negotiable instrument. It cannot confer a better title than what the consignor or the named consignee has on the transferee. After the transfer, the transferee is entitled to take delivery.

7.4 PARTIES' RIGHTS AND LIABILITY

The amended Convention governs the rights and obligations of the consignor, carrier and consignee within its scope. The main concern of the amended Convention is the carrier's liability during the carriage by air. It also states the rights and obligations of the consignor and the consignee.

As an instrument providing a set of unified international rules, the amended Convention may exclude the application of domestic law. This is confirmed by Lord Hope in *Sidhu v British Airways Plc* [1997] AC 430, 453:

> The convention does not purport to deal with all matters relating to contracts of international carriage by air. But in those areas with which it deals—and the liability of the carrier is one of them—the code is intended to be uniform and to be exclusive also of any resort to the rules of domestic law.

The US Supreme Court has come to the same conclusion. It was held in *El Al Israel Airlines Ltd v Tsui Yaun Tseng* (1999) 525 US 155 that:

> Given the Convention's comprehensive scheme of liability rules and its textual emphasis on uniformity, we would be hard put to conclude that the delegates at Warsaw meant to subject air carriers to the distinct non-uniform liability rules of the individual signatory nations.

Further, the amended Convention or the Warsaw Convention prevails if there is conflict between the contract of carriage and the convention applicable in the case, unless it provides otherwise. For example, the carrier's liability for the loss of or damage to the goods cannot be reduced or restricted by the contract. But the rights and obligations with respect to disposition and delivery of cargo under Articles 12, 13 and 14 are subject to express provisions in the air waybill.

Since the main concern of the Warsaw regime is carrier's liability to consignor or consignee, it does not govern the legal relations between the consignor and consignee. Article 15 of the amended Convention states explicitly that Articles 12, 13 and 14 do not affect the legal relations between the consignor and consignee or other parties who derive their rights from them. The rights and obligations between the carrier and other parties are also not covered by the amended Convention. If the negligence of the carrier causes damage to another aircraft, the owner of the aircraft is, subject to any relevant legislation, entitled to sue the carrier in the law of negligence.

7.5 CARRIER'S RIGHTS AND OBLIGATIONS

The Warsaw regime provides a legal regime different from the common law system to determine the rights and obligations of carriers, consignor and consignee. At common law, the carrier is liable only if it is proved to be at fault. Once the claimant proves the carrier was in breach of contract, the carrier is liable for all the reasonably foreseeable losses suffered by the claimant. Generally, the carrier is subject to unlimited liability. Under the Warsaw regime, the carrier is in principle liable for any damage, loss and delay if the claimant can prove it occurred during carriage by air. The carrier is not liable only if any of the special defences can be proved. In return for this presumed liability, the carrier's liability is generally limited. Liability becomes unlimited only in certain specified circumstances.

Right to be indemnified

The carrier's right to get compensation or indemnity from the consignor is associated with the consignor's related obligations under the amended Convention. For example, the carrier is entitled to indemnity for:[6]

6. See *Nantong Angang Garments Co Ltd v Hellmann International Forwarders Ltd* [2012] HKEC 782 (CA), the Court held that it is clear that the letter of indemnity was intended and understood by the parties that Hellmann was to be indemnified for the consequence of the carriage of the garments to Hong Kong; it is a common practice for carrier to request for a letter of indemnity.

- any damage suffered as a result of the irregularity, incorrectness or incompleteness of the particulars and statements furnished by the consignor under Article 10, or
- the expenses occasioned by the consignor's order of disposal under Article 12.

Presumed liability for loss, damage and delay

Under the Warsaw regime, the carrier is generally liable for the loss of and damage to the cargo during the carriage by air and for delay.[7] In the event that several successive carriers perform an international carriage, the liability of different carriers becomes an issue. Article 30 states that each carrier who accepts the cargo is deemed to be a party to the contract of carriage in so far as the contract deals with that part of the carriage performed under the carrier's supervision. A carrier who accepts the cargo from another carrier is, for the part under his supervision, liable as the carrier under the amended Convention.

For loss, damage or delay, the consignor may always sue the first carrier. The consignee may, on the other hand, sue the last carrier. If the carrier who performed the carriage during which the damage was caused is known, the consignor or consignee may also take an action against them. These carriers are jointly and severally liable to the claimant.

The time at which the damage manifests itself is irrelevant. For example, the amended Convention or the Warsaw Convention applies in a case where a computer malfunctions six months after the carriage if the malfunctioning results from a latent defect caused by the fault of the carrier during the carriage by air.

Carriage by air

The term 'carriage by air' comprises:

> the period during which the . . . cargo is in the charge of the carrier, whether in an aerodrome or on board an aircraft, or in the case of a landing outside an aerodrome, in any place whatsoever.

If the plane lands outside an aerodrome, the cargo is in the course of carriage by air so long as it is in the charge of the carrier.

7. See Articles 18(1) and 19 of the amended Convention.

> *Manohar Gangaram Ahuja v Hill & Delmain (Hong Kong Ltd)* [1993] HKLY 181
>
> The consignor, M, concluded a contract with D for the carriage of goods by air from Hong Kong to Brussels, Belgium. The air waybill stated the place of destination as Brussels. In fact, the goods were carried by air from Hong Kong to Schipol Airport, Amsterdam, Netherlands and then by land from Schipol Airport to Brussels Airport. The goods were misdelivered at Brussels Airport. The amended Warsaw Convention applied to Hong Kong, the Netherlands and Belgium. M sued D for damages. D contended that the misdelivery occurred in the course of 'carriage by air' and thus the amended Warsaw Convention applied and set the limit of the carrier's liability.
>
> The court held that the amended Warsaw Convention governed. The goods were still during 'carriage by air' when the misdelivery took place.

The period of carriage by air does not extend to other modes of carriage performed outside an aerodrome. However, if such a carriage is done in the performance of a carriage by air contract and for the purpose of loading, delivery or transhipment, the presumption is that any loss or damage is occasioned during the carriage by air, unless it is proved to the contrary. Sometimes, the claimant wants to prove the damage was not caused during carriage by air. If it can be proved the cargo was in fact damaged when it was being transported outside the airport, the common law rules apply and the carrier is subject to unlimited liability.

The operation of the rules is illustrated by the following example. A carrier was required, under the carriage by air contract, to deliver the cargo by truck to the named consignee in Causeway Bay after it had arrived in Hong Kong from Egypt. When the cargo was being delivered to the consignee, it was damaged in a traffic accident in Central caused by the negligence of the carrier. The damage is presumed to be caused by an occurrence taking place during the carriage by air. The carrier's liability is subject to the amended Convention unless the consignor can prove the damage was in fact caused in the traffic accident.

In charge of cargo

It is a matter of fact whether the carrier is in charge of the cargo.[8] Generally, the cargo is in the charge of the carrier if the carrier has it in possession or control. Carriers are in charge of the cargo if they keep it for the purposes of safekeeping, custody or care.

8. *Swiss Bank Corp. v Brink's-Mats Ltd* [1986] 2 Lloyd's Rep 79.

Aerodrome

The old-fashioned term 'aerodrome' means 'airport'.[9] The version of the Warsaw Convention adopted in the United States in fact uses the term 'airport'. Giving the term a strict interpretation by reference to the geographical limit of airport, a US court held in *Victoria Sales Corp v Emery Air Freight Inc* 917 F.2d 705 (1990) that cargo in the carrier's warehouses outside the airport perimeter were not in the airport.

Aircraft

The definition of 'carriage by air' also refers to 'aircraft'. The term 'aircraft' is not defined in this context. It is common sense that it must cover aeroplanes and helicopters. It is less clear whether it covers hovercrafts. Since the CAO does not define the term, the general definition given by the Interpretation and General Clauses Ordinance (Cap 1) should apply if no contrary intention appears. Under section 3 of the Interpretation and General Clauses Ordinance, 'aircraft' is defined as:

> any machine that can derive support in the atmosphere from the reactions of the air.

This is a very wide definition. Helicopters, gliders, hovercrafts and even balloons are clearly included. This domestic definition may not apply because the amended Convention is meant to provide a set of unified rules with respect to international carriage. It is instructive to look into the definition given by the International Civil Aviation Organization (ICAO) under the Chicago Convention on International Civil Aviation:

> Any machine that can derive support in the atmosphere from the reactions of the air other than the reactions of the air against the earth's surface.

This definition is similar to but narrower than the one given by the Interpretation and General Clauses Ordinance. Obviously, it does not cover hovercrafts. In view of the similar natures of the Chicago Convention and the amended Convention, the better view is that the ICAO definition is preferable.

9. *Rolls Royce Ltd v Heavylift-Volga Dnepr Ltd* [2000] 1 All ER (Comm) 796.

Delay

Under Article 19 of the amended Convention, the carrier is generally liable for damage occasioned by delay in the carriage by air. The Warsaw regime is not relevant to delays caused by events which did not take place in carriage by air.

Defence

Carriers are prima facie liable for any loss, damage or delay caused by incidents occurring during carriage by air. They must prove one of the two special defences provided in Article 20 if they want to escape liability.

One defence is that they and their servants or agents have taken all necessary measures to avoid the loss, damage or delay. Another defence is that it was impossible for carriers or their servants or agents to take such measures. Although the term 'all necessary measures' should not be given a strictly literal interpretation, it connotes a standard of skill and care higher than the standard of reasonableness applied in the law of negligence. Carriers cannot escape liability by merely proving that they were not negligent. A party who has taken 'some' reasonable measures is generally not negligent. But this is not enough for the carrier to escape liability under the Warsaw regime. All measures reasonably necessary in the circumstances must be taken. It was held in the US case of *Manufacturers Hanover Trust Co v Alitalia Airlines*, 429 F Supp 964 (1977) that

> [the first limb of] Article 20 requires of defendant proof, not of a surfeit of preventatives, but rather, of an undertaking embracing all precautions that in sum are appropriate to the risk, i.e., measures reasonably available to defendant and reasonably calculated, in cumulation, to prevent the subject loss.

The following case is instructive.

> *Rugani v KLM*, City Court, New York County, January 20, 1954
> A carrier placed some expensive furs in his store in an airport before the carriage. For security purposes, the carrier had an unarmed guard on duty. The furs were stolen in an armed robbery.
>
> The court held that the carrier had not taken all necessary measures, because the guard was unarmed and therefore not able to provide effective protection against armed robbers.

To establish the second limb of the defence, the carrier must prove that the damage was 'inevitable, or at least which no human precaution or foresight would have prevented'. This is even more difficult to prove. For all practical purposes, it is not as important as the first limb. One rare case in which the defence of impossibility was successfully pleaded is about the loss of some carpets from an aircraft stranded at Kuwait Airport after the Iraqi invasion in 1990. The Frankfurt Higher Regional Court held that it was impossible for the carrier to take any measures to prevent the loss because no one could have contemplated the invasion.[10]

Error in navigation under the Warsaw Convention

Under Article 20(2) of the Warsaw Convention, error in piloting, in the handling of the aircraft or in navigation provides another defence to the carrier. The carrier is required to prove

(a) damage was occasioned by negligent pilotage or negligence in the handling of the aircraft or in navigation; and
(b) in all other respects, carriers and their servants or agents have taken all necessary measures to avoid the damage.

The operation of the defence is illustrated in a case about an air accident in Hong Kong. It was held in *American Smelting & Refining Co v Philippine Airlines*, 4 AVI 17,413 1954 that

> The proof adduced upon the trial conclusively established that [the carrier] took all possible precautions to ensure the safety of the flight and to avoid the crash of the aircraft. The record shows that [the carrier] properly equipped, loaded and fuelled the plane, supplied an airworthy and duly licensed aircraft, a licensed and qualified pilot and crew who were given all necessary maps, charts and instructions, and that [the carrier] adopted sound flight procedures in accordance with all applicable rules and regulations. The credible evidence proves that the crash of [the carrier's] plane was caused by a combination of factors, including negligent piloting, faulty and erroneous instructions from Kai-Tak Airport control tower, possible failure of the pilot to obey instructions from the control tower and/or to follow [the carrier's] established landing procedure, poor weather conditions, and a dangerous landing field and surrounding terrain. [In the circumstances, the carrier may rely on the defence of negligent pilotage provided by Article 20 of the Warsaw Convention].

10. *Case No 21U 62/91* (1992) 2 S & B Av R VII/151.

Limits of liability

In return for the carrier's presumed liability, protection is provided by the limits laid down in Article 22. The limits of the carrier's liability are generally determined by the weight of the goods damaged or lost. The limit is normally 250 francs per kilogramme. The term 'francs' is taken as a currency unit consisting of 65.5 milligrammes of gold of millesimal fineness 900. The Hong Kong Monetary Authority may specify in Hong Kong dollars the limits. In the case of judicial proceedings, the limits are determined by referring to the value of gold on the date of judgment but not the date the loss or damage was caused. In general, the limits apply to the total weight of the packages damaged or lost. However, if the lost or damaged packages affect the value of other packages covered by the same air waybill, Article 22(2)(b) of the amended Convention states that the total weight of all such packages is taken into account in calculating the limits. For example, a consignment of a machine has twenty packages. Five packages are covered by one air waybill. Another air waybill covers the other fifteen packages. One of the five packages covered by the first air waybill is lost and the machine cannot work and is, therefore, rendered a total loss. The total weight of the five packages covered by the same air waybill determines the limit applied in the case.

> *Applied Implants Technology Ltd v Lufthansa Cargo AG* (2000) 97(22) LSG 43
>
> A machine was carried by air. One part of the machine was damaged. Until replacement of the damaged part, the machine was unworkable and worthless. The court held that the total weight of the machine covered by the air waybill was the basis for calculating the limit of liability under Article 22(2).

It should be noted that Article 22(2)(b) is found only in the amended Convention. For a claim under the Warsaw Convention, the relevant weight for calculating the limits is the packages actually lost or damaged.[11]

The liability of the carrier cannot be relieved. Similarly, the limits laid down by Article 22 cannot be lowered by the carriage of goods by air contract. Any terms purporting to do so are null and void. The nullity of such terms does not affect the validity of the remaining part of the contract. There is, however, one exception. The carrier may exclude or limit liability with respect to loss or damage resulting from the inherent defect or vice of the cargo. In practice, most carriers include such an exemption clause in their standard terms of contract.

---

11. *Data Card Corp v Air Express International Corp* [1983] 2 All ER 639.

Special declaration

The consignor may raise the limit by making a special declaration of interest in delivery at destination. The declaration must be made at the time the cargo was handed over to the carrier. If the carrier requires a supplementary sum, the declaration is of no effect until the sum is paid. With a valid special declaration, the declared amount becomes the limit. But the general principle that compensation is only for actual loss still applies. The carrier has the burden to prove that the amount claimed is greater than the consignor's actual interest in delivery at destination.

Unlimited liability

The limits prescribed in Article 22 apply in any legal action, regardless of the cause of action on which it is founded. They do not apply if the damage resulted from an act or omission of carriers, their servants or agents done with some degree of culpability. Article 25 states that carriers are not protected if the act or omission is done by any of them

(a) with intent to cause damage; or
(b) recklessly and with knowledge that damage would probably result.

For an act done by a servant or agent, there is one more condition. The carrier is only deprived of the protection if the act is done within the scope of the servant or agent's employment. If the servant or agent's act was done outside the scope of his employment, the carrier may still be protected by Article 22.

In *Nugent v Michael Goss Aviation Ltd* [2000] 2 Lloyd's Rep 222, it was held that 'knowledge' required in the second limb referred to 'actual conscious knowledge'. The test is subjective. To invoke Article 25, it was not enough to prove only 'background knowledge' or 'imputed knowledge'. 'Background knowledge' means knowledge that would be present in the mind of the carrier or the servant if they thought about it. 'Imputed knowledge' is knowledge that the carrier or the servant should have had.

The operation of Article 25 in relation to an under-valued special declaration is considered in the following case.

> *Antwerp United Diamonds BVBA v Air Europe (a firm)* [1995] 3 All ER 424
> The consignor of a consignment of diamonds made a special declaration of interest in delivery at destination of 10,000 Belgian francs (about HK$2,400). The true value of the diamonds was over HK$546,000.

The diamonds were lost during the carriage by air. The consignor sued the carrier for the true value of the cargo, contending that the loss was caused by an act or omission of the carrier, or his servants or agents acting within the scope of their employment, done with intent to cause damage or recklessly and with knowledge that damage would probably result.

The court held that a special declaration made under Article 22(2)(a) was one of the 'limits of liability' referred to in Article 22. Article 25 therefore applied to remove the limit of liability stated in the special declaration.

Wilful misconduct under the Warsaw Convention

Under the Warsaw Convention, the carrier is subject to unlimited liability if the damage is caused by his wilful misconduct. Similarly, the carrier is deprived of the protection of limited liability if the damage is caused by the wilful misconduct of the servant or agent acting within the scope of employment of the wrongdoer. To be wilful misconduct, the act must be misconduct and wilful, ie, far beyond gross or culpable negligence. It refers to acts done regardless of the consequences and with knowledge and appreciation of wrongfulness. Examples of wilful misconduct include flying at a height lower than the set route without good reason and switching off radar or altimeter.

Rustenburg v South African Airways [1977] 1 Lloyd's Rep 564

R consigned two boxes of platinum from South Africa to the USA. The cargo was transferred from one airplane to another at Heathrow, London when one box was stolen by some of the loaders.

The court held that the carriers could not limit their liability and were liable for the full value of the stolen platinum. The theft was committed within the scope of the loaders' employment. It was the willful misconduct of the loaders.

Sometimes it may be rather hard to prove wilful misconduct, as shown in the following case.

Thomas Cook Group Ltd v Air Malta Co Ltd [1997] 2 Lloyd's Rep 399

D entered into a contract of carriage for the consignment of banknotes from London to Malta. As agreed, an agent of the consignor would collect the banknotes from the side of the aircraft at the airport of destination and deliver the cargo to a third party in an armoured vehicle. The relevant authorities never gave consent for the collection of valuable cargo 'planeside'. The agreed arrangement was for the cargo to be escorted by armed Air Force Malta personnel. They were not in position when the banknotes entered the customs examination room in an import cargo warehouse. The cargo was stolen in an armed robbery when it was awaiting customs clearance.

The court held that D was not guilty of wilful misconduct. To determine whether a person entrusted with valuable goods was guilty of wilful misconduct, the court should consider the conduct ordinarily to be expected in the particular circumstances before deciding whether the acts in issue amounted to misconduct. The carrier had to take security measures appropriate to and compatible with the local situation. The next issue was whether any misconduct was wilful. Lastly, it was necessary to see if the wilful misconduct caused the loss of or damage to the goods. In this case, the agent was also at fault. Although D's acts or omissions in this case could be taken as misconduct, they were not wilful.

Action against carrier's servants or agents

The consignor or consignee may choose to sue the servant or agent whose act caused damage to the cargo. If the act was done within the scope of their employment, they are also protected by Article 22, on the condition the carrier himself may invoke the limits in the case. The only exception is where the servant or agent acted, in causing the damage, with intent to cause damage or recklessly and with knowledge that damage would probably result. In such a case, the servant or agent is subject to unlimited liability.

If the carrier, the servants and agents are sued in different legal actions, the aggregate of the amounts recoverable from them cannot exceed the limit applicable in the case. It is not possible to get around the limits by taking separate legal actions against both the carrier and the servant or agent.

Effect of taking delivery

If the person entitled to delivery of the cargo receives it without complaint, this is prima facie evidence the cargo has been delivered in good condition and in accordance with the document of carriage. The consignee may later provide other evidence to prove it had already been damaged.

Owner's right to sue

In a recent case, the court was asked to decide whether only the consignor or consignee is entitled to sue the carrier for loss of or damage to the cargo. Article 15 states that the provisions in the amended Convention relating to the consignor's right of disposal and the consignee's right to take delivery do not affect the mutual relations of third parties whose rights are derived from either of them. At common law, the owner of the cargo or a party who has the cargo in possession may sue the negligent carrier for the loss. The owner's right does not derive from the consignor or consignee. Therefore, Article 15 is not relevant.

In the UK, it was commented, obiter, by Lord Hope in *Abnett v British Airways Plc* [1997] 1 All ER 193 that

> [i]t would seem to be more consistent with the purpose of the [amended Convention] to regard it as providing a uniform rule about who can sue for goods which are lost or damaged during carriage by air, with the result that the owner who is not a party to the contract has no right to sue in his own name.

The UK Court of Appeal, however, came to a different conclusion.

> *Western Digital Corp v British Airways Plc* [2001] 1 All ER 109
>
> WDS consigned some computer equipment to WDN. WDS's forwarder, LEP in Singapore, arranged for the equipment to be carried by an airline, Q Ltd, to the forwarder of WDN, ECF in England. The amended Warsaw Convention and the Guadalajara Convention governed the carriage. LEP acted as the agent of Q Ltd in issuing an air waybill. The air waybill named LEP as consignor and ECF as consignee. The carriage was actually performed by BA but not Q Ltd. The cargo was lost but ECF informed BA that it had been received in a damaged condition. WDS and WDN sued BA for the loss. BA argued, among other things, that WDS and WDN were not entitled to sue.
>
> The court held that WDS and WDN were entitled to sue. Article 13 of the amended Convention conferred rights of suit for loss or damage on the consignor and consignee, whether or not they had interests in the cargo or sustained any actual damage. There was 'no reason to infer an equally general and unstated exclusion of any right of suit by any principal of the consignor or consignee who has really sustained the relevant damage'. Further, there was some flexibility in the identification of the consignor or consignee. Unless there was inconsistency with the conventions, there were strong considerations of commercial sense to give effect to the underlying contractual structure. The principals of persons named in the air waybill could be identified as consignor or consignee. Otherwise, named consignors or consignees might be required to litigate against carriers at peril of liability for costs in lawsuits in which they had no real interest. Similarly, it was against commercial sense for owners of goods to get no compensation only because nominal consignees refused to take action against carriers.

In New Zealand, it is also settled that an owner may sue the carrier in his own name.[12] It is submitted that allowing the owner to sue in their own name is not inconsistent with the purpose of the Warsaw regime and would give business efficacy to the underlying international transaction.

Complaint

Generally, if the person entitled to delivery intends to ask for compensation, a complaint must be made in writing either upon the document of carriage or by separate notice to the carrier. In the case of damage, complaint must be made forthwith after the discovery of damage and, at the latest, within 14 days from the date of receipt of the cargo under the amended Convention. Complaint for delay is to be made within 21 days from the date on which the cargo has been placed at the disposal of the person entitled to delivery.

If no complaint is duly made, no action may be brought against the carrier. However, if the case involved the fraud of the carrier, the above-mentioned time limits do not apply. There is no need to make a complaint if the cargo is totally lost or destroyed. Partial loss is, however, taken as damage and a complaint thus required.[13]

Under the Warsaw Convention, the time limits for making complaint are 7 days for damage and 14 days for delay.

Limitation period

Under the amended Convention, the time limit for taking legal action is 2 years. Time runs from, as the case may be,

- the date of arrival at the destination,
- the date on which the aircraft ought to have arrived, or
- the date on which the carriage stopped.

> *Sidhu v British Airways Plc* [1997] AC 430
>
> D's aircraft landed in Kuwait for refueling after the Iraqi invasion. Passengers on board the aircraft were taken prisoner. They took actions against D in the third year after the event. At common law, the actions were within the three-year limitation period for negligence claims.
>
> The House of Lords held that their actions were barred by Article 24 of the Warsaw Convention (which is, for all practical purposes, identical to Article 24 of the amended Convention). The Warsaw Convention provided an exclusive cause of action. The court was not

12. *Tasman Pulp and Paper Co Ltd v Brambles J. B. O'Loghlen Ltd* [1981] 2 NZLR 225.
13. *Fothergill v Monarch Airlines Ltd* [1980] 2 All ER 696.

unaware of the fact that in certain cases that might deprive the claimants of any remedy. However, the Warsaw regime was established on a give-and-take basis. Carriers gave up their right to exclude or limit liability. In return, passengers and consignors were restricted to claims available under the regime.

(Note: other passengers took similar actions against D in France. The Paris Court of Appeal held that they might take actions outside the scope of the Warsaw Convention and awarded damages to them.)

Forum

The claimant may take action against the carrier in any of the following High Contracting Parties:

- where the carrier is ordinarily resident;
- where the carrier has the principal place of business;
- where the carrier has an establishment by which the contract has been made;
- which is the place of destination.

Normally, a Hong Kong company should be ordinarily resident in Hong Kong or have its principal place of business here. However, an overseas company having a regular place of business here may not be ordinarily resident in Hong Kong. It was held in *Rothmans of Pall Mall (Overseas) Ltd v Saudi Arabian Airlines Corp* [1981] QB 368 that the establishment of a branch office is not, in itself, evidence of ordinary residence. It is a matter of fact in which country the carrier has the principal place of business. In the case of successive carriage, the place of destination refers to the ultimate destination: In *Re Alleged Food Poisoning Incident, Al Zamil v British Airways*[14] Other procedural matters are determined by the governing law of the carriage of goods by air contract.

Milor SRL v British Airways [1996] 3 All ER 537

D performed a carriage for a consignment of jewellery from Milan, Italy to Philadelphia, the USA. The carriage was governed by the Warsaw Convention. After arrival, the cargo was stolen from a bonded warehouse in the airport. P brought an action in the UK where D was ordinarily resident and had its principal place of business. D argued that Article 28 was relevant only to the choice of jurisdiction to commence proceedings, and the law of the court chosen should decide all procedural issues. D contended that the doctrine of forum non

14. US Court of Appeals, 2nd Circuit, 1985; 19 AVI 17,646 1985–6.

conveniens should apply and the case be tried in Pennsylvania, since it was the most convenient forum for evidence to be adduced at the trial.

The court held that P had the right to choose the court for the claim to be heard under Article 28. (Note: in *Feng Zhen Lu v Air China International Corp* (US DC EDNY, 1992; 24 AVI 17,369 1992–5), a US court ordered that the case be tried in the PRC, applying the doctrine of forum non conveniens.)

The carriage of goods contract and any agreement formed before the damage occurred cannot change these rules as to the law to be applied or jurisdiction. Any provision purporting to do so is rendered null and void by Article 32. But the article expressly gives effect to an arbitration clause for the carriage of cargo, provided it is subject to the amended Convention and the arbitration is to take place in one of the jurisdictions allowed by the amended Convention. Since Article 32 is silent on agreement formed after the damage was caused, it is not clear whether such an agreement is valid.

7.6 ACTUAL CARRIER AND THE GUADALAJARA CONVENTION 1961

It is not uncommon for the carrier forming the contract with the consignor to authorize another carrier to perform the whole or part of the carriage. The authorization is often given by way of a sub-contract. If the carriage is an international carriage, the amended Convention still applies. In addition, the Guadalajara Convention also applies to govern the rights and obligations of the carriers. Although the Guadalajara Convention refers only to the Warsaw Convention, it is clear from section 2(1) of the CAO that, the term 'Warsaw Convention' may, in this context, mean the Warsaw Convention or the amended Warsaw Convention, depending on which one applies in the particular case.

The carrier who forms the contract with the consignor is, in the Guadalajara Convention, termed the 'contracting carrier', and the one who performs the carriage is the 'actual carrier'. It should be noted this is not a case of successive carriers since only one carrier forms a contract with the consignor.

Servants and agents

For acts of the servants or agents of the actual carrier done within the scope of their employment, they are deemed to be acts of the servants or agents of the contracting carrier, if they are relating to the part of carriage performed by the actual carrier. The servants and agents of the contracting carrier are also deemed to be the servants or agents of the actual carrier in such circumstances. The effect is that both carriers are liable for the acts of all the servants and agents of the contracting carrier and the actual carrier, if they are done within the scope of their employment and in the carriage performed by the actual carrier.

For the part of carriage not performed by the actual carrier, these rules do not apply.

Liability of contracting and actual carriers

The contracting carrier is liable for the whole of the carriage and the actual carrier solely for the part of carriage they perform. Thus, if the damage is not caused during the part of carriage performed by the actual carrier, only the contracting carrier may be sued. For damage caused during the part of carriage performed by the actual carrier, the claimant may decide to sue the contracting carrier, the actual carrier or both of them. If only one carrier is sued, that one may require the other carrier to be joined in the proceedings so that the court may also order that other carrier to compensate the claimant.

In the case that an action is taken against the actual carrier, the claimant may decide to sue at any place where

- permitted to sue the contracting carrier under the amended Convention;
- the actual carrier is ordinarily resident; or
- the actual carrier has his principal place of business.

Limits of liability

Generally, Article 22 of the amended Warsaw Convention provides the limits of the carriers' liability. Even if the actual carrier, the contracting carrier and the servant whose act caused the damage are separately sued in different legal actions, the aggregate of the amounts recoverable cannot exceed the limit applicable in the case.

The contracting carrier may be liable to a greater extent. For example, if the act causing the damage was done by an agent of the contracting carrier within the scope of his employment and with intent to cause the damage, the contracting carrier is subject to unlimited liability. However, the limits laid down in Article 22 still provide the ceiling of the actual carrier's liability, even though the person doing the act is deemed to be the agent. But if the act is done by a servant of the actual carrier that is deemed to be an act of a servant of the contracting carrier, both carriers are subject to unlimited liability for the damage.

Besides, a special declaration of interest in delivery at destination made by the consignor may have no effect on actual carrier's liability. If the special declaration is valid, the declared value sets the limit of the contracting carrier's liability. However, the special declaration does not bind the actual carrier, unless they agreed to it. Similarly, any additional obligations promised or waivers of rights made by the contracting carrier do not bind the actual carrier without their agreeing to them.

7.7 CONSIGNOR'S RIGHTS AND OBLIGATIONS

Consignors are generally liable for the carrier's loss if it was caused by their fault. Unlike the carrier, their liability is unlimited.

Right of disposal of cargo

Article 12 states that the consignor has the right to dispose of the cargo. Before exercising this right, all obligations under the contract of carriage must have been carried out, such as payment of the agreed freight. There are several ways to exercise this right:

- withdrawing the cargo at the aerodrome of departure or destination;
- stopping it in the course of the journey on any landing;
- calling for it to be delivered to a party other than the consignee; or
- requiring it to be returned to the aerodrome of departure.

The consignor is responsible for the expenses occasioned by the exercise of the right. The carrier must forthwith inform the consignor if it is impossible to carry out the consignor's order of disposition. The consignor's right of disposition is not exercisable if the carrier or other consignors would be prejudiced.

The carrier should require the consignor exercising the right of disposition to produce the part of the air waybill handed to the consignor when the cargo was accepted. If the carrier carries out the order without asking for it, there is liability to the person who is in lawful possession of this part of the air waybill. The carrier may, however, claim against the consignor to recover any such damage.

The consignor's right of disposal ceases at the moment when the consignee's right to take delivery begins. It is revived if the consignee declines to accept the air waybill or the cargo, or cannot be communicated with.

Article 15 states that the consignor's right of disposition may be varied by express provision in the air waybill.

Liability with respect to air waybill

Under Article 10, consignors are responsible for the correctness of the particulars and statements relating to the cargo inserted by them. Consignors are also under a duty to ensure that they are both regular and complete. If carriers suffer any damage by reason of the irregularity, incorrectness or incompleteness of the particulars or statements, they are entitled to seek indemnity from the consignor.

Obligation to provide information

The consignor is also required to furnish all necessary information and documents required by customs, octroi or police before the cargo can be delivered to the consignee. For the carrier's damage occasioned by the absence, insufficiency or irregularity of such information or documents, the consignor is liable unless they are caused by the fault of the carrier or the servants or agents. The consignor cannot escape liability by arguing that the carrier failed to check the information or documents, since Article 16 states that the carrier has no obligation to enquire into their correctness or sufficiency.

7.8 CONSIGNEE'S RIGHTS AND OBLIGATIONS

Subject to the consignor's right of disposition, the consignee has the right to require the carrier to hand over the air waybill and deliver the cargo under Article 13. This right is only exercisable after the goods have arrived at the place of destination and the consignee has paid the charges and complied with the conditions set out in the air waybill. It is the duty of the carrier to give notice to the consignee as soon as the cargo arrives unless the parties agree otherwise. The consignee's right to take delivery may be varied by express terms in the air waybill.

At common law, the doctrine of privity of contract prevents consignees from enforcing the contract of carriage if they are not a party to it. The amended Convention changes the rule to a great extent. The consignee is entitled to take delivery under Article 13. If the cargo was damaged or delayed, the consignee may sue the carrier. Further, in the case that the carrier admits the loss of the cargo or the cargo has not arrived seven days after the date on which it ought to have arrived, the consignee is given a right to enforce against the carrier the rights which flow from the contract of carriage.

Similarly, the common law rule that a party cannot sue in their own name for the loss of another party is changed. By virtue of Article 14, either the consignee or the consignor may, in their own name, sue the carrier in relation to delivery or disposal of the cargo. If the consignee or consignor wants to exercise the right under Article 14, they must first carry out the obligations imposed by the contract.

7.9 GUATEMALA CITY PROTOCOL 1971

In 1971, agreement was reached in Guatemala City on a protocol to amend the Warsaw-Hague Convention 1955. The Guatemala City Protocol 1971 further raises the monetary cap on the carrier's liability in respect of passengers and their luggage, but does not change the relevant provisions in relation to cargo.[15] The Protocol has, however, never entered into force and will, therefore, not be further considered here.

7.10 MONTREAL ADDITIONAL PROTOCOLS NUMBERS 1, 2 AND 3 OF 1975

As a result of developments at the International Monetary Fund (IMF) which led to the demonetization of gold and prevented the member States from setting official prices to gold in relation to currency, three additional protocols were drawn up in Montreal, known as the Montreal Additional Protocols Numbers 1, 2 and 3 of 1975 (hereinafter referred to as 'MAP 1 1975', 'MAP 2 1975' and 'MAP 3 1975', respectively, and as 'MAP 1 to 3 1975', collectively).[16]

MAP 1 to 3 1975 replace the monetary unit of account when referring to the monetary cap on the air carrier's liability from the gold franc to the Special Drawing Right (SDR) established by the IMF and calculated on the basis of a basket of international currencies. MAP 1 to 3 1975 amend the following international air conventions:

15. For injury or death of passengers the carrier's liability was capped at 1,500,000 gold francs, for delay in the carriage of passengers the carrier's liability was capped at 62,500 gold francs and, for loss, damage or delay in respect of passenger luggage, the carrier's liability was capped at 15,000 gold francs per passenger
16. UNCTAD, 'Carriage of Goods by Air: A Guide to the International Legal Framework' (United Nations Conference on Trade and Development, 27 June 2006), UNCTAD/SDTE/TLB/2006/1.

- MAP 1 1975 amends the Warsaw Convention 1929;
- MAP 2 1975 amends the Warsaw-Hague Convention 1955; and
- MAP 3 1975 amends the Warsaw-Hague-Guatemala Convention 1971. However, neither MAP 3 1975, nor the Warsaw-Hague-Guatemala Convention 1971, have entered into force, as they have not been adopted by the required number of States.

7.11 MONTREAL ADDITIONAL PROTOCOL NO. 4 OF 1975

Montreal Additional Protocol No. 4 came into effect on 14 June 1998 and has since been ratified by fifty countries. They include Argentina, Australia, Belgium, Brazil, Canada, Chile, Denmark, Egypt, France, Israel, Italy, Japan, New Zealand, Singapore, Spain, Sweden, the United Kingdom and the United States of America. The PRC and Hong Kong have, at the time of writing, not ratified Montreal Additional Protocol No. 4.

Articles XVII and XIX of Montreal Additional Protocol No. 4 state that, for countries which are not parties to the Warsaw Convention or the amended Convention, ratification of or accession to Montreal Additional Protocol No. 4 has the effect of accession to the amended Convention (as amended by Montreal Additional Protocol No. 4) as well. The amended Convention and Montreal Additional Protocol No. 4 are thus, as between all parties to the latter, to be read and interpreted together as one single instrument.[17] The situation is not so clear for a carriage between a signatory to Montreal Additional Protocol No. 4 (the USA, for example) and Hong Kong, which is not a party to it. It might be argued that the provisions in the amended Convention could apply.[18] The counter-argument is that it would be unfair to a signatory of Montreal Additional Protocol No. 4 if the state does not want to join the amended Convention for some reasons. The better view is that only the Warsaw Convention applies. In practice, the Chief Executive's orders are conclusive as to the application of the conventions to any air transport.

The main changes that Montreal Additional Protocol No 4 brings about are as follows:

- Any other means which would preserve a record of the carriage may, with the consent of the consignor, be used instead of an air waybill. The consignor and the carrier may agree in advance for an electronic

17. Article XV.
18. cf *Re Korean Airlines Disaster*, 1 September 1983 (District Court, District of Columbia) on the application of the Warsaw Convention.

method to be used. If another means is used, the consignor may require the carrier to deliver a receipt to him.

- Non-compliance with the requirements relating to air waybills does not deprive the carrier of the protection of limited liability.
- Whilst the consignor is responsible for the correctness of the particulars and statements inserted in the air waybill, the carrier is responsible for the irregularity, incorrectness or incompleteness of the record preserved or receipt prepared by the carrier.
- For loss of or damage to cargo, the carrier is not liable if it solely results from
  - inherent defect of the cargo;
  - defective packing by persons other than the carrier;
  - an act of war or armed conflict; or
  - an act of public authority carried out in connection with the entry, exit or transit of the cargo.
- Taking all necessary measures or the impossibility of taking such measures is no longer a defence to the carrier, except in the case of damage occasioned by delay.
- The limit of the carrier's liability is 17 Special Drawing Rights per kilogramme.

7.12 THE MONTREAL CONVENTION 1999

The Montreal Convention is the outcome of the attempt of the international community to unify the fragmented Warsaw regime. It supersedes the Warsaw Convention, the amended Convention, the Guadalajara Convention and Montreal Additional Protocol No. 4. However, it brings no drastic changes to the Warsaw regime since it adopts many principles and rules from those instruments to be repealed. Many of the changes brought about by the Montreal Convention are related to international carriage of passengers by air. Thus, the effects of the implementation of the Montreal Convention on international carriage of goods by air are further reduced.

Scope of Application

How to determine the applicable international air[19]

There are different legal regimes which may potentially be applicable to a claim arising from the international carriage of goods by air. Whether one of

19. UNCTAD (above n 49) is referred to for the materials of these sections.

the Warsaw-system conventions (the original Warsaw Convention, adopted in 1929, its amended versions and the Guadalajara Convention 1961 is collectively referred to as 'Warsaw-system conventions')[20] or, alternatively, the Montreal Convention 1999 is applied, is an important and, in practice, often complicated question.

In all cases, the trigger for the application of any one of the international air conventions and its corresponding legal regime is the concept of 'international carriage'. There is a single definition of 'international carriage', which has not been changed in substance by the various amendments to the original Warsaw Convention 1929, or by the most recent Montreal Convention 1999. To determine whether a specific contract of carriage is 'international carriage' governed by one of the international air conventions, there is a two-stage inquiry, which is complex, and in practice often creates considerable difficulty for traders and courts charged with the resolution of disputes.[21]

In simple terms, the process may be summarized as follows. First, it is necessary to determine whether the carriage comes within the technical concept of 'international carriage', defined by reference to the agreed places of departure and destination and any agreed stopping place. Secondly, it is necessary to check that the State/s of departure and destination are Contracting States to the same version of either one of the Warsaw-system conventions, or the Montreal Convention 1999.

The first stage of the inquiry consists of considering the definition of 'international carriage'. Both the Warsaw-system conventions and the Montreal Convention 1999 use similar language to define the term 'international carriage'. They make reference to the 'contract made by the parties',[22] or the 'agreement between the parties',[23] in two distinct situations.

(i) The agreed place of departure and the place of destination are situated within the territories of two Contracting States, whether or not there is a break in the carriage or a transhipment;

(ii) The agreed place of departure and the place of destination are situated within the territory of a single Contracting State, if there is an agreed stopping place within the territory of another State, whether or not this is a Contracting State.

20. ibid 12.
21. ibid.
22. Art 1(2) Warsaw Convention 1929.
23. Art 1(2) Warsaw-Hague Convention 1955, Warsaw-Hague-MAP 4 Convention 1975, Montreal Convention 1999.

Therefore, in order to determine whether a contract for the transport of goods is 'international carriage' governed by any one of the Warsaw-system conventions or the Montreal Convention 1999, it is imperative to study the air waybill or ticket closely to ascertain the agreed places of departure and destination, as well as any agreed stopping place, and to determine whether these meet the requirements set out in (i) or (ii) above. If the requirements are met, the contract is one of 'international carriage' governed by one of the Warsaw-system conventions or the Montreal Convention 1999, as applicable. Otherwise, the contract is not one of 'international carriage' and, therefore, not subject to any of the international air conventions. In these cases, national law and/or the terms of the contract (ie, terms and conditions printed on the air waybill or passenger ticket or incorporated by reference) will be applicable.

In cases where more than one international air convention may be applicable, a second stage is necessary, which consists in identifying the specific legal regime applicable to a contract of 'international carriage'. Thus, it is necessary to determine the 'latest' treaty relationship common to both States. This has also been described as the determination of the 'lowest common denominator'.[24] For illustrative purposes, some examples of the legal regime applicable in various types of contracts of carriage by air are set out below.

- Cairo (Egypt) to Luxor (Egypt):
  Domestic carriage, not covered mandatorily by any of the international air law conventions.
- Cairo (Egypt) to Luxor (Egypt), via Khartoum (Sudan):
  Egypt is a Contracting State to the Montreal Convention 1999. Therefore, the Montreal Convention 1999 is applicable.
- Geneva (Switzerland) to Bangkok (Thailand):
  Thailand is not a Contracting State to any of the Warsaw-system conventions or the Montreal Convention 1999. Switzerland is a Contracting State to the Montreal Convention 1999. Therefore, as the two States are not Contracting States to a common international air convention, national law and/or the terms of the contract of carriage are applicable.
- Kilimanjaro (United Republic of Tanzania) to Jakarta (Indonesia):
  The United Republic of Tanzania is a Contracting State to the Montreal Convention 1999, but Indonesia is not. Both Tanzania and Indonesia

24. Paul Stephen Dempsey, 'International Air Cargo & Baggage Liability and the Tower of Babel' (2004), 36 Geo Wash Int'l L Rev 239, 8.

are Contracting States to the unamended Warsaw Convention 1929. Therefore, the unamended Warsaw Convention 1929 is applicable.

- Mexico City (Mexico) to Sydney (Australia):
  Mexico is a Contracting State to the Montreal Convention 1999, whereas Australia is not. Australia is a Contracting State to the Warsaw-Hague-Guadalajara-MAP 4 Convention 1975, whereas Mexico is not. Both Mexico and Australia are Contracting States to the Warsaw-Hague-Guadalajara Convention 1961. Therefore, the Warsaw-Hague-Guadalajara Convention 1961 is applicable.

- Beijing (People's Republic of China) to Tirana (Albania):
  China is a Contracting State to the Warsaw Convention 1929 and the Warsaw-Hague Convention 1955. Albania is not a Contracting State to any of the Warsaw-system conventions. However, both P.R. China and Albania are Contracting States to the Montreal Convention 1999. Therefore, the Montreal Convention 1999 is applicable.

Once it is determined there is a contract of '*international carriage*' covered by one of the Warsaw-system conventions, or the Montreal Convention 1999, the application of the identified legal regime is both exclusive and mandatory. Exclusive application means the conditions and limits of liability set out in the applicable convention provide the only cause of action in disputes arising out of the 'international carriage' of cargo by air. All the international air conventions provide[25] that 'any action for damages, however founded, can only be brought subject to the conditions and limits' of liability set out in the applicable convention. Therefore, a claimant who commences an action for damages[26] arising out of 'international carriage' of goods by air cannot circumvent the application of the rules laid down in the international air conventions, by pleading another cause of action.

Mandatory application means the parties to the contract of carriage cannot agree to relieve the carrier of liability, or agree to lower limits of liability other than those laid down by the international air conventions.[27] Thus, whilst it is specified that the parties have freedom to contract and to

25. Art 24(1) Warsaw Convention 1929 and Warsaw-Hague Convention 1955. Art 24(2) Warsaw-Hague-MAP 4 Convention 1975 and Art 29 Montreal Convention 1999 add the words 'whether under this Convention or in contract or in tort or otherwise'.
26. It is not entirely clear whether other remedies such as injunctions may be available. One commentator argues that this would be possible, as the authoritative French text of the Warsaw Convention 1929 makes reference to the wider term 'action en responsabilité'. See Clarke and Yates, Land and Air para 3.146.
27. Art 23 Warsaw-system conventions and Art 26 Montreal Convention 1999.

agree to the terms and conditions of their contract,[28] the carrier may not rely on any contractual terms conflicting with the mandatory rules laid down in the Warsaw-system conventions, or the Montreal Convention 1999. Any such term would be null and void.[29] As from 28 June 2004 and, in particular, as between the member states of the European Community, the Montreal Convention prevails over the Warsaw Convention, pursuant to article 55 of the Montreal Convention.[30]

Carriage not covered by the international air conventions

Some types of carriage are either expressly excluded from the scope of application of the international air conventions, or may be excluded if a Contracting State makes a reservation (subject to the conditions laid down) by declaring that the relevant international air convention will not apply in certain circumstances.

(i) The international air conventions do not, in general, apply to the carriage of mail and postal packages.[31]

(ii) The Warsaw Convention 1929 does not apply to carriage performed by way of experimental trial by 'air navigation undertakings' (ie, commercial carriers) with a view to establishing a regular line of air navigation.[32] This is due to the fact that commercial aviation was still developing at the time when the Warsaw Convention 1929 was signed.

---

28. Art 33 Warsaw-system conventions and Art 27 Montreal Convention 1999. Art 27 Montreal Convention 1999 further specifies that the carrier may also waive any defences which are available to him under the Montreal Convention 1999.
29. Art 23 Warsaw-system conventions and Art 26 Montreal Convention 1999. However, note that Art 23(2) Warsaw-Hague Convention 1955 and Warsaw-Hague-MAP 4 Convention 1975 entitles the carrier to contractually exclude liability arising from loss or damage resulting from the inherent defect, quality or vice of the cargo carried. This corresponds to one of the specific *statutory* defences in respect of carriage of cargo that were added by Art 18(3)(a) Warsaw-Hague-MAP 4 Convention 1975 and Art 18(2)(a) Montreal Convention 1999.
30. *Cuadrench More v Koninklijke Luchtvaart Maatschappij NV (KLM)* [2013] 1 Lloyd's Rep 341.
31. It should be noted, however, that there are some differences between the various international air conventions. Art 2(2) Warsaw Convention 1929 provides that it does not apply to 'carriage performed under the terms of any international postal Convention'. Art 2(2) Warsaw-Hague Convention 1955 provides that it does not apply to carriage of 'mail and postal packages'. Art 2 Warsaw-Hague-MAP 4 Convention 1975 and Art 2 Montreal Convention 1999 provide that they do not apply to the 'carriage of postal items', except that the carrier 'shall be liable only to the relevant postal administration in accordance with the rules applicable to the relationship between the carriers and the postal administrations'.
32. Art 34 Warsaw Convention 1929.

(iii) The Warsaw Convention 1929 does not apply to carriage performed in extraordinary circumstances outside the normal scope of an air carrier's business, for example, in cases of carriage of cargo to a territory affected by hostilities. In this context it should be noted, however, that all the other international air conventions, namely, the Warsaw-Hague Convention 1955, the Warsaw-Hague-MAP 4 Convention 1975 and the Montreal Convention 1999 provide differently. Under any of the aforementioned international air conventions, the provisions as to the liability of the air carrier continue to apply, but the provisions relevant to the requirements as to the issuing of an air waybill and as to the particulars to be contained therein, do not apply.[33]

(iv) All the international air conventions, with the exception of the Warsaw Convention 1929, provide[34] that a Contracting State may make a reservation in relation to carriage for a State's military authorities on aircraft, registered in (or leased by)[35] that State the whole capacity of which has been reserved by or on behalf of such authorities; for example, where a State charters aircraft to transport military goods to a war zone in another State. Only a few States have taken advantage of this reservation.[36]

(v) All the international air conventions, provide[37] that whilst they apply to carriage performed by the State or by legally constituted public bodies, a Contracting State may make a reservation[38] so that the relevant international air convention will not apply in relation to carriage performed (and operated) directly by the State (for non-commercial purposes in respect of its functions and duties as a sovereign State).[39]

33. Art 34 Warsaw-Hague Convention 1955 and Warsaw-Hague-MAP 4 Convention 1975, and Art 51 Montreal Convention 1999.
34. Art XXVI Hague Protocol 1955, Art XXI(1)(a) MAP 4 Protocol 1975, and Art 57(b) Montreal Convention 1999.
35. Phrase within brackets added by Art 57(b) Montreal Convention 1999.
36. Warsaw-Hague Convention 1955: reservation made by three countries; Warsaw-MAP 2 Convention 1975: reservation made by two countries; Warsaw-Hague-MAP 4 Convention 1975: reservation made by two countries; Montreal Convention 1999: reservation made by seven countries.
37. Art 2(1) Warsaw-system conventions and Montreal Convention 1999.
38. 'Additional Protocol With Reference to Art 2', Warsaw Convention 1929 (at the very end of the convention), which has not been modified by the subsequent Warsaw-system conventions. Art 57(a) Montreal Convention 1999.
39. Warsaw Convention 1929: reservation made by seven countries; Montreal Convention 1999: reservation made by eight countries.

Form and purpose of the air waybill

The air waybill[40] is by far the most essential document issued in respect of the international carriage of cargo. It evidences the contract or agreement of international carriage between the parties and plays a central role in the liability regime. In current practice, air waybills are usually not negotiable.[41] This is explained by the speed of air transport, meaning there is normally no need for a document which enables sale of goods in transit.[42]

The airline members of IATA83 agreed to introduce a standard form air waybill[43] for international carriage by air of cargo. This has been adopted as the international norm because its layout and wording enables the incorporation of all the particulars required by the various international air conventions. The airline members of IATA have also agreed on alternative form *Conditions of Contract*,[44] printed on the reverse of the standard form air waybill. The Conditions of Contract include the provisions required under the international air law conventions, as well as other terms, applicable in cases where none of the conventions applies or dealing with matters not regulated in the conventions. Terms cover issues such as limitation of the air carrier's liability, the liability of servants and agents of the carrier, written notice of complaint within a specified number of days, time limitation and related matters.[45]

The Conditions of Contract are supplemented by the airline's *Conditions of Carriage* of cargo, which are often contained in a separate booklet or manual, issued by the carrier to interested parties upon request and incorporated into the air waybill by reference. An airline's Conditions of Carriage deal with issues such as when delivery takes place, the handling of perishable goods or dangerous goods, the carrier's rights of disposal in the event

40. The term 'air consignment note' is only used in the Warsaw Convention 1929. The other international air conventions use the more modern term 'air waybill'.
41. The Warsaw Convention 1929 does not deal with the question of the negotiability of the air waybill. Art 15(3) Warsaw-Hague Convention 1955 adds that '[n]othing in this Convention prevents the issue of a negotiable air waybill'. This additional paragraph was deleted from the text of the Warsaw-Hague-MAP 4 Convention 1975, and is not reflected in the text of the Montreal Convention 1999. The fact that air waybills are not negotiable is often expressly stated in the top right hand corner of an air waybill.
42. This is in contrast to the maritime practice of issuing negotiable bills of lading, where the right to possession of the goods on board the ship may be transferred by endorsement and transfer of the document.
43. As set out in Attachment 'A' of IATA Resolution 600a. See Annex 3 for a copy of the standard IATA Air Waybill specifications.
44. IATA Resolution 600b (II).
45. In cases where one of the air law conventions applies, contractual terms that conflict with the mandatory provisions of the respective Convention will be considered null and void.

of non-collection or non-payment of fees and the shipper's obligation with regard to delivery of the cargo, its condition and packing.

The air waybill is the most important cargo document issued by the carrier or its authorized cargo agent and serves several purposes.[46] Most important of these is its evidentiary function. The Warsaw-system conventions and the Montreal Convention 1999 (with minor changes indicated in brackets) provide[47] that the air waybill or cargo receipt is prima facie evidence[48] of the following:

- The conclusion of the contract of carriage and conditions of carriage.
- The receipt of the goods (or acceptance of the cargo) by the carrier and the statements as to the weight, dimensions, packing of the cargo and number of packages.
- The stated quantity, volume and condition of the cargo (as against the carrier); however, only if (a) the carrier, in the presence of the consignor, has checked these and (b) a statement to this effect is included on the face of the air waybill, or if the stated fact relates to the *apparent* condition of the cargo. This means that in the absence of any indication on the face of the air waybill, there is no presumption that the carrier received the cargo in good condition.[49]

Delivery and description of the air waybill

All the international air conventions contain similar provisions on the requirement as to delivery and description of air waybills, except that the two most recent of these, the Warsaw-Hague-MAP 4 Convention 1975 and Montreal Convention 1999, also authorize the use of an electronic record in place of a traditional paper air waybill.

First, the international air conventions provide that '*an air waybill shall be delivered*'[50] in respect of the carriage of cargo. This was more extensively described in the Warsaw Convention 1929 and Warsaw-Hague Convention

46. See *Antwerp Diamond Bank NV v Brink's Inc* [2013] 1 HKLRD 396, air waybills were not documents of title or negotiable. P had not pleaded the existence of 'an established, clear and well-defined custom' treating the air waybills as documents of title and in any event, had not adduced any positive evidence as to such a custom.
47. Art 11 Warsaw-system conventions and Montreal Convention 1999.
48. A presumption is established which, however, may be rebutted by other evidence.
49. In practice, a notice on the face of the air waybill states that: '[i]t is agreed that the goods described are accepted in apparent good order and condition (except as noted)'.
50. Art 5(1) Warsaw-Hague-MAP 4 Convention 1975 and Art 4(1) Montreal Convention 1999, which also specify that electronic air waybills may substitute paper air waybills.

1955; however, the meaning is essentially the same.[51] Secondly, the international air conventions provide that if there is more than one package, the carrier has the right to require the consignor to make out separate air waybills, or cargo receipts.[52] Thirdly, the international air conventions provide[53] that the consignor shall make out the air waybill in three original parts and hand it over to the carrier with the goods, in the following order.

- The first part shall be marked 'for the carrier' and shall be signed[54] by the consignor. This part is for the carrier's file and the signature of the consignor is an acknowledgement that the contents of the air waybill are correct.
- The second part shall be marked 'for the consignee' and shall be signed by the consignor and by the carrier.[55] This part is for the consignee, who may use it to complain to the carrier if the goods are not delivered in good condition, or not delivered at all.
- The third part shall be signed[56] by the carrier and shall be handed to the consignor after the goods have been accepted by the carrier for carriage.[57] This part is to facilitate disposal of the goods in accordance with the consignor's right of disposal during the carriage and prior to delivery to the consignee.

51. Art 5(1) Warsaw Convention 1929 and Warsaw-Hague Convention 1955 provide that '[e]very carrier of goods has the right to require the consignor to make out and hand over to him a document called an [air waybill]; every consignor has the right to require the carrier to accept this document'.
52. Art 7 Warsaw-system conventions and Art 8 Montreal Convention 1999.
53. Art 6 Warsaw-system conventions and Art 7 Montreal Convention 1999.
54. Art 6 (4) Warsaw Convention 1929 and Warsaw-Hague Convention 1955 provide that the signature of the carrier may be stamped, and that of the consignor may be printed or stamped. Art 6(3) Warsaw-Hague-MAP 4 Convention 1975 and Art 7(3) Montreal Convention 1999 provide that the signature of the carrier and that of the consignor may be printed or stamped.
55. Art 6(2) Warsaw Convention 1929 and Warsaw-Hague Convention 1955 include an additional provision that the second part of the air waybill '*shall accompany the goods*'. This has been omitted from the text of the Warsaw-Hague-MAP 4 Convention 1975 and Montreal Convention 1999.
56. Art 6(4) Warsaw Convention 1929 and Warsaw-Hague Convention 1955 provide that the signature of the carrier may be stamped, and that of the consignor may be printed or stamped. Art 6(3) Warsaw-Hague-MAP 4 Convention 1975 and Art 7(3) Montreal Convention 1999 provide that the signature of the carrier and that of the consignor may be printed or stamped.
57. Art 6(3) Warsaw Convention 1929 specifies 'on acceptance of the goods', and the same provision in the Warsaw-Hague Convention 1955 amends this phrase to 'prior to the loading' of the goods on board the aircraft. Reference as to when the carrier signs has been omitted from the Warsaw-Hague-MAP 4 Convention 1975 and Montreal Convention 1999 because the purpose of the carrier's signature is to acknowledge receipt of the cargo, which is fulfilled when the carrier hands the third part of the air waybill to the consignor.

Whilst the international air conventions specify that the consignor makes out the air waybill,[58] in practice it is often made out and completed by the carrier, as agent of the consignor[59] and on the consignor's instructions. However, it is the responsibility of the consignor to ensure the correctness of the particulars and statements contained in the air waybill,[60] and the consignor is ultimately liable for the accuracy of any particulars provided to the carrier.[61]

One of the innovations of the Warsaw-Hague-MAP 4 Convention 1975, reflected in almost identical terms in the Montreal Convention 1999, is that the parties may use simplified electronic records to facilitate shipments instead of paper air waybills. Upon request, the carrier must deliver to the consignor a cargo receipt permitting identification of the consignment and access to the information stored electronically.[62]

The Warsaw-Hague-MAP 4 Convention 1975 includes two additional provisions on the use of electronic records, which have been omitted from the subsequent Montreal Convention 1999. First, the consignor needs to give consent to the carrier to substitute the paper air waybill with an electronic record.[63] Secondly, the carrier may not refuse to accept the cargo for carriage because the points of transit and destination are not equipped for the use of electronic air waybills.[64] It is interesting to note the above two provisions have not been included in the Montreal Convention 1999. This may be due to the fact that electronic storage and retrieval of information are increasingly becoming standard practice in transport transactions.

Particulars to be included

Under the Warsaw Convention 1929, the air waybill must contain the following list of seventeen particulars:[65]

> (a) 'The place and date of its execution;
> (b) The place of departure and of destination;

58. Art 6(1) Warsaw-system conventions and Art 7(1) Montreal Convention 1999.
59. Art 6(5) Warsaw Convention 1929 and Warsaw-Hague Convention 1955, Art 6 (4) Warsaw-Hague-MAP 4 Convention 1975, and Art 7(4) Montreal Convention 1999.
60. Art 10(1) Warsaw-system conventions and Montreal Convention 1999.
61. Art 10(2) Warsaw-system conventions and Montreal Convention 1999.
62. Art 5(2) Warsaw-Hague-MAP 4 Convention 1975, and Art 4 (2) Montreal Convention 1999.
63. Art 5(2) Warsaw-Hague-MAP 4 Convention 1975.
64. Art 5(3) Warsaw-Hague-MAP 4 Convention 1975.
65. Art 8 Warsaw Convention 1929.

(c) The agreed stopping places, provided that the carrier may reserve the right to alter the stopping places in case of necessity, and that if he exercises that right the alteration shall not have the effect of depriving the carriage of its international character;
(d) The name and address of the consignor;
(e) The name and address of the first carrier;
(f) The name and address of the consignee, if the case so requires;
(g) The nature of the goods;
(h) The number of the packages, the method of packing and the particular marks or numbers upon them;
(i) The weight, the quantity and the volume or dimensions of the goods;
(j) The apparent condition of the goods and of the packing;
(k) The freight, if it has been agreed upon, the date and place of payment, and the person who is to pay it;
(l) If the goods are sent for payment on delivery, the price of the goods, and, if the case so requires, the amount of the expenses incurred;
(m) The amount of the value declared in accordance with Article 22(2);[66]
(n) The number of parts of the [air waybill];
(o) The documents handed to the carrier to accompany the [air waybill];
(p) The time fixed for the completion of the carriage and a brief note of the route to be followed, if these matters have been agreed upon;
(q) A statement that the carriage is subject to the rules relating to liability established by [the Warsaw Convention 1929].

The Warsaw Convention 1929 provides that failure to include the particulars listed in (a) to (i) and (q) above in the air waybill, or acceptance of goods by the carrier without an air waybill, leads to loss of the carrier's right to limitation of liability.[67]

The Warsaw-Hague Convention 1955 reduces to only three the list of particulars that the air waybill must contain,[68] namely:

(a) An indication of the places of departure and destination;
(b) If the places of departure and destination are within the territory of a single [Contracting State], one or more agreed stopping places

66. Art 22(2) Warsaw Convention 1929 provides that when the consignor hands over the package to the carrier, the consignor may make a special declaration of the value of the package at delivery and pay a supplementary sum, if required.
67. Art 9 Warsaw Convention 1929.
68. Art 8 Warsaw-Hague Convention 1955.

being within the territory of another State, an indication of at least one such stopping place;

(c) A notice to the consignor to the effect that, if the carriage involves an ultimate destination or stop in a country other than the country of departure, the Warsaw Convention may be applicable and that the Convention governs and, in most cases, limits the liability of carriers in respect of loss of or damage to cargo.

Particulars (a) to (c) above, are designed to draw the parties' attention to facts indicating that the contract is one of *international carriage*, as defined, and thus subject to the Warsaw-Hague Convention 1955, and to put on notice any *successive* carriers. In addition, the notice required under (c) draws attention to the fact that in cases where the Warsaw-Hague Convention 1955 applies, there is a monetary cap limiting the carrier's liability. This notice is fundamental to the carrier's right to limited liability: if the air waybill does not include the notice, or if cargo is loaded on board the aircraft without an air waybill, the carrier is deprived of the monetary cap limiting his liability.[69]

> *Fujitsu Computer Products Corporation v Bax Global Inc.*[70]
> Under a contract of carriage Bax agreed with a company in the Fujitsu group to carry by air a consignment of hard disk drives from Manila to Glasgow. Fujitsu Computer Products was the shipper and Fujitsu Europe was the consignee. The contract of carriage was governed by the Warsaw Convention, as amended by the Hague Protocol, as set out in schedule 1 to the Carriage by Air Act 1961. Bax subcontracted the carriage to Emirates, the airline. By a House Air Way Bill (HAWB) Bax acknowledged the shipment of the consignment. The consignment was stolen and Bax became liable to one or other of the companies in the Fujitsu group (the claimants) subject to proof of title.
>
> The Court held that the air waybill did not comply with article 8(c) and that the limitation of liability under article 22 was not open to Bax. The HAWB did not contain a notice. Warsaw-Hague required there to be what was recognisable as "a notice", i.e., a discrete form of words warning the reader of the potential applicability of the Convention and its effect, namely to govern and limit liability.

The Warsaw-Hague-MAP 4 Convention 1975 and the Montreal Convention 1999 simplify even further the list of particulars that the air waybill, or cargo receipt must contain,[71] namely:

---

69. Art 9 Warsaw-Hague Convention 1955. For an interesting case on the important issue of what the notice requirement under Art 8(c) entails, see *Fujitsu Computer Products Corporation v Bax Global Inc.* [2005] EWHC 2289 (Comm); [2006] 1 Lloyd's Rep 367; [2006] All ER (Comm) 211 (at para 19 of the judgment).
70. [2006] 1 Lloyd's Rep 367.
71. Art 8 Warsaw-Hague-MAP 4 Convention 1975 and Art 5 Montreal Convention 1999.

(a) An indication of the places of departure and destination;
(b) If the places of departure and destination are within the territory of a single [Contracting State], one or more agreed stopping places being within the territory of another State, an indication of at least one such stopping place; and
(c) An indication of the weight of the consignment.

The required particulars under (a) and (b) are the same as under the Warsaw-Hague Convention 1955. The statement required under (c) serves as a basis for the calculation of the monetary cap limiting the air carrier's liability. More importantly, however, the Warsaw-Hague-MAP 4 Convention 1975 and Montreal Convention 1999 provide[72] that absence of any or all of the particulars, or failure to deliver an air waybill, does *not* deprive the air carrier of the monetary cap limiting liability. This represents an important change to the position under the earlier legal instruments.

As is evident from the above, the Montreal Convention 1999 preserves and modernizes the benefits achieved under the Warsaw-Hague-MAP 4 Convention 1975. Thus, the formerly cumbersome particulars to be included in air waybills have been simplified, the use of electronic air waybills to facilitate shipments is expressly envisaged, and any 'penalty' for a carrier's failure to comply with the air waybill requirements has been removed.

Consignor's right of disposal and consignee's right to delivery

The international air conventions confer on the consignor a right of disposal of the goods during the carriage by air and before the goods are delivered to the consignee. The consignor may withdraw the goods at the airport of departure or destination, or may stop the goods in the course of carriage at any landing, or may name a person (other than the consignee named in the air waybill) as the person to whom the goods should be delivered. Naturally, the consignor may not prejudice the carrier or other consignors in the exercise of this right, and the consignor must repay any expenses occasioned as a result.[73] The right of disposal of the consignor ends when the right of the consignee to delivery of the cargo at destination begins, except if the consignee declines to accept the goods, or cannot be communicated with.[74]

To exercise the right of disposal, the consignor must produce the third part of the air waybill signed by the carrier and handed over to the consignor

72. Art 9 Warsaw-Hague-MAP 4 Convention 1975 and Montreal Convention 1999.
73. Art 12 Warsaw-system conventions and Montreal Convention 1999.
74. ibid.

after the carrier has accepted the cargo for carriage. The carrier must inform the consignor forthwith if it is impossible to carry out the consignor's orders. However, if the carrier complies with the consignor's directions, the carrier must request the consignor produce the third part of the air waybill. If the carrier does not require the production of the third part of the air waybill, the carrier is liable to any person lawfully in possession of that part of the air waybill and has suffered damage as a result. The carrier may, of course, make a claim against the consignor to recover any resulting loss.[75]

Unless the consignor has exercised right of disposal of the goods, the consignee has the right to the delivery of the goods[76] on arrival at the place of destination.[77] The carrier needs to give notice to the consignee as soon as the goods arrive. At delivery, the consignee must pay any charges due and comply with the conditions of carriage set out in the air waybill.

If the carrier admits that the goods are lost, or if the goods have not arrived after seven days from the date on which they ought to have arrived, the consignee is entitled to enforce his rights under the contract of carriage, namely, seek damages against the carrier.[78] Further, all the international air conventions confer a right of action on the consignor and the consignee, who may each enforce their respective right of disposal and right to delivery of the goods, whether acting in their own interest, or in the interest of another.[79]

> *Western Digital Corporation and ors. v British Airways Plc*[80]
>
> The action arose out of a shipment of hard discs which the second claimant Western Singapore despatched to the third claimants Western Netherlands. The freight forwarders LEP issued two house air bills covering the seven pallets which made up the shipment. These air bills named the shipper as Western Singapore and the consignee as Express Cargo Forwarding. At the same time, LEP issued a Qantas Air Waybill naming LEP as shipper and Express Cargo Forwarding as consignee. Neither air waybill was negotiated. However, three out of the seven pallets were never delivered.

75. Art 12 Warsaw-system conventions and Montreal Convention 1999.
76. Art 12(4) and Art 13(1) Warsaw-system conventions and Montreal Convention 1999.
77. Art 13 Warsaw-system conventions and Montreal Convention 1999.
78. ibid.
79. Provided they carry out the 'obligations imposed by the contract'; Art 14 Warsaw-system conventions and Montreal Convention 1999. It should be noted, however, that the scope of this right is not uncontroversial. See further Clarke and Yates, Land and Air at para 3.92. Under English common law, the owner of the goods may be entitled to sue the carrier independently of any rights derived from either the consignor or the consignee; see *Western Digital Corporation and ors. v British Airways Plc.* [2001] QB 733 (CA).
80. [2000] 2 Lloyd's Rep 142 (CA); [2001] QB 733 (CA).

> The Court held that the second and/or third claimants had a properly arguable case for saying, under common law principles, that they were a party to a contract for carriage by air evidenced by the Qantas air waybill. Under common law principles this would not be a case of agency for an undisclosed principal.
>
> Nothing in the Warsaw Convention as amended required the naming of a consignor or consignee in the air waybill or explicitly restricted the concept of consignor or consignee to someone so named. The cargo interests should be able to intervene and to sue for loss and damage as a principal on a contract evidenced by an air waybill issued to his shipping or customs agent appeared to considerably outweigh any argument based on the supposed inconvenience or uncertainty that this might involve for air carriers. The claimants' action in respect of international carriage by air was thus barred and fell to be dismissed; the cross-appeal succeeded.

Finally, subject to any rights and obligations that the consignor and the consignee may have towards each other or towards any third parties, they may vary the aforementioned provisions of the international air conventions, by express reference recorded in the air waybill.[81] In practice, such derogation from the provisions of the international air conventions is, however, exceptional.

Air carrier's liability for loss, damage or delay

The *raison d'etre* of all the international air conventions and thus of central importance are the provisions on the air carrier's liability, which apply mandatorily and may not be contractually modified to the benefit of the carrier. The common features of the liability regime under the international air conventions can be summarized as follows.

First, in case of damage or delay to cargo, the person entitled to delivery or the claimant must complain in writing to the carrier after the discovery of the damage, and within a specified number of days from the date of receipt in case of damage,[82] or from the date on which the cargo should have been delivered in case of delay.[83] This is to provide the carrier with the opportunity

81. Art 15 Warsaw-system conventions and Montreal Convention 1999.
82. Art 26(2) Warsaw Convention 1929 stipulates seven days from the date of receipt in case of damage. Art 26(2) Warsaw-Hague Convention 1955 and Warsaw-Hague-MAP 4 Convention 1975, as well as Art 31(2) Montreal Convention 1999 stipulate fourteen days from the date of receipt in case of damage.
83. Art 26(2) Warsaw Convention 1929 stipulates fourteen days from the date on which the cargo should have been delivered in case of delay. Art 26(2) Warsaw-Hague Convention 1955 and Warsaw-Hague-MAP 4 Convention 1975, and Art 31(2) Montreal Convention 1999 stipulate twenty-one days from the date on which the cargo should have been delivered in case of delay.

to investigate the facts and circumstances of the damage or delay, collect and retain the necessary documents and information, and assess his potential liability. Failure to complain within the specified number of days is prima facie evidence that the goods have been delivered in good condition and in accordance with the documents of carriage. More importantly, failure to complain will prevent the claimant from subsequently bringing an action against the carrier, except in cases of fraud on the part of the carrier.[84]

> *Western Digital Corporation and ors. v British Airways Plc.*[85]
> The Court held that as to the cross-appeal, there must be within the time stated a complaint which must at least embrace the damage to which the subsequent action related; such complaint as was made within the relevant time was specifically limited to physical damage to identified items and did not embrace the loss of such items, the subject of this action; it indicated a problem about the condition of the identified items but not about their arrival; and art 26(2) was not complied with; the claimants' action fell to be dismissed on that ground.

Secondly, a claimant has two years to bring a court action, or arbitral proceedings[86] claiming damages against the carrier, from the date of arrival of the goods at their destination, or from the date on which the aircraft ought to have arrived, or from the date on which the carriage stopped.[87] If more than one point is applicable, the latest in date is relevant.[88] After the period of two years, the claimant's right to bring an action against the carrier is extinguished.[89] This means the claimant's right is lost. As succinctly put by a New South Wales court, it is '*non-existent . . . finished, gone forever*'.[90]

Thirdly, the claimant needs to prove the extent of loss as damages are only payable in respect of the actual loss suffered.

If the three conditions listed above have been complied with (ie, written complaint within the specified notice period, action within the two-year time limit, and proof of loss), then the carrier is prima facie liable for the loss

84. Art 26 Warsaw-system conventions, and Art 31 Montreal Convention 1999. For an illustrative case, where it was held that a claim against a carrier was barred for failure to notify a complaint within the prescribed time limit, see *Western Digital Corporation and ors. v British Airways Plc.* [2001] QB 733 (CA).
85. [2000] 2 Lloyd's Rep 142 (CA); [2001] QB 733 (CA).
86. Art 32 Warsaw-system conventions and Art 34 Montreal Convention 1999. Contractual arbitration agreements are permitted in relation to the carriage of cargo.
87. Art 29 Warsaw-system conventions and Art 35 Montreal Convention 1999, which also provide that the method of calculating the period of limitation, ie, whether a year means twelve calendar months or 365 days, or whether parts of a day are disregarded, is determined by the law of the court seized of the case.
88. *All Transport v Seaboard World Air Lines*, 349 NYS 2d 277 (1988).
89. Art 29 Warsaw-system conventions and Art 35 Montreal Convention 1999.
90. *Proctor v Jetway* [1982] 2 NSWLR 264, 271.

of or damage to cargo and for delay during the time the cargo is in the charge of the carrier. However, the carrier is entitled to rely on specified defences, which need to be proved in order to be exonerated, wholly or partly, from liability. The liability of the air carrier for delay, loss, or damage to cargo is limited to a maximum amount per kilogramme, (also called monetary cap). In certain circumstances of serious misconduct, under some of the international air conventions, the carrier loses the benefit of the monetary cap limiting liability for delay, loss, or damage to cargo.

Presumed liability of the carrier for loss or damage during carriage by air

A fundamental tenet of all the international air conventions is the presumed liability of the air carrier for all loss or damage during air carriage. Thus, the claimant whose goods are lost or damaged does not need to prove that the carrier was at fault. In this respect, the relevant provisions of the international air conventions are substantially the same, with minor semantic differences for the Montreal Convention 1999, which are here indicated within brackets, as appropriate.

All the international air conventions provide that the air carrier is liable if '*the occurrence (event) which caused the damage . . . took place during the carriage by air*',[91] even if the substantive consequential damage occurred later.[92] The period of '*carriage by air*' is defined,[93] in all the international air conventions, as the period during which the goods are '*in the charge of the carrier*'. The additional phrase '*whether in an airport*[94] *or on board an aircraft, or, in the case of landing outside an airport, in any place whatsoever*', which is included in the relevant provisions of the Warsaw-system conventions, was omitted from the text of the Montreal Convention 1999.[95]

---

91. Art 18(1) Warsaw Convention 1929 and Warsaw-Hague Convention 1955, Art 18(2) Warsaw-Hague-MAP 4 Convention 1975, Art 18(1) Montreal Convention 1999.
92. See Shawcross, Air Law, para 589 citing the discussion in *Nowell v Qantas Airways Ltd* 22 Avi 18, 071 (WD Wash, 1990). Further, Shawcross in the same paragraph states that Art 18(1) Warsaw-Hague Convention 1955 [and the other international air conventions] contemplate the award of consequential damages, such as loss of expected profit or the cost of hiring replacement items, but the precise scope of recovery is not specified in the international air conventions and will therefore be determined by the appropriate conflict rules of the court before which an action is brought.
93. cf Art 18(2) Warsaw Convention 1929 and Warsaw-Hague Convention 1955, Art 18(4) Warsaw-Hague-MAP 4 Convention 1975, Art 18(3) Montreal Convention 1999.
94. Art 18(2) Warsaw Convention 1929 and Warsaw-Hague Convention 1955 use the term 'aerodrome', whereas Art 18(4) Warsaw-Hague-MAP 4 Convention 1975 uses the more modern term 'airport'. The difference is semantic.
95. Art 18(3) Montreal Convention 1999.

Therefore, the central question for determining the liability of the carrier during the '*carriage by air*' is whether or not the goods are in the '*charge of the carrier*'. The carrier must be in a position to control the situation and protect the goods. In the United Kingdom, it has been held that the goods must be effectively in the '*safe-keeping, custody, [and] care*'[96] of the carrier. The air waybill, or cargo receipt may be decisive in determining when the cargo first came into the carrier's charge. Further, the period of the carrier's responsibility should normally end when the goods have been delivered to the consignee.[97]

> *Swiss Bank Corp. and Others v Brink's-MAT Ltd and Others*[98]
> The three plaintiff banks ordered £825,000 Bank of England bank notes from the fourth defendants, Midland Bank. Midland Bank at the request of the plaintiffs but in their own name, engaged the first defendants, Brink's-Mat to arrange the carriage of the notes by air to Zurich. Brink's-Mat indicated that the carriage would be by the second defendants Swissair and they prepared an air waybill on Swissair blank forms which they held in the office. The third defendants KLM were the ground handling agent and Swissair's cargo handling agents at Heathrow, and it was to KLM's warehouse that the notes were taken for delivery.
>
> The Court held that KLM accepted the consignment when each consignment having been placed on the scales and weighed and found to conform with the document was moved off the scale and placed on the trolley; Brink's-Mat, when the loss occurred had the CS consignment in their actual custody and control but not the consignments owned by SBC and UBS. It is further held that the consignments did not come into the charge of Swissair/KLM when the van drove into the warehouse; Swissair/KLM became in charge of the goods at the precise moment when they ceased to be in the custody and control of Brink's-Mat; upon the occurrence which caused the loss, the SBC and UBS consignments were, and the CS consignment was not in the charge of Swissair/KLM.

96. *Swiss Bank Corp. and Others v Brink's-MAT Ltd and Others* [1986] 2 Lloyd's Rep 79.
97. Often goods are delivered to third parties, in which case there are issues of agency to consider. Whether the goods are in the carrier's charge when they are subject to customs procedures, depends on the facts and circumstances of each case, see Shawcross, Air Law, para 601, and cited case law. For example, in a French case (cited in fn 4) it was held that the goods were no longer in the charge of the carrier, when the retention of cargo in a warehouse during a strike of customs officers lasted for five months, *Societé National Air France v Societé Arlab* (Aix-en-Provence CA, 29 November 1983), (1985) 39 RFDA 478. In a German case (cited in fn 5) it was held that confiscation of goods by the authorities brought the period of carriage by air to an end, Landgericht Hamburg (64 O 36) (1988).
98. [1986] 2 Lloyd's Rep 79.

The term '*airport*' has been interpreted as including the terminals and other buildings within the airport's premises. Whether or not a freight-handling or storage facility outside the airport perimeter is part of the airport is a matter of interpretation, and may depend on the law of the court where a claim is brought. A United States court held[99] that, notwithstanding the commercial realities, the term airport excluded a warehouse located less than a quarter of a mile beyond the airport limits.

All the international air conventions specify[100] that the period of '*carriage by air*' does not include '*carriage by land, by sea, or by river performed outside an airport*'. However, a presumption is established (which may be rebutted by other evidence) that the damage resulted during the '*carriage by air*' in cases where there is '*carriage by land, by sea, or by river (inland waterway) performed outside an airport*', in the '*performance of a contract for carriage by air*', and for the '*purpose of loading, delivery or transhipment*' of the cargo. Therefore, in the absence of contrary evidence, the claimant is not required to prove exactly where the damaging event occurred, and it is presumed the damage occurred during air carriage. However, if there is evidence the damage was occasioned by an event outside an airport, and thus clearly not during the carriage by air, the liability of the carrier will not be governed mandatorily by an international air convention.

Whether carriage by land or water performed outside an airport is for the '*purpose of loading, delivery or transhipment*', and thus triggers the rebuttable presumption that the damage occurred during the '*carriage by air*' is a question of fact and depends on the circumstances of each case. Information stated on the air waybill may be decisive. For example, when the air waybill specifies delivery to the consignor's own address, surface carriage from the airport to that address will be for the purpose of delivery.[101] However, when onward surface carriage is arranged after receipt of the goods has been acknowledged by the consignee at the airport of destination, it will not be treated as carriage in the '*performance of a contract for carriage by air*' and for the '*purpose of loading, delivery or transhipment*'.[102]

Therefore, if it is envisaged that the carriage by air will involve some ancillary transport by other modes, the parties to the contract should

99. *Victoria Sales Corp. v Emery Air Freight Inc.*, 917 F 2d 705 (2nd Cir 1990), 22 Avi 18, 502.
100. Art 18(3) Warsaw Convention 1929, Warsaw-Hague Convention 1955, Art 18(5) Warsaw-Hague-MAP 4 Convention 1975, and Art 18(4) Montreal Convention 1999 are substantially the same, except for minor semantic differences indicated within brackets for the Montreal Convention 1999.
101. *Jaycees Patou Inc v Pier Air International Ltd*, 714 F Supp 81 (SDNY, 1989), 21 Avi 18, 496.
102. *Compagnie Trans World Airlines v Guigui* (Cour de Cass 17 March 1966), (1966) 20 RFDA 333.

include a corresponding reference in the air waybill, to ensure application of any of the international air conventions throughout.

It should be noted that the international air conventions provide[103] that in case of '*combined carriage performed partly by air and partly by any other mode of carriage*', the provisions of the relevant international air convention '*apply only to the carriage by air*'. Furthermore, in the case of combined carriage, the parties may insert '*in the document of air carriage conditions relating to other modes of carriage, provided that the provisions of the [relevant international air convention] are observed as regards the carriage by air*'.[104] The parties may also stipulate in the air waybill that the same conditions and liability limits of the international air conventions also govern surface portions of the through combined air/surface transport.[105]

> *Siemens Ltd v Schenker International (Australia) Pty Ltd & Another*[106]
> Siemens Australia bought equipment from its German parent (Siemens Germany). For many years, Siemens Germany had used Schenker Germany to carry its goods to Australia. Schenker Australia caused the consignment to be placed on a truck at the airport for delivery to Schenker Australia's bonded warehouse. After the truck left the airport but before arrival at the warehouse, a portion of the consignment was damaged through the negligence of the truck driver employed by Schenker Australia.
>
> The Court held that since the damage occurred outside the physical boundary of the airport, it did not take place during the carriage by air within the meaning of Art 18.1 of the Convention. Accordingly, Schenker Germany was not liable for damage thereunder and the limitation of its liability under Art 22.2(a) did not arise.
>
> As cl 4 operated only in respect of carriage to which the Convention did not apply, 'carriage' in cl 4 had a different meaning from Art 18 of the Convention. Hence, the waybill provided for the transportation of goods otherwise than through carriage by air. The Schenker companies and Siemens Australia contemplated delivery at the warehouse not at the airport.
>
> Decision of the Supreme Court of New South Wales (Court of Appeal) affirmed.

Such provisions would be enforceable, absent other mandatory law applicable to the through combined air/surface transport that is inconsistent with such provisions. If, however, nothing is specified in the air waybill, an air

103. Art 31(1) Warsaw-system conventions and Art 38(1) Montreal Convention 1999.
104. Art 31(2) Warsaw-system conventions and Art 38(2) Montreal Convention 1999.
105. See *Siemens Ltd v Schenker International (Australia) Pty Ltd & Another* [2004] HCA 11, a decision by the highest court in Australia.
106. [2004] HCA 11.

carrier who may routinely operate road vehicles in an integrated intermodal movement, may find a road liability regime (and not the air liability regime) would apply to the road part of the carriage.[107]

> *Quantum Corporation Ltd v Plane Trucking Ltd*[108]
> A consignment of hard disk drives (the goods) were sent by the first claimant in Singapore to the second claimant in Dublin. The third and fourth claimants were freight forwarders and it was common ground that the second defendant Air France agreed to carry the goods from Singapore to Dublin. At Charles De Gaulle airport the three pallets were unloaded from the aircraft. Air France used a sub-contractor, the first defendant Plane Trucking, to carry the goods from Paris to Dublin. Plane Trucking collected two of the pallets (the goods) and loaded them onto a trailer vehicle for carriage by road to Dublin. While in the course of being carried by road in England towards Holyhead the goods were stolen.
>
> The Court held that CMR should be applied to the carriage by road from Paris to Dublin. The present contract was for carriage by road within art 1(1) of CMR in relation to the roll-on, roll-off leg from Charles de Gaulle, Paris to Dublin; Air France's own conditions were to the extent that they would limit Air France's liability overridden accordingly; it was open to the claimants to seek to show that under CMR art 29 there was wilful misconduct or equivalent default disentitling Air France to limit its liability for the loss which occurred during the road transit; that issue would be remitted to the Commercial Court; the appeal would be allowed.

The Montreal Convention 1999 expressly adds an important qualification to the provision that in cases of combined carriage the provisions of the relevant international air convention '*apply only to the carriage by air*'.[109] Namely, '*if a carrier, without the consent of the consignor, substitutes carriage by another mode of transport for the whole or part of a carriage intended by the agreement between the parties to be carriage by air, such carriage by another mode of transport is deemed to be within the period of carriage by air*'.[110] Thus, in these cases, the Montreal Convention 1999 would apply, even if it were established the damage occurred in fact during land transport outside an

107. See *Quantum Corporation Ltd v Plane Trucking Ltd* [2002] 2 Lloyd's Rep 25; [2002] 1 WLR, a decision by the English Court of Appeal, where CMR 1956 was held to apply to a road carriage segment of transport, overriding the airline's own contractual conditions limiting liability.
108. [2002] 2 Lloyd's Rep 25 (CA).
109. Art 38(1) Montreal Convention 1999 is stated to be subject to Art 18(4) Montreal Convention 1999.
110. Last sentence of Art 18(4) Montreal Convention 1999.

airport.[111] This provision is in line with case law decided under the Warsaw-system conventions. Namely, where cargo is carried by land for reasons connected with the carrier's operational convenience only, such carriage will be considered to be for the purpose of delivery,[112] and therefore, come within the scope of the international air conventions.

Presumed liability of the carrier for delay

In contrast to some other international transport conventions, the international air conventions provide expressly for liability in case of delay of the goods. If the goods are delayed for an unreasonable amount of time,[113] or if the delay (even if not unreasonable in its duration) causes the destruction or loss of the cargo, the cargo owner does not need to prove that the carrier was at fault. Subject to giving written notice of complaint within the prescribed time-limits,[114] the international air conventions provide[115] that the air carrier is liable for damage occasioned by delay in the carriage of goods or cargo.

It should be noted there are no special rules on the monetary limitation of liability in cases of delay and the rules applicable to loss of or damage to cargo apply equally to damage caused by delay.

Defences available to the air carrier

Whereas there is a presumption the air carrier is liable for any loss of or damage to cargo and for delay, the air carrier disposes of a number of narrowly defined defences to be exonerated, wholly or partly, from liability. The burden of adducing the necessary evidence to prove the defences is on the carrier.

The defences, examined in more detail below, are the following:

---

111. The question arises, however, as to what amounts to 'consent of the consignor', in particular whether tacit consent may be assumed in certain circumstances. This would depend on the circumstances of the case.

112. See *Cie UTA v Ste Electro-Enterprise* (Cour de Cass, 31 January 1978) (1979) 33 RFDA 310.

113. The word 'delay' is generally thought to mean that, in the absence of any express contract, a carrier is only bound to perform the carriage within a reasonable time having regard to all the circumstances of the case.

114. The time-limit for written notice of complaint in case of delay is calculated from the date of receipt. Art 26 Warsaw Convention 1929 stipulates fourteen days. Art 26 Warsaw-Hague Convention 1955 and Warsaw-Hague-MAP 4 Convention 1975, and Art 31 Montreal Convention 1999 stipulate twenty-one days.

115. Art 19 Warsaw-system conventions and Montreal Convention 1999.

(a) defence of 'all necessary measures';
(b) defence of 'negligent pilotage'; and
(c) defence that the claimant was 'contributory negligent'.

In relation to the carriage of cargo, four further specific defences may apply, namely:

(a) inherent defect, quality or vice of the goods;
(b) defective packing of the goods;
(c) act of war; and
(d) act of public authority.

'All necessary measures'

The international air conventions (subject to the change introduced by the Montreal Convention 1999 which is indicated within brackets) provide[116] that the air carrier is liable, unless it can be proved a) that the carrier and servants and agents took '*all necessary measures*' (or, '*all measures that could reasonably be required*') to avoid the damage, or b) that '*it was impossible . . . to take such measures*'.

The Warsaw-system conventions use the phrase '*all necessary measures*', which has been construed by the courts to mean the carrier should prove '*all reasonably necessary measures*'[117] were taken. This interpretation is reinforced by the wording of the Montreal Convention 1999 replacing the phrase '*all necessary measures*' with the phrase '*all measures that could reasonably be required to avoid the damage*'. In practice, the defence is quite difficult to establish and is, therefore, rarely successfully invoked in litigation.

> *J. J. Silber v Islander Trucking*[118]
> A consignment of goods, the property of the plaintiffs, was lost when the trailer in which they were being carried by the defendants (Islander) from Reggio Calabria to Paris was seized by armed robbers while parked at a motorway tollgate near Salerno. The plaintiffs claimed damages in respect of this loss. The defendants however, sought to rely

116. Art 20(1) Warsaw Convention 1929 and Warsaw-Hague Convention 1955 make reference to 'agents', whereas Art 20 Warsaw-Hague-MAP 4 Convention 1975 and Art 19 Montreal Convention 1999 make reference to 'servants and agents'.
117. *Swiss Bank Corp. and Others v Brink's-MAT Ltd and Others* [1986] 2 Lloyd's Rep 79, 96–97. The requirement has also been construed by courts to be more akin to the duty of 'utmost care' required of road carriers under the CMR 1956 (La Convention relative au Contrat de Transport International de Marchandises par Route/The Convention on the International Carriage of Goods by Road), see *J.J. Silber v Islander Trucking* [1985] 2 Lloyd's Rep 243.
118. [1985] 2 Lloyd's Rep 243.

on the International Carriage of Goods by Road Convention (CMR) art 17 (2).

The Court held that (1) art 17 (2) set a standard somewhere between a requirement to take every conceivable precaution, however extreme, within the limits of the law and a duty to do no more than act reasonably in accordance with prudent current practice; the words 'could not avoid' were to be treated as comprising the rider 'even with the utmost care'; and in deciding whether this criterion was satisfied in the individual case, the Court would need to look at all the circumstances and not simply at any constraints which might be imposed by the criminal law on steps which would have prevented the loss. There would be judgment for the plaintiffs.

If the carrier is not successful in establishing '*all reasonably necessary measures*' were taken by the carrier or the servants and agents, the carrier may still be exempt from liability if it can be proven '*it was impossible* [for them] *to take such measures*'. For example, when a flight is prevented by an unexpected natural disaster, such as a typhoon or a volcanic eruption, the carrier would not be liable, as it would be plainly impossible for a carrier to take measures to prevent such an event.[119]

It is worth pointing out that the Warsaw Convention 1929 and the Warsaw-Hague Convention 1955 provide[120] the carrier may use the defence of '*all necessary measures*' in respect of a claim for loss, damage, or delay to cargo. In contrast, the Warsaw-Hague-MAP 4 Convention 1975 and Montreal Convention 1999 restrict the availability of the defence of '*all necessary measures*' (or, '*all measures that could reasonably be required*') to a claim for damage caused by delay in the carriage of cargo.[121] It should, however, also be noted that under the Warsaw-Hague-MAP 4 Convention 1975 and the Montreal Convention 1999, four specific defences have been added in respect of carriage of cargo, two of which (namely, '*inherent defect, quality or vice*' and '*defective packing*') are similar to those included in other international transport conventions.

Negligent pilotage

The Warsaw Convention 1929 provides[122] that the carrier is excused from liability if it can be proven the damage to the goods was caused by '*negligent pilotage or negligence in the handling of the aircraft or in navigation*', and that

119. *DeVera v Japan Airlines*, 24 Avi 18, 317 (SD NY, 1994).
120. Art 20(1) Warsaw Convention 1929 and Warsaw-Hague Convention 1955.
121. Art 20 Warsaw-Hague-MAP 4 Convention 1975 and Art 19 Montreal Convention 1999.
122. Art 20(2) Warsaw Convention 1929.

in all other respects the carrier and the agents '*have taken all necessary measures to avoid the damage*'.

The defence of '*negligent pilotage*' is an unusual example of exemption from liability in cases of negligence. In view of the considerable technological progress in air navigation equipment since the Warsaw Convention 1929, the defence of '*negligent pilotage*' became unnecessary and difficult to justify. Thus, the defence of '*negligent pilotage*' has not been reproduced in the other Warsaw-system conventions or the Montreal Convention 1999, and is only available in cases where the unamended Warsaw Convention 1929 applies.

Contributory negligence of the claimant

Under all of the international air conventions, the carrier is wholly or partly relieved from liability if it is proved negligence on the part of the claimant caused or contributed to the loss, damage, or delay in question. There are, however, some differences between the relevant provisions in the various international air conventions. The Warsaw Convention 1929 and Warsaw-Hague Convention 1955 provide[123] that the carrier may be exonerated wholly or partly from liability if it is proven the damage '*was caused by or contributed to by the negligence of the injured person*'. In the context of carriage of cargo, the negligence must be attributable to the claimant, ie, the consignor or the consignee.

The Warsaw-Hague-MAP 4 Convention 1975 and the Montreal Convention 1999 make two main additions to the text of the Warsaw Convention 1929 and Warsaw-Hague Convention 1955. They provide that the carrier may be exonerated wholly or partly from liability if it is proved the damage was caused or contributed to by the negligence, or 'other wrongful act or omission' of the person claiming compensation, or 'the person from whom he derives his rights'. The additional phrase of 'other wrongful act or omission' appears to refer to acts which are deliberate rather than negligent.[124] Examples of such acts are false declarations of weight or content giving rise to delay in customs clearance, or a knowing failure to provide documents essential to avoid seizure of the cargo upon arrival by customs.[125] The question of whether there was, in fact, any contributory negligence and the question of how losses in these cases are to be attributed as between

123. Art 21 Warsaw Convention 1929 and Warsaw-Hague Convention 1955.
124. Art 21(2) Warsaw-Hague-MAP 4 Convention 1975 and Art 20 Montreal Convention 1999.
125. *KLM v Tannerie des Cuirs*, Paris 06.06.2001, BTL 2001.664.

the claimant and the defendant, depend upon the facts of each case and are determined by the law of the court before which a case is brought.[126]

Specific defences in relation to carriage of cargo

As was stated above, under the Warsaw-Hague-MAP 4 Convention 1975 and the Montreal Convention 1999, the carrier can rely on the defence of '*all necessary measures*' only with regard to damage caused by *delay* in the carriage of cargo. However, in relation to claims arising from loss of or damage to cargo, the Warsaw-Hague-MAP 4 Convention 1975 and the Montreal Convention 1999 provide[127] that the carrier disposes of four additional specific defences, namely:

(a) 'inherent defect, quality or vice of that cargo;
(b) Defective packing of that cargo performed by a person other than the carrier or [his] servants or agents;
(c) An act of war or an armed conflict;
(d) An act of public authority carried out in connection with the entry, exit, or transit of the cargo'.

The first two specific defences, namely, '*inherent defect, quality or vice*' and '*defective packing*' are also encountered in other transport conventions. Whilst interpretation of the defences may vary between jurisdictions, some degree of uniformity in interpretation has developed.

As concerns the English view, a leading commentator states, by way of summary: '[a]n inherent vice (or defect) is a defect in cargo which by its development through ordinary processes within the cargo itself tends to the injury or destruction of that same cargo, to such an extent that it does not survive the normal rigours of the journey in question and remain suitable for use in commerce for a reasonable time after the end of the journey.'[128]

Noten v Harding[129]
The plaintiffs imported industrial leather gloves from their principal supplier who was a manufacturer in Calcutta (Artonex). Four

126. The Warsaw Convention 1929 and Warsaw-Hague Convention 1955 make express reference to the law of the court seized of the case to determine whether and to which extent the carrier is exonerated from liability. This was not reproduced in the text of the Warsaw-Hague-MAP 4 Convention 1975 and Montreal Convention 1999.
127. Art 18(3) Warsaw-Hague-MAP 4 Convention 1975 and Art 18(2) Montreal Convention 1999.
128. For an interesting discussion of the concept and a review of earlier case-law, see *Noten v Harding* [1990] 2 Lloyd's Rep 283 (CA), a marine insurance case.
129. [1990] 2 Lloyd's Rep 283 (CA).

> shipments of gloves were made, three in 1982 and the fourth in 1983. The 1982 shipments were insured with the defendants (Lloyd's underwriters) on the terms of the Institute Cargo Clauses (All Risks) which included the following exception of inherent vice or nature of the subject matter insured. In each case the gloves were found on out-turn to be wet, stained, mouldy and discoloured.
>
> The Court held that there was no doubt that the goods suffered damage in that the gloves were mouldy and mildewed and the boxes and wrapping in which they were carried were sodden and collapsed; the real or dominant cause of that damage was that the gloves when shipped contained excessive moisture, given the conditions in which they were, and were expected to be carried.
>
> It is further held that if the factual cause of the damage to these gloves had been correctly identified then that was an excepted peril under these policies; the goods deteriorated as a result of their natural behaviour in the ordinary course of the contemplated voyage without the intervention of any fortuitous external accident or casualty; the damage was caused because the goods were shipped wet; the cause of the damage fell within the excepted peril of inherent vice or nature of the goods.

'Packing is defective, if its state is such that the particular goods are unable to withstand the dangers of normal transit of the kind contemplated by the particular contract of carriage.'[130] The last two specific defences, '*an act of war*' or '*an act of public authorities*', refer to instances of armed hostilities and the enforcement of customs, excise, trade, embargo or quarantine regulations, respectively.

There is one important difference between the Warsaw-Hague-MAP 4 Convention 1975 and the Montreal Convention 1999 in respect of these four defences. Under the Warsaw-Hague-MAP 4 Convention 1975, the air carrier needs to prove the destruction, loss of, or damage to cargo resulted '*solely*' from one or more of the above four defences. A literal interpretation of the word '*solely*' would mean the carrier's four specific defences would be very difficult to prove. An English court held[131] that the right approach was to '*look at the adventure as a whole*' and the judge concluded it was sufficient for a successful defence by the carrier that the inherent vice in the goods was the '*dominant cause of their deterioration*'.

> *Winchester Fruit Ltd v American Airlines Inc.*[132]
>
> The claimants were the consignees of peaches and nectarines which were carried from Asuncion in Paraguay to London Heathrow via

130. Note that in English law packing is considered as part of the cargo.
131. *Winchester Fruit Ltd v American Airlines Inc.* [2002] 2 Lloyd's Rep 265 (276).
132. [2002] 2 Lloyd's Rep 265 (276).

> Sao Paulo and New York in November, 1996 under the defendants' airway bill. The claimants alleged that the goods deteriorated during the carriage by air and sought to recover damages under the Warsaw Convention.
>
> The Court held that on the facts and the evidence the defendant took the steps which they did in the ordinary course of events so that it was impossible to identify a relevant failure on their part which could be said to represent an 'occurrence'; any criticism which might have been directed was met by the 72 hour provision; the provision warned the consignor that insofar as perishable goods might, simply by the ordinary passage of time, deteriorate before the expiry of 72 hours the risk was vested in him. On the facts the claimants failed to establish that there was any relevant occurrence within the meaning of art 18 and their claim failed.

In the corresponding provision of the Montreal Convention 1999, the word '*solely*' has been removed. Thus, evidence the carrier was or should have been aware, for example, of defective packing, may not necessarily defeat the defence in question altogether.

Financial limitation of carrier's liability

The international air conventions limit the air carrier's liability for loss, damage, or delay to cargo to a certain maximum amount (also called monetary cap).

General

The monetary cap limiting the liability of the air carrier for cargo is essentially the same under all the international air conventions. However, under the Warsaw Convention 1929 and the Warsaw-Hague Convention 1955, it is expressed in the monetary unit of gold francs (250 gold francs per kilogramme). Under all the other, more recent international air conventions,[133]

133. Warsaw-MAP 1 Convention 1975, Warsaw-Hague-MAP 2 Convention 1975, Warsaw-Hague-MAP 4 Convention 1975, and Montreal Convention 1999. See *GKN Westland Helicopters Limited and Another v Korean Air* [2003] 2 Lloyd's Rep 629, the Korean Air offer was made more than six months after the accident and more than six months before the commencement of proceedings; it was, therefore, not an offer to which Art 22(4) applied. *Rolls Royce Plc and Another v Heavylift-Volga Dnepr Ltd and Another* [2000] 1 Lloyd's Rep 653, RR were entitled to recover the amount of damage which they had suffered only up to the limit specified in the Convention. 2000 *Applied Implants v Lufthansa Cargo* 2 Lloyd's Rep 46; *Applied Implants Technology Ltd and Others v Lufthansa Cargo AG and Others* [2000] 2 Lloyd's Rep 46, the relevant limit on the agreed facts was to be calculated by reference to the combined weight of all the packages covered by the House Air Waybill, namely 11,675 kilogrammes.

it is expressed in the monetary unit of SDR (17 SDR per kilogramme). The main advantage of using the monetary unit of the SDR, rather than the gold franc, is that it creates certainty for carriers and cargo interests as to the value of any compensation that may be payable. The exchange rate of the SDR in relation to major currencies is published daily by the IMF.[134] As the SDR is calculated by reference to a basket of generally stable currencies, the use of the SDR as a monetary unit of account safeguards against the erosion of the liability limits by inflation.

In contrast, in respect of the Warsaw Convention 1929 and Warsaw-Hague Convention 1955, the difficult issue arises which exchange rate is to be used when converting gold francs to today's currency. This may vary greatly depending on the jurisdiction where a dispute is resolved.[135]

> *SS Pharmaceutical Co Ltd v Qantas Airways Ltd*[136]
>
> The second plaintiffs consigned pharmaceutical products to the plaintiffs. The goods were carried by the defendants from Melbourne, Australia for delivery to the first plaintiffs in Tokyo, Japan. The cartons were each marked with a stencilled umbrella to denote that the goods in the carton would be damaged if exposed to water. The weather forecast for that day indicated showers with occasional thunderstorms. On arrival at Tokyo the goods were surveyed and it was found that the cartons were torn and wet on both sides of the front, back and top.
>
> The Court held in the particular circumstances in Australia the SDR as an appropriate conversion rate did not apply; to accept the last official gold price as a unit of conversion neither satisfied the intention of the framers of the Convention nor paid sufficient regard to the changed circumstances; the last official price fixed in Australia was fixed in the circumstances of this country with scant regard to other countries prices; it no longer had any relevance to today's values in a financial market place; the only gold price that could be used was the only price currently in existence, the free market price; effect could be given to the framers of the Convention only by holding that the market price of gold should be the appropriate conversion mechanism.
>
> On the evidence the defendants' conduct was reckless; there was clear knowledge of the likelihood of damage to specially vulnerable cargo in the weather conditions then obtaining; the plaintiffs had satisfied the test propounded by art 25 and the defendants were not entitled to limit their liability.

134. See <www.imf.org>.

135. For discussion of different possible approaches, see *SS Pharmaceutical Co Ltd v Qantas Airways Ltd* [1989] 1 Lloyd's Rep 319 (NSW SC), affirmed without reference to this issue at [1991] 1 Lloyd's Rep 288 (NSW CA). See *ESPADA Sancheq and Others v Iberia Lineas Aereas De Espana SA* [2013] 1 Lloyd's Rep 411.

136. [1989] 1 Lloyd's Rep 319.

The Warsaw Convention 1929 makes reference to the '*French franc consisting of 65.5 milligrams gold of millesimal fineness 900*'[137] (which was the actual currency in use in France between 1928 and 1937). It also provides that the amount of gold francs (ie, 250 gold francs) '*may be converted into any national currency in round figures*'.[138] The original cargo limit of 250 gold francs per kilogramme amounted to some US$10, at the rates of exchange prevailing in 1929.[139]

The Warsaw-Hague Convention 1955 adds that the conversion of 250 gold francs into national currency other than gold '*shall, in case of judicial proceedings, be made according to the gold value of such currencies at the date of the judgment*'.[140] Therefore, many countries enacted legislation prescribing the equivalent, in their national currency, of the Warsaw Convention 1929 limits. However, often that legislation has not been revised to take account of inflation over the past fifty or so years. Even when a country does have an official value of its currency in terms of gold, that official value may deviate significantly from the free market value. This creates additional uncertainty for the various parties involved in the contract of carriage. In order to ensure greater stability of the liability limits, some Contracting States to the unamended Warsaw Convention 1929, or the Warsaw-Hague Convention 1955 (or States whose trading partners include such States), have adopted the SDR as the relevant unit of account, at the exchange rate used in MAP 1 1975 and MAP 2 1975, ie, one SDR equalling fifteen gold francs, and have enacted national legislation to this effect. However, this practice is not uniform.

Special declaration of value

Under the international air conventions, the liability of the carrier is limited to a sum of 250 gold francs[141] or 17 SDR per kilogramme,[142] unless the consignor has made, at the time when the package was handed over to the carrier, a special declaration of the value[143] and has paid a supplementary

137. Art 22(4) Warsaw Convention 1929.
138. ibid.
139. See Shawcross, Air Law, para 106 stating that 125,000 gold francs amounted to some US$5,000 at the rates of exchange prevailing in 1929.
140. Art 22(5) Warsaw-Hague Convention 1955.
141. Art 22(2) Warsaw Convention 1929 and Warsaw-Hague Convention 1955.
142. Art 22(2)(b) Warsaw-Hague-MAP 4 Convention 1975 and Art 22(3) Montreal Convention 1999.
143. The phrase 'special declaration of value' used in Art 22(2) Warsaw Convention 1929 is replaced by the phrase 'special declaration of interest' in Art 22(2) Warsaw-Hague Convention 1955, Art 22(2)(b) Warsaw-Hague-MAP 4 Convention 1975, and Art 22(3)

sum, if so required. If a special declaration of value has been made, the carrier's liability may not exceed the declared sum, except if the carrier proves the consignor has declared a value greater than the actual value of the package at delivery.

In commercial practice, carriers commonly provide in their conditions of contract for an acknowledgment by the consignor that they has had the opportunity to make a special declaration of the value of the goods at delivery and identifying as the special declaration the entry on the air waybill of a 'declared value for carriage'.

Calculation of the limit

The financial limitation of the carrier's liability is calculated on a per kilogramme basis. Under the Warsaw Convention 1929, where only a part of a consignment of goods is lost, damaged or delayed, the carrier's liability is calculated by reference to the actual weight of the lost, damaged, or delayed goods.[144] All the other international air conventions contain an additional provision which is more advantageous to the cargo owner. Namely, if the loss, damage or delay of a part of the cargo, or of an object contained therein, affects the value of other packages covered by the same air waybill, the total weight of such package or packages shall also be taken into consideration in determining the limit of liability. In other words, the weight to be taken into consideration in determining the amount to which the carrier's liability is limited is not only the total weight of the package or packages lost, damaged, or delayed, but also the weight of those other packages whose value has been affected as a result.[145]

Carrier may lose the benefit of the financial limitation of his liability in certain circumstances: 'wilful misconduct' or 'recklessness'

Under the Warsaw Convention 1929 a carrier may not rely on the monetary cap limiting their liability in cases where the carrier or any of the

---

Montreal Convention 1999. The difference in wording is semantic. The phrase 'special declaration of value' will be used throughout this report.

144. The actual gross weight of the goods will provide the relevant data, unless the terms of the air waybill provide otherwise. See *CPH International Inc. v Phoenix Assurance Co of New York* (SD NY, 1994), where the effect of the air waybill was that the 'chargeable weight' (which was five times higher than the gross weight), used to determine the transportation charge, was to be relied upon.

145. Art 22(2)(b) Warsaw-Hague Convention 1955, Art 22(2)(c) Warsaw-Hague-MAP 4 Convention 1975 and Art 22(4) Montreal Convention 1999.

agents acting within the scope of their employment are guilty of '*wilful misconduct*'.[146] This also affects cases where a special declaration of value at delivery has been made.[147] In these cases, the carrier's liability will, therefore, not be limited to the declared value.

'*Wilful misconduct*' is not defined in the Warsaw Convention 1929, but would seem to require a degree of intention or subjective recklessness.[148] According to the relevant provision,[149] the same consequences arise in cases of '*wilful misconduct*', or '*such default* . . . [by the carrier or his agents acting within the scope of their employment] *as, in accordance with the law of the Court seized of the case is considered to be equivalent to wilful misconduct*'. Therefore, the Warsaw Convention 1929 leaves the determination of whether or not the carrier or the agents acting within the scope of their employment are guilty of the relevant misconduct to the law of the court before which a case is brought.[150]

> *The Thomas Cook Group Ltd v Air Malta Company Ltd*[151]
>
> The defendants, entered into a contract for international carriage of a consignment of banknotes from London Heathrow to Luqa Airport, Malta. The banknotes were being transported in the course of a contract for the sale of currency by the first plaintiffs TCGL to BOV. The third plaintiffs now Guardforce International (UK) Ltd. (Group 4 Airborne) had been employed by the TCGL and the second plaintiffs TCFM to collect the consignment from Air Malta and to deliver it to the BOV as the ultimate consignee to whom the property would have passed on delivery at the bank. The carriage of the banknotes was subject to the unamended Warsaw Convention which was set out in Schedule 2 to the Carriage by Air Acts (Application of Provisions) Order, 1967. In the event the consignment was off loaded at Luqa Airport. There

146. Art 25 Warsaw Convention 1929.

147. See, for instance, *Antwerp United Diamonds BVBA and the Excess Insurance Co Ltd v Air Europe* [1993] 4 All ER 469 and [1995] 3 All ER 424 (CA).

148. The term '*wilful misconduct*' is used in other international conventions, (for example, in Art 29 CMR), and has been interpreted as requiring either intention or subjective recklessness. In other words, the carrier must know of the risk and acts or fails to act regardless of the consequences. For example, in one case (*Agrippina v KLM* [Milan, 20 February 1964] (1967) 6 Dirito Aereo 170) gold jewellery valued at US$5,000 was stolen at Kennedy Airport, New York. There was evidence of wilful misconduct in that the packages had been left unguarded for approximately one hour. However, wilful misconduct does not include negligence, even gross negligence. See further, Shawcross, Air Law, paras 666 to 680, and Clarke and Yates, *Land and Air*, para 3.150.

149. Art 25 Warsaw Convention 1929.

150. For the approach taken by English courts in examining the issue, see *The Thomas Cook Group Ltd v Air Malta Company Ltd* [1997] 2 Lloyd's Rep 399 at 407; *Goldman v Thai Airways International Ltd* [1983] 3 All ER 693 at 698 (CA).

151. [1997] 2 Lloyd's Rep 399.

> was an armed robbery and the consignment was stolen from a room within the import cargo warehouse while awaiting customs clearance.
>
> The Court held that the starting point in considering whether the acts or omissions of a person entrusted with goods of another amounted to wilful misconduct was an enquiry about the conduct ordinarily to be expected in the particular circumstances; the next step was to ask whether such acts or omissions were properly to be regarded as 'misconduct'; it was next necessary to consider whether the misconduct was wilful and what did and what did not amount to wilful misconduct; and finally it had to be considered whether the wilful misconduct caused the loss of or damage to the goods.
>
> The acts or omissions of Air Malta were not so far outside the range of conduct ordinarily to be expected in the particular circumstances as to be properly regarded as 'misconduct'; and any misconduct on the part of Air Malta could not be considered as wilful. The presence of the CIT vehicle with its armed AFM escort would have a very significant deterrent; it was for Group 4 Malta to arrange this; they failed to do so and this was not the fault of Air Malta.
>
> Air Malta did not have a defence under art 20; and Air Malta's liability was limited by art 22(2); it was not necessary to consider the defendant's alternative case under art 21 as to whether the plaintiffs had caused or contributed to the loss. Both actions failed.

The Warsaw-Hague Convention 1955 sought to clarify the meaning of the term '*wilful misconduct*' and replaced it[152] with the phrase '*act or omission* [of the carrier or the servants or agents acting within the scope of their employment] . . . *done with intent to cause damage or recklessly and with knowledge that damage would probably result*'.[153]

152. Art 25 Warsaw-Hague Convention 1955.

153. There is a great deal of case law on this question, much in relation to personal injury claims, rather than cargo claims, and views vary between different jurisdictions. In determining whether there has been any relevant misconduct, US courts appear to focus more on an 'objective' view of the circumstances, see *Ospina v TWA*, 24 Avi 17, 109 (2 Cir, 1992); *Cortes v American Airlines Ltd*, 177 F 3d 1272, 1291 (11 Cir, 1999). In contrast, English courts have tended to focus more on the actual awareness of the person, see *Horobin v BOAC* [1952] 2 All ER 1016 at 1022 and *Gurtner v Beaton* [1993] 2 Lloyd's Rep 369 at 387 (CA), but it appears that this view may be changing; see, for instance, the discussion of the question by Auld, LJ and Dyson, J in *Nugent and Killick v Michael Goss Aviation Ltd and Others* [2000] 2 Lloyd's Rep 222 at 227 (CA) and 232–33. See also *Ford v Malaysian Airline Systems Berhad* [2014] 1 Lloyd's Rep 301 (CA), the circumstances in which the injection was administered by the doctor could not be characterized as 'unusual' for the purposes of article 17.1; *Buckley v Monarch Airlines Ltd* [2013] 2 Lloyd's Rep 235; *Barclay v British Airways Plc* [2009] 1 Lloyd's Rep 297 (CA), the term 'accident' in art 17.1 contemplated a distinct event, not being any part of the usual, normal and expected operation of the aircraft, which happened independently of anything done or omitted by the passenger. The causative event had to be 'external' to the passenger; *Ong v Malaysian Airline System Bhd* [2008] 3 HKLRD 153 (CA), dismissing the appeal on

Gurtner v Beaton[154]

A Cessna 310 twin-engined aircraft crashed into Franklyden hillside about eight nautical miles west of Dundee airport. The plane was one of two light aircraft carrying a party of 11 Swiss nationals to a curling engagement in Scotland. Two of the passengers in YI were killed and the three others were severely injured. The pilot (the fifth defendant) was also injured.

The Court held that as to the plaintiffs' appeal, the words of arts 25 and 25A of the Amended Warsaw Convention focused attention on the act or omission; the articles meant that the pilot had to do or omit to do something 'with knowledge that damage would probably result' from that act or omission; there was no basis for a finding that in the moments just before the crash the pilot knew that damage would probably result from flying too low; he thought wrongly that he was over low ground; the appeal and cross-appeal would be dismissed.

Nugent and Killick v Michael Goss Aviation Ltd and Others[155]

Mr. Matthew Harding died in the crash of a helicopter in which he was travelling from Bolton to London as a passenger. The helicopter was owned by the third defendant Polo Aviation Ltd. and piloted by Mr. Michael Goss who was also killed in the crash and whose executor was the second defendant Janine Goss. The claimants, Margaret Rose Nugent and Mark John Killick as executors of the estate of Mr. Harding deceased claimed damages under the Fatal Accidents Act, 1976.

The Court held that whichever test of knowledge was applied it was not possible on the pleadings and other material before the Court, taking the most favourable view from the claimants' point of view, to draw an inference that the pilot or his employers knew that damage would probably result from their conduct; the appeal would be dismissed.

There was no basis for concluding that those who agreed the text of art 25 intended to include background knowledge but exclude other knowledge which was not present to the mind of the actor at the time of the act or omission; it was a requirement of art 25 that the act or omission should be done recklessly and with knowledge that damage would probably result; and although background knowledge could not be rejected on the ground that it added nothing to recklessly, it was sufficient for recklessness that a person should act regardless of the

---

the basis that the Convention applied even though the claim did not fall within art.17; *Deep Vein Thrombosis and Air Travel Group Litigation* [2006] 1 Lloyd's Rep 231; [2005] UKHL 72, art 17 did not apply to DVT cases where all that could be said was that the cramped seating arrangements in the aircraft were a causative link in the onset of the DVT.

154. [1993] 2 Lloyd's Rep 369 (CA).

155. [2000] 2 Lloyd's Rep 222 (CA).

> possible consequences of his act; and what art 25 required was that there should be knowledge of the probable consequences.
>
> Those who drafted art 25 did not intend that anything less than actual conscious knowledge would suffice; the article was not just concerned with death of or injury to passengers but also with damage to or loss of baggage and cargo; the article had to be construed consistently over its whole field of operation, and there was nothing in the language of art 25 or the travaux preparatoires to indicate that it was intended to include some and not all categories of knowledge not present to the mind at the time of the act or omission; the appeal would be dismissed.

A case relevant to damage to cargo,[156] where the issue of interpretation of this phrase arose was decided in Australia.[157] The carrier's agents could see marks on the cargo indicating it should be stored in the dry, as well as note the poor state of the plastic wrapping. Moreover, it was raining at the time and apparent that a typical Sydney summer thunderstorm was likely. Nonetheless, the carrier's agents left the cargo uncovered in the open, exposed to the storm. The court found the agents of the carrier '*must have known that such "deplorably bad handling" of the cargo would probably result in damage to the cargo*'. As a result, the carrier was deprived of the right to limitation of liability.[158]

> *Ericsson Ltd and Ericsson Mobile Communications AB v KLM Royal Dutch Airlines and ors.*[159]
>
> P issued writs against Ds. The claims arose out of thefts of a consignment of mobile phones which were brought into Hong Kong by the airline carrier, D1, and were stolen from the air cargo warehouse at the airport. D1 was the 'actual carrier' of the phones within the meaning of the Amended Warsaw Convention (the AWC) and the Guadalajara Convention (the GC); D2 was a division of D1 and had no legal personality; D3 was the 'contracting carrier' within the meaning of the Conventions; D4 was the local transport agent/freight forwarder and was responsible for picking up the phones and transporting them;

156. Reported decisions involving injury or death of passengers should be read with caution as some courts, especially in the USA, regarding the monetary cap on the carrier's liability for passengers as too low, have been more ready than most to make a finding of 'wilful misconduct'.

157. *SS Pharmaceuticals v Qantas* [1991] 1 Lloyd's Rep 288 (NSW CA).

158. See also *Ericsson Ltd and Ericsson Mobile Communications AB v KLM Royal Dutch Airlines and ors.* [2005] HKEC 2121, a decision at first instance by the High Court of Hong Kong. There, cargo had been stolen with 'inside involvement' of the cargo handling company. It was held that the contracting carrier, the actual carrier, as well as the cargo handlers, as agents, were deprived of the right to limit their liability under the Warsaw-Hague Convention 1955 and the Gudalajara Convention 1961, because the theft amounted to an act done 'with intent to cause damage'.

159. [2005] HKEC 2121.

D5 provided the cargo handling services at the airport, and provided its services to D1 under a written agreement; and D6 provided document handling and administrative services to airlines including D1. D4, in the normal course of events, acted like an agent of D1. Common law actions were pursued against D4–6. Both thefts were an 'inside job' in that they involved an employee of D5. Ds resisted P's claims.

The Court held, finding for P against D1, D3 and D5, but ordering that D5 indemnify D3 and D1, that By virtue of s. 3 of the Carriage by Air Ordinance (Cap. 500), the AWC, as amended by the Hague Convention of 28 September 1955, was given the force of law in Hong Kong. Similarly, by virtue of s 10(1) of the Ordinance, the GC was given the force of law. Since these were international conventions, they were to be given a purposive rather than a narrow, literal interpretation.

The effect of a finding of '*wilful misconduct*', both under the Warsaw Convention 1929 and the Warsaw-Hague Convention 1955, is that the carrier loses the benefit of the monetary cap limiting liability. The carrier will however, not be liable beyond any actual loss proven by the claimant.[160]

It is important to note under the Warsaw-Hague-MAP 4 Convention 1975 and Montreal Convention 1999 the financial limitation of the carrier's liability is 'unbreakable', as the relevant provisions on the effects of serious misconduct[161] do not apply to the carriage of cargo. This is a major change simplifying the settlement of claims related to cargo and avoids lengthy and costly litigation. However, it also means a consignor may have a greater incentive to provide a declaration of value for inclusion in the air waybill. Otherwise, even in cases of intentional misconduct of the carrier, any compensation would be limited to 17 SDR per kilogramme.

Liability of servants or agents of the carrier

The carrier is generally accountable for the acts or omissions of the servants or agents, in particular, his employees. However, a separate question is whether or not servants or agents of the carrier may be sued in their own right by a cargo claimant. This depends on many factors, such as the available evidence, the financial solvency of the servants or agents of the carrier, and issues of jurisdiction (ie, may vary according to the law applicable in

160. Please note that there are no punitive damages under the Warsaw-system conventions or the Montreal Convention 1999. See Art 29 Montreal Convention 1999, and *Re Air Disaster at Lockerbie*, 928 F. 2d 1267 (2 Cir, 1991).

161. Arts 25 and 25 A Warsaw-Hague-MAP 4 Convention 1975 only apply to passengers and baggage and not to cargo. Similarly, Art 22(5), and Art 30(3) Montreal Convention 1999, do not apply to cargo.

the place where a claim is instituted). The courts in some States do not allow a direct action by the cargo claimant against the servants or agents of the carrier, as they are not parties to the contract of carriage, and therefore, are not under any contractual obligation. However, there may be liability of such parties in delict or tort. All the international air conventions, with the exception of the Warsaw Convention 1929, contain a specific provision,[162] which states '*[i]f an action is brought against a servant or agent of the carrier* [for damage, loss, or delay to cargo under the international air conventions] . . . *such servant or agent* . . . *shall be entitled to avail himself (themselves) of the limits of liability (of the conditions and limits of liability)*[163] *which the carrier* . . . *is entitled to invoke* . . .'. This is subject to one condition, namely the servant or agent must prove '*he acted within the scope of his employment*'.

The meaning of '*servants and agents*' is a question of national law, which may be interpreted differently in various jurisdictions. Usually, the term 'servants' refers to those with whom the carrier has entered into a contract of employment, and the term 'agents' refers to independent contractors with whom the carrier has contracted for the performance of a specified task.[164] To benefit from the monetary limitation of liability under the conventions, the servants or agents need to prove they were '*acting within the scope of their employment*'. This is again an issue for national law and has been interpreted differently by courts in various jurisdictions.[165] Strike action is usually considered to be outside the scope of one's employment,[166] whereas theft of goods by a cargo loader has been held to be within the scope of employment.[167]

162. Art 25A(1) Warsaw-Hague Convention 1955 and Warsaw-Hague-MAP 4 Convention 1975, Art 30(1) Montreal Convention 1999.
163. Words within brackets reflect the text of Art 30(1) of the Montreal Convention 1999.
164. For the purposes of the international air conventions, arguably, a distinction needs to be made between agents performing services in furtherance of the contract of carriage, such as those performed by certain airport security services, and agents whose role does not involve activities directly related to the carriage of cargo (for example those engaged in the maintenance or repair of the carrier's aircraft).
165. For instance, where an employee profits from knowledge of valuable cargo acquired during working hours to steal after hours, the question arises as to whether he acted within the scope of his employment. The answer to this question may not only depend on the specific facts of the case, but also on the views adopted by courts in the relevant jurisdiction. See, French Cour de Cassation in *Saint-Paul Fire Co v Air France*, 22 July 1986, (1986) 40 RFDA 428 and *Air France v United Commercial Agencies Ltd*, (1988) 1 S & B Av R VII/293 (Cour de Cass, 12 January 1988). See also Federal Court of Appeal of Canada in *Swiss Bank v Air Canada* (1982) 129 DLR (3rd) 85, 104–105.
166. *OLG Stuttgart*, 24.02.93, TranspR 1995, 74.
167. See *Rustenburg Platinum Mines Ltd v South African Airways and Pan American World Airways Inc.* [1977] 1 Lloyd's Rep 564 at 576, where the court held that 'it was clearly part of [the cargo loader's] duty to take reasonable care of the package during the operation of

> *Rustenburg Platinum Mines Ltd v South African Airways and Pan American World Airways Inc.*[168]
>
> The plaintiffs were the owners, consignors and consignees of platinum which were flown into London Airport (Heathrow) from South Africa by the first defendants, (SAA), and were to be flown to Philadelphia, Pennsylvania, U.S.A., by the second defendants (Pan Am). It was accepted that the platinum was stolen after the box had been loaded into the aircraft but before the aircraft took off.
>
> The Court held that 'wilful misconduct' involved a person doing or omitting to do that which was not only negligent but which he knew and appreciated was wrong and was done or omitted regardless of the consequences and it could not be said in this case that there was wilful misconduct by any servant other than in the theft of the box. The box of platinum was probably stolen by a Pan Am loader at a time when that loader was the only person left since it was clearly part of his duty to take reasonable care of the platinum during the operation of loading and stowing it on the aircraft the theft by the loader was within the scope of his employment. Judgment for the plaintiffs.

If an action is brought against servants or agents of the carrier and the carrier, '*the aggregate of the amounts recoverable*'[169] from the carrier, and the servants or agents, is limited to the monetary cap limiting the carrier's liability, ie, 17 SDR per kilogramme or the amount of any special declaration of value at delivery. Under the Warsaw-Hague Convention 1955,[170] the servants and agents lose the benefit of the monetary cap limiting their liability '*if it is proved that the damage resulted from an act or omission of the servant or agent done with intent to cause damage or recklessly and with knowledge that damage would probably result*'. Therefore, if the servants and agents of the carrier are guilty of '*wilful misconduct*' or '*recklessness*', they face potentially unlimited liability, subject of course to the claimant proving the loss and the amount thereof.[171] Under the Warsaw-Hague-MAP 4 Convention 1975 and the Montreal Convention 1999, in relation to the carriage of cargo, the monetary cap limiting the liability of the servants and agents is 'unbreakable',[172] in the same way as it is 'unbreakable' for the carrier.

---

loading and stowing it on the aircraft'. The decision was confirmed at appeal, see [1979] 1 Lloyd's Rep 19.

168. [1977] 1 Lloyd's Rep 564 (QB) at 576; [1979] 1 Lloyd's Rep 19 (CA).
169. Art 25A (2) Warsaw-Hague Convention 1955 and Warsaw-Hague-MAP 4 Convention 1975, Art 30 (2) Montreal Convention 1999.
170. The Warsaw Convention 1929 does not include the specific provisions on the liability of servants or agents of the carrier added by Art 25A Warsaw-Hague Convention 1955.
171. See, for instance, *Ericsson v KLM* above.
172. Art 25A(3) Warsaw-Hague-MAP 4 Convention 1975, and Art 30(3) Montreal Convention 1999 do not apply in relation to carriage of cargo.

Liability of 'successive', and 'actual' carriers

In principle, the international air conventions regulate the contractual liability, ie, the liability arising out of '*international carriage*' as defined, according to the '*contract made by the parties*' (or '*agreement between the parties*').[173] The party undertaking to perform the carriage in accordance with the contract of carriage is referred to as the carrier, or the '*contracting*' carrier. As has been stated above, '*successive*' carriage performed by several '*contracting*' carriers, even if performed under multiple air waybills, is treated as undivided carriage if the carrier and the consignor considered it as such.[174] An example would be where a consignor contracts with one carrier ('*contracting*; carrier) for carriage from point A to point B, and from point B to point C, but it is agreed from the outset that the last stage from point B to point C is to be performed by another '*successive*' carrier.

A '*successive*' carrier needs to be distinguished from an '*actual*' carrier, to whom a '*contracting*' carrier sub-contracts the performance of the carriage, or part thereof. An '*actual*' carrier, in contrast to a '*successive*' carrier, is not a party to the contract of carriage. Therefore, the question arises as to the liability of such an '*actual*' carrier, if damage occurs during the part of the carriage sub-contracted and performed by the '*actual*' carrier.[175]

The Guadalajara Convention 1961 was concluded to extend the rights and obligations of a carrier to any sub-contracting '*actual*' carrier. As stated earlier, the Guadalajara Convention 1961 is supplementary to both the Warsaw Convention 1929 and the Warsaw-Hague Convention 1955. Its provisions have also been incorporated, largely unchanged, as chapter V in the Montreal Convention 1999.

Definitions

The terms '*contracting*' and '*actual*' carriers are defined[176] in the Guadalajara Convention 1961 as follows:

173. Art 1(2) Warsaw-system conventions, and words within brackets Art 1(2) Montreal Convention 1999.
174. Art 1(3) Warsaw-system conventions and Montreal Convention 1999.
175. The practice of subcontracting the performance of carriage to another carrier came about through the need to market as wide a network of routes as possible, through agreements between carriers, called code-sharing. Under a code-share agreement, two air carriers agree they will each use their unique two-letter code (allocated to each airline by IATA), to sell cargo space on flights operated by the other carrier. Code-sharing is different from 'blocked-space arrangements', whereby an airline allocates cargo space on the flight of another airline. It is also different from 'inter-line agreements', whereby two or more carriers agree to mutually accept air waybills issued by the other.
176. Art I Guadalajara Convention 1961 and Art 39 Montreal Convention 1999.

- '*contracting*' carrier is a person who '*as a principal*[177] *makes an agreement for carriage*' governed by the international air conventions with a consignor or with a person acting on the carrier's behalf; and
- '*actual*' carrier is another person who '*performs the whole or part of the carriage*' contemplated by the agreement between the '*contracting*' carrier and the consignor by virtue of authority from the '*contracting*' carrier. Such authority shall be presumed in the absence of proof to the contrary.

In the definition of '*actual*' carrier, it is further specified that an '*actual*' carrier is not a '*successive*' carrier.

- '*Successive*' carriage occurs when carriage is undertaken by two or more carriers, '*regarded by the parties as a single operation*',[178] but agreed either in the form of a single contract or under a series of contracts.

Therefore, '*successive*' carriage is divided into separate and successive stages, both in terms of time and place, which are identifiable from the outset, ie, when the contract is made. What is important is the intention of the parties at the time the contract of carriage was agreed on.

Some illustrative examples of '*successive*' carriage are the following:

- A consignor contracts with a carrier for carriage from point A to point B, and from point B to point C, and the carrier's timetables (which form part of the contract of carriage) indicate that part of the journey is to be performed by another carrier.[179]
- A consignor contracts with a carrier for carriage from point A to point B, and from point B to point C, but, for marketing reasons, the carriage is identified as a single flight bearing a joint designator code[180] identifying the two carriers involved in the carriage.
- A consignor contracts with a carrier for carriage from point A to point B, and from point B to point C, but, for marketing reasons, the carriage is identified as a single flight bearing the designator code of only one of the carriers, and the carrier's timetables (which form part of the contract of carriage) make the use of two carriers clear to the consignor.

In cases of '*successive*' carriage, each carrier is '*deemed to be one of the parties to the contract of carriage [for the part of the carriage] performed under its*

177. It is important that the 'contracting' carrier is a party who concludes a contract of carriage as a principal and not as an agent for another carrier.
178. Art 1(3) Warsaw-system conventions and Montreal Convention 1999.
179. *Haldimann v Delta Airlines Inc.*, 168 F 3d 1324 (DC Cir 1999).
180. A unique two-letter code allocated to each airline by IATA.

supervision'.[181] Thus, '*successive*' carriers are deemed to be '*contracting*' carriers. In contrast, '*actual*' carriers are not parties to the contract of carriage, as their involvement in the performance of the carriage is not agreed on and evident from the outset. A relevant scenario for '*actual*' carriage could be the following:

- Under a code-sharing agreement, an airline with designator code AA (airline AA) is operating freight services from Asia to Europe, and another airline with designator code BB (airline BB) is operating freight services from Europe to the Americas. The two airlines agree airline AA will sell cargo space on the freight services operated by airline BB from Europe to the Americas, *by using its own designator code, AA*. By the same token, airline BB will sell cargo space on the freight services operated by airline AA from Asia to Europe, *by using its own designator code, BB*.

The distinction between '*successive*' and '*actual*' carriage is important, because the liability provisions applicable under the international air conventions are different, depending on the type of carriage identified.

Liability of 'successive' carriers

All of the international air conventions state (in the relevant definition provisions)[182] that carriage to be performed by several '*successive*' carriers is deemed to be '*one undivided carriage*' if the parties from the outset regarded it as a single operation, whether one or more air waybills were issued. In addition, '*successive*' carriage '*does not lose its international character*' because one or more stages are to be performed entirely within the territory of the same State.[183]

The above provisions are particularly important in deciding whether the '*successive*' carriage is '*international carriage*', as defined, and thus governed by the international air conventions. For example, a consignor contracts with carrier AA for carriage from Seoul (Republic of Korea) to Anchorage (USA), from Anchorage (USA) to Chicago (USA), and from Chicago (USA) to San Francisco (USA). The last stage (Chicago to San Francisco) is to be performed by carrier BB. The carriage from Seoul to Anchorage to Chicago to San Francisco is deemed to be '*one undivided carriage*'. Further, whilst the Chicago to San Francisco stage is performed entirely within the territory of the same State (USA), the carriage is nonetheless international carriage. For

181. Art 30(1) Warsaw-system conventions and Art 36(1) Montreal Convention 1999.
182. Art 1(3) Warsaw-system conventions and Montreal Convention 1999.
183. ibid.

the purposes of determining the carriage is '*international*' it does not matter one or more air waybills have been issued.

The international air conventions provide that each carrier who accepts cargo (in the example above, carrier AA, and carrier BB) is subject to the rules of the applicable international air convention, '*and is deemed to be one of the contracting parties to the contract of carriage in so far as the contract deals with that part of the carriage which is performed under his (its) supervision*'.[184] Therefore, the provisions of the applicable international air convention apply to each '*successive*' carrier, if the cargo comes effectively into their possession,[185] and if they have performed the carriage.[186]

In terms of who may be sued, the international air conventions provide[187] that the *consignor* has a right of action against the first '*successive*' carrier, who is usually identified as the first carrier in the air waybill and is often the one who actually contracted with the consignor. Further, '*the consignee who is entitled to delivery*' has a right of action against the last '*successive*' carrier. Moreover, if the stage of the carriage during which the '*destruction, loss, damage, or delay* [of the goods] *took place*' can be identified, the consignor and the consignee, may each take action against the '*successive*' carrier who actually performed the carriage. All the international air conventions provide[188] that '*these carriers will be jointly and severally liable*' to the consignor or consignee. In other words, if one of the aforementioned '*successive*' carriers is successfully sued, that carrier is liable for the totality of the loss or damage, but that one may be entitled to take recourse against another '*successive*' carrier.

Liability of 'actual' carriers

As was stated above, '*actual*' carriers are neither '*contracting*' nor '*successive*' carriers and are thus not parties to the contract of carriage. The regime applicable to '*actual*' carriers is contained in the Guadalajara Convention 1961, and chapter V of the Montreal Convention 1999, incorporating the respective provisions largely unchanged. In respect of the liability of the '*actual*' carrier, the Guadalajara Convention 1961 and the Montreal Convention 1999 provide[189] that if the '*actual*' carrier performs the whole or part of the carriage, both the '*actual*' and '*contracting*' carriers are liable.

184. Art 30(1) Warsaw-system conventions, and Art 36(1) Montreal Convention 1999.
185. See *Wright v TACA International Airlines* (1984) 2 S & B Av R VII/119 (Belize CA, 1984).
186. See *Emery Air Freight Corpn v Nerine Nurseries Ltd* [1997] 3 NZLR 723 (NZ, CA).
187. ibid.
188. Art 30(3) Warsaw-system conventions, and Art 36(3) Montreal Convention 1999.
189. Art II Guadalajara Convention 1961 and Art 40 Montreal Convention 1999.

The '*actual*' carrier is liable for the part of the carriage performed and the '*contracting*' carrier is liable for the entire carriage contemplated in the contract. Therefore, if the damage to the cargo occurred during the part of the carriage performed by the '*actual*' carrier, the claimant may sue the '*actual*' and/or the '*contracting*' carriers, either jointly or separately.

If the claimant sues only one carrier, that carrier is entitled to require the other carrier to be joined in the proceedings, in accordance with the law of the court seized of the case.[190] Thus, the '*contracting*' carrier cannot decline liability on the basis the damage occurred during the part of the carriage performed by the '*actual*' carrier. As between themselves, the '*actual*' and '*contracting*' carriers may agree as to their respective rights and obligations, including the right of recourse or indemnification.[191] Further, the Guadalajara Convention 1961 and the Montreal Convention 1999 provide[192] that the limit of liability under the international air conventions applies to each one of them individually and to their respective servants and agents acting within the scope of their employment. Moreover, the aggregate damages awarded against the '*actual*' carrier, '*contracting*' carrier, or their servants and agents may not exceed the highest amount the claimant would recover by suing either the '*actual*' carrier or the '*contracting*' carrier.

The '*actual*' and '*contracting*' carriers are, in principle, accountable for each other's acts and omissions[193] (and that of their respective servants or agents),[194] in respect of the part of the carriage performed by the '*actual*' carrier. However, an '*actual*' carrier is not accountable for any potential '*wilful misconduct*' or '*recklessness*' on the part of the '*contracting*' carrier and would, therefore, not lose the right to monetary limitation of liability.[195] By the same token, if the '*contracting*' carrier by special agreement assumes obligations over and above the limit of 17 SDR per kilogramme for cargo (for example, where the consignor has made a special declaration of value at delivery),[196] these will not be binding on the '*actual*' carrier, unless they

190. Art VII Guadalajara Convention 1961 and Art 45 Montreal Convention 1999.
191. Art X Guadalajara Convention 1961 and Art 48 Montreal Convention 1999, which specifies that each carrier has the 'right of recourse or indemnification' against the other.
192. Art VI Guadalajara Convention 1961 and Art 44 Montreal Convention 1999.
193. Art III Guadalajara Convention 1961 and Art 41 Montreal Convention 1999.
194. Art III Guadalajara Convention 1961 and Art 41 Montreal Convention 1999. Art V Guadalajara Convention 1961 and Art 43 Montreal Convention 1999 also provide that servants and agents of the 'actual' or 'contracting' carriers, acting within the scope of their employment, benefit from the same limits of liability (or may lose such benefit) applicable to the carrier whose servant or agent they are.
195. Note, however, that under the Warsaw-Hague-MAP 4 Convention 1975 and Montreal Convention 1999, the limits of liability are in any event 'unbreakable' in relation to carriage of cargo.
196. Art 22(2) Warsaw-system conventions and Art 22(3) Montreal Convention 1999.

have agreed to them.[197] Thus, an '*actual*' carrier's liability in relation to the carriage of cargo is capped at 17 SDR per kilogramme, unless (a) agreed otherwise or (b) the carrier is (or his servants or agents are) guilty of any relevant misconduct.[198] The other provisions of the Guadalajara Convention 1961, and the Montreal Convention 1999, in relation to carriage performed by the '*actual*' carrier are to a large extent similar to the provisions already discussed in relation to the liability of air carriers.[199]

Liability of consignor to the air carrier

The international air conventions mainly deal with the air carrier's liability in case of loss, damage or delay of cargo, and the potential claimants are either the consignor or the consignee. However, the consignor has certain obligations related to the particulars included in the air waybill and the completion of required customs formalities, which may give rise to liability.

First, all the international air conventions provide that the consignor is responsible for the correctness of the particulars and statements relating to the cargo which the consignor inserts,[200] or provides for insertion[201] in the air waybill, cargo receipt, or other record, as applicable. In commercial practice, air waybills are usually completed by the air carrier on behalf of the consignor, and on the basis of information furnished by the consignor or on his behalf. This is reflected in the wording of the relevant provision in the Warsaw-Hague-MAP 4 Convention 1975 and Montreal Convention 1999. The Montreal Convention 1999 additionally includes a new sentence, which specifies that the consignor remains responsible for the correctness of the particulars in situations where the person acting on the consignor's behalf is also the agent of the carrier.

Under the Warsaw Convention 1929, the consignor is '*liable for all damage suffered by the carrier or any other person*' due to the irregularity,

197. Art III Guadalajara Convention 1961 and Art 41 Montreal Convention 1999.
198. See, for instance, *Ericsson v KLM*.
199. See Art IV Guadalajara Convention 1961 and Art 42 of Montreal Convention 1999: notice of complaint may be addressed to the 'actual' or 'contracting' carrier, except that orders or instructions by the consignor to the carrier in the exercise of the consignor's right to dispose of the cargo prior to delivery may only be addressed to the 'contracting' carrier. See also Art VIII Guadalajara Convention 1961 and Art 46 of Montreal Convention 1999: provide for an additional jurisdiction, namely an action may be brought before the court of a country where the 'actual' carrier is ordinarily resident, (or 'domiciled', in the case of the Montreal Convention 1999) or has his principal place of business. This is in addition to the four places in which an action may be brought against the 'contracting' carrier, pursuant to Art 28 Warsaw-system conventions and Art 33 Montreal Convention 1999.
200. Art 10(1) Warsaw Convention 1929 and Warsaw-Hague Convention 1955.
201. Art 10(1) Warsaw-Hague-MAP 4 Convention 1975 and Montreal Convention 1999.

incorrectness or incompleteness of the particulars and statements relating to the goods, the consignor inserts in the air waybill.[202] Under the other international air conventions, the position is similar, but this is expressed in terms of an obligation to '*indemnify the carrier against all damage*', including third party liability, due to the '*irregularity, incorrectness or incompleteness*' of such particulars and statements.[203] It is important to note that liability of the consignor to the carrier, which may potentially be considerable, is not subject to any monetary limit.[204]

The Warsaw-Hague-MAP 4 Convention 1975 and Montreal Convention 1999 add a further provision[205] requiring that the '*carrier shall indemnify the consignor against all damage*', including third party liability, due to the '*irregularity, incorrectness or incompleteness of the particulars and statements inserted by the carrier or on his behalf*' in the cargo receipt or other electronic record. This would include statements inserted by the carrier, for instance, as to relevant stopping places.[206] Secondly, all the international air conventions provide[207] that the consignor must furnish such information and such documents as are necessary to meet the formalities of customs, police, and any other public authorities before the goods may be delivered to the consignee. In addition, under the Warsaw Convention 1929 and the Warsaw-Hague Convention 1955, the consignor must also attach the necessary documents to the air waybill.[208]

This is particularly important where dangerous goods are to be carried.[209] The consignor must furnish the operator of the aircraft with a dangerous goods transport document, which must describe the dangerous goods as required by the latest edition of the Technical Instructions for the Safe Transport of Dangerous Goods by Air published by ICAO.[210]

202. Art 10(2) Warsaw Convention 1929.
203. Art 10(2) Warsaw-Hague Convention 1955, Warsaw-Hague-MAP 4 Convention 1975 and Montreal Convention 1999.
204. For example, if the cargo consignment includes illegal substances, subject to fines, and the seizure and release of the aircraft is made conditional on payment or forfeiture of the aircraft, see *Air Canada v United Kingdom* (1995) 20 EHRR 150.
205. Art 10(3) Warsaw-Hague-MAP 4 Convention 1975 and Montreal Convention 1999.
206. *American Home Assur. Co v Jacky Maeder* (Hong Kong Ltd), 999 F Supp 543, 548 (SDNY, 1998).
207. Art 16(1) Warsaw-system conventions and Montreal Convention 1999.
208. Art 16(1) Warsaw Convention 1929 and Warsaw-Hague Convention 1955.
209. In future, this is also likely to become increasingly relevant in relation to the documentary requirements arising from international security regulations.
210. The Technical Instructions amplify the basic provisions of Annex 18 to the Convention on International Civil Aviation 1944 (as amended) — The Safe Transport of Dangerous Goods by Air. All Contracting States of ICAO are required to take the necessary measures to achieve compliance with the ICAO Technical Instructions. See further <www.icao.org>.

In addition, airline operators also require compliance with their own IATA 'Dangerous Goods Regulations'. The consignor is liable to the carrier for any damage occasioned by the absence, insufficiency or irregularity of any such information or documents, unless the damage is due to the fault of the carrier or the servants and agents.[211] The carrier is under no obligation to enquire into the correctness or sufficiency of such information or documents. Again, the liability of the consignor is not subject to any limit. Ultimately, the consignor bears the responsibility for obtaining and providing the correct relevant information and documentation being in the best position to do so, and as a result is liable to the carrier if the information or documentation is incorrect or insufficient.

Jurisdiction

Issues of international jurisdiction are of great practical importance and often give rise to litigation, as the place where legal proceedings are commenced (the 'forum')[212] may affect the substantive rules applicable to the liability of the air carrier. Ultimately, it is the law of the forum (including its conflict of law rules) that determines which, if any, of the international air conventions may be applicable. Moreover, questions of procedure are governed by the law of the forum[213] as are many issues that are not dealt with in the international air conventions. These include important matters, such as the measure and extent of compensatory damages (remoteness and quantum), the question of whether there has been any wilful misconduct/recklessness of the carrier or contributory negligence of the claimant and how liability should be apportioned.

The international air conventions provide[214] that an action for damages may only be brought in a limited number of jurisdictions in places connected with the carrier, but also likely to be convenient for the claimant. In relation to cargo, the claimant has the option to bring an action for damages in the territory of one of the Contracting States to the applicable international air convention before the competent court at one of the following four places:

a. Where the carrier is '*ordinarily resident*' (the word '*domiciled*' is used in the Montreal Convention 1999), usually the place of incorporation; or

211. Although this is not expressly noted, on the basis of general principles the burden of proving such fault would be on the claimant.
212. Normally, this refers to the country where legal proceedings are brought. However, in federal States, there may be a number of different jurisdictions.
213. Art 28(2) Warsaw-system conventions and Art 33(4) Montreal Convention 1999.
214. Art 28(1) Warsaw-system conventions and Art 33(1) Montreal Convention 1999.

b. Where the carrier has a '*principal place of business*', usually the operational headquarters; or
c. Where the carrier has '*an establishment* [the word "*business*" is used the Montreal Convention 1999] *by which the contract has been made*', usually where the air waybill is issued; or
d. The place of destination, usually designated in the air waybill.

All the international air conventions provide[215] that any '*clause . . . and all special agreements entered into before the damage occurred by which the parties purport to infringe the rules laid down by [the applicable international air convention], whether by deciding the law to be applied, or by altering the rules as to jurisdiction, shall be null and void*'. Thus, exclusive contractual jurisdiction agreements are not permitted. However, in respect of carriage of cargo, arbitration clauses are permitted, provided that the arbitral proceedings are brought in one of the four optional jurisdictions specified, and provided the arbitrator or arbitration tribunal apply the provisions of the international air conventions.[216]

215. Art 32 Warsaw-system conventions and Art 49 Montreal Convention 1999.
216. Art 32 Warsaw-system conventions and Art 34 Montreal Convention 1999, which is clearer on this point.

8

Land Carriage Law (Road and Rail)

8.1 INTRODUCTION

The carriage of goods by land is the most ancient of all forms of carriage. Many goods are carried by rail, truck or lorry. As the demand for door-to-door service increases, many carriers carry on their business as couriers. In Hong Kong, it is generally the case goods produced in factories locally or in Mainland China are delivered by trucks to the Hong Kong International Airport at Chek Lap Kok or the container terminal at Kwai Chung for consignment to overseas customers by air or by sea. An effective and reliable service of domestic carriage of goods by land is essential if Hong Kong is to remain an international trade centre and logistic hub.

Liabilities of land carriers in general

Subject to any applicable legislation, the rights and obligations of a carrier are determined by the contract of carriage formed with his client. How the contract is drafted has an important bearing on the carrier's liabilities.

> *Orient Overseas Container Inc v Regal Motion Industries Ltd*[1]
> N formed a contract of carriage with R. Under the contract, R was required to transport some refrigerated chickens from Hong Kong to Shenzhen. The goods were packed in a container provided by P. R's driver was involved in a traffic accident which caused extensive damage to P's container. The issue was whether R might escape liability

1. [1994] 1 HKLR 282.

by referring to an exemption clause attached to the cargo receipt issued by R. The clause stated that '*[T]he owner of goods shall be responsible for the damage done to the goods during loading and unloading; the consignor shall purchase its own insurance. If during the consignment the goods are damaged due to any traffic accident, theft and burglary, fire and water flood, driver's negligence and other disaster not resistible by human beings, our company shall not be responsible and the transportation charges are still payable*'. [emphasis added]

The court held that R was liable. On its face, the clause exempted only liabilities in relation to damage to the goods. Even if it also covered damage to the container, its validity was determined by the reasonableness test laid down by the Control of Exemption Clauses Ordinance. Taking account of the short duration of the contract, the inability of other parties to have any meaningful control over the carriage, and the infeasibility of the risk of damage to the container being covered by insurance, the exemption clause was not reasonable and thus not valid against N.

When carriers are voluntarily in possession of the goods, they are also liable as bailees: see *Dense Billion Ltd v Hui Ting Sung*.[2]

Wing Hing (Tangs) Fabrics Mfg Co Ltd v Ever Reach Freight Ltd[3]

P formed a contract with E for the carriage of certain bales of cotton first from Shenzhen to Hong Kong, then from Hong Kong back to Shenzhen, and finally from Shenzhen to Sweden. E subcontracted with F for the carriage from Shenzhen to Hong Kong. The goods were lost in Shenzhen. P claimed against F, as a bailee for reward, for the loss.

The court held that F was liable. Where a bailee failed to deliver goods, the settled rule was that he had the burden of proving that it was not due to his fault. The basic duty of a bailee was to take reasonable steps to protect the goods or the bailor's title to them. In the circumstances, F should at least have notified P.

A carrier may also be liable for the loss of other parties in the law of negligence. Often, the carrier is responsible for negligent driving of carrier employees.

Lam Shui-tsin v James Tong[4]

A carrier (D) drove his lorry round a corner when P was standing on the part of the footway outside the railings which were installed to prevent pedestrians from straying on to the carriageway. Part of D's

2. [1996] 2 HKLR 107.
3. [1991] HKLY 59.
4. [1974] HKLR 357.

lorry overhung the footway and P was knocked down. P sued D for personal injuries, claiming that D was negligent.

The court held that D was liable. D was negligent in encroaching the lorry upon the footway. P, as a pedestrian, was entitled to be on any part of the footway and therefore not guilty of contributory negligence.

Generally, carriers are under a duty to take reasonable care in providing services. They should make sure, among other things, their employees are competent, vehicles are in a stage of reasonable repair and appropriate equipment is used to handle the cargo. Carriers are vicariously liable for their employees' acts done within their course of employment. They are, however, not liable for damage caused by a latent defect of the vehicle. For the defence of latent defect to succeed, the carrier must prove the nature of the defect and that a reasonable examination could not have detected the defect. It is not easy to discharge this burden of proof.[5] Similarly, a carrier is not liable for failure to use special equipment to load or unload cargo if the consignor had not informed the carrier of the equipment to be used and it was an unreasonable expectation in the circumstances to be aware of the need.

As a road user, a carrier delivering goods by road should observe the Road Users' Code and the Code of Practice for the Loading of Vehicles issued by the Director of Highways under the Road Traffic Ordinance (Cap 374). Non-compliance with the codes is prima facie evidence of negligence.[6] However, the court may not hold a pedestrian guilty of contributory negligence even though failing to comply with the Road Users' Code.

A railway should properly maintain its railway tracks and a functional system of signalling. In relation to training and supervision of employees, a railway is under a duty to see that 'all persons connected with the carrying and with the means and appliances of the carrying, with the carriages, the road, the signalling, and otherwise, shall use care and diligence, so that no accident shall happen'.[7] In the event two trains collide or a train runs off the lines, the railway is prima facie negligent. The railway may, however, prove the accident was caused by, say, a latent defect in the rolling stock or the negligence of others.[8]

Carriers must also comply with any statutory obligations imposed upon them. Whilst breach of contract and negligence carry only civil liabilities, breach of statutory duty often results in criminal prosecution. For example,

5. See *Henderson v H.E. Jenkins & Sons* [1970] AC 282; *Pearce v Round Oak Steel Works* [1969] 1 WLR 595; *Ritchie v Western Scottish M.T. Co* (1935) SLT 13.
6. Section 109(5).
7. *Wright Midland Ry* (1873) LR 8 Ex 137, 140.
8. See *Readhead v Midland Ry* (1869) LR 4 QB 379; *Latch v Rumner Ry* (1858) 27 LJ Ex 155.

if the consignor sends the goods for consignment in an unsafe container, the carrier who has the container in possession as a bailee may incur criminal liability for the use of the container under the Freight Containers (Safety) Ordinance (Cap 506). The Ordinance gives effect to the International Convention for Safe Containers 1972 and lays down statutory requirements for the safe use of containers. Under the Ordinance, a bailee who, after being served with a notice prohibiting the use of an unsafe container and continues to use it, commits an offence punishable by a fine at level 2 and to imprisonment for 3 months.[9]

In Hong Kong, a carrier may also incur criminal liability if the goods that are being carried are for the purpose of smuggling.

> *R v Wong Yin Chung*[10]
>
> W was a carrier and formed a contract with one customer, X, for the transport of some TV sets to a waterfront. A fishing vessel berthed at the dock when W's lorry reversed towards the water's edge. The TV sets were being loaded onto the vessel when police officers who had been waiting in ambush rushed out and arrested W and other men. It was an offence to assist with the carrying of any article (such as the smuggled TV sets in this case) the carriage of which is restricted. The law also stated that 'in circumstances that give rise to a reasonable suspicion that there is intent on the part of [a] person to evade a restriction or prohibition or to assist another person to evade a restriction or prohibition, [he] will be presumed to have such intent in the absence of evidence to the contrary'. W claimed that he had not been aware that the TV sets were being smuggled until they were put on board the vessel.
>
> The court held that the presumption applied and W was guilty of the offence.

In the *Wong Yin Chung* case, the contract is illegal as performed because it is for an illegal purpose. The carrier may sue for remuneration for service provided before being aware the cargo was to be smuggled. However, the court will not entertain any claim relating to performance done with knowledge of the illegal element.

8.2 DOMESTIC CARRIAGE OF GOODS BY LAND

Consignment note

When the goods are sent to the carrier, a consignment note is usually issued as a receipt for the goods. A consignment note is also evidence of the

9. Section 21.
10. [1993] HKLY 331.

underlying contract of carriage. Further, it may contain the terms of the contract. However, consignment notes are not documents of title. Possession of a consignment note does not equal possession of the goods covered by it.[11]

Rights of consignor and consignee under the contract

In relation to a domestic carriage of goods by land, whether the consignor or consignee is a party to the contract requires further consideration. In the uncommon event that the contract of carriage states explicitly that both the consignor and the consignee may sue under the contract, the entitlement to sue is not an issue. If the contract is silent on this point, one view is that the consignor forms the contract of carriage on behalf of the consignee if the consignor and the consignee are not the same party.[12] It follows that the consignor forms the contract of carriage as the agent of the consignee. If there is, especially at the time of the contract, evidence to show the consignor is the owner of the goods, this view is inappropriate and the consignor should be taken as a party to the contract. The consignee is, in such a case, not entitled to sue under the contract.

A consignor who is an agent of the consignee is generally not able to enforce it. On the other hand, if the consignor is a party to the contract, it may be argued the consignee is nevertheless entitled to sue under an implied contract, ie, one similar to the implied contract recognized in *Brandt v Liverpool Brazil & River Plate SN Co*[13] for carriage of goods by sea. However, the court has become increasingly reluctant to accept the existence of such an implied contract.[14] It is, therefore, not advisable for the consignee to rely solely on the implied contract.

If there is an underlying sale of goods contract between the consignor (the seller) and the consignee (the buyer), the consignee may instead sue under the sale of goods contract if unable to sue and the consignor refuses to sue for loss of or damage to the goods under the contract of carriage. This is possible only if the seller is under an obligation to deliver the goods to the buyer. The sale of goods contract determines whether it is the duty of the consignor to deliver or of the consignee to take possession of the goods.[15]

It is also possible for the sale of goods contract to provide that the seller (the consignor) is under an obligation to sue for the benefit of the buyer (the

11. *L. & S.W. Ry and G.N. Ry v Biship* (1898) 42 SJ 255.
12. *Stephenson v Hart* (1828) 4 Bing 476 and *Heugh v L.N.W. Rly* (1870) LR 5 Ex 51.
13. [1924] 1 KB 575.
14. *The Aramis* [1989] 1 Lloyd's Rep 213.
15. See section 31 of the Sale of Goods Ordinance (Cap 26).

consignee) under the contract of carriage. In refusing to do so, the seller is in breach of the sale of goods contract.

Rights to sue of consignor and consignee in the law of negligence

The rule is that a party who is the owner or in possession of the goods is entitled to sue for damage or loss.[16] If the consignee is not the owner or not in possession of the goods at the time the damage or loss occurs, the carrier cannot be sued in tort. Even if the seller is entitled to sue, full compensation of the loss may not be possible. For example, pure economic loss is generally not recoverable in the law of negligence. For example, if the loss of the cargo causes the closure of the consignee's factory, the consignee cannot sue for the loss of profit resulting from the closure.

Carriage of goods by rail

The Kowloon-Canton Railway Corporation (the KCR) is a public authority established under the Kowloon-Canton Railway Corporation Ordinance (Cap 372). The Ordinance specifies the powers and authorities of the KCR. The KCR is required to conduct its business on commercial principles. The Ordinance also declares the KCR is not a common carrier. That means the KCR has the right of refusal. Carriage of goods is part of the business of the KCR as a railway. It may provide services for the consignment of goods from Hong Kong to other regions or countries. The ordinance also authorizes the KCR, in association with carriers outside Hong Kong, to provide services of through carriage of goods. The contracts formed by the KCR and its customers determine their respective rights and obligations.

On 2 December 2007 the Rail Merger Ordinance came into effect. The Ordinance expressly empowered KCRC to grant a service concession to Mass Transit Railway Corporation Limited (MTR) and expanded the scope of the MTR's franchise to enable it to take up the operation of KCR's transport services.

8.3 INTERNATIONAL CARRIAGE OF GOODS BY LAND

Although Mainland China is the only region that a lorry or train from Hong Kong may enter, it is possible for a Hong Kong carrier to enter into a

16. *Leigh and Sillavan v Aliakmon* [1986] AC 785.

contract of international carriage of goods by land to, say, Poland. In fact, goods have been transported from Hong Kong to Russia by rail. As shown by *Wing Hing (Tangs) Fabrics Mfg Co Ltd v Ever Reach Freight Ltd,*[17] it is also possible to consign goods from Hong Kong to a European country by truck.

The two main conventions governing carriage of goods by land are the Convention on the Contract for the International Carriage of Goods by Road (CMR, the acronym in French for *Convention Relative au Contrat de Transport International de Marchandises par Route*) and the Convention concerning International Carriage by Rail (COTIF).

Since Hong Kong and the PRC are not parties to the CMR and the COTIF, international carriage of goods by land departing from or arriving at Hong Kong is, from a legal point of view, similar to domestic carriage if the governing law of the contract is the law of Hong Kong. If the contract of carriage is governed by the law of another country, for example Germany or Russia, it is possible one of the international conventions applies to part of the carriage outside Hong Kong. Further, parties doing business in Hong Kong with an international dimension may be involved in international carriage outside Hong Kong. It is, therefore, desirable for practitioners in the logistics industry to have some knowledge of the two international conventions.

It should also be mentioned that the PRC became a party to the International Agreement on Carriage of Goods by Rail (IACGR) in 1954. Other contracting states to the IACGR include Albania, Bulgaria, Hungary, Mongolia, North Korea, Poland, Rumania, Russia and Vietnam. The IACGR specifies, among other things, the rights and obligations of railway authorities, consignors and consignees in relation to international carriage of goods by rail through the contracting states. The IACGR does not apply to Hong Kong.

Convention on the contract for the international carriage of goods by road

The CMR came into force in 1961. Contracting parties to the CMR are mainly European countries including Austria, Belgium, Denmark, France, Germany, Gibraltar, Guernsey, Hungary, the Isle of Man, Italy, Luxembourg, the Netherlands, Norway, Poland, Portugal, the Russian Federation, Sweden, Switzerland and the United Kingdom.

17. [1991] HKLY 59.

The CMR applies 'to every contract for the carriage of goods by road in vehicles for reward, when the place of taking over the goods and the place designated for delivery, as specified in the contract, are situated in two different countries, of which at least one is a Contracting country'. The place of taking over the goods and the place designated for delivery determine whether the CMR applies. It applies to, for example, a carriage of goods contract by truck from Hong Kong to Moscow if the governing law of the contract is Russian law. In addition, the CMR may be voluntarily incorporated into any carriage of goods by land. If it is so adopted, it forms part of the contract and thus regulates the rights and obligations of the parties.

Under the CMR, the carrier is generally liable for any loss of or damage to the goods occurring between the time of taking over by the carrier and the time of delivery. The carrier is also liable for delay in delivery. The carrier is, however, not liable if the loss or damage is caused by

- the fault of the claimant,
- inherent vice, or
- unavoidable circumstances.

The carrier is presumed to be not liable in certain specified circumstances. The specified circumstances include

- the use of open unsheeted vehicles agreed by the parties,
- defective condition of packing in the case of goods liable to wastage,
- handling or loading of the goods by the sender,
- goods particularly exposed to loss or damage through breakage or decay,
- insufficiency of marking, and
- carriage of livestock.

The presumption is rebutted if the claimant proves that the loss, damage or delay was not attributable to the specified circumstances.

Normally, the liability of the carrier is limited. For loss of or damage to goods, the amount cannot generally exceed the amount payable in the case of total loss. Subject to this, the limit of liability is 8.33 units of account per kilogramme of gross weight. For delay, the compensation cannot exceed the carriage charges. The limits may, however, be raised by agreement. Besides, the carrier is not entitled to limit his liability if the loss, damage or delay is caused by his wilful misconduct.

A claim must be brought within one year, generally from the day of delivery. Where there is wilful misconduct, the limitation period is extended to three years.

Convention concerning international carriage by rail

The COTIF joins together three conventions on carriage by rail, namely the revised Berne Rail Conventions ('the CIM' and 'the CIV') and the Additional Convention ('the CAV'). Contracting states are mainly European countries and include Austria, Belgium, Bulgaria, Croatia, Denmark, Finland, France, Germany, Greece, Iran, Iraq, Luxembourg, Norway, Poland, Slovakia, Syria, Spain, Turkey, Ukraine, the United Kingdom and Yugoslavia.

The COTIF governs only carriage taking place exclusively over railway lines registered under the Convention. For carriage wholly or partly over non-listed railways, the COTIF does not apply. The COTIF applies only to international carriage by rail, with the departing and arriving stations in two different contracting states.

For carriage of goods by rail, the CIM is relevant. For the CIM to apply, a through-consignment note covering the whole carriage must be issued. The particulars contained in a consignment note include the name of the destination station, the names and addresses of the consignee and consignor, a description of the goods, the quantity and packages of goods, and a detailed list of required customs documents. The consignor may specify the route in the consignment note. If no route is specified, the railway may choose any route which is most advantageous. The consignor may also indicate in the consignment note the amount of carriage charges that he undertakes to pay. By paying a collection fee, the consignor may ask the railway to collect a 'cash on delivery payment'. The amount to be collected must be entered into the consignment note.

The consignor is generally required to prepare a duly completed consignment note for each consignment. The consignor is responsible for the correctness of the particulars on the consignment note. Generally, a consignment note cannot cover more than one wagonload. The original is sent with the goods. A duplicate is returned to the consignor. If the railway considers it appropriate, the goods may be checked and examined to verify the particulars on the consignment note. The consignor or consignee should be invited to attend a content examination. If none of them can attend, two witnesses independent of the railway must be present, unless the local legislation states otherwise.

The railway is generally under an obligation to carry all goods that are presented as complete wagonloads. The obligation to accept is dispensed with if the carriage of the goods will cause delay or require special means of handling the railway station does not have. On the other hand, the railway must refuse to accept goods which are prohibited in any states through which they would be carried. If such goods are inadvertently accepted, the carriage must immediately stop when the railway realizes the mistake. The matter should then be dealt with by the local police. For dangerous goods, the relevant regulations and conditions must be complied with.

The consignor is responsible for any necessary packing. The loading and unloading are governed by the relevant rules applying in the forwarding station, subject to any special agreement between the consignor and the railway contained in the consignment note. In loading the goods, the consignor is liable in the case of overloading. Delivery is effected when the railway hands over the consignment note and delivers the goods to the consignee at the destination station.

The railway is generally liable for the loss of or damage to the goods after their acceptance for carriage and before delivery. The railway is not liable if it is proved that the loss or damage was caused

- in unavoidable circumstances the consequences of which it was unable to prevent,
- by the wrongful act or neglect of the claimant,
- by his instructions,
- or by inherent vice of the goods.

Further, there are certain special risks on which the railway may rely. Examples of the special risks are

- carriage in open wagons under certain conditions,
- inadequate packing,
- defective loading by the consignor,
- completion by the consignee,
- goods inherently liable to loss or damage,
- and incorrect description of goods.

Even though the railway can prove one or more of the special risks apply, it is still liable if the claimant can show the loss was not in fact caused by the special risks. The railway's liability is generally limited. For loss of goods, the limit is 17 SDR per kilogramme of gross weight. The limits for damage are based on the loss in value of the goods. Compensation for damage cannot exceed the amount recoverable for loss of the damaged goods. If the loss or damage is caused by the transit period being exceeded, the railway is liable up to a limit of four times the carriage charges. If the goods have not been delivered within 30 days after the expiry of the transit period, the party entitled to take delivery may consider the goods lost. On receipt of compensation for the goods in such a case, the party may request the railway to notify him without delay if the goods are discovered within one year. The above-mentioned limits may be raised by a declaration of a special interest in delivery. If the loss or damage is caused by an act of the railway, done with intent to cause such loss or damage or recklessly and with knowledge that such loss or damage will probably result, the railway is not able to invoke the limits of liability. Generally, a claim against the railway must be made within one year.

It should be noted that a protocol amending the COTIF ('the 1999 Protocol') was signed by most of the contracting states in Vilnius on 3 June 1999. The amended COTIF came into force in 2006. Some of the major changes made by the 1999 Protocol are:

- English is introduced as a working language, in addition to French and German.
- The revised CIM applies to contracts of carriage of goods by rail for reward if the place of taking over of the goods and the place designated for delivery are situated in two different Member States. Where the place of taking over and the place of delivery are within two different states, the CIM also applies if one of the states is a Member State and the parties to the contract agree to be bound by it.
- Instead of 'railway', 'carrier' is now the focus.
- The Regulations concerning the International Haulage of Private Owners' Wagons by Rail ('the RIP') is no longer in force.
- Railway vehicles may always be carried as goods.
- The consignment note contains more particulars.
- If a carriage performed by several successive carriers is governed by a single contract, all successive carriers become parties to the contract by taking over the goods with the consignment note. Each of them is responsible for the whole carriage until delivery.
- If the carrier entrusts the performance of the carriage, in whole or in part, to a substitute carrier, the original carrier is still liable for the whole carriage. For the part of the carriage performed by the substitute carrier, both carriers are jointly liable.

9

Multimodal Carriage Law

9.1 INTRODUCTION

Nowadays, demand for door-to-door service is on the increase. Exporters favour door-to-door service because it provides a one-shop solution for delivering goods to virtually anywhere in the world. As door-to-door service usually has a higher profit margin than unimodal carriage, many carriers have extended their business to provide door-to-door service. Door-to-door service is also bolstered by the wide use of containers. The standardization of containers makes multimodal transport of goods (also called combined or intermodal transport) much easier. Regardless of the numbers of transhipments or changes in modes of transport, all a consignor needs to do is to form one contract with one carrier. Under the contract, the carrier collects the cargo in the consignor's warehouse and delivers it to the final destination. Although door-to-door service is often provided by carriers, other practitioners in the logistics industry may also do so. In practice, freight forwarders, consolidators, shipping agents and other intermediaries often act as multimodal transport operators.

Depending on the intention of the parties, the carriage may be unimodal or multimodal under a door-to-door service contract. For unimodal transport, the choice of mode of transport is simple. If it is practicable, goods are usually carried by truck or rail. If not, air transport is chosen if prompt delivery is required and the goods are of high value. Since containers are heavy, containerized goods are not normally used in air transport. Air transport is also appropriate if the cargo cannot endure the much longer sea transport. Since sea transport is much cheaper than air transport, consignors prefer sea transport if at all possible. In practice, most containerized goods are transported by sea. The trend is, however, that multimodal transport is increasingly used because it provides more flexibility to carriers and is more customer-oriented. Since air transport is expensive but fast and sea transport slow but cheap, it is rather uncommon for door-to-door carriage to have both an air leg and a sea leg.

Regulatory framework

If the carriage is unimodal, the carrier is, as discussed in other chapters, subject to the relevant international convention. The general view is that a sea or air carrier does not act as a multimodal transport operator if, in addition to performing the unimodal carriage of goods, they provide ancillary services such as collecting the cargo at the premises of the customer or delivering it to the consignee at the place of destination. For example, Article 1 of the UN Convention on International Multimodal Transport of Goods (1980) ('the Multimodal Convention') states that

> [t]he operations of pick-up and delivery of goods carried out in the performance of a unimodal transport contract, as defined in such contract, shall not be considered as international multimodal transport.

For multimodal carriage of goods, until now no international convention is in force. The rights and obligations of the multimodal service provider are determined by the contract formed with his client, subject to the law of Hong Kong. In particular, applicable conventions governing international carriage of goods are most relevant. Since different legal regimes govern different modes of transport, the law is rather complicated in relation to multimodal carriage.

> *Quantum Corp Ltd v Plane Trucking Ltd*[1]
> A forwarder arranged for the carriage of a consignor's goods from Singapore to Dublin. An air carrier agreed to perform the carriage. After the plane landed at Charles De Gaulle airport, Paris, France, the sub-contractor of the carrier carried the goods by road to Dublin. During the carriage by road, an employee of the sub-contractor stole some goods. The consignor claimed against the carrier and the sub-contractor for the loss. The consignor argued that the Convention for the International Carriage of Goods by Road 1956 ('CMR') applied, since international carriage by road formed part of the whole carriage.

1. [2001] All ER (Comm) 916.

> The consignor also contended, in the alternative, that the amended Warsaw Convention governing international carriage by air applied. Under both the CMR and the amended Warsaw Convention, the carrier could not limit liability in this case. The air carrier claimed that either the carriage was not governed by any international conventions or the amended Warsaw Convention as further amended by Protocol No 4 of Montreal 1975 applied. If the court accepted any one of the two arguments, the air carrier was entitled to limit liability in this case.
>
> The court held that the contract was a single contract for carriage from Singapore to Dublin. The parties intended to carry the goods by road from Paris to Dublin. Under the contract, the air carrier was not obliged to perform this part of carriage. If it wished to do so, the carriage might be by air. The contract could not be described as an international carriage of goods by road contract. According to the terms of the contract, the Warsaw regime did not govern the theft. The general conditions of the air carrier thus applied. The air carrier was therefore entitled to limit its liability.

In general, the Carriage by Air Ordinance (Cap 500) and the Carriage of Goods by Sea Ordinance (Cap 462) apply to the sea and air legs respectively. For the sea leg, the Hague/Visby Rules are given in the force of law in Hong Kong. Similarly, the Warsaw Convention or the Montreal Convention applies to an international carriage of goods by air. For carriage of goods by land, the legal relationship between the carrier and the shipper is determined by common law since no international convention applies to Hong Kong. Rights and obligations relating to other services, such as the collection, grouping, consolidating and regrouping of goods, are also defined by the contract. In short, if there is no applicable international convention or mandatory law for any stage of the carriage, the terms of the contract govern. Often, the parties agree to apply one international convention to other parts of the carriage not governed by any convention. In such a case, the terms of the contract must be carefully examined to determine which legal regime governs. The following cases show the complexity of the issue.

> *The 'OOCL Bravery'*[2]
>
> The consignor formed a contract with the carrier for the delivery of some goods, door-to-door from Wisconsin to the Netherlands. The carrier issued a through bill of lading for the goods. The bill stated that the goods were to be transported on board the ship 'OOCL Bravery' from Montreal, Canada to Antwerp, Belgium and the place of delivery was the consignee's warehouse in the Netherlands. Clause 4 of the bill stated that '*[e]ach stage of the transport shall be governed according*

2. [2000] 1 Lloyd's Rep 394.

to any law and tariffs applicable to such stage ' and Clause 23 read: 'CLAUSE PARAMOUNT. *All carriage under this Bill of Lading to or from the United States of America shall have effect subject to the provisions of COGSA . . . which shall be deemed to be incorporated herein . . . Except as otherwise provided herein COGSA . . . shall govern the Goods before loading on board and after discharge from the Vessel and while subject to this Bill of Lading.*' After the ship arrived at Antwerp, the carrier sub-contracted a trucking company to carry the goods by road to the consignee's warehouse. As a result of the negligence of the trucking company, the goods were stolen. The carrier argued that the lower limit prescribed by the Convention on the Contract for the International Carriage of Goods by Road ('CMR') applied.

The court held that Clause 4 was of no effect and the carrier was subject to the higher limit stated in the COGSA (the USA equivalent to the Carriage of Goods by Sea Ordinance). The COGSA had been voluntarily extended to cover the time when the goods were stolen. It nullified any clauses in the bill that would have the effect of lessening the carrier's liability.

Finagra (UK) Ltd v OT Africa Line Ltd[3]

A multimodal transport operator carried goods for a consignor, first by sea from Lagos, Nigeria to Rotterdam, the Netherlands and then by road to Amsterdam, the Netherlands. The goods were damaged during the carriage. The stage in which the damage had been caused was unknown. The whole carriage was covered by bills of lading. Clause 5 was the liability clause. Clause 5(B) provided as follows:

> 'Where the Carriage . . . is Combined Transport then, save as is otherwise provided in this Bill of Lading, the [multimodal transport operator] shall be liable for loss or damage occurring during Carriage to the extent set out below:
>
> (1) Where the stage of Carriage where loss or damage occurred is not known, the Hague Rules . . . apply . . .
>
> (4) Time bar Special Provisions for Combined Transport . . .
>
> . . .
>
> (b) Subject to any provisions of this clause 5 to the contrary, the [multimodal transport operator] shall be discharged of all liability under this Bill of Lading unless suit is brought . . . within nine months after delivery of the Goods . . .'

The limitation period stated in the Hague Rules was twelve months. The consignor sued the multimodal transport operator for the damaged

3. [1998] All ER (D) 296.

goods within one year but after nine months. The multimodal transport operator contended that the action was time barred.

The court held that the limitation period under the Hague Rules applied. The limitation period article was one of the provisions referred to in the proviso '[s]ubject to any provisions of this clause 5 to the contrary'. The nine-month period applied only outside the scope of the Hague Rules.

9.2 INTERNATIONAL CONVENTION

It is desirable to have a unified system for the regulation of the rights and obligations of multimodal transport operators. The Multimodal Convention shows the efforts of the international community to develop such a system. It was adopted by the UN at Geneva on 24 May 1980. So far, the attempt is not very successful. Only a few countries have ratified it. It seems it will take some time for the Multimodal Convention to come into effect. Although it is not yet in force, it governs multimodal transport if the parties agree to incorporate it into the contract. In such a case, its provisions take effect as terms of the contract.

The Multimodal Convention is subject to international conventions on unimodal carriage or national law. The multimodal transport operator is presumed to be liable for any loss, damage or delay caused by an occurrence which takes place while the goods are in his charge. The limit of liability for loss or damage is generally 920 units of account per package or 2.75 units of account per kilogramme, whichever is the higher. If there is no sea transport in the whole carriage, a higher limit of 8.33 units of account per kilogramme applies. Where the loss is localized, the higher limit prescribed in the relevant unimodal convention or national law applies. The limit for loss resulting from delay is generally 2.5 times the freight for the delayed goods. The liability cannot, however, exceed the total freight paid under the contract. The limit for total loss is adopted as the limit for aggregate liability.

9.3 TRANSPORT DOCUMENTS

If the carriage is unimodal, the usual transport documents are issued. For example, air carriers and sea carriers will issue air waybills and bills of lading respectively.

It is permissible for a multimodal transport operator to issue different transport documents covering different segments of the transport. In practice, a multimodal transport document is issued for the whole transport. A multimodal transport document may be in many forms. It could be an air waybill if the multimodal transport operator is an air carrier. Article 31 of the amended Convention explicitly provides that 'conditions relating to other modes of carriage' may be inserted in an air waybill.

If issued in the form of a bill of lading, the multimodal transport document is usually called a combined transport bill of lading. For example, a forwarder may issue a FIATA Combined Transport Bill of Lading.

A multimodal transport document may be negotiable or non-negotiable. Sometimes it may not be easy to tell whether or not a transport document is negotiable.

> *The Chitral*[4]
> Two sets of bills of lading were issued with regard to the goods. One was in negotiable form. The other bill (the one in issue) named A as the consignee. Although the printed box for naming the consignee specifically provided 'if order state notify party', no notify party was named. The phrase 'unto the above-mentioned consignee or to his or her assigns' appeared elsewhere in the bill. The goods were damaged in transit. In the ensuing legal action, one issue was whether the bill in issue was negotiable.
>
> The court held that the bill was not negotiable. With the words 'if order state notify party', it might be used as a straight consigned bill or a negotiable bill. In the commercial world, the general view was that the words 'or order or assigns' were required to make the bill negotiable. Without such words and no notify party identified, the bill in issue was not a negotiable instrument. The phrase 'unto the above-mentioned consignee or to his or her assigns' did not have the effect as if the words 'or order' had been added. Otherwise, the bill might not be issued as a straight consigned bill.

Multimodal transport document as document of title

A multimodal transport document is a receipt of the cargo and evidence of the contract. It may also be a document of title. At common law, the custom of merchants determines whether a particular document is a document of title. It is a document of title if there is a custom that its holder transfers the property in the goods by transferring the document to the transferee.[5] The shipped bill of lading is the most well-known document of title. However, whether a 'received for shipment' bill of lading is a document of title is still not clear.[6] In relation to door-to-door service, a multimodal transport

4. [2000] 1 Lloyd's Rep 529.
5. *Lickbarrow v Mason* (1794) 5 TR 683, 101 ER 380.
6. cf *The Marlborough Hill* [1921] AC 444; *Diamond Alkali Export Corp v Fl. Bourgeois* [1921] 3 KB 443; *Ishag v Allied Bank International, Fuhs and Kotalimbora* [1981] 1 Lloyd's Rep 92.

document in the form of bill of lading is often issued as a 'received for shipment' bill. Its status as a document of title is therefore unclear. Until the court rules on this point, the better view is that such a document is, in the absence of a custom required by *Lickbarrow v Mason*, not a document of title. To avoid any possible dispute, it is advisable for the consignor to change a 'received for shipment' bill to a 'shipped' bill by notation once the cargo is on board the ship.

It is settled that an air waybill issued by an air carrier is not a document of title. In the Australian case, *The Cape Comorin*,[7] it was held that a bill of lading issued by a forwarder is not a document of title. If the multimodal transport operator is an air carrier or a freight forwarder, the multimodal transport document is unlikely to be taken as a document of title.

Through bill of lading

Sea carriers usually issue through bills of lading as multimodal transport documents. For example, such a document is issued if the shipper sends the cargo to the sea carrier's loading depot for grouping. If the transport involves only successive carriages by sea, an 'ocean through bill of lading' may be issued.

The use of through bills of lading is widely recognized. For example, Article 19 of the ICC Standard Documentary Credit Forms (UCP 600)[8] states such a bill is acceptable by the advising bank as a valid transport document, provided it is issued and signed by a multimodal transport operator. The bill is acceptable even though the credit requires an ocean bill of lading. Besides, a multimodal transport document issued by forwarders is also acceptable under UCP 600, provided that they act as a multimodal transport operator or an agent of the operator.

Where a through bill of lading is issued, the multimodal transport carrier usually undertakes to collect the goods at the shipper's factory or warehouse, to consolidate them later with goods of other shippers, to carry them to the final destination and to deliver them to the consignee. Multimodal transport operators may not themselves perform the whole or certain parts of the transport. For example, forwarders acting as a multimodal transport operators may collect the goods themselves at the premises of the consignor and deliver them to the consignee when the goods have arrived at the destination, whilst sub-contracting an air carrier to perform the air transport.

7. (1991) 24 NSWLR 745.
8. UCP 600 came into effect on 1 July 2007 with 39 articles.

Sometimes it is not clear whether the party concluding with the consignor acts as a principal or an agent of a carrier. Where the carriage is not performed by the multimodal transport operator, the situation becomes more complicated. The basic rule is that in forming a contract of carriage with a client, a multimodal transport operator is at law a carrier. The parties actually performing the carriage are only the operator's agents.

Kuehne & Nagel (Hong Kong) Ltd v Yuen Fung Metal Works Ltd[9]

The consignor, YF, formed a contract with KN for the carriage of aluminium goods from Hong Kong to Germany. The goods were stuffed in containers and to be carried in three stages: (1) by sea to Russia, (2) by the Trans-Siberian Railway and (3) by trucks across Poland to Germany. The carriage was performed by a company closely related to KN. One contractual document called 'Shipping Order/Dock Receipt' contained the following term:

> *'Carriage of cargo is subject to the terms and conditions of the Carrier's Through Bill of Lading and the applicable tariff as well as our shipping trading conditions printed on the reverse side.'*

Three original through bills of lading were issued in KN's own name. The following was one of the conditions stated in the bills:

> *'Delivery will be made upon surrender of one original of this Bill of Lading, duly receipted.'*

When the goods arrived at the border between Poland and Germany, they were delivered to the customer of YF without the surrender of any bill of lading. The customer refused to pay for the goods or return them to YF. YF sued KN for the loss, claiming that KN was a carrier. KN contended that it was not a carrier in that it had no means of transport of its own and that it carried on the business of international freight forwarder. Besides, KN tried to rely on the exemption clause in the contract.

The court held that KN was liable. First, KN was a carrier with respect to the carriage. The terms of the contract determined KN's status. In the circumstances, KN undertook to carry the goods to the destination. In addition, the only through bills of lading were issued in KN's name. Second, the delivery of the goods to YF's customer without surrender of an original bill amounted to a fundamental breach and KN was therefore not able to rely on the exemption clause.

9. [1979] HKLR 526.

Container bill of lading

Container bills of lading are issued by sea carriers providing door-to-door service if the goods are carried in containers. At law, there is no difference between a container bill and an ordinary bill of lading. They are subject to the Carriage of Goods by Sea Ordinance (Cap 462). For tackle-to-tackle liability, the Hague Rules or Hague/Visby Rules therefore apply. For other stages of the entire transport, the terms of the contract govern.

The practice is that if the cargo is less than a full container load (LCL), it may be sent to a freight forwarder or consolidator for grouping. The cargo is loaded in a container with the goods of other shippers. Where the carrier receives goods at a container freight station or in a container at the shipper's premises, the container bill of lading is a 'received for shipment' bill. For full container load (FCL), the carrier usually sends an empty container to the premises of the shipper for loading. The container with an FCL is then sent to the carrier. A container bill of lading is often issued in such a case. Depending on the situation, a 'shipped on board' bill may be issued. No matter what type of transport document is issued, the facts of the case determine the liability of the carrier under the bill.

> *The Esmeralda I*[10]
> Z, a cutlery manufacturer in Brazil, wanted to send products to A, Z's sole distributor in Australia. Z asked D, the carrier, to supply an empty container for stuffing at Z's premises. After the packing, Z sealed the container with a seal provided by D. The bill of lading was a 'clean on board' bill. It stated that there were 437 boxes in the container. It bore on its face the words 'FCL/FCL'. Under the heading 'Number and Kind of packages, description of goods', the following words were inserted:
>
> > *'01–container 20' with 437 cardboard boxes, containing: "cutlery and leaflets and posters"'.*
>
> The phrase '*said to contain–packed by shippers*' was stamped along the margin of the bill. The bill also had this clause:
>
> > *This bill of lading shall be prima facie evidence of the receipt by the carrier . . . of the total amount of containers packages or other units . . . specified on the face hereof . . .*
>
> At the premises of A, A broke the seal and opened the container. It was found that 118 cartons were missing. A sued D for the loss of the goods, contending that D was estopped from denying the receipt of 437 boxes.

10. [1988] 1 Lloyd's Rep 206.

The court held that D was not liable. The words 'FCL/FCL' had two meanings. The first 'FCL' meant that the shipper packed the container, whilst the second indicated that the importer (ie, A in this case) would unpack the container. It was obvious that D could not have checked the content of the container. The phrase 'said to contain–packed by shippers' made it clear that the carrier made no representation as to the accuracy of the statement that there were 437 boxes in the container. Besides, the court found it as a fact that the goods were stolen before the container was sent to D.

Winkenson Impex Co Ltd v Haverton Shipping Ltd[11]
W formed a contract with a carrier, D, for the carriage of a consignment of goods from Hong Kong to Nigeria on a door-to-door basis. W packed the goods in a container. The goods were covered by four bills of lading. The bills had the following remark:

> '*Shipper's load and count. Containerized cargo. Freight prepaid.*'

The container arrived safely at the port of destination. It was stolen when it was transported to a bonded warehouse somewhere away from the port for unstuffing. In an action against D, W relied on the *prima facie* presumption that the bills were receipt of the goods described in them. W also provided various invoices and packing lists which he claimed to have received from the supplier to prove that the goods had actually been loaded.

The court held that W's action failed. The presumption did not apply in this case since it was impossible for D to check the content of the container. W could not prove that the goods as described in the bills had actually been shipped. Very little weight, if at all, could be given to the invoices and packing lists for this purpose.

9.4 INTERNATIONAL RULES

The ICC contributes to facilitate multimodal transport by preparing two sets of international rules. These are:

- The ICC Uniform Rules for a Combined Transport Document (1975 Revision) (the ICC Rules); and
- The UNCTAD/ICC Rules for Multimodal Transport Documents (1992) (the UNCTAD/ICC Rules).

Both the ICC Rules and the UNCTAD/ICC Rules are not mandatory rules. Either set of rules have to be incorporated into a contract to take effect.

11. [1985] HKLR 141.

Formally, the ICC Rules have been repealed by the UNCTAD/ICC Rules and are no longer in force. In practice, the ICC Rules are still incorporated in some old forms of transport documents. Newer versions adopt, on the other hand, the UNCTAD/ICC Rules. For example, the Combidoc adopted by the Baltic and International Maritime Conference and the FIATA Multimodal Transport Bill of Lading 1992 (FBL 92) are based on the ICC Rules and the UNCTAD/ICC Rules respectively. The fact is that the UNCTAD/ICC Rules have a wider scope of application. With the disuse of old transport documents in the foreseeable future, general acceptance of the UNCTAD/ICC Rules is expected.

Since the two sets of rules may be found in transport documents, it is desirable to examine each of them in detail for practical reasons.

ICC Rules

The ICC Rules apply to a 'contract concluded for the performance and/or the procurement of performance of combined transport of goods which is evidenced by a combined transport document'. The intention of the parties to have a combined transport as expressed in the contract is decisive. If it is so stated in the contract and the actual transport is unimodal, the ICC Rules still apply. Under the ICC Rules, transport by sea and by inland waterway are two different modes of transport. Transport by rail and by road are also different.

Combined transport is defined as:

> the carriage of goods by at least two different modes of transport, from a place at which the goods are taken in charge situated in one country to a place designated for delivery situated in another country.

As a set of rules issued by the International Chamber of Commerce, the ICC Rules are subject to national law or applicable international conventions on unimodal carriage.

Combined transport document

For the ICC Rules to apply, a combined transport document (CTD) must be issued. The party issuing a CTD is the combined transport operator (CTO). All parties having or acquiring an interest in the CTD are subject to the ICC Rules.

A CTD may be negotiable or non-negotiable. If it is negotiable, it must bear the heading 'Negotiable combined transport document issued subject to Uniform Rules for a Combined Transport Document (ICC Publication No 298)'. For a non-negotiable CTD, the heading is 'Non-negotiable combined transport document issued subject to Uniform Rules for a Combined Transport Document (ICC Publication No 298)'. A negotiable CTD must be made out to order or to bearer. The CTO is discharged of obligation to deliver the goods if, in good faith, the goods have been delivered against surrender of an original of the CTD. A non-negotiable CTD must indicate a named consignee. The CTO's obligation to deliver is discharged by delivering the goods to the consignee or the agent entitled to take delivery.

The CTD must indicate clearly the quantity, weight, volume or marks of the goods. This is prima facie evidence of the CTO's taking in charge of the goods described in it. If the CTD is negotiable, it becomes conclusive evidence when it has been transferred to a third party acting in good faith. If there is reasonable suspicion that some particulars are inaccurate or do not have reasonable means of checking, the CTO may enter reservations. The CTD is not evidence of matters subject to the reservations.

CTO's responsibilities in general

In issuing a CTD, the CTO accepts responsibility for performing or procuring the performance of the whole transport to deliver the goods. The ICC Rules govern the CTO's liability for the period from the moment of taking over the goods to the time of delivery.

Generally, the CTO is presumed liable for the loss or damage but not liable if there is proof the loss or damage is caused by any of the events or matters listed in Rule 12:

- Act or omission of the consignor or consignee or person (other than the CTO) acting on behalf of either of them or from whom the CTO took the goods in charge
- Insufficient or defective packing or marks
- Handling, loading, stowage or unloading of the goods by the consignor or consignee or their agents
- Inherent vice of the goods
- Strike, lockout, stoppage or restraint of labour, the consequences of which cannot be avoided by the CTO's exercise of reasonable diligence
- Any cause or event the CTO could not avoid and the consequences of which could not be prevented by the exercise of reasonable diligence

The CTO is not required to prove one or more of the above events or matters in fact caused the loss or damage. If it can be shown the loss or damage could have been caused by any of them, it is presumed it was so caused. However, the claimant may adduce evidence to show the loss or damage was in fact not caused by any of such causes or matters.

The CTO is presumed liable as a carrier for the loss of or damage to the goods, and for delay.[12] Although the liability is limited, there is no limit on the aggregate liability. If no limit on aggregate liability is prescribed by any applicable convention or law, a claimant who sues the CTO and the party whose act caused the damage in different actions may obtain damages in total exceeding the stated limit.

If the damage or loss results from an act or omission of the CTO done with intent to cause damage or recklessly and with knowledge damage would probably result, the CTO is subject to unlimited liability.

The CTO is also liable for the acts or omissions of the agents or servants done within their scope of employment. In using other independent sub-contractors for the performance of the transport, the CTO is liable for their acts.

Unlocalized loss

If the stage of transport in which the occurrence causing the loss is not known, the loss is unlocalized. Generally, the CTO's liability is not governed by international conventions on unimodal transport in such a case. Under the ICC Rules, the claimant is entitled to claim the value of the lost or damaged goods at the place of destination and time of delivery. The CTO's liability cannot, however, exceed 30 francs per kilogramme of gross weight of the lost or damaged goods. This is the same as the original limit laid down in the Hague/Visby Rules. Because the unit of gold francs has fallen into disuse, the parties often adopt another unit, say 2 SDR per kilogramme, as the basic limit. Since the limit is calculated with reference to the weight of the goods, the number of packages is irrelevant. If the consignor has declared in the CTD a higher value of the goods and it is accepted by the CTO, the declared value becomes the limit.

Localized loss

If the stage of transport in which the loss was caused is known, the loss is localized. In such a case, rule 13 states that the applicable international convention governing the mode of carriage or other mandatory law determines

12. See rule 5.

the CTO's liability. It does not matter that no unimodal transport contract had been concluded or no transport document required by the convention issued. For example, where the goods were carried by sea from Hong Kong to New York when the loss was caused, the Carriage of Goods by Sea Ordinance, and therefore, the Hague Rules apply, as if a carriage of goods by sea contract for the voyage had been formed and a bill of lading issued. Similarly, if the damage occurred during carriage by air, the Carriage by Air Ordinance, and therefore, the Warsaw regime govern. If no international convention or law compulsorily applies in the way described above, the parties may agree to apply the relevant convention.

If the localized damage was caused during carriage by inland waterways performed by a sub-contractor, the sub-contract formed by the CTO and the sub-contractor governs, provided the carriage is not governed by any international convention and the CTD provides for the application of the sub-contract.

In any other cases, the rules applying to unlocalized damage determine the CTO's liability.

Delay

The CTO may be liable for delay in delivery. There is liability if the stage of transport in which the delay was caused is known and the law or applicable international convention provides for such liability. Subject to any mandatory rule, the CTO's liability cannot exceed the amount of freight for that stage of transport.

If the delay is more than 90 days, the party entitled to take delivery may, in the absence of evidence to the contrary, treat the goods as lost.

Limitation of servants' liability

The ICC Rules do not automatically extend the benefit of limitation of liability to the CTO's servants or agents. However, the multimodal transport contract may extend the benefit to them.

Consignor's liability

Under the ICC Rules, the consignor is under an obligation to provide accurate and sufficient particulars to be inserted in the CTD. The consignor is deemed to guarantee the accuracy and adequacy of the description, marks, weight and volume of the goods that are furnished. If the CTO suffers any loss as a result of the inaccuracy or inadequacy of the particulars, the consignor may be asked for indemnity.

If the goods are of a dangerous nature, the consignor must comply with any requirements laid down by the law or applicable conventions. Further, there is a duty to inform the CTO in writing of the exact nature of the dangerous goods before the goods are taken in charge by the CTO. In the event the consignor fails to do so and the CTO is not aware of the dangerous nature of the goods or the necessary precautions to be taken, the goods may, without compensation, be unloaded, destroyed or rendered harmless when they become or are deemed to be a hazard to life or property. Further, the consignor is responsible for all the loss, damage, delay or expenses suffered or incurred by the CTO.

The burden of proving the CTO knew the exact nature of the dangerous goods rests on the person entitled to the goods. If the consignor cannot discharge the burden, the CTO is not liable for the destruction of or damage to the goods.

Claims

A party who intends to claim against the CTO for loss of or damage to the goods must give notice in writing to the CTO or representative at the place of delivery before the removal of the goods into the custody of the person entitled to delivery. If the damage is not apparent, the notice may be given within 7 days of the removal. No notice is required if the goods are treated as lost for undue delay.

Action against the CTO must be taken within 9 months after:

- the delivery of the goods;
- the date when the goods should have been delivered; or
- the date the goods are treated as lost for long delay.

UNCTAD/ICC Rules

Introduced jointly by UNCTAD and the ICC on 1 January 1992, the UNCTAD/ICC Rules are based on the Multimodal Convention. It is hoped the anticipated wide acceptance of the UNCTAD/ICC Rules will lead to the coming into effect of the Multimodal Convention. The UNCTAD/ICC Rules consist of 13 rules. They do not provide a complete set of rules governing multimodal transport. Matters which are not covered by the UNCTAD/ICC Rules include freight and charges, routeing, stowage, both-to-blame collision, liens, general average, and governing law. Thus, other terms are required to supplement the UNCTAD/ICC Rules for the regulation of the legal relations between multimodal transport operator and consignor.

Application

The UNCTAD/ICC Rules may be incorporated into any carriage of goods contract. The incorporation may be made in writing, orally or by any other means. Being a set of rules for multimodal transport, Article 1 states explicitly they may apply to either a unimodal or multimodal transport contract. The issue of transport documents is not a condition for the application of the UNCTAD/ICC Rules. Since most rules are concerned with the liabilities of multimodal transport operator (MTO) and the functions of multimodal transport document (MTD), they may not easily be interpreted if the UNCTAD/ICC Rules are applied to a unimodal transport contract.

If incorporated, the UNCTAD/ICC Rules prevail over other terms of the contract. However, terms which increase the responsibility or obligations of the MTO are not invalidated by the UNCTAD/ICC Rules.

Multimodal transport document and EDI

To cater for the advent in communication technology and the continuing use of traditional transport documents, the UNCTAD/ICC Rules are prepared for both the issue of MTD and the use of electronic data interchange (EDI). In the hand of the consignor, the MTD or the equivalent EDI message is prima facie evidence of the MTO's taking in charge of the goods as described in it. When the MTD has been transferred or the EDI message transmitted to the consignee who in good faith has relied and acted on it, it becomes conclusive evidence and no other evidence may be admitted to disprove the truth of the information contained in it. However, the MTD or EDI message has no evidentiary effect at all if it contains expressions such as 'shipper's weight, load and count' or 'shipper-packed container'.

Liability of consignor

The consignor is deemed to guarantee the accuracy, at the time the goods were taken in charge by the MTO, of all particulars relating to the general nature and any dangerous character of the goods, their marks, number, weight, volume and quantity that were furnished for insertion in the MTD. If the particulars provided are inaccurate or inadequate and the MTO as a result suffers loss, the consignor is liable to indemnify the MTO. There is liability even if the consignor has transferred the MTD or transmitted the EDI message to another party.

Responsibilities of MTO

The UNCTAD/ICC Rules prescribe the MTO's responsibilities and liability for the period from the time of taking charge of the goods to their delivery.

The term 'taking in charge' is given a restricted meaning. It means that 'the goods have been handed over to and accepted for carriage by the MTO'. In accepting the goods for, say, the purpose of warehousing or regrouping, it may be argued the MTO does not take them in charge under the UNCTAD/ICC Rules.

The primary responsibility of the MTO is to perform or procure the performance of the multimodal transport contract to ensure delivery of the goods. 'Delivery' is defined as:

- the handing over of the goods to the consignee;
- the placing of the goods at the disposal of the consignee in accordance with the multimodal transport contract, law or usage; or
- the handing over of the goods to the authorities pursuant to law.

The MTD may take many forms. It may be negotiable or non-negotiable. The form of the MTD determines how the delivery should be done. If it is negotiable and issued 'to bearer', the goods may be delivered to any person who surrenders one original of the MTD. For a negotiable MTD issued 'to order', a holder of a duly endorsed original is entitled to take delivery. Where the MTD is issued in a negotiable form to a named person, generally only that person may take delivery upon proof of identity and surrender of an original of the MTD. However, if such an MTD has been transferred 'to order' or in blank, any holder of it may take delivery of the goods. For a non-negotiable MTD, the goods must be delivered to the named consignee upon proof of identity. In the case no MTD is issued, the MTO must deliver the goods to the person as instructed by the consignor or a person entitled to give such instruction under the multimodal transport contract.

The MTOs are responsible for the acts and omissions of their servants or agents done within the scope of their employments. MTOs are also liable for the conduct of other parties of whose services they make use for the performance of the multimodal transport contract.

Liability of MTO

The MTO is prima facie liable for any loss of or damage to the goods and delay in delivery if the occurrence causing such loss, damage or delay took place while the goods were in the MTO's charge. The MTO is not liable if there is proof that the loss, damage or delay was not caused by the fault or neglect of

- the MTO;
- servants or agents of the MTO within the scope of their employments, or
- other persons of whose services were used for the performance of the multimodal transport contract.

Generally, delay in delivery occurs when the goods have not been delivered within the agreed time for delivery. If there is no agreed time for delivery, the MTO is liable for delay if the goods have not been delivered after the expiry of the time in which it is reasonable to require a diligent MTO to make delivery. If there is a delay of more than 90 days, the goods may, in the absence of evidence to the contrary, be treated as lost.

In relation to carriage by sea or inland waterways, special defences are available to the MTO. There is no liability if the loss, damage or delay is caused by:

- the act, neglect, or default of the master, mariner, pilot or servants of the carrier in the navigation or in the management of the ship; or
- fire, unless caused by the actual fault or privity of the carrier.

Where the damage is caused by the unseaworthiness of the ship, the MTO is not liable if it can be proved that, at the commencement of the voyage, due diligence was exercised to make the ship seaworthy.

Limits of liability

Different limits are prescribed for localized and unlocalized damage. If the stage of transport in which the damage was caused is known, the damage is 'localized' and the associated loss, 'localized loss'. If the stage is unknown, the damaged and loss are unlocalized.

If the damage is localized, the limit of the MTO's liability is determined by the international convention or other mandatory law applied to that stage.

For unlocalized damage, the liability of the MTO cannot generally exceed 666.67 SDR per package or 2 SDR per kilogramme of gross weight of the lost or damaged goods, whichever is the higher. But if the multimodal transport does not include, in accordance with the multimodal transport contract, any carriage by sea or carriage by inland waterways, the MTO is subject to a higher limit of 8.33 SDR per kilogramme. The number of packages or other shipping units enumerated in the MTD is used to calculate the limit. The limit may be raised by the consignor's declaration made before the goods have been taken in charge by the MTO and inserted in the MTD.

For delay in delivery or consequential loss or damage other than loss of or damage to the goods, the freight agreed in the multimodal transport contract sets the limit of the MTO's liability. Further, the limits for total loss of the goods are the ceilings of the aggregate liability of the MTO.

The MTO is not entitled to limit liability if the claimant proves that the loss, damage or delay resulted from a personal act or omission of the MTO done with the intent to cause such damage, or recklessly and with knowledge that such damage would probably result.

The parties employed by MTOs and their servants and agents are also entitled to the benefit of limitation of liability.

Notice and time bar

If the consignor intends to sue the MTO, it is required to give notice in writing when the goods are handed over to the consignee. In the event that the loss or damage is not apparent, the notice may be given within 6 consecutive days after the handing over of the goods. If no notice is given, the handing over of the goods is prima facie evidence of the delivery by the MTO of the goods as described in the MTD.

Generally, action against the MTO must be brought within 9 months after, as the case may be, the delivery of the goods, the date when the goods should be delivered or the date when the goods are treated as lost for long delay. The parties may agree to extend the time bar.

APPENDIX I

UNCTAD/ICC Rules for Multimodal Transport Documents*

1. *Applicability*

1.1 These Rules apply when they are incorporated, however this is made, in writing, orally or otherwise, into a contract of carriage by reference to the "UNCTAD/ICC Rules for multimodal transport documents", irrespective of whether there is a unimodal or a multimodal transport contract involving one or several modes of transport or whether a document has been issued or not.

1.2 Whenever such a reference is made, the parties agree that these Rules shall supersede any additional terms of the multimodal transport contract which are in conflict with these Rules, except insofar as they increase the responsibility or obligations of the multimodal transport operator.

2. *Definitions*

2.1 '**Multimodal transport contract**' means a single contract for the carriage of goods by at least two different modes of transport.

2.2 '**Multimodal transport operator**' (MTO) means any person who concludes a multimodal transport contract and assumes responsibility for the performance thereof as a carrier.

2.3 '**Carrier**' means the person who actually performs or undertakes to perform the carriage, or part thereof, whether he is identical with the multimodal transport operator or not.

2.4 '**Consignor**' means the person who concludes the multimodal transport contract with the multimodal transport operator.

2.5 '**Consignee**' means the person entitled to receive the goods from the multimodal transport operator.

2.6 '**Multimodal transport document**' (MT document) means a document evidencing a multimodal transport contract and which can be replaced by electronic data interchange messages insofar as permitted by applicable law and be,

(a) Issued in a negotiable form or,
(b) Issued in a non-negotiable form indicating a named consignee.

* ICC Publication No 481 – ISBN 92.842.1126.3
Published in its official English version by the International Chamber of Commerce
Copyright © 1992 – International Chamber of Commerce (ICC), Paris
Available from: *ICC Publishing* SA., 38 Cours Albert 1er, 75008 Paris, France.

2.7 '**Taken in charge**' means that the goods have been handed over to and accepted for carriage by the MTO.

2.8 '**Delivery**' means:

(a) The handing over of the goods to the consignee, or
(b) The placing of the goods at the disposal of the consignee in accordance with the multimodal transport contract or with the law or usage of the particular trade applicable at the place of delivery, or
(c) The handing over of the goods to an authority or other third party to whom, pursuant to the law or regulations applicable at the place of delivery, the goods must be handed over.

2.9 '**Special Drawing Right**' (SDR) means the unit of account as defined by the International Monetary Fund.

2.10 '**Goods**' means any property including live animals as well as containers, pallets or similar articles of transport or packaging not supplied by the MTO, irrespective of whether such property is to be or is carried on or under deck.

3. *Evidentiary effect of the information contained in the multimodal transport document*

The information in the *MT document* shall be *prima facie* evidence of the taking in charge by the MTO of the goods as described by such information unless a contrary indication, such as "shipper's weight, load and count", "shipper-packed container" or similar expressions, has been made in the printed text or superimposed on the document. Proof to the contrary shall not be admissible when the *MT document* has been transferred, or the equivalent electronic data interchange message has been transmitted to and acknowledged by the consignee who in good faith has relied and acted thereon.

4. *Responsibilities of the multimodal transport operator*

4.1. *Period of responsibility*
The responsibility of the MTO for the goods under these Rules covers the period from the time the MTO has taken the goods in his charge to the time of their delivery.

4.2 *The liability of the MTO for his servants, agents and other persons*
The multimodal transport operator shall be responsible for the acts and omissions of his servants or agents, when any such servant or agent is acting within the scope of his employment, or of any other person of whose services he makes use for the performance of the contract, as if such acts and omissions were his own.

4.3 *Delivery of the goods to the consignee*
The MTO undertakes to perform or to procure the performance of all acts necessary to ensure delivery of the goods:

(a) When the *MT document* has been issued in a negotiable form "to bearer", to the person surrendering one original of the document, or
(b) When the *MT document* has been issued in a negotiable form "to order", to the person surrendering one original of the document duly endorsed, or
(c) When the *MT document* has been issued in a negotiable form to a named person, to that person upon proof of his identity and surrender of one original document; if such document has been transferred "to order" or in blank the provisions of (b) above apply, or
(d) When the *MT document* has been issued in a non-negotiable form, to the person named as consignee in the document upon proof of his identity, or
(e) When no document has been issued, to a person as instructed by the consignor or by a person who has acquired the consignor's or the consignee's rights under the multimodal transport contract to give such instructions.

5. Liability of the multimodal transport operator

5.1 *Basis of Liability*
Subject to the defences set forth in Rule 5.4 and Rule 6, the MTO shall be liable for loss of or damage to the goods, as well as for delay in delivery, if the occurrence which caused the loss, damage or delay in delivery took place while the goods were in his charge as defined in Rule 4.1, unless the MTO proves that no fault or neglect of his own, his servants or agents or any other person referred to in Rule 4 has caused or contributed to the loss, damage or delay in delivery. However, the MTO shall not be liable for loss following from delay in delivery unless the consignor has made a declaration of interest in timely delivery which has been accepted by the MTO.

5.2 *Delay in delivery*
Delay in delivery occurs when the goods have not been delivered within the time expressly agreed upon or, in the absence of such agreement, within the time which it would be reasonable to require of a diligent MTO, having regard to the circumstances of the case.

5.3 *Conversion of delay into final loss*
If the goods have not been delivered within ninety consecutive days following the date of delivery determined according to Rule 5.2, the claimant may, in the absence of evidence to the contrary, treat the goods as lost.

5.4 *Defences for carriage by sea or inland waterways*
Notwithstanding the provisions of Rule 5.1. the MTO shall not be responsible for loss, damage or delay in delivery with respect to goods carried by sea or

inland waterways when such loss, damage or delay during such carriage has been caused by:

- Act, neglect, or default of the master, mariner, pilot or the servants of the carrier in the navigation or in the management of the ship,
- Fire, unless caused by the actual fault or privity of the carrier.

However, always provided that whenever loss or damage has resulted from unseaworthiness of the ship, the MTO can prove that due diligence has been exercised to make the ship seaworthy at the commencement of the voyage.

5.5 *Assessment of compensation*
5.5.1 Assessment of compensation for loss of or damage to the goods shall be made by reference to the value of such goods at the place and time they are delivered to the consignee or at the place and time when, in accordance with the multimodal transport contract, they should have been so delivered.

5.5.2 The value of the goods shall be determined according to the current commodity exchange price or, if there is no such price, according to the current market price or, if there is no commodity exchange price or current market price, by reference to the normal value of goods of the same kind and quality.

6. Limitation of liability of the multimodal transport operator

6.1 Unless the nature and value of the goods have been declared by the consignor before the goods have been taken in charge by the MTO and inserted in the *MT document*, the MTO shall in no event be or become liable for any loss of or damage to the goods in an amount exceeding the equivalent of 666.67 SDR per package or unit or 2 SDR per kilogramme of gross weight of the goods lost or damaged, whichever is the higher.

6.2 Where a container, pallet or similar article of transport is loaded with more than one package or unit, the packages or other shipping units enumerated in the *MT document* as packed in such article of transport are deemed packages or shipping units. Except as aforesaid, such article of transport shall be considered the package or unit.

6.3 Notwithstanding the above-mentioned provisions, if the multimodal transport does not, according to the contract, include carriage of goods by sea or by inland waterways, the liability of the MTO shall be limited to an amount not exceeding 8.33 SDR per kilogramme of gross weight of the goods lost or damaged.

6.4 When the loss of or damage to the goods occurred during one particular stage of the multimodal transport, in respect of which an applicable international convention or mandatory national law would have provided another limit of liability if a separate contract of carriage had been made for that particular stage of transport, then the limit of the MTO's liability for such loss or damage shall be determined by reference to the provisions of such convention or mandatory national law.

6.5 If the MTO is liable in respect of loss following from delay in delivery, or consequential loss or damage other than loss of or damage to the goods, the liability of the MTO shall be limited to an amount not exceeding the equivalent of the freight under the multimodal transport contract for the multimodal transport.

6.6 The aggregate liability of the MTO shall not exceed the limits of liability for total loss of the goods.

7. Loss of the right of the multimodal transport operator to limit liability

The MTO is not entitled to the benefit of the limitation of liability if it is proved that the loss, damage or delay in delivery resulted from a personal act or omission of the MTO done with the intent to cause such loss, damage or delay, or recklessly and with knowledge that such loss, damage or delay would probably result.

8. Liability of the consignor

8.1 The consignor shall be deemed to have guaranteed to the MTO the accuracy, at the time the goods were taken in charge by the MTO, of all particulars relating to the general nature of the goods, their marks, number, weight, volume and quantity and, if applicable, to the dangerous character of the goods, as furnished by him or on his behalf for insertion in the *MT document*.

8.2 The consignor shall indemnify the MTO against any loss resulting from inaccuracies in or inadequacies of the particulars referred to above.

8.3 The consignor shall remain liable even if the *MT document* has been transferred by him.

8.4 The right of the MTO to such indemnity shall in no way limit his liability under the multimodal transport contract to any person other than the consignor.

9. Notice of loss of or damage to the goods

9.1 Unless notice of loss of or damage to the goods, specifying the general nature of such loss or damage, is given in writing by the consignee to the MTO when the goods are handed over to the consignee, such handing over is *prima facie* evidence of the delivery by the MTO of the goods as described in the *MT document*.

9.2 Where the loss or damage is not apparent, the same *prima facie* effect shall apply if notice in writing is not given within 6 consecutive days after the day when the goods were handed over the consignee.

10. Time-bar

The MTO shall, unless otherwise expressly agreed, be discharged of all liability under these Rules unless suit is brought within 9 months after the delivery of the goods, or the date when the goods should have been delivered, or the date when in accordance with Rule 5.3, failure to deliver the goods would give the consignee the right to treat the goods as lost.

11. Applicability of the rules to actions in tort

These Rules apply to all claims against the MTO relating to the performance of the multimodal transport contract, whether the claim be founded in contract or in tort.

12. Applicability of the rules to the multimodal transport operator's servants, agents and other persons employed by him

These Rules apply whenever claims relating to the performance of the multimodal transport contract are made against any servant, agent or other person whose services the MTO has used in order to perform the multimodal transport contract, whether such claims are founded in contract or in tort, and the aggregate liability of the MTO of such servants, agents or other persons shall not exceed the limits in Rule 6.

13. Mandatory law

These Rules shall only take effect to the extent that they are not contrary to the mandatory provisions of international conventions or national law applicable to the multimodal transport contract.

APPENDIX II

ICC Uniform Rules for a Combined Transport Document (1975)*

General Provision

Rule 1

(a) These Rules apply to every contract concluded for the performance and/or procurement of performance of combined transport of goods which is evidenced by a combined transport document as defined herein.

These Rules shall nevertheless apply even if the goods are carried by a single mode of transport contrary to the original intentions of the contracting parties that there should be a combined transport of the goods as defined hereafter.

(b) The issuance of such combined transport document confers and imposes on all parties having or thereafter acquiring an interest in it the rights, obligations and defences set out in these Rules.

(c) Except to the extent that it increases the responsibility or obligation of the combined transport operator, any stipulation or any part of any stipulation contained in a contract of combined transport or in a combined transport document evidencing such contract, which would directly or indirectly derogate from these Rules shall be null and void to the extent of the conflict between such stipulation, or part thereof, and these Rules. The nullity of such stipulation or part thereof shall not affect the validity of the other provisions of the contract of combined transport or combined transport document of which it forms a part.

Definitions

Rule 2

For the purpose of these Rules:

(a) **Combined transport** means the carriage i of goods by at least two different modes of transport, from a place at which the goods are taken in charge situated in one country to a place designated for delivery situated in a different country.

(b) **Combined transport operator** (CTO) means a person (including any corporation, company or legal entity) issuing a combined transport document.

* ICC Publication No 298 – out of print.
Published in its official English version by the International Chamber of Commerce

The 1975 rules are no longer in force. Since the rules are still used in old forms of transport documents, they are included here only as reference.

Where a national law requires a person to be authorised or licenced before being entitled to issue a combined transport document, then combined transport operator can only refer to a person so authorised or licenced.

(c) **Combined transport document** (CT Document) means a document evidencing a contract for the performance and/or procurement of performance of combined transport of goods and bearing on its face either the heading "Negotiable combined transport document issued subject to Uniform Rules for a Combined Transport Document (ICC Publication N° 298)" or the heading "Non-negotiable combined transport document issued subject to Uniform Rules for a Combined Transport Document (ICC Publication N° 298)".

(d) **Different modes of transport** means the transport of goods by two or more modes of transport, such as transport by sea, inland waterway, air, rail or road.

(e) **Delivery** means delivering the goods to or placing the goods at the disposal of the party entitled to receive them.

(f) **Franc** means a unit consisting of 65.5 milligrammes of gold of millesimal fineness 900.

Negotiable document

Rule 3

Where a CT document is issued in negotiable form:

(a) it shall be made out to order or to bearer;
(b) if made out to order it shall be transferable by endorsement;
(c) if made out to bearer it shall be transferable without endorsement;
(d) if issued in a set of more than one original it shall indicate the number of originals in the set;
(e) if any copies are issued each copy shall be marked "non-negotiable copy";
(f) delivery of the goods may be demanded only from the CTO or his representative, and against surrender of the CT document duly endorsed where necessary;
(g) the CTO shall be discharged of his obligation to deliver the goods if, where a CT document has been issued in a set of more than one original, he, or his representative, has in good faith delivered the goods against surrender of one of such originals.

Non-negotiable document

Rule 4

Where a CT document is issued in nonnegotiable form:

(a) it shall indicate a named consignee;

(b) the CTO shall be discharged of his obligation to deliver the goods if he makes delivery thereof to the consignee named in such non-negotiable document, or to the party advised to the CTO by such consignee as authorised by him to accept delivery.

Responsibilities and liabilities of the CTO

Rule 5

By the issuance of a CT document the CTO:

(a) undertakes to perform and/or in his own name to procure performance of the combined transport including all services which are necessary to such transport from the time of taking the goods in charge to the time of delivery, and accepts responsibility for such transport and such services to the extent set out in these Rules;
(b) accepts responsibility for the acts and omissions of his agents or servants, when such agents or servants are acting within the scope of their employment, as if such acts and omissions were his own;
(c) accepts responsibility for the acts and omissions of any other person whose services he uses for the performance of the contract evidenced by the CT document;
(d) undertakes to perform or to procure performance of all acts necessary to ensure delivery;
(e) assumes liability to the extent set out in these Rules for loss of or damage to the goods occurring between the time of taking them into his charge and the time of delivery, and undertakes to pay compensation as set out in these Rules ins respect of such loss or damage;
(f) assumes liability to the extent set out in Rule 14 for delay in delivery of the goods and undertakes to pay compensation as set out in that Rule.

Rights and duties of the parties

Rule 6

In addition to the information specifically required by these Rules, the parties shall insert in a CT document such particulars as they may agree to be commercially desirable.

Rule 7

The consignor shall be deemed to have guaranteed to the CTO the accuracy, at the time the goods were taken in charge by the CTO, of the description, marks, number, quantity, weight and/or volume of the goods as furnished him, and the consignor shall indemnify the CTO against all loss, damage and expense arising or resulting from inaccuracies in or inadequacy of such particulars.

The right of the CTO to such indemnity shall in no way limit his responsibility and liability under the CT Document to any person other than the consignor.

Rule 8

The consignor shall comply with rules which are mandatory according to the national law or by reason of international Convention, relating to the carriage of goods of a dangerous nature, and shall in any case inform the CTO in writing of the exact nature of the danger before goods of a dangerous nature are taken in charge by the CTO and indicate to him, if need be, the precautions to be taken.

If the consignor fails to provide such information and the CTO is unaware of the dangerous nature of the goods and the necessary precautions to be taken and if, at any time, they are deemed to be a hazard to life or property, they may at any place be unloaded, destroyed or service incidental thereto.

The burden of proving the CTO knew the exact nature of the danger constituted by the carriage of the said goods shall rest upon the person entitled to the goods.

Rule 9

The CTO shall clearly indicate in the CT document, at least by quantity and/or weight and/or volume and/or marks, the goods he has taken in charge and for which he accepts responsibility.

Subject to paragraph 1 of this Rule, if the CTO has reasonable grounds for suspecting that the CT document contains particulars concerning the description, marks, number, quantity, weight and/or volume of the goods which do not represent accurately the goods actually taken in charge, or if he has no reasonable means of checking such particulars, the CTO shall be entitled to enter his reservations in the CT document, provided he indicates the particular information to which such reservations apply.

The CT document shall be prima facie evidence of the taking in charge by the CTO of the goods as therein described. Proof to the contrary shall not be admissible when the CT Document is issued in negotiable form and has been transferred to a third party acting in good faith.

Rule 10

Except in respect of goods treated as lost in accordance with Rule 15 hereof, the CTO shall be deemed prima facie to have delivered the goods as described in the CT document unless notice of loss of, or damage to, the goods, indicating the general nature of such loss or damage, shall have been given in writing to the CTO or to his representative at the place of delivery before or at the time of removal of the goods into the custody of the person entitled to delivery thereof under the CT document, or, if the loss or damage is not apparent, within seven consecutive days thereafter.

Liability for Loss or Damage

A. Rules applicable when the stage of transport where the loss or damage occurred is not known.

Rule 11

When in accordance with Rule 5(e) hereof the CTO is liable to pay compensation in respect of loss of, or damage to, the goods and the stage of transport where the loss or damage occurred is not known:

(a) such compensation shall be calculated by reference to the value of such goods at the place and time they are delivered to the consignee or at the place and time when, in accordance with the contract of combined transport, they should have been so delivered;
(b) the value of the goods shall be determined according to the current commodity exchange price or, if there is no such price, according to the current market price, or, if there is no commodity exchange price or current market price, by reference to the normal value of goods of the same kind and quality.
(c) compensation shall not exceed 30 francs per kilogramme of gross weight of the goods lost or damaged, unless, with the consent of the CTO, the consignor has declared a higher value for the goods and such higher value has been stated in the GT document, in which case such higher value shall be the limit.

However, the CTO shall not, in any case, be liable for an amount greater than the actual loss to the person entitled to make the claim.

Rule 12

When the stage of transport where the loss or damage occurred is not known the CTO shall not be liable to pay compensation in accordance with Rule 5(e) hereof if the loss or damage was caused by:

(a) an act or omission of the consignor or consignee, or person other than the CTO acting on behalf of the consignor or consignee, or from whom the CTO took the goods in charge ;
(b) insufficiency or defective condition of the packing or marks;
(c) handling, loading, stowage or unloading of the goods by the consignor or the consignee or any person acting on behalf of the consignor or the consignee;
(d) inherent vice of the goods;
(e) strike, lockout, stoppage or restraint of labour, the consequences of which the CTO could not avoid by the exercise of reasonable diligence;
(f) any cause or event which the CTO could not avoid and the consequences of which he could not prevent by the exercise of reasonable diligence;

(g) a nuclear incident if the operator of a nuclear installation or a person acting for him is liable for this damage under an applicable international Convention or national law governing liability in respect of nuclear energy.

The burden of proving that the loss or damage was due to one or more of the above causes or events shall rest upon the CTO.

When the CTO establishes that, in the circumstances of the case, the loss or damage could be attributed to one or more of the causes or events specified in (b) to (d) above, it shall be presumed that it was so caused. The claimant shall, however, be entitled to prove that the loss or damage was not, in fact, caused wholly or partly by one or more of these causes or events.

B. Rules applicable when the stage of transport where the loss or damage occurred is known.

Rule 13

When in accordance with Rule 5(e) hereof the CTO is liable to pay compensation in respect of loss or damage to the goods and the stage of transport where the loss or damage occurred is known, the liability of the CTO in respect of such loss or damage shall be determined:

(a) by the provisions contained in any international Convention or national law, which provisions:
   (i) cannot be departed from by private contract, to the detriment of the claimant, and
   (ii) would have applied if the claimant had made a separate and direct contract with the CTO in respect of the particular stage of transport where the loss or damage occurred and received as evidence thereof any particular document which must be issued in order to make such international Convention or national law applicable ; or
(b) by the provisions contained in any international Convention relating to the carriage of goods by the mode of transport used to carry the goods at the time when the loss or damage occurred, provided that:
   (i) no other international Convention or national law would apply by virtue of the provisions contained in sub-paragraph (a) of this Rule, and that
   (ii) it is expressly stated in the CT Document that all the provisions contained in such Convention shall govern the carriage of goods by such mode of transport; where such mode of transport is by sea, such provisions shall apply to all goods whether carried on deck or under deck; or
(c) by the provisions contained in any contract of carriage by inland waterways entered into between the CTO and any sub-contractor, provided that:
   (i) no international Convention or national law is applicable under sub-paragraph (a) of this Rule, or is applicable, or could have been made

applicable, by express provision in accordance with subparagraph (b) of this Rule and that

(ii) it is expressly stated in the CT Document that such contract provisions shall apply; or

(d) by the provisions of Rules 11 and 12 in cases where the provisions of subparagraphs (a), (b) and (c) above do not apply.

Without prejudice to the provisions of Rule 5(6) and (c), when, under the provisions of the preceding paragraph, the liability of the CTO shall be determined by the provisions of any international Convention or national law, this liability shall be determined as though the CTO were the carrier referred to in any such Convention or national law. However, the CTO shall not be exonerated from liability where the loss or damage is caused or contributed to by the acts or omissions of the CTO in his capacity as such, or his servants or agents when acting in such capacity and not in the performance of the carriage.

Liability for Delay

Rule 14

The CTO is liable to pay compensation for delay only when the stage of transport where a delay occurred is known, and to the extent that there is liability under any international Convention or national law, the provisions of which:

(i) cannot be departed from by private contract to the detriment of the claimant;

(ii) would have applied if the claimant had made a separate and direct contract with the CTO as operator of that stage of transport and received as evidence thereof any particular document which must be issued in order to make such international Convention or national law applicable.

However, the amount of such compensation shall not exceed the amount of the freight for that stage of transport, provided that this limitation is not contrary to any applicable international Convention or national law.

Miscellaneous Provisions

Rule 15

Failure to effect delivery within 90 days after the expiry of a time limit agreed and expressed in a CT Document or, where no time limit is agreed and so expressed, failure to effect delivery within 90 days after the time it would be reasonable to allow for diligent completion of the combined transport operation shall, in the absence of evidence to the contrary, give to the party entitled to receive delivery the right to treat the goods as lost.

Rule 16

The defences and limits of liability provided for in these Rules shall apply in any action against the CTO for loss of, damage, or delay to the goods whether the action be founded in contract or in tort.

Rule 17

The CTO shall not be entitled to the benefit of the limitation of liability provided for in Rule 11 hereof if it is proved that the loss or damage resulted from an act or omission of the CTO done with intent to cause damage or recklessly and with knowledge that damage would probably result.

Rule 18

Nothing in these Rules shall prevent the CTO from including in the CT document provisions for protection of his agents or servants or any other person whose services he uses for the performance of the contract evidenced by the CT document, provided such protection does not extend beyond that granted to the CTO himself.

Time-bar

Rule 19

The CTO shall be discharged of all liability under these Rules unless suit is brought within nine months after

(i) the delivery of the goods, or,
(ii) the date when the goods should have been delivered, or
(iii) the date, when in accordance with Rule 15, failure to deliver the goods would, in the absence of evidence to the contrary, give to the party entitled to receive delivery the right to treat the goods as lost.

10

Marine Insurance Law

The use of warranties originated in marine insurance in the Middle Ages.[1] Then, as now, policyholders were in a position of power: they knew material facts about the marine adventure and could also alter the risk once cover had been provided. It, therefore, became common practice for the insurer to require the policyholder do or refrain from doing something, with severe consequences for the policyholder if the term was breached.

10.1 WARRANTY

S 33(1): 'A warranty . . . means a promissory warranty, that is to say, a warranty by which the assured undertakes that some particular thing shall or shall not be done, or that some condition shall be fulfilled, or whereby he affirms or negatives the existence of a particular state of facts.'

Materiality to the risk and exact compliance

S 33(3): 'A warranty . . . is a condition which must be exactly complied with, whether it be material to the risk or not.'

A warranty does not have to be material to the risk. A warranty must be *exactly* complied with. Every policy in which an express warranty is inserted is a conditional contract, to be binding if the warranty be literally complied

1. The twelfth and early thirteenth centuries. See further Baris Soyer, *Warranties in Marine Insurance* (2nd ed 2006) at para 1.11 and following; J Hill, *O'May on Marine Insurance and Practice* (1993) 1.

with, but not otherwise. There is a material distinction between a warranty and a representation. A representation may be equitably and substantially answered, but a warranty must be strictly complied with.

In *De Hahn v Hartley*[2] a vessel was warranted to sail with a crew numbering at least fifty. Having safely completed the first part of the voyage from Liverpool to Anglesey with a crew of only forty-six, they picked up six more crew in Anglesey. The ship was captured and lost off the coast of Africa. The warranty was held to be broken. Ashurst J remarked that the 'very meaning of a warranty is to preclude all questions whether it has been substantially complied with; it must be literally so.'

In *Overseas Commodities Ltd v Style*[3] was insurance on canned pork against all risks, 'warranted all tins marked by manufacturers with a code for verification of date of manufacture'. It transpired that, although some of the tins were correctly marked, others were not. Many of the tins were condemned whilst the balance was sold below the market price. It was held that the policy could not be severed into as many contracts as there were tins and, as there were a substantial number of tins not marked, a breach of warranty of the whole policy had occurred.

A warranty must relate to the risk. A promise to pay the premium by a particular date is a normal contractual obligation but not a promissory insurance warranty.

10.2 EXPRESS WARRANTIES

S 35: '(1) An express warranty may be in any form of words from which the intention to warrant is to be inferred. (2) An express warranty must be included in, or written upon, the policy, or must be contained in some document incorporated by reference into the policy.'

An express warranty, contrasted with an implied warranty, is a condition of the contract expressed and set forth in the policy.

'An express warranty does not exclude an implied warranty, unless it be inconsistent therewith.' (See S 35(3).) For example, where a voyage policy covering a ship incorporates an express warranty requiring the ship to sail by a certain date, it does not mean the ship can sail in an unseaworthy condition to comply with the express warranty. On the other hand, section 35 provides that an express warranty takes precedence over an implied warranty where it is inconsistent with the demands of the implied warranty.

2. (1786) 1 TR 343.
3. [1958] 1 Lloyd's Rep 546.

The use of the word 'warranty' is by no means decisive.[4] Merely calling something a 'warranty' is not enough as the term is 'always used with the greatest ambiguity in a policy'.[5] In *HIH Casualty & General Insurance Co Ltd v New Hampshire Insurance Co*, Lord Justice Rix provided this guidance:

> It is a question of construction, and the presence or absence of the word 'warranty' or 'warranted' is not conclusive. One test is whether it is a term which goes to the root of the transaction; a second, whether it is descriptive or bears materially on the risk of loss; a third, whether damages would be an unsatisfactory or inadequate remedy.[6]

Warranty of neutrality

S 36: '(1) Where insurable property, whether ship or goods, is expressly warranted neutral, there is an implied condition that the property shall have a neutral character at the commencement of the risk, and that, so far as the assured can control the matter, its neutral character shall be preserved during the risk.

(2) Where a ship is expressly warranted 'neutral' there is also an implied condition that, so far as the assured can control the matter, she shall be properly documented, that is to say, that she shall carry the necessary papers to establish her neutrality, and that she shall not falsify or suppress her papers, or use simulated papers. If any loss occurs through breach of this condition, the insurer may avoid the contract.'

Implied warranty of nationality

S 37: 'There is no implied warranty as to the nationality of a ship, or that her nationality shall not be changed during the risk.'

However, a statement that a vessel is an 'American ship' constitutes a warranty that the vessel is of that nationality[7] and also that the vessel has on board all documentation a vessel of that nationality ought to carry.

Warranty of good safety

S 38: 'Where the subject-matter insured is warranted "well" or "in good safety" on a particular day, it is sufficient if it be safe at any time during that day.'

4. *Barnard v Faber* [1893] 1 QB 340.
5. *Roberts v Anglo Saxon Insurance Company* (1927) 2 Lloyd's Rep 550.
6. [2001] 2 Lloyd's Rep 161; [2001] EWCA Civ 735.
7. *Baring v Claggett* (1802) 3 Bos & Pul 201.

In *Blackhurst v Cockell*,[8] a policy on a ship contained a warranty stating 'Lost or not lost. Warranted well on December 9'. The policy was subscribed between 1 p.m. and 3 p.m. on December 9. The vessel was, in fact, lost at about 8 a.m. on that day. It was held that the warranty was complied with by the fact that the vessel was safe at some time on the day when the underwriters subscribed the policy, although she was not safe at the particular hour when they subscribed it. The warranty protected them against all losses before that day, but not against losses on that day.

Warranted 'to sail'

An express warranty may require the vessel to sail on or before a certain date. A question may then arise as to what constitutes a sailing within the meaning of such a warranty. The test is whether there was a clear intention on the part of the master, when the vessel left her moorings, to proceed directly on the voyage. The motive of moving the vessel from her moorings must be looked to, and that motive supplies the test whether or not the warranty has been complied with.[9]

'To sail from'

A warranty to sail from a specified port on a certain date will not have been complied with unless the vessel has actually left the precincts of the port named. This warranty is absolute. If the ship is ready to leave port and is prevented by bad weather or accident, this is beside the question. If she has not quitted the port by the date specified, the policy is void by reason of the breach of warranty.

Thus, in *Bond v Nutt*,[10] where a ship was warranted 'to have sailed on or before August 1', Lord Mansfield said: 'Had she or had she not sailed on or before that day? No matter what cause prevented her, if the fact is that she had not sailed, though she stayed behind for the best reasons, the policy was void.'

The use of express warranties is common in marine hull insurance practice, particularly in time policies. For example, in clause 10 of the International Hull Clauses 01/11/03 (IHC 01/11/03): 'Unless and to the extent otherwise agreed by the Underwriters in accordance with Clause 11,

8. (1789) 3 Term Rep 360.
9. *Sea Insurance Co v Blogg* [1989] 2 QB 398.
10. (1777) 2 Cowp 601.

the vessel shall not breach any provisions of this insurance as to cargo trade or locality.'

The 'disbursement' warranty[11] in the various sets of Institute hull clauses is, also, an express warranty but breach of this warranty is *not* held covered by the Institute hull clauses.

Effect of 'held covered' provisions

Where express warranties are concerned a breach can often be resolved by immediate advice to the underwriter and payment of an additional premium. This particularly applies where the assured has been offered a reduced premium in return for giving the undertaking incorporated in the warranty. In some cases, the assured is held covered only acceptance of a change in the insurance conditions as well as paying the additional premium.

10.3 IMPLIED WARRANTIES

Warranty of seaworthiness

S 39: '(1) In a voyage policy there is an implied warranty that at the commencement of the voyage the ship shall be seaworthy for the purpose of the particular adventure insured.

(2) Where the policy attaches while the ship is in port, there is also an implied warranty that she shall, at the commencement of the risk, be reasonably fit to encounter the ordinary perils of the port.

(3) Where the policy relates to a voyage which is performed in different stages, during which the ship requires different kinds of or further preparation or equipment, there is an implied warranty that at the commencement of each stage the ship is seaworthy in respect of such preparation or equipment for the purposes of that stage.

(4) A ship is deemed to be seaworthy when she is reasonably fit in all respects to encounter the ordinary perils of the seas of the adventure insured.'

The implied warranty of seaworthiness applies only to 'voyage' policies, but is applied to both hull insurance contracts and cargo insurance contracts. The term is a relative one, and seaworthiness is that condition in which a ship should be to enable her to encounter whatever perils of the sea a ship of her kind, and laden as she is, may fairly be expected to encounter in performing the voyage concerned.

11. IHC, cl 24, especially cl 24.2.

A ship is not unseaworthy if she is properly found in all requisite appliances, although, through the negligence of the persons responsible, such appliances are not used.

Thus, in *Steel v State Line SS Co*,[12] where a vessel sailed on her voyage with a porthole insecurely fastened, through which the sea entered and damaged the cargo, Lord Blackburn took the view that if the porthole was in a place where it would in practice be left open from time to time, but yet was capable of being speedily shut when occasion required, the vessel could not be said to be unfit to encounter the perils of the voyage. And if, when weather threatened, it was not shut, that would be negligence of the crew and not unseaworthiness.

The burden of proving a breach of the implied warranty of seaworthiness lies on the insurers where they allege it.

Seaworthiness in hull time policies

'In a time policy there is no implied warranty that the ship shall be seaworthy at any stage of the adventure, but where, with the privity of the assured the ship is sent to sea in an unseaworthy state, the insurer is not liable for any loss attributable to unseaworthiness.' (See S 39(5).)

In *Russell v Provincial Insurance*,[13] the insured's motor boat was insured under a time policy against *inter alia* 'stress of weather, stranding, sinking'. She sank on a voyage from Burnham-on-Crouch to Southend whilst cutting across a sandbank. When the assured claimed for a total loss, one of the defences raised by the insurers was that she was unseaworthy, and that the assured had not kept her in a proper state of repair and seaworthiness as required by the policy. The court held that this defence failed, for she was seaworthy, and that she sank as a result of a genuine casualty. The court explained that there was evidence that the boat had been recently repaired. The boat might strike something harder than the sand.

In *Thomas v Tyne and Wear SS Freight Insurance Asso.*,[14] a vessel was insured under a time policy. The assured was privy to her being sent to sea with an insufficient crew, but did not know of a defect in her hull. The ship was lost on account of such defect. It was held that the assured could recover on the policy because the loss had not been caused by the particular seaworthiness to which the assured was privy.

12. (1877) 3 App Cas 72.
13. [1959] 2 Lloyd's Rep 275, QBD (Commercial Court).
14. (1917) 117 LT 55.

There is no implied warranty of seaworthiness in a time policy. Therefore, in a policy which incorporates the Institute Time Clauses Hulls (ITC Hulls), there is no implied warranty requiring the ship to be seaworthy at any time during the currency of the policy.

Seaworthiness in cargo policies

Unless the policy conditions provide otherwise, the implied warranty of seaworthiness applies equally to the overseas vessel which is to carry the goods insured by a cargo policy. This was practical in the days then merchants carried goods in their own vessels. Nowadays, few cargo shippers could control the seaworthiness of the ships that were to carry their goods. Accordingly, it has become the practice for cargo underwriters to waive breach of the warranty of seaworthiness except in cases where the assured or their servants are privy to such seaworthiness.

To protect their interest, however, cargo underwriters insuring goods by open cover usually incorporate a 'classification' clause in the contract allowing them to charge an additional premium for goods carried by ships of a lower standard than is outlined in the clause.

Implied warranty of cargoworthiness

S 40: '(1) In a policy on goods or other movables there is no implied warranty that the goods or movables are seaworthy.

(2) In a voyage policy on goods or other movables there is an implied warranty that at the commencement of the voyage the ship is not only seaworthy as a ship, but also that she is reasonably fit to carry the goods or other movables to the destination contemplated by the policy.'

That is to say, the holds and other storage spaces in the ship shall be fit to carry the goods safely. It is unlikely the cargo assured will have control over the fitness of the ship to carry their goods so it has become the practice for cargo underwriters to waive the breach of this warranty except where the assured or their servants are privy to the unfitness of the vessel at the time the goods are loaded therein.

Implied warranty of legality

S 41: 'There is an implied warranty that the adventure insured is a lawful one, and that, so far as the assured can control the matter, the adventure shall be carried out in a lawful manner.'

As a general rule, the legality or otherwise of an adventure is determined according to the common and statute laws of Hong Kong. If the adventure to be performed is wholly or partly illegal according to Hong Kong law, the contract of insurance would be affected.

The implied warranty of legality is qualified with the term 'so far as the accused can control the matter'. The case of *Pipon v Cope*,[15] concerning barratry, is apt to illustrate this point. The crew members had committed repeated acts of smuggling on three consecutive voyages. The shipowner could and should have taken positive steps (for example, by replacing the ship with a new crew member) to prevent the repeated acts of smuggling.

The warranty of legality is an exception to the rule in section 34(3) that it cannot be waived by the insurer.

Effect and excuse of breach

No defence for breach and irremediable

With the exception of the 2 excuses laid down in section 34, there is clearly no defence for a breach of an express or implied warranty.

S 34(2): 'Where a warranty is broken, the assured cannot avail himself of the defence that the breach has been remedied, and the warranty complied with, before loss.'

For example, a ship is insured under a time policy from 1 January to 31 December. The policy conditions incorporate an express warranty restricting navigation to within specified geographical limits, breach of this warranty not being held covered by the policy conditions. During June, the insured vessel sails beyond the specified limits, without the assured obtaining approval from underwriters, but returns to policy limits without incident. During September, whilst sailing within the policy limits, the ship is damaged by an insured peril. However, no claim would attach to the policy for the damage to the ship because the underwriter had been automatically discharged from all liability under the policy as from the date in June when the ship sailed beyond the policy limits.

Excuses

S 34(1): 'Non-compliance with a warranty is excused when, by reason of a change of circumstances, the warranty ceases to be applicable to the

15. (1808) 1 Camp 434.

circumstances of the contract, or when compliance with the warranty is rendered unlawful by any subsequent law.'

S 34(3): 'A breach of warranty may be waived by the insurer.'

Thus, in *Weir v Aberdeen*,[16] a vessel was insured 'at and from London to Bahia'. She sailed from London in an unseaworthy condition by reason of her being overloaded. She was forced to put in at Ramsgate to unload part of the cargo, and then, with the consent of the insurers, proceeded on her voyage and was lost. It was held that the breach of the implied warranty of seaworthiness had been waived, and that the insurers were liable for the loss.

In *Samuel v Dumas*,[17] a vessel was insured under a time policy which contained a warranty that the assured was entitled to insure freight for 6 months for $13,750. In fact, the freight was insured for 6 months for $27,500. The vessel was lost, and when a claim was made under the policy, the insurers repudiated liability for breach of warranty. The House of Lords held there had been a breach of warranty but the breach had been waived by the insurers so they could not rely on it. The insurers were prevented from taking advantage of it by the circumstances they were a party to the excessive insurance on freight which constituted the breach.

Often the policy contains a clause concerning the waiver of a breach of warranty by the insurer as long as the assured is willing to pay an additional premium.

Effect of breach

S 33(3): 'If it be not so complied with, then, subject to any express provision in the policy, the insurer is discharged from liability as from the date of the breach of warranty, but without prejudice to any liability incurred by him before that date.'

No action is required from an insurer, being automatically discharged from all liability under the contract as from the date of the breach. Effectively, this means that an assured who ignored a breach of warranty and continued trading would be uninsured as from, and including, the date of the breach; with the only remedy being to advise the insurer immediately after becoming aware of the breach, requesting continuance of cover (continuance being at the discretion of the insurer unless the breach was held cover by a policy condition).

16. (1819) 2 B & Ald 320.
17. *P Samuel & Co Ltd v Dumas* [1924] All ER Rep 66.

Warranties in consumer insurance: Other protections

For consumers there are other statutory and regulatory safeguards. These include Financial Services Authority rules, the Unfair Terms in Consumer Contracts Regulations 1999 (UTCCRs), and the Consumer Insurance (Disclosure and Representations) Act 2012.

Furthermore, the Financial Ombudsman Service is not bound by the strict letter of the law, and has jurisdiction to hear complaints from consumers and microbusinesses. Below, we consider these additional protections.

Basis of the contract clauses

If a prospective policyholder signs a statement on a proposal form stating that the answers given form the 'basis of the contract', this has the effect of converting all the answers into warranties. This gives the insurer additional remedies should one of the statements be untrue. An insurer may only avoid the contract if a misrepresentation is 'material' under section 20 of the Marine Insurance Act. A basis of the contract clause goes beyond section 20, and allows the insurer to avoid liability for any inaccuracy, however unimportant.[18]

The case of *Dawsons Ltd v Bonnin*[19] demonstrates the effect of a basis of the contract clause. A furniture removal firm in Glasgow (Dawsons) took out insurance for one of its removal lorries. The proposal form included the following clause: 'which proposal shall be the basis of this contract and be held as incorporated herein'. This converted all the answers on the form into warranties. Dawsons filled out the form, and gave its business address in central Glasgow. When it was asked where the lorry would normally be parked, it inadvertently wrote 'above address'. In fact, the lorry was usually parked in the outskirts of Glasgow. The lorry was destroyed in a fire and Dawsons made a claim. At court, it was argued that Dawsons' mistake about the address did not add to the risk and arguably had reduced it. The court held that this did not matter: the insurance company was entitled to refuse to pay all claims under the policy.

18. Law Commission (2012) *Insurance Contract Law: The Business Insured's Duty of Disclosure and the Law of Warranty*, Consultation Paper No 204, p 135. The content of this chapter makes references to the above Consultation Paper No 204.
19. [1922] 2 AC 413; 1922 SC (HL) 156.

The law of warranties has attracted criticisms for many years.[20] In 1980, the Law Commission commented that 'it seemed quite wrong that an insurer should be entitled to demand strict compliance with a warranty which was immaterial to the risk'.[21] Similarly, it seems unjust that an insurer should be entitled to reject a claim for any breach, 'no matter how irrelevant that breach to the loss'.[22]

Although consumers are now protected against basis of the contract clauses, they remain a problem for business policyholders. In 1996, in *Unipac (Scotland) Ltd v Aegon Insurance Co (UK) Ltd*,[23] the Court of Session confirmed that in business insurance, where a basis of the contract clause was in place, an insurer may refuse a claim for any inaccuracy on a proposal form, including those which were immaterial.

In an international context, the UK law on warranties seems unbalanced, tending to favour the insurer over the policyholder. In the common law world, most jurisdictions have moved away from the UK approach. Although both Australia and New Zealand originally adopted statutory law equivalent to the Marine Insurance Act 1906, both have now enacted reforms. In Canada, the Supreme Court has limited the effect of a breach of warranty to situations where the breach is material to the particular type of loss.[24] In the USA, insurance law is left to individual states.[25] Many states have introduced statutory reform.

Proposals of the Law Commission

The proposals are aimed at remedying the particular problems caused by sections 33 and 34 of the Marine Insurance Act 1906. The three main proposals are:

---

20. See, for example, Dr Baris Soyer, 'Reforming Insurance Warranties: Are We Finally Moving Forward?' *Reforming Marine & Commercial Insurance Law* (2008); Sir Andrew Longmore, 'Good Faith and Breach of Warranty: Are We Moving Forwards or Backwards?' (2004) LMCLQ 158. See also the National Consumer Council, *Insurance Law Reform: The Consumer Case for Review of Insurance Law* (May 1997), and Kenneth Reid and Reinhard Zimmermann (eds), *A History of Private Law in Scotland* (2000), vol 2, 360–61 contributed by ADM Forte.
21. 'Insurance Law: Non-Disclosure and Breach of Warranty' (1980) Law Com No 104, para 6.9(a).
22. Above, para 6.9(b).
23. (1996) SLT 1197.
24. See *Century Insurance Company of Canada v Case Existological Laboratories Ltd (The Bamcell II)* [1984] 1 *Western Weekly Reports* 97.
25. See *Wilburn Boat Co v Fireman's Fund* Ins 348 US 310 (1955).

(1) *To abolish basis of the contract clauses.* Insurers may still use warranties of past or present fact, but they should be included specifically in the contract.
(2) *To treat warranties as suspensive conditions.* A breach of warranty would suspend the insurer's liability, rather than discharge it. Where the breach is remedied before the loss, the insurer must pay the claim.
(3) *To introduce special rules for terms designed to reduce the risk of a particular type of loss, or the risk of loss at a particular time or in a particular location.* For these terms, a breach would suspend liability in respect only of that type of loss (or a loss at that time or in that place). Thus, the breach of a warranty to install a burglar alarm would suspend liability for loss caused by an intruder but not for flood loss. Similarly, a failure to employ a night watchman would suspend the insurer's liability for losses at night but not for losses during the day.

In consumer insurance, the proposed consequences of breach could not be excluded by a contract term. In business insurance, the parties would be able to contract out of these provisions, provided they did so in clear, unambiguous terms and brought the term to the attention of the other party.

Basis of the contract clauses have been criticized for many years. Few policyholders understand their effect. In 1986, the Statement of General Insurance Practice barred their use. They are not thought to treat customers fairly and the FOS would almost certainly reject an attempt by an insurer to rely on them. The draft Bill abolishes basis of the contract clauses. This brings the law into line with recognized good practice.

10.4 CONTRACT OF INDEMNITY, UTMOST GOOD FAITH AND INSURABLE INTEREST

Introduction

Among different types of insurance, marine insurance is one of the earliest types to be found in the market. It is believed that a form of mutual insurance was practised in China around 4000 BC, but there is no evidence of private underwriters offering cover at that time.[26] The first statute in marine insurance in the United Kingdom was the Marine Insurance Act 1745. Judicial decisions in courts were consolidated and codified into the Marine Insurance Act 1906, which is still in force in the United Kingdom. The law of marine insurance in the Hong Kong Special Administrative Region of

26. See Marine Insurance Training Services (Great Britain) (ed), *Introduction to Marine Insurance: Training Notes for Brokers* (Witherby 1987) Section 1.

The People's Republic of China (HKSAR) is mainly governed by the Marine Insurance Ordinance (Cap 329) which came into force on 9 June 1961. The Ordinance is concerned only with marine insurance, but many of its provisions are presumed to be equally applicable to other types of insurance in Hong Kong:

> *Pan Atlantic Insurance Co Ltd v Pine Top Insurance Co Ltd*[27]
> Although the issues arise under a policy of non-marine insurance it is convenient to state them by reference to the Marine Insurance Act 1906 since it has been accepted in argument, and is indeed laid down in several authorities, that in relevant respects the common law relating to the two types of insurance is the same, and that the Act embodies a partial codification of the common law.

Insurance is one of the arrangements by which risks in connection with operations can be transferred to a professional risk carrier, i.e., an insurance organization or insurance company, in a sophisticated community like Hong Kong. Parties which are involved in transport or logistics, like shippers, consignors, consignees, shipowners, airlines, charterers, logistics operators, freight forwarders, warehouse operators, barge owners, road hauliers, rail operators or inland waterway operators, may transfer their risks arising from operation of freight to a third party or risk carrier.

Contract of indemnity

One of the main features of the law of marine insurance is that a policy is a contract of indemnity with which many other principles of marine insurance are connected. Section 2(1) of the Marine Insurance Ordinance on 'Marine insurance defined' states:

> A contract of marine insurance is a contract whereby the insurer undertakes to indemnify the assured, in manner and to the extent thereby agreed, against marine losses, that is to say, the losses incident to marine adventure.

The rights and liabilities of underwriters and assureds are governed by this fundamental principle, i.e., a contract of indemnity.

> *Castellain v Preston* (1883) 11 QBD 380 at p 386
> Mr. Justice Brett remarked: 'The contract of insurance contained in a marine or fire policy is a contract of indemnity, and of indemnity only, and this contract means that the assured, in case of a loss against which

27. [1995] AC 501, at p 518 by Lord Mustill.

the policy has been made, shall be fully indemnified, but shall never be more than fully indemnified.'

A contract which adopts the form of marine insurance may not necessarily be a contract of marine insurance and indemnity if a contract of marine insurance is to be against marine damage or loss.

> *Re County Commercial* [1922] 127 LT 20
>
> The risk, which was insured against by certain policies of insurance or re-insurance issued on marine insurance forms and containing the p.p.i. (policy proof of interest) and f.i.a. (full interest admitted) clauses, was a total loss in the event of peace not being declared on or before a certain date.
>
> The court held that the provisions of the Marine Insurance Act 1906 were not applicable to the peace policies; they were insurances within the meaning of the Life Assurance Act 1774 and the assured had no interest or insurance by way of gaming or wagering with s 1 of the 1774 Act. They were therefore illegal and no premiums paid on them were recoverable.

Not a perfect contract of indemnity

A contract of marine insurance is in principle a contract of indemnity, but in practice it may not compensate with perfect indemnity an assured who has suffered loss or damage. This means that the suffering party cannot be fully compensated, because one may not be able to measure damages and/or loss perfectly. An assured may recover 100 per cent of the sum insured from the underwriters on the policy, but may not then be able to buy a similar property (or similar subject matter) in an open market with all the proceeds just recovered. One reason may be that the value of the subject matter fluctuates or a subject matter of similar nature is not available in the market. It is not a perfect indemnity, as underwriters would not indemnify the assured by delivering a similar ship or a similar subject matter in case the subject matter is lost.

For example, a shipowner bought a ship on January 1999 at a value of US$10 million and insured it for US$10 million. The freight market improved and subsequently the market value of the ship increased to US$12 million on September 1999. The ship then sank in an accident. The shipowner received from underwriters the sum insured by the policy of US$10 million, but the shipowner could not go to the market to buy back a similar ship with the claim proceeds, i.e., US$10 million.

In *British and Foreign Insurance Co Ltd v Wilson Shipping Co Ltd* [1921] 1 AC 188, Lord Sumner said: 'In practice contracts of insurance by no means always result in a complete indemnity, but indemnity is always the basis of the contract.'

Utmost good faith

The heading of section 17 of the Marine Insurance Ordinance is 'Insurance is uberrimae fidei'. The phrase 'uberrimae fidei' is explained in Osborn's *Concise Law Dictionary* as 'of the fullest confidence'. It further says that a contract is said to be uberrimae fidei when the promise is bound to communicate to the promisor every fact and circumstance which may be influential in deciding to enter into the contract or not. Contracts of insurance of every kind are of this class. The Marine Insurance Ordinance section 17, on 'Insurance is uberrimae fidei', states:

> A contract of marine insurance is a contract based upon the utmost good faith, and, if the utmost good faith be not observed by either party, the contract may be avoided by the other party.
>
> *Container Transport International Inc v Oceanus Mutual Underwriting Association (Bermuda)* [1984] 1 Lloyd's Rep 476 at p 525, CA.
>
> Lord Stephenson stated that: '. . . It is enough that much more than an absence of bad faith is required of both parties to all contracts of insurance.'

In contrast, ordinary commercial contracts start from the premise of *caveat emptor*—let the buyer beware. Though they are often subject to implied terms, such as those set out in the Sale of Goods Act 1979, ss 13, 14 and 15, which may have the indirect effect of requiring that the seller disclose defects. A party to a contract *caveat emptor* must not misrepresent facts, but is under no obligation to disclose facts about which it is not asked.[28]

The principle of utmost good faith is applicable to all classes of insurance and it involves the duty of disclosure of assured, agent of assured, underwriter and underwriter's agent.

> *Carter v Boehm* (1766) 3 Burr. 1905, 1909
>
> Lord Mansfield stated: 'Insurance is a contract upon speculation. The special facts upon which the contingent chance is to be computed lie most commonly in the knowledge of the insured only; the underwriter trusts to his representation and proceeds upon confidence that he does not keep back any circumstance in his knowledge to mislead the underwriter into a belief that the circumstance does not exist, and to induce him to estimate the risk as if it did not exist. . . . Good faith

28. Law Commission and Scottish Law Commission, *Insurance Contract Law Issue Paper 1 Misrepresentation and Non-Disclosure* (2006).

> forbids either party, by concealing what he privately knows to draw the other into a bargain from his ignorance of the fact, and his believing the contrary.'

The consequence is that the duty tends to be unevenly heavier on the applicant for insurance than on the underwriter or insurer. In many areas, the balance of knowledge in relation to risks between insured and insurer has drastically altered in the last 250 years since the comments made by Lord Mansfield.

If a party to an insurance contract does not comply with the duty of utmost good faith, the other party may avoid the contract from the beginning of the contract (i.e., *ab initio*) on an all or nothing basis. The contract is treated as if it never existed and it is applicable to breach of fraudulent, negligent or innocent nature.

> *Black King Shipping Corporation and Wayang (Panama) SA v Mark Ranald Massie (The 'Litsion Pride')* [1985] 1 Lloyd's Rep 437
>
> The vessel was chartered and later nominated to make a voyage from Europe to Bandar Khomeini, a port in the Persian Gulf which was at the time by far the most dangerous port in the Gulf, attracting additional premium (A.P.) at a very substantial rate.
>
> The court held that, even on a view most favourable to the owners, on the facts it would still have been impossible to acquit them of breach of duty of utmost good faith having regard to their culpable failure to communicate to the underwriters the facts which they knew well before 15 July, such as the ETA in the war zone, the likely duration of the vessel's stay and the nomination of Bandar Khomeini as the discharge port; all these matters were highly material both in order to enable the underwriters to fix the premium and to enable them to consider reinsurance.
>
> '[The word] "avoidance" in s 17 of the Marine Insurance Act 1906 meant avoidance ab initio; s 17 provided that the policy may be avoided, not that it must be avoided and here in the case of post-contract breach it was open to the underwriters simply to defend the claim without avoiding the policy.'

The English Law Commission issued a report[29] on the issues of non-disclosure and breach of warranty of insurance law in October 1980. It concluded that the law relating to non-disclosure and breach of warranty is undoubtedly in need of reform, and this reform has been too long delayed. Unfortunately, the Report excluded marine, aviation and transport (MAT)

29. Insurance Law, Non-disclosure and Breach of Warranty (1980) Law Com No 104, para 1.21(a), p 10.

insurance from its recommendations. The MAT insurances are covered by Issues Paper 1.[30]

In The *'Captain Panagos'* D.P. [1986] 2 Lloyd's Rep 511, Mr. Justice Evans held that any breach of the post-contractual duty of utmost good faith in relation to the making of claims also breaks an implied term of the contract. The House of Lords held in *Banque Financiere de La Cite SA v Westgate Insurance Co Ltd* [1991] 2 AC 249 that in relation to the pre-contractual duty of good faith the only remedy for breach lies in avoidance under section 17 of the Marine Insurance Act, and that there is no right to damages.

The issue of duration of the duty of utmost of good faith was discussed in *Manifest Shipping & Co Ltd v Uni-Polaris Insurance Co Ltd and La Reunion Europeene (The 'Star Sea')* [1997] 1 Lloyd's Rep 360, in which Leggatt LJ said:

> It was common ground between assured and underwriters that the duty of utmost good faith continues to subsist after the making of the contract. There is less common ground as to the content of the duty or as to the remedy for breach of the duty. Whatever its origin, there is force in the argument that the scope of the duty of utmost good faith will alter according to whether underwriters have to make a decision under the policy or the assured decides to make a claim, and may also be affected according to the stage of the relationship at which the scope of the duty becomes material.

Disclosure by assured

Before a contract of insurance is concluded, an assured is usually required to provide for underwriters information of the risks concerned in the form of a proposal or otherwise. In section 18(1) of the Marine Insurance Ordinance, on 'Disclosure by assured', it states:

> Subject to the provisions of this section, the assured must disclose to the insurer, before the contract is concluded, every material circumstance which is known to the assured, and the assured is deemed to know every circumstance which, in the ordinary course of business, ought to be known by him. If the assured fails to make such disclosure, the insurer may avoid the contract.

The reason that the Ordinance requires the assured in section 18(1) to disclose to the insurer (or underwriters) before the conclusion of the

30. Law Commission and Scottish Law Commission, *Insurance Contract Law Issue Paper 1 Misrepresentation and Non-Disclosure* (2006).

contract every material circumstance is that it may not be possible for the insurer to know, even though going themselves or sending an agent to find out, every material circumstance. If insurers were required to find out every material circumstance by themselves, the costs incurred in respect of investigation would be high and would finally go back to the assured, who might have to pay a much higher premium. In *Container Transport International Inc v Oceanus Mutual Underwriting Association (Bermuda)* [1984] 1 Lloyd's Rep 476, Stephenson LJ stated: 'It seems to me essential to bear in mind two things, each of which stems from the need for equality between those bargaining in the marine insurance market. . . . The first is that the insured is the one who knows most of what the underwriter needs to know but does not know; the second that, though the underwriter must trust the insured to give it him, he in his turn must be trusted not to abuse the help and protection given him by the duty the law imposes on the insured to disclose and represent truly all that a prudent underwriter needs to know, and so turn the duty into a means of avoiding a contractual liability which he ought in fairness to honour. This the [Act of 1906] recognizes by making the duty to observe the utmost good faith mutual in section 17 and by providing the exceptions of circumstances which need not be disclosed that are to be found particularly in section 18(3)(b) and (c).'

It is clear that the assured is liable to the duty of disclosure to insurer. The decision of the Court of Appeal in relation to the duty of disclosure of insurer towards assured was upheld by House of Lords majority in *Banque Financiere de La Cite SA v Westgate Insurance Co Ltd* [1991] 2 AC 249. The Marine Insurance Ordinance stipulates the assured must disclose to the insurer before the contract is concluded every material circumstance, and the issue arises as to when a contract of marine insurance is deemed to be concluded. In section 21 of the Marine Insurance Ordinance, on 'When contract is deemed to be concluded', it states:

> A contract of marine insurance is deemed to be concluded when the proposal of the assured is accepted by the insurer, whether the policy be then issued or not; and, for the purpose of showing when the proposal was accepted, reference may be made to the slip or covering note or other customary memorandum of the contract.

The proposal of insurance (sometimes called 'Application of Insurance' in marine cargo) may have to be in writing and signed by the assured in accordance with the details of the form designed by the underwriters, or some underwriters accept a proposal in free form submitted by an insurance broker or intermediary. The important part is at what time the proposal of the assured is accepted by the insurer, while the date of the issue of the slip, cover note, policy, other customary memorandum and even the inception of the risk are not relevant. In *Economides v Commercial Union* [1997] 3 All ER 636, the court held that it does not mean that assureds are required to investigate matters outside their knowledge, as these are for the insurer alone to uncover in reaching an underwriting decision.

> *Success Insurance Ltd v George Kallis (Manufacturers) Ltd* [1981] HKLR 616, CA
>
> The respondent bought denim material from a Hong Kong supplier for delivery to Limassol. The insurance policies covered all risks and were subject to the Institute Cargo Clauses. The goods were shipped on another vessel to Keelung under a separate bill of lading and stored in a customs godown for three months. The goods were then shipped on a further vessel for Limassol but were lost en route. One of the issues arising was whether the policies were void for non-disclosure.
>
> The court held that there was no duty upon the Hong Kong supplier nor upon the respondent to seek out information as extraordinarily prudent businessmen, as opposed to being deemed to know information available in their office which should be noted and acted upon. Hence there was no case for non-disclosure.

The court further stated in *Success Insurance Ltd v George Kallis (Manufacturers) Ltd* [1981] HKLR 616: 'It is one thing to say that an assured is to be deemed to know information which has been sent to his office and which, in the ordinary course of business, ought to have been noted and acted upon (e.g. the casualty slip in *London General Insurance Co Ltd v General Marine Underwriters Association Ltd* [1921] I KB 104) but quite another to say that he must go out and look for information, even though an extraordinarily prudent man might do so. The present case is stronger in favour of the Assured than was *Australia and New Zealand Bank Ltd v Eagle Wharves Ltd* [1960] 2 Lloyd's Rep 241, where the information not disclosed related to the manner in which a company's own operations were performed.'

It may come as a surprise to many policyholders, particularly consumers, that they are expected to disclose information about which they have not been asked. Professor Malcolm Clarke criticized the current position:

> Applicants in England may complete the form with scrupulous care, but still find that there was something else material to prudent insurers which, apparently, the particular insurer did not think to ask about but which, nonetheless, the applicant was expected to think of and disclose.[31]

31. MA Clarke, *Policies and Perceptions of Insurance Law in the Twenty-first Century* (OUP 2005) 103.

What circumstance is deemed to be material?

An assured or underwriter may want to find out what circumstance is deemed to be material. The Marine Insurance Ordinance section 18(2), on 'Disclosure by assured', states:

> Every circumstance is material which would influence the judgment of a prudent insurer in fixing the premium, or determining whether he will take the risk.

In *Container Transport International Inc v Oceanus Mutual Underwriting Association (Bermuda) Ltd* [1984] 1 Lloyd's Rep 476, it was held that a circumstance was material only if its disclosure would have decisively influenced the mind of a prudent insurer. The issue is the hypothetical, not the actual or particular insurer. In other words, whether the actual or particular insurer was or was not induced by the undisclosed fact or misrepresentation to enter into the contract was considered irrelevant.

> *Container Transport International Inc v Oceanus Mutual Underwriting Association* (Bermuda) Ltd [1984] 1 Lloyd's Rep 476
>
> The first plaintiffs Container Transport International Inc (C.T.I.) carried on a worldwide business in the hiring out of containers for use in ocean transport. C.T.I. needed insurance for containers on leases incorporating 'Damage Protection Plan'. C.T.I. changed underwriters several times because of poor claims experience, and in due course the insurance was placed with the defendants Oceanus, who came on risk. Oceanus' solicitors wrote avoiding the policy on the grounds of misrepresentation and non-disclosure.
>
> On appeal by the defendants Oceanus, it was held by the Court of Appeal that an insurer was entitled to avoid a contract under s 18 (1) of the Marine Insurance Act 1906 if there was undisclosed before the contract was concluded any circumstance which a prudent insurer would take into account when reaching his decision whether or not to accept that risk or what premium to charge; the yardstick was the prudent insurer and not the particular insurer; here the representation made was a representation which was material and untrue within s 20.

The 'decisive influence' test was rejected by the majority of the House of Lords in *Pan Atlantic Insurance Co Ltd and Another v Pine Top Insurance Co Ltd* [1994] 2 Lloyd's Rep 427. '"Influence the mind" is not the same as "change the mind" . . . it should be observed that the expression used is "influence the judgment of a prudent insurer in" determining whether he will take the risk. The expression clearly denotes an effect on the thought processes of the insurer in weighing up the risk, quite different from words which might have been used but were not, such as "influencing the insurer to take the risk".' The House of Lords stated that the Court was at liberty to give to the definition of materiality in sections 18(2) and 20(2) an interpretation which accorded with the natural and ordinary meaning of the words used, the underlying obligation of good faith and the practicalities of the situation. It further states that a circumstance may be material for the purposes of an insurance contract (whether marine or non-marine), even though had it been fully and accurately disclosed it would not have had a decisive effect on the prudent underwriter's decision whether to accept the risk and if so at what premium. There is to be implied in the Marine Insurance Act 1906 a requirement that a material misrepresentation will only entitle the insurer to avoid the policy if it induced the making of the contract, and that a similar conclusion must be reached in the case of a material non-disclosure.

> *Pan Atlantic Insurance Co Ltd and Another v Pine Top Insurance Co Ltd* [1994] 2 Lloyd's Rep 427, HL
>
> The plaintiffs were the reassured and the defendants were the reinsurers under a Casualty Account Excess of Loss reinsurance. The contract covered losses occurring in the calendar year 1982. It was a renewal of similar contracts on which the defendants were reinsurers covering losses in the calendar years 1980–1981. The contract wording was signed on 29 Oct, 1982. The issues for decision were whether there was non-disclosure of a material part of the loss record on this contract i.e. on the 1977–1979 years and whether in relation to the record for the years 1980 and 1981 there were additional losses which should have been disclosed and or whether there was a misrepresentation in relation to the losses that occurred in relation to those two years.
>
> The House of Lords held that there was to be implied in the Marine Insurance Act 1906 a requirement that a material misrepresentation would only entitle the insurer to avoid the policy if it induced the making of the contract; a similar conclusion was to be reached in the case of non-disclosure: the decisive influence test would be rejected. Appeal dismissed.

In *Great Northern Shipping Co Ltd v American International Assurance Co Ltd* [1952] HKLR 267, an employee of the insurance company was aware that the voyage was to be to North China. The Court stated it was of the greatest importance to insurers to know the flag the ship was flying because of the blockade of the coast by Nationalist gunboats. The Court was satisfied the defendants would only have accepted insurance of the ship under the Chinese flag on a voyage to North China, if at all, on very high rates and not on the rates quoted.

Zeus Tradition Marine Ltd v Bell (The 'Zeus V') [2000] 2 Lloyd's Rep 587

The plaintiffs' yacht *Zeus V* was a motor sailing vessel which had been rebuilt and refitted during the period 1990 to 1992. The vessel was brought into commission in May 1992. The Institute Yacht Clauses were incorporated and the conditions set out in the certificate of insurance included the following: 'Subject to survey including valuation by independent qualified surveyor prior to commencement of in commission period.' On 20 April 1993, the *Zeus V* caught fire off the island of Sikinos in the Aegean Sea. It was towed to a position off Ios where the fire caused it to become a total loss.

The claim was rejected by the underwriters on the grounds that, inter alia, there had been non-disclosure of the material fact that the vessel had been put into commission during the immediately previous period of insurance from 1 April 1992 to 31 March 1993 without the owners having first provided to underwriters a condition survey and without underwriters having been informed that she was to go into commission.

The court held that if the plaintiff had complied with the subject condition, the bringing of the vessel into commission following an inadvertent omission to send a favourable and sufficiently comprehensive report to underwriters was incapable of amounting to a fact material to the risk on the grounds of the assured's lack of good faith, and once it was accepted that the starting point was an oversight by the assured rather than deception and once it was assumed that the report in question was not unfavourable, there was no logical basis for the underwriters to take this conduct into account in deciding whether to accept the risk; the underwriters would not have been entitled to avoid the policy for non-disclosure.

One issue in *The 'Zeus V'* [2000] 2 Lloyd's Rep 587 decided by the Commercial Court of the Queen's Bench was criticized by the Court of Appeal as incorrect. The Court of Appeal held that since the judge predicated his conclusion that the survey by SVL was insufficiently comprehensive to satisfy the reasonable requirements of the underwriters for the purposes of the clause upon reasoning which was regarded as incorrect, that conclusion seemed to be irremediably flawed and the appeal would be allowed on the issue of construction.

Sealion Shipping Ltd v Valiant Insurance Co (The 'Toisa Pisces') [2012] 1 Lloyd's Rep 252

The claimants (C) claimed an indemnity under a marine insurance policy with the defendant insurers (U) for loss of hire of their vessel. In 2004 the vessel's starboard motor failed and the whole motor was replaced. During repairs the vessel was off-hire for 10 days. In 2005 there were vibrations in the port motor and it was repaired. C had claimed on its hull and machinery insurance for both of those incidents. In 2008 C obtained loss of hire insurance with U, who noted, from conversations with C's brokers, that the vessel had an excellent hull record and no major business interruption. In 2009 the vessel suffered a port motor breakdown and was replaced, It was off-hire for a total of almost three months.

The Court held the judgment for the claimants. (1) Although disclosure of the other hull claim was good broking practice, it did not make a hull claim material if it was not otherwise material. If the hull claims were immaterial they remained immaterial even if one was mentioned and not the other. The materiality of the hull incidents was linked to the extent to which they caused loss of hire, which had to be appraised in light of the conclusion on undisclosed periods of off-hire. The undisclosed loss of hire for 10 days in 2004 was not material when compared to the 21-day excess in the 2008 policy. That conclusion was consistent with a reasonable commercial approach; it was not a particularly long period of off-hire, it was nearly four years before U's policy, it did not result in a claim, and it did not come close to the excess period. The assertion that there had been no major business interruption was true for the same reasons as the undisclosed off-hire periods, and the excellent hull record comment was both a statement of opinion and made in good faith.

(2) The standard of care required of C was one of negligence, and 'want of due diligence' meant a lack of reasonable care. The failures alleged by U did not amount to negligence. There was a proper opportunity to investigate the problem on inspection, and there was no reason to doubt the adequacy of that inspection when the stator was found to be secure and the rotor was diagnosed as the cause. There was no reason to modify or repair the stator since no problem was identified.

In *The 'Toisa Pisces'*, marine insurers could not avoid liability under a loss of hire policy for machinery breakdown where the vessel owners' failure to disclose both of two previous hull claims and a 10-day off-hire period had not been material when compared to the 21-day period of excess provided for by the policy. A proviso that the insurer would not be liable for machinery claims resulting from the insured's failure to exercise 'due diligence' meant a lack of reasonable care by the insured.

Circumstances that need not be disclosed in the absence of inquiry

There are circumstances which need not be disclosed if the underwriter does not inquire about them. See the Marine Insurance Ordinance section 18(3), which lists:

(a) Any circumstance which diminishes the risk;
(b) Any circumstance which is known or presumed to be known to the insurer. The insurer is presumed to know matters of common notoriety or knowledge, and matters which an insurer in the ordinary course of his business, as such, ought to know;
(c) Any circumstance as to which information is waived by the insurer;
(d) Any circumstance which it is superfluous to disclose by reason of any express or implied warranty.

A particular circumstance may vary in different cases in connection to the issue of materiality due to the fact that it is not a question of law but a question of fact. The Marine Insurance Ordinance section 18(4) on 'Disclosure by assured' states:

> Whether any particular circumstance, which is not disclosed, be material or not is, in each case, a question of fact.

This means that a circumstance which may have been considered to be material in one case, may not be material in another case. For example, the condition of the insulation in a cargo hold may be held to be material for a reefer ship, but may not be so in another reefer ship. Section 18(5) of the Marine Insurance Ordinance on 'Disclosure by assured' states:

> The term 'circumstance' includes any communication made to, or information received by, the assured.

Fames Yachts Ltd v Thames and Mersey Marine Insurance Co Ltd [1977] 1 Lloyd's Rep 206
Builders' risk policy. The assured failed to disclose:
i. The fact that a local authority had refused to give him permission to use his boatyard for any industrial purpose because it was a high hazard risk; and
ii. The fact that he was in financial difficulties and would be less able to maintain the equipment used in the yard and thereby would increase the hazard.

The court held that these facts were material and the insurers were entitled to avoid liability.

K/S Merc-Scandia XXXXII v Certain Lloyd's Underwriters Subscribing to Lloyd's Policy No. 25T 105487 and Ocean Marine Insurance Co Ltd and Others (The 'Mercandian Continent') [2000] 2 Lloyd's Rep 357
The claimants were the owners of the vessel and the defendants were Lloyd's underwriters and companies who subscribed to liability policies in favour of a ship repairing and building company. The policies incorporated the ship repairers' liability clauses of 1 April 1985. The

vessel was repaired at Port of Spain during April and May 1988 but shortly after the vessel left the yard her engine exploded and severe damage was done to the vessel. It was accepted that the explosion and the consequent damage were due to the negligent repair work done by the assured.

The assured then produced to the solicitors appointed by the insurers a forged document which the assured thought would assist the solicitors. The forgery was discovered and the insurers avoided liability for breach of the duty of utmost good faith by the assured.

The court held that the deliberate and culpable misrepresentation, which was genuine, was immaterial as it only concerned the issue of jurisdiction and had no legal relevance to the assured's claim on the policy itself. The duty of utmost good faith did not attach to the acts of the assured in producing it or the act was not a breach of the duty that gave the insurers the right to avoid the policy. Even if the misrepresentation by the assured was deliberate and culpable it was to be treated as irrelevant if it did not relate directly to the 'risque' of the insurer or the liability of the insurer for a claim under the policy, but only to some collateral matter such as the jurisdiction in which the claim against the insured was to be determined.

Representation

A representation is a statement or assertion of fact made by one party to a contract to the other, before or at the time of the contract, of some matter or circumstance relating to it. But a representation of belief, expectation, or intention may be a representation of fact. Both misrepresentation and non-disclosure are frequently pleaded together by insurer while the law of non-disclosure becomes more predominant.[32]

The Marine Insurance Ordinance, section 20 on 'Representations Pending Negotiation of Contract', states:

> (1) Every material representation made by the assured or his agent to the insurer during the negotiations for the contract, and before the contract is concluded, must be true. If it be untrue, the insurer may avoid the contract.
>
> (2) A representation is material which would influence the judgment of a prudent insurer in fixing the premium, or determining whether he will take the risk.
>
> (3) A representation may be either a representation as to a matter of fact, or as to a matter of expectation or belief.
>
> (4) A representation as to a matter of fact is true, if it be substantially correct, that is to say, if the difference between what is represented

32. J Birds and N J Hird, *Birds' Modern Insurance Law* (6th ed, 2004) 101.

and what is actually correct would not be considered material by a prudent insurer.

(5) A representation as to a matter of expectation or belief is true if it be made in good faith.

(6) A representation may be withdrawn or corrected before the contract is concluded.

(7) Whether a particular representation be material or not is, in each case, a question of fact.

A false or fraudulent misrepresentation is one made with knowledge of its falsehood, and intended to deceive. A fraudulent misrepresentation is actionable as a tort. A negligent misrepresentation is one made with no reasonable grounds for believing it to be true. An innocent misrepresentation is one made with reasonable grounds for believing it to be true, as where an honest mistake is made.

When a person has been induced to enter into a contract by misrepresentation, they may in general either:

1. Affirm the contract and insist on the misrepresentation being made good, if that is possible; or
2. Rescind the contract if it is still executory, and if all parties can be restored to their original positions; or
3. Bring an action for damages; or
4. Rely upon the misrepresentation as a defence to an action on the contract.

A contract may be rescinded on the ground of misrepresentation even if innocent. Specific performance will not be decreed if a definite untrue representation has been relied on.

Inversiones Manria SA v Sphere Drake Insurance plc, The Dora [1989] 1 Lloyd's Rep 69, QB

The policy was on a yacht. The assured made a representation that they were producing yachts at the rate of about four per annum. Representation was intended to indicate to the insurers the scale of their business.

The Court held that the representation was untrue, and was material, and, therefore, the insurers could avoid liability.

Recommendations for changes in consumer insurance contracts

The result of the Statements of Practice issued by the Association of British Insurers, the Conduct of Business (COB) and Insurance Conduct of Business

(ICOB) Rules in the Financial Services Authority (FSA) Handbook and, above all, the FOS scheme (the dispute resolution service of the Financial Ombudsman Service) mean that consumer insurance is subject to an additional set of rules that should go some way to improving the consumer's position so far as misrepresentation or non-disclosure is concerned.

There are reasons of substance that may make it sensible to treat consumer insurance separately. Consumers might be thought to deserve more favourable treatment under the law than, say, large businesses because:[33]

> (1) Consumers will almost always contract on standard policy terms which they lack the bargaining power to alter.
>
> (2) Consumers are unlikely to have access to the level of expert advice which will be available to large businesses whether in-house or externally.
>
> (3) Consumers are typically less able to understand the technicalities of insurance contract law particularly if it does not meet their reasonable expectations. FOS has pointed out that consumers may not even be aware there is an issue of law to be considered.
>
> There are two problems for consumers: firstly, they do not know there are issues of insurance contract law and secondly, in the unlikely event that they do, they are less likely to understand them than a business advised by its broker or in-house specialists.
>
> This problem is compounded by the fact that expert broker or intermediary advice is hard to find for consumers and there is little understanding of the need to obtain this given the ignorance of the pitfalls in insurance contract law. In this regard, it is important to bear in mind that the reasonable expectations of the consumer may well be based upon their expectation of the law in relation to all, or nearly all, other consumer transactions they may enter into. Applying for an insurance contract may be the only transaction to acquire goods or services they enter into in any year in which they are expected in law to assume duties and responsibilities that are not known or explained to them and, where they are not required to seek expert assistance for their own benefit.

Where misrepresentation or non-disclosure is alleged, the ABI (Association of British Insurers) Statements and FSA Rules may be of assistance to a well-informed consumer. However, it is the FOS which ultimately offers the consumer the most valuable protection. In particular, it addresses the harshness of the law in three important ways:[34]

33. Issue Paper 1, p 31.
34. Issue Paper 1, p 43.

(1) The FOS treats consumers as if there is no residual duty of disclosure. Neither the Statements nor the Rules have this effect.
(2) The FOS will require an insurer to settle claims on a proportionate basis where there has been 'inadvertent' misrepresentation and the insurer, had it been aware of the truth, would simply have accepted the application at an increased premium. In contrast, both Statements and Rules allow avoidance for 'negligent' misrepresentation.
(3) The FOS does not offer an insurer any remedy in cases of innocent misrepresentation. It appears to take this line regardless whether there has been a claim or not whereas corresponding provisions in the Statements and Rules only apply where a claim has been made.

The Law Commission's 1980 report and the others that followed it identified a number of ways in which the law is unfair. It is not suggested that insureds are being cheated; the market for insurance is highly competitive and no doubt they get value for money. What seems unfair about the law is that it defeats the reasonable expectations of consumers. Insurance is intended to be an effective risk-transfer mechanism, bringing peace of mind to the purchaser. The consumer exchanges the risk of a loss of unknown amount for the payment of a known premium. This process performs a valuable function in enabling consumers to plan their financial affairs prudently. The present system is unsatisfactory in four ways:[35]

(1) It fails to prevent some insurers avoiding policies inappropriately. Insurers may repudiate claims for negligent misstatements that have nothing to do with the claim in issue.
(2) It is inaccessible and obscure. To understand how the current law works, insurers and consumer advisers need to work through multiple layers of law, self-regulation, regulation and ombudsman case law. Not only do the rules differ but the meaning of some of the individual rules, and thus the differences intended, are unclear. It is hardly surprising that insurers often do not understand the rules that govern this area, and attempt to avoid policies inappropriately.
(3) Consumers are deprived of a genuine choice between the FOS and the courts. Except in limited circumstances, no well-advised consumer would choose to take a case to court, because a court would be forced to apply the full rigour of the law.
(4) It requires the FOS to exercise undue discretion. Instead of using its discretion to supplement the law, the FOS has been required to develop its approach almost from first principles. The FSA Rules provide little assistance as COB and ICOB do not and are

35. Issue Paper 1, p 49.

not intended to deal with anything like the breadth and extent of disputes that arise in insurance policies.

The Law Commission concluded that the time has come to state clearly what rights and obligations a consumer insured should have. It believes that a new statute should replace the current patchwork by a single system that is coherent and clear, and that is fair in that it meets the reasonable expectations of both consumers and insurers.[36]

Policy

The policy is deemed to be concluded at the time when the slip is accepted and initialized by the underwriters. The Marine Insurance Ordinance section 21 on 'When contract is deemed to be concluded' states:

> A contract of marine insurance is deemed to be concluded when the proposal of the assured is accepted by the insurer, whether the policy be then issued or not; and, for the purpose of showing when the proposal was accepted, reference may be made to the slip or covering note or other customary memorandum of the contract.

> *Eagle Star Insurance Company Ltd v Spratt* [1971] 2 Lloyd's Rep 116
> The plaintiffs, an insurance company, reinsured certain risks with (inter alia) Lloyd's underwriters. On re-negotiation of terms between a committee of underwriters and the plaintiffs, the underwriters approved terms of settlement, enclosing a draft addendum and signing for the plaintiffs. The signatures of the leading underwriters were obtained thereto by the placing brokers. The placing brokers then took to Lloyd's Policy Signing Office (L.P.S.O.) (*inter alia*) (i) the original slips initialled by all underwriters; (ii) statements of the proposed terms initialled by all underwriters; (iii) the addenda initialled by the leading underwriters. The addenda were stamped and signed by L.P.S.O. on behalf of all underwriters. The defendant subsequently asserted that he was not bound by any settlement with the plaintiffs and purported to avoid the policies. The plaintiffs commenced an action against the defendant, claiming a declaration that he was so bound, and, on the preliminary issue, whether the defendant was bound by an agreement alleged to have been made on his behalf with the plaintiffs.
>
> The Court of Appeal held that, according to the practice of Lloyd's, an agreement is made when the slip is signed, and that, therefore, there was an agreement made by S.-P. on 25 June, and since S.-P. did have the ostensible authority of the defendant to conclude such an agreement, that was binding on the defendant. Appeal allowed.

36. Issue Paper 1, p 50.

Under section 21 of the Marine Insurance Ordinance, a contract of insurance may exist separately of an insurance policy as a contract of marine insurance is deemed to be concluded when the proposal of the assured is accepted by the insurer, whether the policy be then issued or not. A contract of marine insurance is not valid unless the same is expressed in a policy of marine insurance, despite the fact that the slip in connection with the contract of insurance has properly been initialled by underwriters, because only the policy itself may be sued upon in the court. The Marine Insurance Ordinance section 22, 'Contract must be embodied in policy', states:

> Subject to the provisions of any Ordinance, a contract of marine insurance is inadmissible in evidence unless it is embodied in a marine policy in accordance with this Ordinance. The policy may be executed and issued either at the time when the contract is concluded, or afterwards.

In *Eide U.K. Ltd v Lowndes Lambert Group Ltd, The Sun Tender* [1998] 1 Lloyd's Rep 389, Lord Justice Phillips at p 397 states: 'The significance of this provision lay in other provisions of the Act which required the policy to be stamped—provisions repealed by the Finance Act, 1959. . . . The Court would normally be expected to insist upon the production of a policy where a claim was made under it. Quite apart from this it was generally considered that the underwriter could not be required to pay under a policy unless the policy was produced. Thus, in 1906, the possessory lien appeared a highly effective security, for it prevented the assured from recovering under a policy until a lien secured by the policy was discharged.' It is unclear whether section 22 of the Marine Insurance Act 1906 has the effect of rendering a contract of marine insurance unenforceable if the assured cannot produce the policy.

In *Swan Cleland's Graving Dock and Slipway Co v Maritime Insurance Co* [1907] 1 KB 116, Mr. Justice Channel dismissed a defence plea that the underwriters were not liable because the plaintiff, an assignee, was unable to produce the policy. This decision, coupled with the subsequent abolition of the stamping requirement, has had the effect of reducing the value of the possessory lien on the policy, although production of the policy is sometimes a contractual condition precedent to the underwriters' liability (*Eide U.K. Ltd v Lowndes Lambert Group Ltd, The Sun Tender* [1998] 1 Lloyd's Rep 389.

A policy must contain the name of the assured or the agent. If the agent effects the policy, it is only the person or persons who is/are intended to be covered that will be able to claim under the policy. As a bill of lading can be resold in international trade, a seller or the agent of a seller may not exactly know who will finally take delivery of the goods at the destination. It may be the case that the party who collects the goods may not be the party on the policy. See section 23 of the Marine Insurance Ordinance, on 'What policy must specify', which states:

> A marine policy must specify the name of the assured, or of some person who effects the insurance on his behalf.

In *National Oilwell (UK) Ltd v Davy Offshore Ltd* [1993] 2 Lloyd's Rep 582, an issue arose as to whether the plaintiffs (NOW), whose name did not appear on the policy, were a party to the contract of insurance in respect to the subrogation clause. The subrogation clause stipulated:

> Underwriters agree to waive rights of subrogation against any Assured and any person, company or corporation whose interests are covered by this policy and against any employee, agent or contractor of the Principal Assureds or any individual, agent, firm affiliate or corporation for whom the Principal Assureds may be acting or with whom the Principal Assureds may have agreed prior to any loss to waive subrogation, including but not limited to helicopters, supply boats etc., existing installation(s) and tugs and/or insurers. The foregoing shall not apply in respect of operations not connected with the project. (*National Oilwell (UK) Ltd v Davy Offshore Ltd* [1993] 2 Lloyd's Rep 582 at p 591)

There can be no doubt that NOW as supplier to the defendants (DOL) of the equipment under the agreement was a sub-contractor and was a company with whom DOL, the assured under the policy, had entered into an agreement or contract in connection with the subject matter of the insurance and/or any works, activities or preparations connected with that subject matter. It, therefore, fell within the description of 'Other Assureds' under the definition of 'Assured' in the insurance conditions of the policy. That is to say, NOW qualified for inclusion in the class of other assureds. The waiver of subrogation clause by the words 'against any assured and any person company or corporation whose interests are covered by this policy' confined the effect of the waiver to claims under the policy for losses which were insured for the benefit of the party claimed against. The waiver clause operated consistently with the commercial purpose of that contract if its meaning was confined to the waiver of claims based on losses insured for the benefit of NOW, i.e., pre-delivery losses; and, in as much as the subrogated claims advanced against NOW were based on losses arising in relation to particular items of equipment after delivery to DOL of that equipment, the waiver clause did not preclude, or provide a defence in respect of, such claims.

A marine policy should be signed by the insurer or an agent on behalf of the insurer. It is sufficient for a corporation to use the corporate seal on a marine policy. The Marine Insurance Ordinance section 24(1), on 'Signature of insurer', states:

> A marine policy must be signed by or on behalf of the insurer, provided that in the case of a corporation the corporate seal may be sufficient, but nothing in this section shall be construed as requiring the subscription of a corporation to be under seal.

In *Touche Ross & Co v Colin Baker* [1992] 2 Lloyd's Rep 207, Lord Mustill at p 209 stated: 'Every underwriting member of Lloyd's incurs liability to the assured on a basis which is several not joint. The concluding words of each policy express this ancient principle—We, the Underwriters . . . hereby bind ourselves each for his own part and not one for another . . . in respect of his due proportion only . . . and it is now embodied in section 8(1) of Lloyd's Act, 1982.'

In *PT Buana Samudra Pratama v Maritime Mutual Insurance Association (NZ) Ltd*,[37] the tug was insured under a policy of marine insurance by three underwriters: M, Aegis and Axa as the lead underwriter. Under the policy it was agreed 'to follow Axa in respect of all decisions, surveys and settlements regarding claims within the terms of the policy, unless these settlements are to be made on an ex gratia or without prejudice basis'. The policy permitted the vessel to be employed in operations which entailed cargo loading or discharging at sea from or into another vessel. The Court held that on the correct construction of a 'follow the leader' clause in a policy of marine insurance it obliged the following market to follow settlements both as to liability and quantum and also applied where it was said that there had been an antecedent breach of warranty.

> *PT Buana Samudra Pratama v Maritime Mutual Insurance Association (NZ) Ltd*[38]
>
> A tanker owned by a related company ran aground and the applicant tug owner (B) sent the tug to tow it to Indonesia for tank cleaning and repairs. En route the tug ran aground and was declared a constructive total loss. Axa paid B's claim on the policy. The respondent insurer (M) rejected any liability because, in breach of warranty, the tug had been engaged to provide towage and/or salvage services to the tanker.
>
> The Court held that the application was refused. The wording indicated that all settlements would be followed, save for those expressly

37. [2011] EWHC 2413 (Comm).
38. ibid.

stated, and that the process would include issues of both liability and quantum. To impose a limited obligation to follow settlements only as to quantum would require much clearer language than that found in the clause.

Designation of subject matter

The Marine Insurance Ordinance section 26, on 'Designation of subject-matter', states:

(1) The subject-matter insured must be designated in a marine policy with reasonable certainty.
(2) The nature and extent of the interest of the assured in the subject-matter insured need not be specified in the policy.
(3) Where the policy designates the subject-matter insured in general terms, it shall be construed to apply to the interest intended by the assured to be covered.
(4) In the application of this section, regard shall be had to any usage regulating the designation of the subject-matter insured.

Under section 26 of the Marine Insurance Ordinance, it is required that the subject matter should be designated in the policy with reasonable certainty. It is important, therefore, that the assured should describe the subject matter clearly and sufficiently for underwriters to understand the nature of the risks. General terms such as ship, freight, goods, hull and machinery may be used, but it is essential, particularly in the case of goods, that the assured should not mislead or confuse underwriters. A mis-description may lead to underwriters being able to avoid the policy on the ground of either non-disclosure (see s 18) or misrepresentation (see s 20). In the absence of fraud or illegality, the assured would be able to recover the premium paid, but would not be entitled to recover or claim under the policy.

Anglo-African v Bayley [1969] 1 Lloyd's Rep 268
An insurance was effected against all risks on plaintiffs' goods (unused 20 year old army surplus leather jerkins in bales). The plaintiffs disclosed that the goods were 'new' government surplus leather jerkins. During the policy period, 245 bales were lost and the plaintiffs claimed under the policy.

The court held that, on the evidence, there was no doubt that the underwriters were not told that the goods were government surplus or that they were at least 20 years old, and the description of the goods in this policy would not have put a prudent underwriter on inquiry as to the precise nature of the goods.

Overseas Commodities v Style [1958] 1 Lloyd's Rep 546

Where on two 'All risks' cargo policies, they included a term 'Warranted all tins marked by manufacturers with a code for verification of date of manufacture', many tins were not marked in the stipulated manner.

The court held that that the use of the word that 'for verification' meant that the tins must be marked in a manner which would identify the actual date of manufacture, and that there was a breach of warranty in that a substantial number of the tins were not marked in a code which enabled the true date of manufacture to be established.

Where a general description is used, usage or custom may be relied upon to show what is intended to be covered.

Denoon v Home and Colonial Assurance Co (1872) LR 7 CP 341

The plaintiff's ship was employed for the conveyance of 360 coolies [Cooly or coolie, Which means an unskilled native labourer in Eastern countries] and 1,200 bags of rice from Calcutta to Mauritius. The plaintiff effected a marine insurance and declared the interest to be on freight valued at 20,00l.

The intention of the plaintiff in effecting this insurance was to insure the freight of the rice only, but was not communicated to the defendants. No binding custom of trade limiting the meaning of the term freight was proved, but the most frequent course in the insurance business, where the freight of coolies is intended, is to describe it as freight of coolies, or passage money of coolies, or by some other term distinguishing it from freight of merchandise; the rate of the premium differs for the insurance of passage money of coolies and freight of goods.

The ship was wrecked, and there was a total loss of the rice and consequently of the freight of the rice, but the coolies, with the exception of twelve, were saved, and their passage money, which was payable on arrival, paid.

The court held that in the present case, the term 'freight' did not include passage money and consequently there was a total loss of the freight insured by the policy.

Under section 26(3), it is provided that where the policy describes the subject matters in general terms the policy is to be understood as applying to the interest intended by the assured. What this means is that the policy will cover the actual interest the assured has in the subject matters when entering into the policy.

The requirement for a policy in writing was originated in preventing tax evasion. In 1795, a stamp duty was imposed on marine insurance and it did not tax the insurance contract as such, but the document, which had to be stamped. To make sure that the parties paid the duty, a succession of statutes declared insurance contracts to be invalid without a formal stamped policy.[39] This stamp duty was abolished in 1970 (Finance Act 1970, Schedule 1(2)(b)) but section 22 remains. This introduces a technicality with potentially drastic consequences for the insured. It provides that where there is no marine policy a person is not allowed to prove the terms of the contract in court, and therefore, cannot prove their right to make a claim. This appears to have little effect in practice. Insurers will pay claims whether the insured has a written policy document or not. However, there is always the possibility an insurer could invoke section 22 to the disadvantage of an insured. Given that the rationale for section 22 has disappeared, it is hard to defend its continued existence and a reform should be seriously considered. The provision only applies to marine insurance. In other forms of insurance, an insurance contract does not need to be in any particular form. The cases show that section 22 does not require an insured to possess a policy to bring a claim. He or she simply needs to prove that the policy has been executed.[40] But what if a policy has not been executed? The courts have resisted attempts to circumvent section 22 by arguing that the insurer is in breach of a collateral contract to execute a policy.[41]

The Marine Insurance Act 1906 makes a distinction between a 'policy' and a 'slip' which is the initial document creating the contract.[42] It allows the court to look at the slip, but only where there is a 'duly stamped policy'.[43] The Act does not define a 'policy', though it does require the policy specifies the name or agent of the insured, and is signed by the insurer. It is argued whether nowadays a slip would be enough to fulfil the need for a policy, provided it meets these requirements. It may be sufficient, but the law is far from clear. The insurance industry has implemented various initiatives to improve the way contracts are agreed, culminating in two significant developments: the Contract Certainty Code of Practice,[44] and the Market Reform Contract. The uncertainty surrounding section 22 has the potential to undermine these initiatives.[45]

39. Stamp Act 1795, s 22; Stamp Act 1891, s 93. See *Warwick v Slade* (1811) 3 Camp 127.
40. *Swan and Cleland's Graving Dock and Slipway Co v Maritime Insurance Co and Croshaw* [1907] 1 KB 116.
41. *Fisher v Liverpool Marine Insurance Co* (1872–73) LR 8 QB 469; Law Commission and Scottish Law Commission (2010) *Reforming Insurance Contract Law Issue Paper 9 The Requirement for a Formal Marine Policy: Should Section 22 Be Repealed?*
42. *Ionides v Pacific Fire & Marine Insurance Co* (1871) LR 6 QB 674.
43. Marine Insurance Act 1906, s 89.
44. 'Contract Certainty Code of Practice: Principles and Guidance', June 2007.
45. Law Commission and Scottish Law Commission, *Reforming Insurance Contract Law Issue Paper 9 The Requirement for a Formal Marine Policy: Should Section 22 Be Repealed?* (2010).

10.5 MARINE POLICIES AND ELECTRONIC COMMERCE

In 2001, the Law Commission said that a policy had to be tangible to satisfy section 22.[46] This is because the 1906 Act gives the broker a lien over the policy, and a lien can only attach to a tangible object.[47] The Law Commission took the view that this does not necessarily prevent an electronic document from meeting the requirements for a policy. A policy may exist, even if it is in a form the broker cannot take a lien over. Another difficult issue is what is meant by the insurer's signature under section 24(1). In 2001, the Law Commission took a broad view. It thought signatures could include digital signatures, scanned manuscript signatures and typed names. Since 2001, several cases have supported the view that typing one's name into a document will suffice, provided the necessary authenticating intention is evinced.[48] However, until a ruling by the higher courts, the issue is not beyond doubt.[49]

Valued policy

Under section 27 of the Marine Insurance Ordinance a policy may be valued or unvalued, and it is provided that a valued policy is a policy which sets out the agreed value of the subject matters as in section 27(2). The Marine Insurance Ordinance section 27, on 'valued policy', states:

> (1) A policy may be either valued or unvalued.
> (2) A valued policy is a policy which specifies the agreed value of the subject-matter insured.
> (3) Subject to the provisions of this Ordinance, and in the absence of fraud, the value fixed by the policy is, as between the insurer and assured, conclusive of the insurable value of the subject intended to be insured, whether the loss be total or partial.
> (4) Unless the policy otherwise provides, the value fixed by the policy is not conclusive for the purpose of determining whether there has been a constructive total loss.

Where no such agreed value appears on the policy it will be an unvalued policy. It is possible in the case of a valued policy for the sum insured to be less than the agreed value. In such a case, assureds will be found to be under-insured and by section 81 of the Marine Insurance Ordinance assureds will

46. Law Commission, Electronic Commerce: Formal Requirements in Commercial Transactions, December 2001.
47. *Tappenden v Artus* [1964] 2 QB 185 at p 195.
48. *Mehta v J Pereira* [2006] EWHC 813 (Ch); *Orton v Collins* [2007] EWHC 803 (Ch); *Lindsay v O'Loughnane* [2010] EWHC 529 (QB).
49. ibid.

be held to be their own underwriters in respect of the difference. Normally, the insured value and the agreed value correspond with each other.

> *Lewis v Rucker* (1761) 2 Burr 1167
> Where it was held that an agreement by the parties as to the basis of indemnity did not render the contract of insurance void as a gaming or wagering contract.

For example, a shipowner purchases a vessel at a sum of US$1,000,000 on a sale and purchase contract. The shipowner may agree with underwriters that 100 per cent of the value of the vessel is US$1,000,000 or US$1,150,000, which may include other transaction costs to acquire the vessel. Underwriters may always query the actual value of the vessel purchased by the shipowner, who should prove the price paid is a market price.

> *Kyzuna Investments Ltd v Ocean Marine Mutual Insurance Association (Europe)* [2000] 1 Lloyd's Rep 505
> The claimants acquired the yacht. A proposal of insurance was accepted and the policy that covered the yacht provided, inter alia, the sum insured for hull machinery, gear, equipment etc., dinghy and/or boat and flatacraft gave a total sum insured.
>
> The court held that if the whole of the clauses were read together and the proposal form considered, there was nothing that pointed to the intention of the parties that the sums stated in the schedule were to be the agreed value of the yacht and her equipment; the words 'sum insured' ordinarily indicate a ceiling on recovery in an unvalued policy; there was nothing which displaced this ordinary meaning and the policy did not specify, in accordance with s 27(2) of the Marine Insurance Act 1906, the agreed value of the yacht. The policy was an unvalued policy.

It is important to note that an agreed value should appear on the policy if the parties intend to have a valued policy. In Kyzuna Investments, the court stated that the draftsman of the part of the policy specifically drafted by the defendant underwriters referred in the definition of special equipment to 'the sum insured' and not an insured value. The main insuring clause expressed the agreement of the underwriters to indemnify 'up to the amounts and/or limits contained herein'. The words actually used in the schedule were 'the sum insured' and far from meaning the value had been agreed. It represents the ceiling on recovery in respect of the liability of the underwriters.

The references in the Institute Clauses to an insured value were references in standard clauses applicable only if there was an agreed value. The reference in the clause defining a constructive total loss to the insured value of the vessel could not, in the face of the terms of the general insuring clause and the special equipment clause, mean the policy as a whole was to be read as a valued policy. The parties had not made their intention clear that the sum in the schedule was an agreed value.

The court further stated it was common for underwriters to insist on a valuation and the vessel concerned was an unusual vessel. The underwriters wanted some independent valuation before considering insuring her, and they were not thereby indicating they were prepared to agree to a valued policy. Although the proposal contained the words 'value to be insured', this was not an indication that the value so stated would be agreed as the insured value by the underwriters.

Where 100 per cent of the value of the vessel is agreed between the underwriters, say the subject matter is US$1,000,000, the next step is that the shipowner may decide whether to insure 100 per cent of the value with the underwriters or only part of the value, say 80 per cent. Where the shipowner insures only 80 per cent of the value, the agreed value appearing on the policy is US$800,000. The balance, 20 per cent, of the value of the vessel will be insured by the assured, i.e., the shipowner. We call it self-insurance or under-insurance. Insofar as over-valuation is concerned, it is possible for an underwriter to claim that gross over-valuation is in fact either a non-disclosure of the true value or a misrepresentation of the true value. However, in cases of gross over-valuation, it is possible for underwriters to claim there has been fraud on the part of the assured.

> *Ionides v Pender* [1872] 27 LT 244
>
> The adventure was expected to be enormously profitable, the profit being variously estimated at from 80 to 125 per cent. To secure these profits, it was admitted that the goods had been overvalued to the extent of 25 to 30 per cent, and there were heavy insurances of commissions. The shipment was lost in fine weather in mid-ocean without any known cause.
>
> The court held that an insurance on profit must be taken to mean possible profit. Excessive valuation is almost conclusive evidence of a fraudulent intent. The slips mentioned that profits were to be insured 'however high they might be'. No further notice of the over-insurance was given to the underwriter. The jury found that the over-valuations were excessive and material, and were concealed from the underwriters.

The more likely area for claiming non-disclosure by underwriters in respect of the over-valuation of a ship arises in the case of smaller vessels such as yachts or pleasure boats, in that underwriters may not have sufficient information to know the real value of the craft. In the case of commercial merchant vessels, underwriters should have sufficient knowledge to know whether the vessel is over-valued.

> *Slattery v Mance* [1962] 1 Lloyd's Rep 60
> The assured had earlier tried to sell his yacht at the sum (£2,250) less that given (£4,500) in the policy. The yacht was destroyed by fire as a total loss and the insurer denied liability.
> The court held that the representation as to the value of the yacht made by him in the proposal form was untrue and that the representation was material but not made fraudulently.

It is possible for the court to refuse to deal with a case where the differences between the agreed value and the real value of the subject matters is such as to show it is in reality a wagering policy (see s 4 of the Marine Insurance Ordinance). If underwriters wish to void a policy on the ground of unreal valuation, then they must specifically plead a defence by way of fraud or non-disclosure, misrepresentation or even illegality. In fact, in relation to merchant vessels, over-valuation is an accepted practice. It, of course, results in a higher premium, knowing the regard of the assured and, also under the Institute Clauses it is more difficult to show a constructive total loss. Clause 19 of the Institute Time Clauses Hulls clause 280 dated 1/10/83 provides that the repaired value of the vessel is deemed to be the insured value.

Generally, the sum insured or the insured value of a vessel or subject matter remains the same depending on the length of the policy. Underwriters will sometimes accept changes of the value of the vessel during the policy period. Insurance markets do not insist the agreed value has to continually reflect the market value of a vessel, which fluctuates within a policy period. Under the common law, without applying the Institute Clauses, the actual value of the vessel when repaired is referred to, and if the cost of repair exceeds that sum then it would be a constructive total loss.

> *Helmville Ltd v Yorkshire Insurance Company Ltd, The Medina Princess* [1965] 1 Lloyd's Rep 361
> Damage was sustained by the plaintiffs' ship during the voyage from Bremen to China laden with a cargo of flour, and the ship was insured under a valued policy containing the Institute Time Clause Hull. The ship was insured for a value of £350,000, and at the time when the policy commenced it was probably worth £65,000, and at the time of loss, £62,500. The damaged value of the ship was nil.
> The court held that (i) the plaintiffs had failed to prove that the Medina Princess was a constructive total loss, and that plaintiffs had, thus far, established a partial loss amounting to £152,713; (ii) that

> the Medina Princess had not been repaired or sold during the risk and therefore, prima facie, the measure of indemnity was the depreciation (not exceeding the reasonable cost of repairs) arising from the unrepaired damage (Marine Insurance Act 1906, s 69(3)), but, as the damaged value of the Medina Princess was virtually nil, the indemnity would be limited to the reasonable cost of repairs.

The policy contained a constructive total loss clause, and in order to prove a constructive total loss it was necessary to show the cost of repair would exceed £350,000. Had the common law operated under section 60 of the Marine Insurance Act, then the vessel in the above case would have undoubtedly been a constructive total loss. Under section 27(3) of the Marine Insurance Ordinance, it is provided that where there is an agreed value in the policy, then in the absence of fraud the value is conclusive as between the assured and the underwriters. This is to prevent underwriters from pleading non-disclosure, misrepresentation, breach of warranty or other possible defence against the assured. Under section 29(4) of the Marine Insurance Ordinance, in the case of a floating policy, the value of the goods shipped is not conclusive if the valuation is not declared before the loss of the subject matter. The Marine Insurance Ordinance section 29(4), on 'Floating policy by ship or ships', states:

> Unless the policy otherwise provides, where a declaration of value is not made until after notice of loss or arrival, the policy must be treated as an unvalued policy as regards the subject-matter of that declaration.

Under section 75(2) of the Marine Insurance Ordinance, underwriters are given the right to show in respect of a claim that the assured had a partial interest or no interest at all in the subject matter, or that the whole or part of the subject matter was never at risk. The Marine Insurance Ordinance section 75(2), on 'General provisions as to measure of indemnity', states:

> Nothing in the provisions of this Ordinance relating to the measure of indemnity shall affect the rules relating to double insurance, or prohibit the insurer from disproving interest wholly or in part, or from showing that at the time of the loss the whole or any part of the subject-matter insured was not at risk under the policy.

Under section 27(4) of the Marine Insurance Ordinance, unless the policy provides otherwise, the value agreed in the policy is not conclusive in determining a constructive total loss under section 60. However, Clause 19 of the Institute Time Clause Hull alters this provision. Under section 32(2)(c), in dealing with double insurance assureds must give credit for any money received under other policies in respect of the same subject matters and risks, and they are bound by the value stated in the policy under which they now claim. If they have already received money in excess of the value stated in the policy which they are claiming under, they will not receive anything under another policy even though the real value of the subject matter may be far greater than the value stated in the particular policy.

Unvalued policy

Where the policy merely states the amount of the sum insured without including an agreed value, it is an unvalued policy. In such a case, reference must be made to section 16 of the Marine Insurance Ordinance, which deals with the measure of an insurable value. It is almost inevitable that on an unvalued policy, in the event of loss by perils insured against, the assured will be found to be either over-insured or under-insured. In the case of over-insurance under an unvalued policy, a proportionate part of the premium is returnable under section 84(3)(e) of the Marine Insurance Ordinance. The Marine Insurance Ordinance section 84(3)(e), on 'Return for failure of consideration', states:

> Where the assured has over-insured under an unvalued policy, a proportionate part of the premium is returnable.

In the case of under-insurance under an unvalued policy, then section 81 provides that the assureds are their own underwriters in respect of the under-insurance. Note that this under-insurance percentage must be taken into account in respect of claims, salvage, general average or recovery under the sue and labour provisions, because the assured is the underwriter in respect of the part of under-insurance. The Marine Insurance Ordinance section 81, on 'Effect of under insurance', states:

> Where the assured is insured for an amount less than the insurable value or, in the case of a valued policy, for an amount less than the policy valuation, he is deemed to be his own insurer in respect of the uninsured balance.

Under section 28 of the Marine Insurance Ordinance, it is set out that the unvalued policy is one which does not give the value of the subject matter but merely provides the limit of recovery to the sum insured.

> *Continental Illinios National Bank & Trust Co of Chicago v Bathurst, The 'Captain Panagos D.P.'* [1985] 1 Lloyd's Rep 625
>
> The shipowners insured the vessel under a hull policy in the sum of US$9,000,000, and they obtained a mortgage from a bank which

insured the same vessel in respect of the bank's interest as mortgagees for the sum of US$6,859,200. The vessel agrounded in the Red Sea, caught fire and suffered damage. An issue arose as to whether the amount was due under the mortgagees' interest policy to be ascertained by reference to the amount which would have been recoverable under the hull policy, i.e. US$9,000,000, subject to an upper limit of US$6,859,200.

The court held that (1) the general tenor (general meaning) and shape of the mortgagees' interest policy spoke more of an insurance against physical damage to (and liability of) the ship than against financial damage to the mortgagees' interest, and the value stated in the hull policy was irrelevant; (2) the purpose of the 'insured amount' in the mortgagees' interest policy was not to provide a basis for calculation of the claim but to fix a figure which was to act as both a point of reference for the calculation of the premium and as an upper limit to the recovery. There was nothing in the policy showing an intention to prescribe that the vessel was deemed to have an agreed value equal to the amount insured.

Insurable interest

It is not true that anyone who can pay may buy a policy or enter into a contract of insurance. It is stated in section 4 of the Marine Insurance Ordinance, on 'Avoidance of wagering or gaming contracts', that:

> (1) Every contract of marine insurance by way of gaming or wagering is void.
>
> (2) A contract of marine insurance is deemed to be a gaming or wagering contract–
>
> (a) Where the assured has not an insurable interest as defined by this Ordinance, and the contract is entered into with no expectation of acquiring such an interest; or
>
> (b) Where the policy is made 'interest or no interest', or 'without further proof of interest than the policy itself', or 'without benefit of salvage to the insurer', or subject to any other like term:
>
> provided that, where there is no possibility of salvage, a policy may be effected without benefit of salvage to the insurer.

An 'insurable interest' means that an assured will suffer loss or incur a liability if the subject matter insured is lost, damaged, detained or fails to arrive on time. If the assured does not have an insurable interest in the subject matters insured or has no expectation of acquiring such an interest, then the policy is deemed to be one considered as a contract of gaming or wagering and is void.

An 'insurable interest' is defined in section 5 of the Marine Insurance Ordinance. It states:

> (1) Subject to the provisions of this Ordinance, every person has an insurable interest who is interested in a marine adventure.
> (2) In particular a person is interested in a marine adventure where he stands in any legal or equitable relation to the adventure or to any insurable property at risk therein, in consequence of which he may benefit by the safety or due arrival of insurable property, or may be prejudiced by its loss, or by damage thereto, or by the detention thereof, or may incur liability in respect thereof.

An assured will benefit from the marine adventure if it goes well or will suffer loss if there is any loss, damage or detention of the subject matter insured, or if incurring liability in respect of the subject matter. An assured may insure against liability to third parties arising out of a marine adventure, e.g., collision damage caused by the ship of the assured to a terminal or buoy of a third party. As to a right of an assured to insure an equitable interest in a marine adventure, see:

> *Samuel v Dumas* [1924] AC 431
> During the currency of the marine policy, the ship was scuttled by the master and crew, or some of them, with the connivance of the owner, but without any connivance or complicity on the part of the mortgagee. In an action on the policy by the plaintiffs on behalf of the mortgagee, the court held that the mortgagee had an insurable interest in the ship.

Although an assured may have suffered a loss by reason of a peril insured against, and although the loss has taken place within the period covered by the policy, there will be no right of recovery if the nature of the assured's interest falls within the provision of section 4 of the Marine Insurance Ordinance relating to gaming and wagering, nor if the assured cannot satisfy the definition of insurable interest as in ss 5–6 of the Marine Insurance Ordinance and the general scope of insurable interest contained in ss 7–14.

> *Seagrave v Union Marine Insurance* [1866] LR 1 CP 305, at 326
> Willes J stated: 'The general rule is clear, that to constitute interest insurable against a peril, there must be an interest such that the peril would, by proximate effect, cause damage to the assured.'

> *Lucena v Craufurd* (1806) 2 Bos & PNR 269, at 302
> Lawrence J stated: 'To be interested in the preservation of a thing, is to be so circumstanced with respect to it as to have benefit from its

existence prejudice from its destruction.' It has also been said the prospect of loss or gain not founded on any right or liability or in respect of the subject matters insured is not insurable.

When an assured must have an insurable interest

By section 6 of the Marine Insurance Ordinance, it is provided that the assured must have an insurable interest in the subject matters insured at the time of loss. It does not affect the right of the assured to claim against an underwriter whether the assured has an insurable interest or not at the time when the insurance is effected. It is possible but very rare the assured effects a policy covering a ship while not the owner of the ship. In a sale and purchase of a ship, the purchaser will have an insurable interest in the ship from the time of acquiring actual ownership. An underwriter normally accepts the risks of a hull and machinery policy on the basis of the identity of shipowner unless the underwriter and shipowner mutually agree to include the loss statistics of a ship, which is technically operated by a ship manager, under the record of the shipowner. It is common practice for a potential shipowner to arrange an insurance contract before taking delivery of the ship.

> *Piper v Royal Exchange* (1932) 44 Ll L Rep 103
>
> A yacht in Norway was bought 'as she lies' by the plaintiff, and was at the risk of the seller until she arrived in London. The plaintiff effected a policy in respect of her and claimed against the insurers for damage which she had suffered by stranding. The insurers counterclaimed (which they had paid in respect of damage suffered by her on the voyage from Norway to London) on the ground that he had no insurable interest in the yacht. The court held that the counterclaim succeeded on this ground.

The Marine Insurance Ordinance section 6, on 'When interest must attach', states:

> (1) The assured must be interested in the subject-matter insured at the time of the loss though he need not be interested when the insurance is effected:
>
> Provided that where the subject-matter is insured 'lost or not lost', the assured may recover although he may not have acquired his interest until after the loss, unless at the time of effecting the contract of insurance the assured was aware of the loss, and the insurer was not.
>
> (2) Where the assured has no interest at the time of the loss, he cannot acquire interest by any act or election after he is aware of the loss.

Cepheus Shipping Corporation v Guardian Royal Exchange Assurance Plc (The 'Capricorn') [1995] 1 Lloyd's Rep 622

The plaintiffs were the registered owners of the reefer vessel and the defendants were party to loss of hire insurance. Section 2 of the general conditions of the insurance stipulated, *inter alia*, that it covers loss due to the vessel being wholly or partly deprived of her earning capacity as a consequence of damage sustained. Return of premium was provided if the vessel has been laid up in a safe port . . . shall be returned depending on whether or not repairs . . . of the vessel have been carried out. At Falmouth the crankshaft was replaced and the generator repaired between 6 June and 7 October 1986 while the vessel lay at a lay-up berth. The plaintiffs claimed under the policy in respect of 60 days' loss of time.

The defendants argued that the policy was not to be read as covering loss which the vessel would have sustained, damage or no damage, because she would anyway have been out of the market. They submitted that the vessel was due to be and would have been laid up at Falmouth throughout the low season of 1986 and that the plaintiffs had no insurable interest within s 6 of the Marine Insurance Act 1906.

The court held that under s 6 of the Marine Insurance Act 1906, the plaintiffs' insurable interest in the subject matter insured (freight and income from trading) must have existed at the time of loss. The loss of earnings which the insurance contemplated would have been sustained over the early months of the off-season period. It was common ground that questions of insurable interest fell to be considered on a continuing or day-to-day basis, but even at the date of the accident it was to all intents and purposes clear that the charterers would not exercise their off-season option and the plaintiffs' intention on that basis was to lay her up. It was not necessary to decide that parties may never agree on or value such an insurable interest but there was a distinct unlikelihood about their so doing, and they had not done so in this case.

There is an exception where the policy covers the subject matter on the basis of 'lost or not lost' as the proviso of section 6(1) of the Marine Insurance Ordinance. The assured may recover under such a policy although not acquiring an interest after the loss of the subject matter. The pre-requisite of the right of the assured to recover is that the assured is not aware of the loss at the time of the formation of the policy. The loss stipulated in section 6(1) is expressed in general terms and it seems that it is applicable in cases of partial loss or total loss.

A person who has an insurable interest in the subject matters may assign the benefit of a policy to a third party after a total loss of the subject matters. It would then not be possible for any interest in the subject matters to be transferred to the third party, because the person does not have any interest in the subject matters to transfer as the subject matters have ceased to exist. The value of section 6(1) would only work in cases where there is a partial loss of the subject matters before the third party acquires an interest. In other words, assureds cannot recover in a total loss under a policy effected by them, whether 'lost or not lost' or otherwise.

If assureds have no interest at the time of loss, they cannot acquire an interest by any act or election on their part after they are aware of the loss. The effect of section 6 of the Marine Insurance Ordinance must not be confused with the provisions of Rule 1 of the Rules of Construction. Rule 1 of the Rules of Construction focuses on when the risk attaches as the loss has occurred before the formation of the contract subject to the principle of 'lost or not lost'. Section 6 of the Marine Insurance Ordinance relates to circumstances in which the interest of the assured must attach in order for the assured to recover from the insurer.

In the section of the Marine Insurance Ordinance, Rule 1 of the Rules for Construction of Policy on 'Lost or not lost' states:

> Where the subject-matter is insured 'lost or not lost', and the loss has occurred before the contract is concluded, the risk attaches unless, at such time the assured was aware of the loss, and the insurer was not.

The underwriter may have the defence in claim handling that the assured had no insurable interest, or was only partly interested in the subject matters insured in accordance with section 75(2) of the Marine Insurance Ordinance.

Gaming and wagering contracts

Section 4 of the Marine Insurance Ordinance, on 'Avoidance of wagering or gaming contracts', states:

> (1) Every contract of marine insurance by way of gaming or wagering is void.
>
> (2) A contract of marine insurance is deemed to be a gaming or wagering contract
>
> (a) Where the assured has not an insurable interest as defined by this Ordinance, and the contract is entered into with no expectation of acquiring such an interest; or
>
> (b) Where the policy is made 'interest or no interest', or 'without further proof of interest than the policy itself', or 'without benefit of salvage to the insurer', or subject to any other like term:

> Provided that, where there is no possibility of salvage, a policy may be effected without benefit of salvage to the insurer.

Where an assured effects a policy, it is required to have an insurable interest at that time, or at least an expectation of acquiring such an interest in the subject matters insured. If the assured effects a policy without having such interest or expectation of an interest, the policy is a gaming or wagering policy. By section 4(1) of the Marine Insurance Ordinance, a gaming or wagering policy is void.

10.6 POLICY AND BROKER

Attachment and termination of risk

Attachment and termination of risk is commonly used to denote the period of liability of an underwriter under a contract of insurance. When a policy begins, the risks of the contract of marine insurance attach. Attachment of risk is sometimes described as inception of risk, which carries a similar meaning. Three basic questions are asked by an underwriter in relation to a submitted claim:

1. Did the assured have insurable interest?
2. Was the loss and/or damage proximately caused by the insured peril?
3. Did the loss and/or damage occur during the currency of the risk?

It is no use for an assured to be able to show the loss complained of occurred as a result of perils insured against if in fact that loss falls outside the period of the policy or the underwriters can show the whole or part of the subject matter is never at risk in the policy.

> *George Kallis (Manufacturers) Ltd* v *Success Insurance Ltd* [1985] 2 Lloyd's Rep 8, PC
>
> The buyers, George Kallis (Manufacturers) Ltd, who were manufacturers of jeans in Cyprus agreed to buy denim from a Hong Kong firm, Wantex Traders, on c.i.f. Limassol terms. In order to comply with the requirements of the contract Wantex arranged for the goods to be shipped on 'Ta Shun'. A bill of lading was issued showing the port of discharge as Limassol. The goods were not shipped on board 'Ta Shun' on the bill of lading date or indeed on that date on any other ship. However, Wantex used those bills to negotiate the credit sometime in August 1976.
>
> On 16 August 1976, the goods were included among 957 packages shipped on a vessel named 'Ta Hung'. They were then transshipped in Keelung on 'Intellect' to Cyprus. On 27 November, 'Intellect' caught

> fire and the goods suffered water damage and were thus actually totally lost.
>
> The court held that the goods in question had never been on board that ship bound for that destination; in these circumstances it was impossible to assert that the risk ever attached when the goods left the warehouse in Hong Kong. The goods were never appropriated by a contract of carriage to the insured voyage since the insured voyage was from Hong Kong to Limassol under a shipped on board bill of lading on Ta Shun; the appeal would be dismissed.

Section 43 of the Marine Insurance Ordinance relates to the acquaintance of risk to the port of departure, and in 'Alteration of port of departure' states:

> Where the place of departure is specified by the policy, and the ship instead of sailing from that place sails from any other place, the risk does not attach.

> *Success Insurance Ltd v George Kallis (Manufacturers) Ltd* [1981] HKLR 616, CA
>
> The respondent bought denim material from a Hong Kong supplier for delivery to Limassol. The insurance policies covered all risks and were subject to the Institute Cargo Clauses. The goods were shipped on another vessel to Keelung under a separate bill of lading and stored in a Customs Godown for three months. The goods were then shipped on a further vessel for Limassol but were lost en route. One of the issues that arose was whether the goods were ever on risk and whether the cover was terminated before loss.
>
> The court held that it was however necessary to look at the main object of the insurance policies and this was to cover the risks of a marine adventure, being a voyage from Hong Kong to Limassol. In the circumstances, the voyage which actually took place was fundamentally different from the marine adventure covered by the policies and hence, by virtue of s 44 of the Marine Insurance Ordinance, the risk never attached. Since the risk never attached, Clause I of the Institute Cargo Clauses was of no assistance to the respondent.

Section 44 of the Marine Insurance Ordinance, on 'Sailing for different destination', states:

> Where the destination is specified in the policy, and the ship, instead of sailing for that destination, sails for any other destination, the risk does not attach.

A risk does not attach if a ship departs from a different port or sails for a different destination from that of a policy. An assured may approach the underwriter for amendment if it is known to the assured before the voyage begins. The underwriter may or may not agree to the amendment. If the underwriter agrees to the amendment an additional premium may be requested where the risk is different from that originally anticipated by the underwriter. Under section 75(2) of the Marine Insurance Ordinance the underwriter may at any time show the whole or part of the subject matter is never at risk. Section 75(2) of the Marine Insurance Ordinance, on 'General provisions as to measure of indemnity', states:

> Nothing in the provisions of this Ordinance relating to the measure of indemnity shall affect the rules relating to double insurance, or prohibit the insurer from disproving interest wholly or in part, or from showing that at the time of the loss the whole or any part of the subject-matter insured was not at risk under the policy.

Voyage policy

Section 25 of the Marine Insurance Ordinance, on 'voyages and time policies', states:

> Where the contract is to insure the subject-matter 'at and from', or from one place to another or others, the policy is called a 'voyage policy', and where the contract is to insure the subject-matter for a definite period of time the policy is called a 'time policy'. A contract for both voyage and time may be included in the same policy.

It is rare to have a voyage policy on a vessel in respect of hull and machinery unless it is the last voyage for scrap or exceptional circumstances. See *Argo Systems FZE v Liberty Insurance Pte Ltd* (MV 'Copa Casino') [2011] EWCA Civ 1572, Court of Appeal in which the respondent had purchased a floating casino for scrap. The casino was to be towed from Alabama to India as a dead ship. The vessel was insured for the voyage under a voyage policy issued by an insurer. The policy incorporated the Institute Voyage Clauses (1983 version) and provided cover for the total loss of the vessel caused by perils of the sea.

It is usual to be specific as to the time of attachment or termination of a voyage policy. Rule 2 of the Rules for Construction of Policy on 'From' states:

> Where the subject-matter is insured 'from' a particular place, the risk does not attach until the ship starts on the voyage insured.

In order to satisfy this, the vessel should have the moorings all let go and cast off, or anchor aweighed and underway; and be intended to proceed on the voyage. Merely moving to an anchorage waiting for documentation would not fulfill the requirement of the attachment of the risk. Rule 3 of the Rules for Construction of Policy, on 'at and from', states:

> (1) Where a ship is insured 'at and from' a particular place, and she is at that place in good safety when the contract is concluded, the risk attaches immediately.
> (2) If she be not at that place when the contract is concluded, the risk attaches as soon as she arrives there in good safety, and, unless the policy otherwise provides, it is immaterial that she is covered by another policy for a specified time after arrival.
> (3) Where chartered freight is insured 'at and from' a particular place, and the ship is at that place in good safety when the contract is concluded, the risk attaches immediately. If she be not there when the contract is concluded, the risk attaches as soon as she arrives there in good safety.
> (4) Where freight, other than chartered freight, is payable without special conditions and is insured 'at and from' a particular place, the risk attaches pro rata as the goods or merchandise are shipped; provided that if there be cargo in readiness which belongs to the shipowner, or which some other person has contracted with him to ship, the risk attaches as soon as the ship is ready to receive such cargo.

If a vessel is at a particular place when the policy is concluded and is in good safety, then the 'at and from' policy attaches immediately. Where a vessel is not at the port of departure when the contract is concluded, then the policy will attach when the vessel arrives in good safety at that place. This will be so even if the vessel is covered by an earlier policy which provides a specific period of cover after arrival at the particular place.

Section 38 of the Marine Insurance Ordinance deals with good safety. Insofar as the subject matter is warranted 'well' or 'good safety' on a particular date, it is sufficient that the subject matter be safe at any time on the particular date. The Marine Insurance Ordinance section 38, on 'Warranty of good safety', states:

> Where the subject-matter insured is warranted 'well' or 'in good safety' on a particular day, it is sufficient if it be safe at any time during that day.

Insofar as the ship is concerned, what is required is that the vessel exists as a ship, even though it may be damaged.

> *Parmeter v Cousins* (1809) 2 Camp 235,
> Where the vessel was insured at and from, but when she arrived off the commencing port she was severely damaged by a storm, i.e. in a

leaky condition and unfit to take on cargo. The only way to keep her afloat was by continuous pumping, and after 24 hours in which she was being constantly damaged by the storm, the vessel was blown out to sea and lost. The court held that the risk had never attached.

As to the termination of risk in the case of a vessel, there is little guidance in the Marine Insurance Ordinance, although the Lloyd's policy provided cover for 24 hours after arriving in good safety.

Samuel v Royal Exchange (1828) 8 B & C 119; 108 ER 987

A vessel was insured from Sierra Leone to London and the insurance was to endure until she had been moored in good safety 24 hours. The vessel arrived and moored near the dock gate. She could not move until 27th, while she had arrived on 18th because of ice. In warping her towards the dock, a rope broke and she grounded and was totally lost.

The court held that the vessel remained at her moorings from 18 to 27 February on account of the ice, and not for want of an order to enter the dock. The plaintiff was entitled to recover because the mooring place was not the place of her ultimate destination. The policy did not expire when she had been there in safety 24 hours. The underwriters were not discharged by the delay.

Where a vessel has the possibility of discharging at several ports, then the policy would be deemed to end at the port of final discharge even though the vessel sails on ballast to a port within the range given on the policy.

Crocker v Sturge [1897] 1 QB 330

The plaintiffs reinsured with the defendants a portion of their risk by a policy which was expressed to be 'at and from Newcastle, N.S.W., to any port or ports place or places in any order on the West Coast of S. America and for 30 days after arrival in the final port however employed'. The ship sailed with a cargo of coal from Newcastle, N.S.W., and proceeded to Valparaiso, on the West Coast of S. America, where she discharged her cargo. She there loaded a small quantity of ballast and sugar and sailed for Talcahuano, another port on that coast, in order to finish loading there a cargo for the U.K. Before reaching Talcahuano she was totally lost, and the loss occurred more than 30 days after her arrival at Valparaiso.

The court held that the expressions in the policy 'port or ports place or places' and 'final port', were not limited to ports or places of discharge and final port of discharge respectively, but must be construed to include ports or places of loading and final port of loading for the voyage to the U.K. The plaintiffs, who had paid the claim of the owner on the original policy, were therefore entitled to recover from the defendants on the policy of reinsurance.

As to insurance 'from' or 'at and from' under section 42 of the Marine Insurance Ordinance, although the vessel need not be at the port of departure at the time when the policy is entered into, there is an implied condition that the adventure may commence within a reasonable time, otherwise the underwriters may avoid the policy. The Marine Insurance Ordinance section 42, on 'Implied condition as to commencement of risk', states:

> (1) Where the subject-matter is insured by a voyage policy 'at and from' or 'from' a particular place, it is not necessary that the ship should be at that place when the contract is concluded, but there is an implied condition that the adventure shall be commenced within a reasonable time, and that if the adventure be not so commenced the insurer may avoid the contract.
> (2) The implied condition may be negatived by showing that the delay was caused by circumstances known to the insurer before the contract was concluded, or by showing that he waived the condition.

Under section 42(2) of the Marine Insurance Ordinance, the implied condition may not arise if the delay is caused by circumstances known to the underwriters before the policy is concluded or the underwriters waive the condition.

> *De Wolf v Archangel Maritime Bank* [1874] LR 9 QB 451
> Where the vessel was insured at and from Montreal, a policy being concluded on 13 July. The underwriters did not make any attempt to discover the position of the vessel at that time, nor did the assured inform the underwriters that the vessel was at sea. The vessel reached Montreal on 13 August; to show that the delay affected the premium chargeable, the underwriters could have avoided the policy.

Under section 43 of the Marine Insurance Ordinance, insofar as the place of departure is stated on the policy the vessel must sail from that place or the risk does not attach. The Marine Insurance Ordinance section 43, on 'Alteration of port of departure', states:

> Where the place of departure is specified by the policy, and the ship instead of sailing from that place sails from any other place, the risk does not attach.

Under section 44 of the Marine Insurance Ordinance, if the destination of the vessel is stated on the policy, the vessel must sail for that destination. It is not sufficient for the vessel to be at the port of departure with no intention of being there and sailing for the port of destination. The Marine Insurance Ordinance section 44, on 'Sailing for different destination', states:

Where the destination is specified in the policy, and the ship, instead of sailing for that destination, sails for any other destination, the risk does not attach.

Time policy

The only difficulty which may arise is probably termination of risk. In general, it is accepted that the date and time of commencement and termination will relate to the country in which the policy is taken out. It follows, therefore, that it is necessary to make provision in a time policy for the possibility of termination coming about as the vessel is still at sea. For example, in Clause 2 of the Institute Time Clauses Hulls, and under the heading of continuation, it is provided that if the vessel is at sea, in distress, in port of refuge or call when the policy expires, then the vessel will be held covered at a pro rata monthly premium until arrival at the port of destination provided advance notice has been given to the underwriter.

Mixed policy

The mixed voyage time policy may present certain difficulties in that the commencement of risk must satisfy the requirement of the policy and insofar as specific voyage(s) are set out. Clause 2 of the Institute Voyage Clauses Hulls allows the vessel to be held covered in the change of voyage subject to such terms and additional premium, and the assured should satisfy requirements of the terms of the policy before the holding cover becomes effective.

M. Almojil Establishment v Malayan Motor and General Underwriters (Private) Ltd, The 'Al-Jubail IV' [1982] 2 Lloyd's Rep 637

The plaintiff bought the vessel, which was then converted. A hull policy on the vessel was effected with the defendants, which provided, inter alia, Warranted: Trading within Persian Gulf but including one delivery from Singapore to Persian Gulf on its own steam sailing on or about 21st April, 1975. The vessel encountered heavy weather and ran aground three times.

The plaintiff claimed under the policy but the defendants denied liability, contending that the policy was a voyage policy and that under s 39 (1) and (5) of the Marine Insurance Act, 1906 unseaworthiness was a defence to the plaintiffs' claim.

The Singapore Court of Appeal held that on the facts, the policy was not a time policy simpliciter but was a 'mixed policy' affording a cover of 12 months and attaching as from and on the voyage from Singapore to the Persian Gulf.

> Compliance with such warranty by the plaintiff was a condition precedent to the risk attaching, and in the circumstances there was no waiver on the part of the defendants of the plaintiff's breach of warranty; the appeal would be allowed.

For example, a policy may cover a vessel for twelve months including voyage(s) between Cardiff and Singapore as from noon 19 July. In this type of policy it is essential that the vessel sails from the nominated port with the intention of proceeding to ports named on the policy. In *The 'Al-Jubail IV'* [1982] 2 Lloyd's Rep 637 the policy was for 12 months and 'as from', but the condition precedent was not satisfied and, therefore, the policy did not attach. The importance of the vessel sailing on the nominated port with intention of proceeding on the policy voyage can be seen from the case.

> *Way v Modigliani* (1787) 2 Term Report 30; 100 ER 17
> The vessel was to sail 'at and from' a port in Newfoundland to Falmouth. The vessel sailed from the port but instead of proceeding on the policy voyage, she went to some fishing banks and later sailed for Falmouth. Under the terms of the policy covered from October 20, the vessel completed the fishing on October 7 but was later lost after October 20 on what would have been the contractual route. The court held that the risk did not attach since the vessel had never sailed for the policy voyage.

In *Way v Modigliani* (1787) 2 Term Report 30, if a ship that is insured from a certain time sails before the time on a different voyage from that insured, the assured cannot recover, even though the ship later resumes the course of the voyage described in the policy and is lost after the day upon which the policy is to have attached.

Goods

Insofar as risks under the policy in respect of goods and moveables, the Ordinance requires the goods to be actually on board and underwriters are not liable if loss occurs between shore and ship during loading. Rule 4 of the Rules for Construction of Policy, on 'From the loading thereof', states:

> Where goods or other movables are insured 'from the loading thereof', the risk does not attach until such goods or movable are actually on board, and the insurer is not liable for them while in transit from the shore to the ship.

Under r 5 of the Rules for Construction of Policy, it covers goods and moveables until they are 'safely landed', which means safely landed in the customary manner or within a reasonable time after arrival even if the goods have not been landed. Rule 5 of the Rules for Construction of Policy, on 'Safely landed', states:

> Where the risk on goods or other movables continues until they are 'safely landed', they must be landed in the customary manner and within a reasonable time after arrival at the port of discharge, and if they are not so landed the risk ceases.

> *Houlder v Merchants Marine* (1886) 17 QBD 354
> A policy of insurance on goods which includes 'all risk of craft until the goods are discharged and safely landed' does not cover the risk to the goods while waiting on lighters at the port of delivery for transhipment into an export vessel.

Where the policy lapses and is terminated, the assured has to take out a separate policy or clause, or a craft clause to cover the transit from the port of discharge to the destination, e.g., warehouse, or known as 'port policy'. A transit clause, formerly known as the 'warehouse to warehouse' clause, covers between-land transits between warehouses and ports at both ends of the journey. Such a clause is necessary in order to cover incidental land risks as well as transit between shore and ship. Section 2(2) of the Marine Insurance Ordinance specifically provides the policy may either expressly or by usage extend to include inland waterways or land risks incidental to sea voyage. On 'Marine insurance defined', it states:

> A contract of marine insurance may, by its express terms, or by usage of trade, be extended so as to protect the assured against losses on inland waters or on any land risk which may be incidental to any sea voyage.

> *Leon v Casey* [1932] 2 KB 576
> A policy contained a warehouse to warehouse clause. The adventure consisted of a journey by land from Cairo to Alexandria and thence by sea on a vessel. In an action upon the policy, the assured alleged that the goods had been damaged by fire in the course of transit by lorry from Cairo to Alexandria. The court held that in respect of a land and sea cover the risk must be substantially a marine adventure.

Under Clause 8 of the Institute Cargo Clauses, this is now known as the Transit Clause. Under Clause 8.1, risks commence from the time the goods leave the warehouse or place of storage or place named in the policy. The risks will terminate under Clause 8.1.1 of the Institute Cargo Clauses on delivery to the consignee or at the final warehouse or place of storage as set out in the policy. As to the port of destination, see

> *Renton v Black Sea and Baltic* [1941] 1 All ER 149
>
> A policy of marine insurance provided that the insurance included risks of non-delivery until discharge at the port of destination and whilst in transit by land or water to the final destination there or in the interior. The insured goods were certain timber which was unloaded at a London dock. The custom of the port was proved to be to stack the timber on the quay alongside the ship from which it was discharged. It was subsequently sorted by the port and then piled in a shed. Until so piled, the consignee was not allowed to touch it.
>
> It was contended that the transit of the timber was not completed until the timber was sorted and piled, and that, because, a claim of some missing after the piling was done. The court held that the transit ended when the timber was unloaded on the quayside, and the cargo owners were accordingly not entitled to recover under the policy.

Difficulties may arise as to the final place of delivery, therefore in Clause 8.1.2 of the Institute Cargo Clauses the risks may be terminated on delivery to any other warehouse or place of storage whether prior to the destination or not and which the assured has elected to use.

Assignment

An assignment in effect means transferring a property. In the course of marine insurance, the assignment refers to the transfer of the rights by the assured under the policy to another party. Under section 15 of the Marine Insurance Ordinance, it is provided that the mere fact the assured has parted with interest in the subject matters assured does not mean the assured's rights under the policy are also assigned. The Marine Insurance Ordinance section 15, on 'Assignment of interest', states:

> Where the assured assigns or otherwise parts with his interest in the subject-matter insured, he does not thereby transfer to the assignee his rights under the contract of insurance, unless there be an express or implied agreement with the assignee to that effect. But the provisions of this section do not affect a transmission of interest by operation of law.

To achieve the assignment of the policy, it is necessary to have an express or implied agreement between the assignee and the underwriters. For example, in a sale of ship where the current shipowner has a time policy in connection to the ship, the benefits of the time policy will not automatically pass to the ship purchaser unless there is an express or implied agreement between the ship purchaser, i.e., assignee, and the underwriters to that effect. Where it is intended among the parties it will be subject to the terms of the particular policy. See

> *Yangtsze Insurance v Lukmanjee* [1918] AC 585
> A policy identified the parcels of logs by their marks, and was expressed, in the usual form, to be made for all persons to whom the goods should appertain in part or in all. The respondent, having paid the price, took delivery of the logs ex ship, but lost a large part due to a gale while they were still afloat. The court held that there was no evidence that the policy was effected on behalf of the respondents, or to cover his interest, and that consequently he could not maintain the suit.

Unless there is a restriction on the assignment, section 50 of the Marine Insurance Ordinance permits the policy to be assigned. The Marine Insurance Ordinance section 50, on 'When and how policy is assignable', states:

(1) A marine policy is assignable unless it contains terms expressly prohibiting assignment. It may be assigned either before or after loss.
(2) Where a marine policy has been assigned so as to pass the beneficial interest in such policy, the assignee of the policy is entitled to sue thereon in his own name; and the defendant is entitled to make any defence arising out of the contract which he would have been entitled to make if the action had been brought in the name of the person by or on behalf of whom the policy was effected.
(3) A marine policy may be assigned by endorsement thereon or in other customary manner.

For the purchaser of insured property to be able to sue under the policy in its own name, it is necessary it intends to have passed to the purchaser and the original assured must have the interest at the time of assignment in order to pass it on. It is important to keep separate the situation where the assured suffers loss during the currency of the policy and that where the assured has parted interest in the subject matter without assigning the policy at or before that time. If loss or damage takes place while the assured has a policy in being benefits may be assigned under the policy regarding loss, unless the policy states otherwise. See

> *Powles v Innes* [1843] 11 M & W 10; 152 ER 695
> A person who assigns away his interest in a ship or goods, after effecting a policy of insurance upon them, and before the loss, cannot sue upon the policy; except as a trustee for the assignee.

Breach of warranty

In contract law, a condition is a very critical term of a contract which goes to the root of the contract. A warranty in contract law is a term carrying lesser importance than a condition. Warranties are extremely important terms in insurance law and the insured faces serious consequences if the insured breaches a warranty of the policy.

A hierarchy of terms imposes different obligations on policyholders and they are:

(1) *Warranties* carry the most severe consequences for policyholders if they are breached. A breach discharges the insurer from any liability under the contract, even if the breach is minor or remedied later. In effect, compliance with a warranty is a condition precedent to liability arising under the policy as a whole.
(2) *Conditions precedent to a claim*. A breach will discharge an insurer from liability to pay a particular claim, but will not affect other possible claims under the policies. Such conditions are most likely to be procedural, requiring (for example) notice of claims.
(3) Clauses which are '*descriptive of the risk*' for which the insurer is liable. These state that the insurer will only cover losses arising in particular circumstances, and if a loss arises in other circumstances, the insurer is not liable. Such terms are sometimes called 'suspensive' conditions, on the basis that they merely suspend liability while a breach taking the risk outside the policy continues. If a policyholder remedies the problem, the insurer resumes liability.

For example, in *Farr v Motor Traders Mutual Insurance*[50] the policyholder insured two taxi-cabs, stating they were only driven for one shift every 24 hours. For a short time, one of the cabs was driven for two shifts while the other was being repaired. The cab was then used for one shift a day in the normal way and a couple of months later was damaged in an accident. The Court of Appeal rejected the insurer's argument that the assured had breached a warranty. Instead, the words were merely 'descriptive of the risk'. This meant that if the cab was driven for more than one shift per day, the risk would no longer be covered, but as soon as the owner resumed one-shift working, the insurer again became liable.

(1) *Innominate terms*, where the remedy for a breach depends on its seriousness. Where the breach is serious, the insurer may repudiate the policy (that is, treat the contract as terminated). Where it is minor, the remedy

50. [1920] 3 KB 669.

would be in damages only. In *Alfred McAlpine Plc v BAI (Run-Off)*,[51] it was suggested that a serious breach of a notification clause may lead to a rejection of the claim while a minor one may not. However, this has been doubted in *Friends Provident Life and Pensions v Sirius International Insurance*.[52]

(2) *Mere terms*, breach of which gives rise to a claim for damages, but which do not affect the insurer's liability to pay claims.[53]

A warranty of a policy is a clause agreed by an underwriter and assured on a contract of insurance and it is so important among other clauses of the marine insurance contract that it is classified as 'warranty'. In practice, a clause which belongs to the class of warranties is usually explicitly remarked as warranty or by starting the clause as 'Warranted . . .'. Special attention should be drawn to the wordings of all warranties prior to acceptance of the contract. As to the nature of warranty, the Marine Insurance Ordinance section 33, on 'Nature of warranty', states:

> (1) A warranty, in the following sections relating to warranties, means a promissory warranty, that is to say, a warranty by which the assured undertakes that some particular thing shall or shall not be done, or that some condition shall be fulfilled, or whereby he affirms or negatives the existence of a particular state of facts.
> (2) A warranty may be express or implied.
> (3) A warranty, as above defined, is a condition which must be exactly complied with, whether it be material to the risk or not. If it be not so complied with, then, subject to any express provision in the policy, the insurer is discharged from liability as from the date of the breach of warranty, but without prejudice to any liability incurred by him before that date.

A term of a policy may be mutually agreed by both parties as a warranty. If it is not expressly stated on the policy that a term of the policy is a warranty, one party may argue it is a warranty while the other party may not agree.

> *Moussi H. Issa NV v Grand Union Insurance Co Ltd* [1984] HKLR 137
> The plaintiff was the holder of two Marine Insurance Policies issued by the defendant and claimed for shortage, apparently due to pilferage, of cargo shipped from Hong Kong to Paramaribo in Surinam, South America. The time for bringing suit against the carrier under the Hague

51. [2000] 1 Lloyd's Rep 437.
52. [2005] 2 Lloyd's Rep 517.
53. Law Commission and Scottish Law Commission, *Insurance Contract Law Issue Paper 2 Warranties* (2006).

> Rules expired and shortly afterwards the defendant rejected the claims under the policies on the ground that the plaintiff failed to comply with the provision in the policies requiring it to ensure that all rights against the carrier were properly preserved and exercised.
>
> (Obiter) The court concluded that the parties did attach such importance to this that they intended to create a warranty. The Bailee Clause was a warranty within the meaning of section 33 of the Marine Insurance Ordinance.

If an insured promises in the form of warranty that certain facts are true but it is not true, the warranty will be broken even if the fact did not change a risk, had no connection to a loss, or the insured was not at fault. The insurer will then be discharged from liability.

An example of a circumstance which has changed may be that an assured warrants sailing with convoy in the time of war or dangerous areas, but later it is found the war is over, or the dangerous areas are clear from anticipated danger or peril. In such a case, the warranty of sailing with convoy may not be applicable to the assured.

The Marine Insurance Ordinance section 34, on 'When breach of warranty excused', states:

> (1) Non-compliance with a warranty is excused when, by reason of a change of circumstances, the warranty ceases to be applicable to the circumstances of the contract, or when compliance with the warranty is rendered unlawful by any subsequent law.
> (2) Where a warranty is broken, the assured cannot avail himself of the defence that the breach has been remedied, and the warranty complied with, before loss.
> (3) A breach of warranty may be waived by the insurer.

> *De Hahn v Hartley* (1786) 1 Term Reports 343
> The insured warranted that the ship 'Juno' would sail from Liverpool with 50 hands or upwards. In fact, when the ship left Liverpool on 13th October 1778 it had only 46 hands, though it picked up six more hands in Anglesea and had 52 hands at the time of the total loss due to capture by King's enemies. The court held that the insurer's liability terminated when the ship left Liverpool. The insurer was not liable for any losses that arose after this date, however caused.

Lord Mansfield, Ch.J. stated that a warranty in a policy of insurance is a condition or a contingency, and unless that be performed, there is no contract. It is perfectly immaterial for what purpose a warranty is introduced; but, being inserted, the contract does not exist unless it be literally complied with. Now in the present case, the condition was the sailing of the ship with a certain number of men; which not being complied with, the policy is void. See Marine Insurance Ordinance, section 34(2).

The form of warranty varies. The use of the word 'warranty' has been said to be indicative but by no means decisive (see *Barnard v Faber* [1893] 1 QB 340). Lord Justice Rix stipulated the form of a warranty in *HIH Casualty and General Insurance Ltd v New Hampshire Insurance Co.*

> *HIH Casualty and General Insurance Ltd v New Hampshire Insurance Co* [2001] 2 Lloyd's Rep 161 at para 101; [2001] EWCA Civ 735
>
> It is a question of construction, and the presence or absence of the word 'warranty' or 'warranted' is not conclusive. One test is whether it is a term which goes to the root of the transaction; a second, whether it is descriptive or bears materially on the risk of loss; a third, whether damages would be an unsatisfactory or inadequate remedy.

The above case concerned film finance insurance, in which the original insured had undertaken to make six films. This was held to be a warranty, even though the word warranty was not used, because it was a fundamental term with a direct bearing on the risk.

The Court of Appeal in *MV 'Copa Casino'*[54] held that section 34(3) of the Marine Insurance Act 1906 provides that an insurer may waive a breach of warranty. Since an insurer is not required to make any election in order to rely on a breach of warranty, the waiver referred to in section 34(3) refers to waiver by estoppel. The insured must demonstrate that: (1) there has been an unequivocal representation, by words or conduct, that the insurer does not, in future, intend to enforce the legal right against the insured; and (2) it has relied on this unequivocal representation in such a way that it would render it inequitable for the insurer to go back on the representation.

> *Argo Systems FZE v Liberty Insurance Pte Ltd (MV 'Copa Casino')*[55]
>
> A vessel sank when she was towed for the last voyage for scrap from the US Gulf to India. After the insurer won in the US courts, a new claim was brought against the insurer in England. In the case, a letter which was sent by the insurer's lawyers after the loss to the assured set out the insurer's reasons for declining liability, but did not refer to the breach of the warranty in question.
>
> The Court of Appeal held that this did not amount to an unequivocal representation because the letter had contained the important words: 'The foregoing is without prejudice to all the remaining terms and conditions of the policy.' It meant that the insurer had reserved the right to raise further defences including the issue of warranty. Since

54. [2011] EWCA Civ 1572.
55. ibid.

the insurer was entitled to rely on the breach of warranty, the insurer's claim for damages for misrepresentation was not pursued.

There is an implied warranty of the assured as a shipowner or ship operator that the ship has to be seaworthy at the time of the commencement of the voyage. See:

> *Quebec Marine Insurance Co v Commercial Bank of Canada* (1870) LR 3 PC234
>
> There was a policy on a ship. The ship was insured 'at and from Montreal to Halifax'. She sailed from Montreal with a defective boiler which was only discovered when she reached salt water. She had to return to Montreal for repairs which were effected. She then sailed again from Montreal, and was lost at the mouth of the St. Lawrence River.
>
> The court held that the insurers were not liable, for the assured could not avail himself of the defence that the breach of warranty of seaworthiness on sailing from Montreal on the first occasion had been remedied.

Normally insurers do not bother to know whether an assured complies fully with those warranties of a policy or not during the currency of the policy. Insurers would as a standard procedure query about the compliance of warranty when a potential claim case is reported to them. If the insurers find out that a warranty has not been complied with, however slight the breach of warranty, or that the damage or loss is not caused by the breach of warranty, the insurers may exercise their right by discharging from liability as from the date of the breach of warranty, but without prejudice to any liability incurred by them before that date. As to warranty of the seaworthiness of a ship in marine insurance, section 39 of the Marine Insurance Ordinance, on 'Warranty of seaworthiness of ship', states:

> (1) In a voyage policy there is an implied warranty that at the commencement of the voyage the ship shall be seaworthy for the purpose of the particular adventure insured.
>
> (2) Where the policy attaches while the ship is in port, there is also an implied warranty that she shall, at the commencement of the risk, be reasonably fit to encounter the ordinary perils of the port.
>
> (3) Where the policy relates to a voyage which is performed in different stages, during which the ship requires different kinds of or further preparation or equipment, there is an implied warranty that at the commencement of each stage the ship is seaworthy in respect of such preparation or equipment for the purposes of that stage.

(4) A ship is deemed to be seaworthy when she is reasonably fit in all respects to encounter the ordinary perils of the seas of the adventure insured.

(5) In a time policy there is no implied warranty that the ship shall be seaworthy at any stage of the adventure, but where, with the privity of the assured the ship is sent to sea in an unseaworthy state, the insurer is not liable for any loss attributable to unseaworthiness.

The Young Shing Insurance v Investment Co Ltd [1921] HKLR 34
The assured under a voyage policy limited to specified perils of the sea sued the underwriters to recover in respect of the total loss of the vessel. The vessel was originally designed for river navigation. The vessel had been overhauled and strengthened for the voyage insured, which was from Hong Kong to Manila. She commenced the voyage on 9 November 1917, but put back into port on 13 November 1917. The coxswain reported on her return that she was too light for the wind and sea. The vessel set out again on 16 November having taken on 40 tons of extra ballast but had not been heard of since. The weather was normal for the time of year, but there was evidence that it was dangerous weather for a vessel of that type.

The court held that the vessel was unseaworthy at the commencement of the voyage, and that there was thus a breach of the warranty of seaworthiness which avoided the contract.

Gompertz Actg CJ in *The Young Shing Insurance v Investment Co Ltd 'Luen On'* [1921] HKLR 34 stated: 'I am clear that the 'Luen On' was not reasonably fit for the perils a vessel of that nature was likely to meet on her voyage. See judgment of Cairns L.C. in *Steel v State Line Steamship Co*. It is true of course that she was not the type of vessel best fitted for navigation of this kind. She was, I take it, designed and built for inland waters: and her certificate is to this effect. It may be impossible to put a vessel of her class into the condition of seaworthiness ordinarily requisite for such a voyage. The condition is not, however, for this reason dispensed with. The duty of the assured is to make her as seaworthy for the voyage as is reasonably practicable for such a vessel by ordinary available means (*Burgess v Wickham* and see Arnould sec. 710).

This duty the assured has, I think, not fulfilled. I see no reason why he should not have satisfied himself before entering on the voyage, by actual experiment, as to the proper ballast for the vessel in a heavy beam sea. He was not entitled to make his experiment on the voyage insured: it was his duty not to enter upon it until he had ascertained her proper trim. He is not allowed to put this responsibility on the insurer, who, if the condition of seaworthiness is not fulfilled, is absolved from liability.'

In the case *Garnat Trading & Shipping (Singapore) Pte Ltd v Baominh Insurance Corp,*[56] it was held there had been a fair presentation to an insurer of the risk of towing a floating dock from Vladivostock to Vung Tau in Vietnam and the dock had been seaworthy at the commencement of the voyage. The towage plan for the voyage permitted towage in conditions up to sea force 5 with a maximum wave height of 3.5m. A number of pontoons were installed on the main deck and lashed to various points on the dock. On the voyage the dock was caught by a tropical storm which unexpectedly changed direction. It encountered waves of up to 10m. The forward pontoon broke free and holed various ballast tanks and might have holed the main deck. The ingress of seawater led to the abandonment of the dock.

> *Garnat Trading & Shipping (Singapore) Pte Ltd v Baominh Insurance Corp*[57]
> The appellant insurer (B) appealed against a decision that it was liable on an insurance policy in respect of a floating dock which was lost in the course of a voyage. B rejected the claim on the insurance by the respondent owners (G) on the grounds of non-disclosure and breach of the warranty of seaworthiness. The judge held that the permissible towage conditions were in fact disclosed to B and that the dock was in fact seaworthy at the commencement of the voyage.
>
> The Court of Appeal held that the appeal was dismissed. (1) The principal director of the first respondent company was an honest witness, and his conclusion as to what had been disclosed was entirely justified. (2) The adventure insured was one where it was contemplated by the parties that there would be a maximum wave height of 3.5m, so that the dock had to be fit in all respects to encounter the ordinary perils of the seas for that adventure, rather than some other voyage. (3) B failed to show that the dock was unseaworthy by reason of the securing arrangements for the bow pontoon, or because all the ballast pumps were not working, or because there were defects in the manhole covers of the dock's main deck and the watertight sub-divisions in the ballast tanks.

The Marine Insurance Ordinance section 41, on 'Warranty of legality', states:

> There is an implied warranty that the adventure insured is a lawful one, and that, so far as the assured can control the matter, the adventure shall be carried out in a lawful manner.

56. [2011] 2 Lloyd's Rep 492.
57. ibid.

A warranty in a contract of marine insurance which stated that the vessel would be laid up in a named port for a specific period did not imply that the lay-up had to be in accordance with the port regulations.

> *Elafonissos Fishing & Shipping Co v Aigaion Insurance Co SA*[58]
> The claimant (E) brought a claim for damages against the defendant insurance company (X). E's fishing vessel had been damaged by a cyclone whilst in a Madagascan port. It had been covered by a policy of insurance underwritten by X. The policy contained a warranty that the vessel would be laid up from November 1 to February 28 in the port.
>
> The Court held that the Judgment was for the claimant, counterclaim dismissed. (1) E had proved on the balance of probabilities that the damage was caused by the vessel colliding with the quayside during the cyclone. The damage was caused by insured perils. (2) On balance, the court was satisfied that at the time of the cyclone the vessel was manned by the chief engineer and at least two other crew members. It could not be satisfied that the captain was on board as he had not given a statement. The state of the engine was largely a matter of speculation and was not established one way or the other. X had therefore not made out its factual case. Further, no port regulations had been produced, and there was evidence that no written regulations existed. In any event, the warranty in the policy was that the vessel would be 'laid up from 1/11/06 until 28/2/07 . . . in the Port of Mahajanga'. Warranties were to be construed narrowly. There was no basis for the implication of further requirements as to compliance with port regulations. If X had wanted such protection, it should have stipulated it in clear terms. E had not breached the warranty.

Basis of the contract clauses

The use of 'basis of the contract' clauses in policy creates difficulty to insured. This is a device whereby potential policyholders are asked to sign a clause at the bottom of the proposal form, declaring they warrant the accuracy of all the answers they have given. The clause usually states that the answers 'form the basis' of the contract. It is well established that such a clause may elevate the answers to contractual terms, even if the terms are not to be found in the policy itself. The use of basis of the contract clauses means that an insurer may avoid liability for an inaccurate answer, even if the answer was not material, and even if it was given innocently.

---

58. [2012] EWHC 1512 (Comm).

Dawson Ltd v Bonnin [1922] 2 AC 413

The insured was asked where a lorry was garaged. They inadvertently gave the firm's place of business in central Glasgow, though the lorry was usually kept on the outskirts. This did not increase the risk in any way (and arguably decreased it). However, when the lorry was destroyed by fire, the insurers argued that the accuracy of the answer had been elevated to a warranty. Given the breach, the insurers were no longer liable. The House of Lords agreed (by a three to two majority) that even though the misrepresentation was immaterial, the insurers were not liable.

There has been widespread criticism of basis of the contract clauses. The 1980 report quoted judicial criticisms of such clauses dating from 1853. In 1908, Lord Justice Moulton said he wished he could 'adequately warn the public against such practices' (see *Joel v Law Union and Crown Insurance Co* [1908] 2 KB 863, at p 885). Despite these criticisms, however, the use of basis of the contract clauses has been upheld as recently as 1996.

Unipac (Scotland) Ltd v Aegon Insurance (1996) SLT 1197

The company answered two questions on the proposal form inaccurately. They said they had been carrying on business for a year, while they had been incorporated for less than a year; and they said they were the sole occupiers of the premises, when they were not. Following a fire, the insurers refused to pay the claim. The policyholders brought an action arguing that they had not warranted the accuracy of the answers, only that they were true to the best of their knowledge and belief. In the absence of a specific warranty, the insurer could avoid liability on the basis of a misrepresentation only if it was material. The Court of Session rejected these arguments, stating that the words used were clear. The court stressed the importance of freedom of contract in ringing terms as the parties agreed to an express warranty with the result that the defenders would have a right to avoid the policy if an answer was untrue whether or not the untrue item was material.

Broker

Only two parties are involved in a marine insurance, underwriters and assured. Historically, however, marine insurance in respect of hull, machinery and cargo has been done through a broker. The broker acts as an agent of the assured and not of the underwriter. It is similar to a Lloyd's broker, who is in the privileged position of being an agent for their assured. Lloyd's syndicates rely on the Lloyd's Policy Signing Office, which acts as their agent for signing policies. As an agent, a broker is subject to the ordinary rules of agency. Thus, the broker must act according to the terms of their express authority. If the assured wishes successfully to restrict the implied or usual authority of their agent, it must be ensured a third party has notice of such restriction.

The most important rule of agency is that the agent must act only for the principal and should not permit personal interests to conflict with those of the principal. Brokers should not act for underwriters unless they make a full disclosure to the assured. The position of a Lloyd's broker was raised in the following case:

> *Anglo African v Bayley* [1969] 1 Lloyd's Rep 268
>
> Where the essence of the argument put forward was that a Lloyd's broker by custom transacted with an underwriter: the court did not arrive at a decision but felt it would be unfair.

There was an attempt to show that the custom for a Lloyd's broker to act as a channel of communication for underwriters was unreasonable. It brought about a direct conflict of interest in which a Lloyd's broker is an agent of the assured and not for a third party, e.g., an underwriter. See

> *North and South Trust Company v Berkeley* [1970] 2 Lloyd's Rep 467
>
> The plaintiffs effected an insurance at Lloyd's through L. Bros., Lloyd's brokers, on some goods, then in an Argentine Customs warehouse, for a voyage from Buenos Aires to Paraguay. The plaintiffs made a claim, through L. Bros., in respect of alleged shortage. The underwriters denied liability on the grounds that the loss occurred prior to inception of risk. The claim was re-submitted through L. Bros., who were instructed, on behalf of the underwriters, to obtain an assessors' report.
>
> The assessors made their report, which was addressed to the underwriters but sent to L. Bros. L. Bros. took copies and handed the original to the underwriters, who again rejected the claim and returned the report to L. Bros.
>
> The plaintiffs brought an action against the underwriters and claimed against L. Bros. for a declaration that the plaintiffs were entitled to (inter alia) delivery up or inspection of the assessors' report. The defendant, a representative underwriter, issued a writ against L. Bros. for an injunction restraining L. Bros. from delivering up or disclosing to the plaintiffs (inter alia) the report.
>
> The court held that although the practice of Lloyd's underwriters in using the Lloyd's broker who placed the insurance as their channel of communication with the assessors was wholly unreasonable and, therefore, incapable of being a legal usage, in this case L. Bros., in so far as they had acted for the underwriters, were not acting in discharge of any duty towards the plaintiffs; therefore, L. Bros. were not under a

duty to pass on to the plaintiffs confidential information received on behalf of the underwriters.

To be acceptable as a custom, it must be legal and reasonable. There would be no problem if the assured has given permission to the broker to act as an intermediary with underwriters. Once brokers are instructed, their duty is to follow the instruction and perform their duty as agents, by informing the assured if any problem arises. Insofar as instructions may have more than one interpretation, the brokers will not be liable if they put a reasonable interpretation upon the instructions given.

Vale v Van Open (1921) 37 TLR 367
Where direction to insure the subject matters against 'all risks' was upheld by court that it was properly carried out where the policy covered 'all marine risks'.

Yuill v Scott Robson [1907] 1 KB 685
Where it was held not sufficient for an all risks policy on cattle to contain a FC&S (war clause).

If a broker properly asks the assured, then the broker would discharge the duty of care to underwriters but must not act in fraudulent manner. A broker effecting a contract of insurance with an underwriter on the instructions of a principal owes no duty of care or skill to the underwriter.

Empress Assurance v Bowring [1906] 11 Com Cas 107
A broker effected a number of open covers for renewal. The premiums were to be the same as those received by the original insurers less a brokerage fee to the firm of the brokers. The insurance company, after learning the facts, disputed the correctness of these deductions and sued the brokers.

The court held that, apart from any question of the correctness of the actual figure, the brokers owed no duty to the insurance company and could not be held liable to such a claim if negligence on their part were proved.

It has long been established at common law that insurance brokers act solely as agents for insured. The mere fact that insurers pay brokerage fees to a broker does not mean that a broker undertakes to perform any obligation on behalf of underwriter.

Hobbins v Royal Skandia Life Assurance Ltd[59]
The second defendant (the Broker) was plaintiff's (P's) independent financial adviser and insurance broker. The Broker arranged for P to

59. [2012] 1 HKLRD 977.

purchase Investment Linked Assurance Scheme products (the ILAS products) from insurers including the first defendant (the Insurer) (the ILAS Agreements). On numerous occasions, the Broker made it known to P that he would not be paying anything for its services or investment advice, but that instead it made money from commissions and fees paid by insurers whose products P purchased. P also signed client agreements (the Client Agreements) covering the ILAS products which acknowledged that they had been explained to P and that the Broker would be paid commission by the Insurer. The ILAS products did not perform as well as P had envisaged and he now sought to set them aside.

The Court held that P's claims were dismissed.

As to the issue of whether the Broker was the Insurer's agent, the Court in *Hobbins v Royal Skandia Life Assurance Ltd*[60] held the Broker was not the Insurer's agent. The contract between the Insurer and the Broker expressly stipulated the Broker was not the Insurer's agent and had no authority to bind the Insurer, and there was no suggestion the Broker had apparent authority so to act. Further, it had long been established at common law that insurance brokers such as the Broker were acting solely as agents for an insured and the mere fact the insurer paid brokerage fees to a broker did not mean the broker was undertaking to perform any obligation on behalf of the underwriter. As to the issue of whether the Broker in breach of its fiduciary, common law or statutory duties to the plaintiff, the Court held that the Broker had not breached its fiduciary, common law or statutory duties to P. It had made adequate disclosure to P by informing P on numerous occasions that it was not charging anything for its services but that it would be earning commission from such of P's business as the Broker placed with insurers (including the Insurer).

Further, the Court stated that P manifested their consent to such arrangement by signing the Client Agreements. To go beyond that and to require the Broker to have disclosed the quantum of commission it expected to receive would be to impose a standard at odds with case law on the prevailing commercial practice among insurance brokers.

As to the issue of whether ILAS Agreements between P and the Insurer illegal, the Court held that the ILAS Agreements were not illegal under section 9(2) of the Prevention of Bribery Ordinance (Cap 201) (PBO). There was 'lawful authority' (consisting of a long line of judicial pronouncements from the 19th century to the present) for the commercial practice that an insurance broker acted as an agent of the insured and not of the insurance company. Consequently, it had long been settled at common law that

60. ibid.

commission paid to an insurance broker by an insurer did not constitute an illegal secret profit unless it was in excess of what was normally paid within the insurance market. There was no evidence that was the case. The Court further stated that while under section 19 of the PBO, the mere fact that a practice was customary within a trade would not constitute a defence to an offence under section 9 of the PBO; here there was more than just a customary practice within the insurance brokerage: the practice had been validated by over a century of judicial authority.

If a broker acts in breach of authority, then the assured will have an action for damages but will not be based specifically under the sum insured on the policy. In certain circumstances difficulties may arise where a broker is acting partly for the assured and partly for the underwriters.

> *Trading & General Investment Corporation v Gault Armstrong & Kemble Ltd, The 'Okeanis'* [1986] 1 Lloyd's Rep 195
>
> The defendants Gault Armstrong & Kemble Ltd (GAK) were Lloyd's brokers, and the plaintiffs were shipowners. GAK were the placing brokers for the London agent of the shipowners. GAK insured 15% of the risk through Italrias for 13 months. Upon renewal Italrias declined and Mutuamar took up the 15%. The ship suffered damage in the currency of the policy and upon renewal Mutuamar declined to renew but Italrias placed a smaller percentage with a different Italian underwriter. By this time engine damage had been agreed and paid by the London underwriters. GAK then sought Italrias for agreement and payment representing Mutuamar's 15%.
>
> The court held that there was little doubt that Italrias should be regarded as agents of the underwriters. It was clear that GAK throughout regarded Italrias as agents for a number of Italian underwriters and not as agents instructed on their own behalf. The signing of the slip and the payment of premium direct to the underwriters were consistent with that relationship, and the fact that GAK pressed Italrias to obtain payment from the underwriters was not in any way inconsistent with their acting as agents for the underwriters.

The issue in *The Okeanis* is whether Italrias acted as sub-brokers and thus agents of GAK or whether they acted as agents of the Italian underwriters, or whether as intermediaries representing both GAK and the Italian underwriter at different stages of the transaction and whether GAK were answerable to the plaintiffs for the amount claimed. The position of a broker may be particularly difficult in respect of disclosure of material circumstances and avoidance of making material misrepresentation. Under section 19 of the Marine Insurance Ordinance, brokers are to disclose every material circumstance which is known to them. These include those which in the ordinary course of business they ought to have known or which ought to have been communicated to them. The Marine Insurance Ordinance section 19, on 'Disclosure by agent effecting insurance', states:

> Subject to the provisions of section 18 as to circumstances which need not be disclosed, where an insurance is effected for the assured by an agent, the agent must disclose to the insurer-
>
> (a) Every material circumstance which is known to himself and an agent to insure is deemed to know every circumstance which in the ordinary course of business ought to be known by, or to have been communicated to, him; and
>
> (b) Every material circumstance which the assured is bound to disclose, unless it comes to his knowledge too late to communicate it to the agent.

Under section 19 of the Marine Insurance Ordinance, there is limited defence if the material circumstances only come to the notice of the assured too late to be communicated to the broker before the conclusion of the policy. As to nature of the disclosure of the broker, see

> *Berger v Pollock* [1973] 2 Lloyd's Rep 442
>
> Four second-hand moulds were insured for the voyage from Australia to England. The bill of lading was claused 'unprotected, second hand and insufficiently packed'. The plaintiffs' brokers declared that the shipment was £20,000 under an open cover. Subsequently a 'signing slip' was issued stating that the moulds were 'unpacked, bound together'. The mould arrived damaged by rust and the defendant repudiated liability on the ground that the moulds were worthless and plaintiffs failed to disclose the material facts, i.e., (i) that the bill of lading was claused, (ii) the history of the moulds, and (iii) the fact that they were over-valued.
>
> The court held that (1) the policy was an unvalued policy under section 28(1) of MIA; (2) the insurable value of the moulds was 5,316.20 under s 16(2); (3) there was an actual total loss of the goods under s 57(3); (4) the existence of the open cover between the brokers and the defendants did not relieve the plaintiffs of their duty under s 18(6) to disclose material facts to the defendants; and (5) the defendants had not proved that the non-disclosure of the clausing of the bill of lading, or of the history of the moulds, or of the over-valuation was material.

A broker may defend its case on the basis of non-disclosure or misrepresentation of its assured, i.e., the principal. A broker may also have certain defences, such as that a policy is illegal, partially illegal, or void on the ground of gaming or wagering policy. However, it has been held that if the assured can show evidence to satisfy the court the underwriters would not have avoided the policy as against the assured but that the avoidance lies solely as a result of the breach of the broker, then the assured may receive damages from the broker. See

> *Everett v Hogg, Robinson & Gardner Mountain (Insurance) Ltd* [1973] 2 Lloyd's Rep 217
>
> The plaintiff, a representative Lloyd's underwriter, instructed the defendant insurance brokers to effect a reinsurance policy. In answer to a question by the reinsurers, the defendants stated that no plastics were used by M. Ltd in their products. In fact this was not true.
>
> The reinsurers repudiated liability under the reinsurance policy on the ground that the answer was untrue. The plaintiff now claimed damages for negligence from the defendants, who contended that (i) the incorrect information had been supplied to them by the plaintiff; and (ii) no loss had been suffered by him because the reinsurers would have been entitled to repudiate liability because he had failed to inform them of adverse claims experience in the year prior to the inception of the reinsurance policy, and would have done so.
>
> The court held that on the evidence, the incorrect information had not been supplied by the plaintiff.

In *Everett v Hogg, Robinson & Gardner Mountain (Insurance) Ltd* [1973] 2 Lloyd's Rep 217, Kerr, J stated: 'Prima facie a plaintiff loses nothing as the result of a defendant's failure to conclude a contract which would have been void. Accordingly, if the plaintiff seeks to assert that he has nevertheless suffered damage in consequence the burden of proof is plainly on him. On the other hand, once a plaintiff has proved that as the result of the defendant's negligence he has lost the benefit of a contract which would have been valid if concluded, but which would have been voidable at the election of the other party, then in my view the burden of proof shifts to the defendant to show that on the balance of probability the plaintiff would in any event have lost all or part of the benefit of the contract as the result of the probable action of the other party'.

The Court stated in *Ground Gilbey Limited, Davey Autos Limited v Jardine Lloyd Thompson UK Limited*[61] that the broker owes the client a duty to draw to the client's attention any onerous or unusual terms or conditions, and should explain to the client their nature and effect. After the risk has been placed, the continuing duty is exemplified by *HIH Casualty & General Insurance Ltd v JLT Risk Solutions Ltd* [2007] 2 Lloyd's Rep 278 in which Longmore LJ stated: 'an insurance broker who, after placing the risk, becomes aware of

61. [2012] 1 Lloyd's Rep 12.

information which has a material and potentially deleterious effect on the insurance cover which he has placed is under an obligation to act in his client's best interest by drawing it to the attention of his client and obtain his instructions in relation to it'.

> *Ground Gilbey Limited, Davey Autos Limited v Jardine Lloyd Thompson UK Limited*[62]
>
> The claimants (O) were the owners of Camden Market in north London and the defendants (B) were their former insurance brokers. The claim arises out of a major fire which occurred in 2008 in the the Canal Market. The cause of the fire was a liquefied petroleum gas (LPG) portable heating appliance which was left on in one of the stalls and which ignited clothing set out for sale in the stall. While the use of LPG heaters was prohibited under the terms of the tenancy agreement with stallholders but the practice was common.
>
> A pre-risk survey took place on 17 March 2005. The insurer (U) stipulated the removal of LPG heaters when U quoted in 2005. B recommended the quote to their principal except for the requirement to remove LPG heaters.
>
> The Court held that A broker owed his client a duty to take reasonable steps to obtain a policy which clearly met his client's needs and was suitable for the client. An aspect of that was that the client should not be exposed to an unnecessary risk of legal disputes with the insurer. The broker owed his client a duty to draw to the client's attention any onerous or unusual terms or conditions, and should explain to the client their nature and effect.

Premium

The premium is the consideration paid by an assured to an underwriter for the transfer of risks or liability from the assured to the underwriters. Unless it is otherwise agreed, the premium is payable under the terms of Schedule 52 of the Marine Insurance Ordinance at the time when the underwriters are ready to issue the policy. The Marine Insurance Ordinance Schedule 52, on 'When premium payable', states:

> Unless otherwise agreed, the duty of the assured or his agent to pay the premium, and the duty of the insurer to issue the policy to the assured or his agent, are concurrent conditions, and the insurer is not bound to issue the policy until payment or tender of the premium.

The words 'unless otherwise agreed' in section 52 of the Marine Insurance Ordinance will cover the case of settlement by account. This is a common

62. ibid.

market practice by which a broker settles with the underwriters periodically for all premiums due and claims payable on policies effected during that period. One of the purposes is to reduce the costs of administration, as transactional costs are rising in Hong Kong and other parts of the world. Some classes of insurance, e.g., car insurance, have been transacted through 'dialing-in' or tele-marketing by insurance companies, which means that a broker is not involved. More classes of insurance are concluded directly between assured and underwriter via the Internet. Many other classes of insurance are effected through brokers who have to add value to the services they provide. Section 53 of the Marine Insurance Ordinance provides that the broker is directly responsible to the underwriters for payment of the premium. Where moneys are payable under the terms of the policy to the assured, the underwriters are directly responsible to the assured for the payment of such moneys. This may be in respect of claim proceeds or return of premium. If the underwriters pay such money to the broker and the broker does not pay the assured, then the underwriters will remain liable to the assured for such moneys. The assured may order the underwriters to pay such moneys to the broker. The Marine Insurance Ordinance section 53, on 'Policy effected through broker', states:

> (1) Unless otherwise agreed, where a marine policy is effected on behalf of the assured by a broker, the broker is directly responsible to the insurer for the premium, and the insurer is directly responsible to the assured for the amount which may be payable in respect of losses, or in respect of returnable premium.
>
> (2) Unless otherwise agreed, the broker has, as against the assured, a lien upon the policy for the amount of the premium and his charges in respect of effecting the policy; and, where he has dealt with the person who employs him as a principal, he has also a lien on the policy in respect of any balance on any insurance account which may be due to him from such person, unless when the debt was incurred he had reason to believe that such person was only an agent.

Effect of receipt of premium on the policy

On the front page of a Lloyd's policy it states that the underwriter has been paid 'the consideration due unto him for this assurance by the assured'. By section 54 of the Marine Insurance Ordinance, the effect of these words is to prevent the underwriter from denying that it has received the premium from the assured, i.e., by this acknowledgment the underwriter is estopped from denying payment. The acknowledgment on a policy is conclusive between the underwriter and the assured, no matter whether the premium has actually been paid or not to the underwriters by the broker, unless it can be shown there has been some fraud in respect of the acknowledgment. Section 54 of the Marine Insurance Ordinance, on 'Effect of receipt on policy', states:

> Where a marine policy effected on behalf of the assured by a broker acknowledges the receipt of the premium, such acknowledgment is, in the absence of fraud, conclusive as between the insurer and the assured, but not as between the insurer and broker.

Section 54 does not prevent the underwriters from recovering the premium from the broker, nor does it prevent the broker from recovering the premium from the assured if the premium has not been paid.

'Premium to be arranged'

Where due to some circumstances or the nature of a policy it is not possible to determine the exact amount of premium to be payable in respect of the liability or risks to be covered by the policy, the assured and underwriter may leave the amount of premium until a later date. Such an arrangement is usually evidenced in the policy by the use of the term 'at a premium to be arranged' or 'held covered subject to additional premium'. By section 31 of the Marine Insurance Ordinance it is stipulated that where a premium is left to be arranged at a later date and this, in fact, is not done, then a reasonable premium is to be payable. The Marine Insurance Ordinance section 31, on 'Premium to be arranged', states:

(1) Where an insurance is effected at a premium to be arranged, and no arrangement is made, a reasonable premium is payable.
(2) Where an insurance is effected on the terms that an additional premium is to be arranged in a given event, and that event happens but no arrangement is made, then a reasonable additional premium is payable.

By section 88 of the Marine Insurance Ordinance, what is 'reasonable' is a question of fact. The Marine Insurance Ordinance section 88, on 'Reasonable time, etc. a question of fact', states:

> Where by this Ordinance any reference is made to reasonable time, reasonable premium, or reasonable diligence, the question what is reasonable is a question of fact.

Section 31 recognizes the custom of providing for certain events which may happen but which are not normally covered in the policy. It is usual to state that an additional premium will become payable as and when the particular event occurs. The premium which is to be paid may be left to be arranged at a later date. The risks in relation to the particular event so agreed with the underwriter may be contingent to the routine operation of the business and the assured may not have to pay the premium before the event occurs. Events covered by way of an additional premium include, for example, trading area, deviation, change of voyage or breach of any particular warranties which may have been given in that policy.

10.7 PERILS

Perils insured against

Under the Institute Clauses, the perils insured against are contained in Clause 6 of the Institute Time Clauses Hulls, Clause 280 dated 1/10/83 and Clause 1 of the Institute Cargo Clauses, i.e., Institute Cargo Clauses (A), Clause 252 dated 1 January 1982, Institute Cargo Clauses (B), Clause 253 dated 1 January 1982 and Institute Cargo Clauses (C), Clause 254 dated 1 January 1982. The Institute Time Clauses Hulls, Clause 280 dated 1/11/95 is not discussed here because the hull clauses referred to in most, if not all, of the legal cases base on the 1983 version. The 1995 version is not widely used in the marine insurance market of Hong Kong.

In all the cases of perils insured against arising under the Institute Time Clauses Hulls, the Clause Paramount is important since the exclusions are overriding in respect of anything in the policy. The exclusions of the Institute Cargo Clauses are contained in Clauses 4, 5, 6 and 7.

The Institute Cargo Clauses (A) is subject to a list of general exclusions, i.e., Clause 4: General Exclusions Clause, Clause 5: Unseaworthiness and Unfitness Exclusion Clause, Clause 6: War Exclusion Clause and Clause 7: Strikes Exclusion Clause. The Institute Cargo Clauses (A) includes peril of piracy back into the Institute Clauses by Clause 6.2, while the same peril is excluded in the Institute Cargo Clauses (B) and (C).

The Institute Cargo Clauses (B) and (C) have similar exclusions in Clauses 4, 5, 6 and 7, except that they both have additional exclusions in Clause 4.7, i.e., '[i]n no case shall this insurance cover deliberate damage to or deliberate destruction of the subject-matter insured or any part thereof by the wrongful act of any person or persons'.

Perils under Institute Time Clauses Hulls

The perils are divided up into two parts in Clause 6 of the Institute Time Clauses Hulls. Clause 6.1 does not contain a proviso as to the use of due diligence by the assured, whereas a proviso to Clause 6.2 expressly provides that the loss or damage arising out of the peril must not result from a want of due diligence on the part of the assured, owners or managers. As to the issue of want of due diligence, see

> *Pacific Queen Fisheries et al. v L. Symes et al, (The 'Pacific Queen')* [1963] 2 Lloyd's Rep 201
>
> Constructive total loss of the vessel Pacific Queen as result of explosion and fire in 1957. The appellants' claim was denied by the appellee insurers of liability, contending, inter alia, (1) that Pacific Queen was, with privity of appellants, sent to sea in unseaworthy condition after (a) alterations to gasoline-carrying capacity and installation of below-deck dispensing arrangements; and (b) inadequate steps to purge the vessel of gasoline and fumes after a 600-gallon spillage on 9 September 1957; (2) that the loss or damage resulted from want of due diligence by the appellants and was excluded by the Inchmaree Clause.
>
> The US Court of Appeals held that the appellants concealed circumstances material to the risk, in that (1) the Pacific Queen was sent to sea in unseaworthy condition; the appellants were privy to state of facts rendering her unseaworthy; and the unseaworthiness was a proximate cause of her loss; (2) loss and damage to the Pacific Queen, assuming that the appellants were negligent, resulted from want of due diligence by the appellants and therefore was excluded by the proviso to the Inchmaree Clause.

As to the issue that the owners and manager of The Pacific Queen did not use due diligence to prevent the loss of the vessel by explosion, the court stated that the gasoline explosion was the proximate cause of the constructive total loss of the vessel. The source of ignition was unknown but the explosion resulted from a want of due diligence by the assured owners and manager to remedy the extremely hazardous condition which existed from the time of the gasoline spill.

Perils of the seas, rivers, lakes or other navigable waters

The onset of winds or waves is never within the control of mankind in history, and perils of the seas are some of the oldest maritime perils in the market of marine insurance. Under Rule 7 of the Rules for Construction of Policy, 'perils of the seas' refers only to 'fortuitous accidents or casualties of the seas' and does not include the ordinary action of wind and waves. The Rules for Construction of Policy Rule 7, on 'Perils of the seas', states:

> The term 'perils of the seas' refers only to fortuitous accidents or casualties of the seas. It does not include the ordinary action of the winds and waves.

Ordinary wear and tear does not fall within the peril and is excluded by section 55(2)(c) of the Marine Insurance Ordinance, which on 'Included and Excluded Losses' states:

> unless the policy otherwise provides, the insurer is not liable for ordinary wear and tear, ordinary leakage and breakage, inherent vice or nature of the subject-matter insured, or for any loss proximately caused by rats or vermin, or for any injury to machinery not proximately caused by maritime perils.

> *Prudent Tankers Ltd SA v The Dominion Insurance Co Ltd (The 'Caribbean Sea')* [1980] 1 Lloyd's Rep 338
> The owner's vessel, the Caribbean Sea, was insured with the defendant underwriters. The vessel sailed from Balboa for Tacoma on 25 May. On 27 May, in the course of her voyage in the Pacific Ocean, off the coast of Nicaragua, although the wind was force 4, the weather fine and the sea and swell moderate, the vessel sank due to the entry of sea water into her engine room.
>
> The court held that the casualty here was not simply attributable to ordinary wear and tear in that the defect upon which the owners relied consisted of the fatigue cracks in the wedge-shaped nozzle which were attributed to two factors: (a) the manner in which the vessel was designed and (b) the effect upon the nozzle of the ordinary working of the vessel. The result of this combination was that the fracture opened up a significant period of time before the end of the life of the vessel and therefore recovery for loss of the vessel consequent upon such a fracture was not excluded by s 55(2)(c) of the Marine Insurance Act 1906, and the defect constituted a latent defect.

The burden of proving fortuity is borne by the assured, who has to prove the adversity of a claim. The claim in relation to an accident may not be an abnormal occurrence, e.g., weather conditions. But 'peril of the sea' includes loss or damage from perils such as ice, rocks and collisions with other vessels on the seas. Loss or damage from stranding is also a peril of the sea. Ingress of water is not a peril of the sea unless the ingress was caused by a peril of the sea, e.g., collision. In other words, underwriters would not indemnify an assured if the damage of cargo is caused by the ingress of water because the vessel is not properly maintained and there is no other peril present, i.e., abnormal action of wind and wave, war risk, scuttling or unseaworthiness. It has been stated that the peril relates essentially to perils of the seas and does not include a peril which occurs merely on the sea which might easily have occurred elsewhere.

> *Thames & Mersey Marine v Hamilton* [1887] 2 AC 484
> For the purposes of navigation, the donkey-engine was being used in pumping water into the main boilers. Owing to a valve being closed which ought to have been kept open, water was forced into and split open the air chamber of the donkey-pump. The closing of the valve was either accidental or due to the negligence of an engineer, and was not due to ordinary wear and tear.
> The court held that whether the injury occurred through negligence or accidentally without negligence, it was not covered by the policy. Such a loss did not fall under the words 'perils of the sea' etc., nor under the general words 'all other perils, losses, and misfortunes that have or shall come to the hurt, detriment or damage of the subject matter of insurance'.

In *Thames & Mersey Marine v Hamilton*, Lord Bramwell stated: 'Definitions are most difficult. . . . I have thought that the following might suffice: "All perils, losses and misfortunes of a marine character, or of a character incident to a ship as such."'As a result of *Thames & Mersey Marine v Hamilton* there came into being what became known as the Inchmaree Clause. This is now contained in the perils set out in Clause 6.2 of the Institute Time Clauses Hulls. The above case was followed in

> *Stott v Martin* [1914] 19 Com Cas 438; 3 KB 1262
> Where the judge described perils of the seas as 'a peril to which the assured would not be exposed if his adventure were not a marine adventure'. Damage was caused by a boiler dropped into a ship's hold. The court held that it was not a peril of the seas.

The form of damage in Stott v Martin was added to the Inchmaree Clause and is part of the perils in Clause 6.2 of the Institute Time Clauses Hulls. It has been held that scuttling by the owner, i.e., wilful sinking of the vessel, is not a peril of the sea. See

> *Samuel v Dumas* [1924] AC 431
> During the currency of the marine policy the ship was scuttled by the master and crew, or some of them, with the connivance of the owner, but without any connivance or complicity on the part of the mortgagee. In an action on the policy by the plaintiffs on behalf of the mortgagee, the court held that the loss of the ship by scuttling was not a loss by a peril of the sea and was not included in the general

words of the policy; it was accordingly not covered by the policy and therefore the action failed.

In *Samuel v Dumas,* the plaintiff was an innocent mortgagee. If the owner is not privy to the scuttling of the vessel and the vessel is wilfully cast away by the action of the crew, this may amount to the peril of barratry. Collision between vessels has been held to be a peril of the seas.[63] It may be confusing to include the peril of collision as a separate peril because it may become too restrictive in a sense that it covers only a collision between two things, both of which are capable of navigation. As to a wider clause, see:

The Monroe [1893] Probate 248
A vessel collided with a wreck and its cargo, and the court held that it was covered under collision.

It may not be easy to determine exactly what the cause of a loss or damage is, and a court may decide that a loss of a vessel is not caused by collision on the balance of probabilities.

The Young Shing Insurance v Investment Co Ltd [1921] HKLR 34
The court held that as to the presumption that she was lost by a peril of the sea, the probabilities were in favour of a loss not by collision, but by a peril insured against.

When the court in *The Young Shing Insurance v Investment Co Ltd* dealt with the plea that the loss was not due to the perils insured against, evidence was given for the defendants as to the likelihood of the vessel having been sunk by collision with a Chinese junk, and as to the probability of such a collision being reported. The court considered that if the vessel had been sunk by collision, the probabilities are in favour of some news of the disaster being available. As to insurance for risks of collision, there is no such cover in the Institute Cargo Clauses (A) but both (B) and (C) provide under Clause 1.1.4 for insurance in respect of loss of or damage to the subject matters reasonably attributable to collision or contact of vessel, craft or conveyance with any external object other than water. There is no identical provision, i.e., 'perils of the seas, rivers, lakes or other navigable waters' in the Institute Cargo Clauses (B) or (C). Institute Cargo Clauses (B), Clause 1.2.3 provides for insurance cover, except as provided in Clauses 4, 5, 6 and 7, loss of or damage to the subject-matter insured reasonably attributable to entry of sea, lake or river water into the vessel, craft, hold, conveyance, container, liftvan or place of storage.

63. See Xantho [1887] 12 AC 503.

Fire and explosions

Peril of fire is generally easy to produce a prima facie case. Cover for loss or damage by fire is provided in Clause 1.1.1 of the Institute Cargo Clauses (B) and (C). Loss or damage that is caused by heating or spontaneous combustion may not be covered by the peril of fire unless it is said to be so in the policy.

Following *The Knight of St. Michael* [1898] Probate 30; 14 TLR 19, fire will not include perils in relation to inherent vice as for example in spontaneous combustion, which may be covered under the HSSC Clause (Heat Sweat and Spontaneous Combustion Clause), as discussed in Pirie v Middle Dock. See:

> *Pirie v Middle Dock* [1881] 44 LT 426
>
> A cargo of coal was shipped and delivered on payment of freight, when a fire broke out spontaneously in the coals. Some of the coal was thrown overboard and the remainder suffered wet damage and was then sold at a port of refuge. Owing to no freight being payable there, the coal realised a larger sum than if it had safely reached its destination. However, the freight upon them was wholly lost.
>
> The court held that the shipowner was entitled to a contribution in general average for the lost freight.

It is possible to cover inherent vice under the 'Heat Sweat and Spontaneous Combustion' Clause. See

> *Soya GmbH Mainz Kommanditgesellschaft v White, (The 'Welsh City')* [1983] 1 Lloyd's Rep 122
>
> The plaintiff bought soya beans from a company c.i.f. Antwerp, Rotterdam, and the company insured under an open cover with the defendant underwriters, including an HSSC. The cargoes was damaged in a heated and deteriorated condition. The question for decision was whether the cause of the damage was inherent vice.
>
> The court held that in the HSSC policy itself 'heat, sweat and spontaneous combustion' were words that were descriptive of the perils insured against, not of the loss occasioned by those perils nor of what caused them to occur. There was no distinction between the words 'heat' and 'sweat'. The words 'sweat' and 'spontaneous combustion' referred to something which could only take place inside the goods themselves, so that those two expressions in their ordinary and natural meaning appeared to be clearly intended to be descriptive of particular kinds of inherent vice, and 'heat' appearing immediately in conjunction with them was apt to include heating of the cargo as a result of some internal action taking place inside the cargo itself.

The court in *Soya GmbH Mainz Kommanditgesellschaft v White* further stated that the phrase 'inherent vice or nature of the subject matter insured' in section 55(2)(c) of the Marine Insurance Ordinance referred to a peril by which a loss was proximately caused. It was not descriptive of the loss itself and meant the risk of deterioration of the goods shipped as a result of their natural behaviour in the ordinary course of the contemplated voyage without the intervention of any fortuitous external accident or casualty. The risk was prima facie excluded from a policy of marine insurance unless the policy otherwise provided.

In *The 'Cendor MOPU'*,[64] a marine insurance policy on cargo dated 5 July 2005, which incorporated the Institute Cargo Clauses (A) of 1 January 1982. One of the main issues is three legs of the oil rig were broken. The Supreme Court formally overruling the decision in *Mayban v General Insurance BHD v Alstom Power Plants Ltd.*[65] The *Mayban* case found that inability of a cargo to withstand the ordinary perils of the seas amounted to inherent vice. It would be commercially unacceptable if cover for loss arising as a result of the interaction of perils of the seas and the nature of the goods were reduced to situations where the conditions of the sea were not reasonably foreseeable.

> *The 'Cendor MOPU'*[66]
>
> An oil rig 'Cendor MOPU' was purchased by the respondents (the assured under the policy) for conversion. The insurance covered the loading, carriage and discharge of the oil rig on a towed barge from the US to Malaysia. The tug and barge arrived north of Cape Town for further inspection and some repairs were carried out on the legs. The starboard leg broke off and fell into the sea later in the voyage. The following evening the forward leg broke off, and then later the port leg broke off. The loss resulted from metal fatigue in the three legs. The insurers rejected the claim for the loss of the legs and alleged that the loss was the inevitable consequence of the voyage.
>
> The Supreme Court affirmed the decision of the Court of Appeal.
>
> The two inter-related questions are whether the loss was proximately caused by a peril insured against, namely perils of the seas, or whether cover is excluded because the failure occurred as a result of the inherent vice in the rig. Inherent vice is a damage to the cargo because of its natural behaviour and without the intervention of any fortuitous external accident or casualty. The Court therefore held that there had been an external fortuity involved, i.e., the actions on the

64. *Syarikat Takaful Malaysia Berhad v Global Process Systems Inc and Another (The 'Cendor MOPU')* [2011] 1 Lloyd's Rep 560, SC.
65. [2004] 2 Lloyd's Rep 609.
66. ibid.

sea, the rolling and pitching of the barge in the sea conditions encountered during the voyage. Judgment for the assured.

The Supreme Court in the above case further explained that 'Perils of the seas' was defined in the Marine Insurance Act 1906, in 'Rules for construction of policy' contained in schedule 1 as referring 'only to fortuitous accidents or casualties of the seas'. It did not include 'the ordinary action of the winds and waves'. Section 55(2) (c) of the 1906 Act (and the Institute Cargo Clauses) made clear that ordinary wear and tear caused by the sea (or otherwise) was something for which the insurer did not provide cover. The purpose of insurance was to afford protection against contingencies and dangers which might or might not occur; it could not properly apply to a case where the loss or injury must inevitably take place in the ordinary course of things. The word 'ordinary' attached to 'action', not to 'wind and waves', so that if the action of the wind or sea was the proximate cause of the loss, a claim lay under the policy notwithstanding that the conditions were within the range which could reasonably have been anticipated.

The cases made it clear that perils of the seas were not confined to cases of exceptional weather or weather that was unforeseen or unforeseeable. Entry of seawater was not automatically a peril of the seas. A fortuitous external accident or casualty, whether identified or inferred, was necessary, but it did not need to be associated with extraordinary weather. It was not a necessary answer to a claim for loss by perils of the sea that the loss only occurred because the vessel was unseaworthy.

When the assured claims regarding fire, proof of how the fire started does not have to be shown and there is no duty to prove anything other than loss by fire. See

Schiffshypothekenbank Zu Luebeck AG v Norman Philip Compton (The 'Alexion Hope') [1988] 1 Lloyd's Rep 311, CA

A serious fire occurred in the engine room of the vessel at Shahjah. The owners gave notice of abandonment to the underwriters as a constructive total loss. The plaintiff mortgagees issued a writ against the hull underwriters, who denied liability, contending that damage was caused by the wilful misconduct of the shipowner. The issue was whether fire was an 'occurrence' within the policy.

The Court of Appeal held that (1) the word 'occurrence' should be given its ordinary everyday meaning of a happening or event but one which was not due to ordinary wear and tear or to the ordinary progression of natural causes; the fire which 'occurred' on 23 November 1982 was a happening or event: it was not due to ordinary wear and tear and it was an occurrence within the meaning of the policy; (2) 'fire' in a marine policy was not confined to an accidental or fortuitous fire;

it included a fire started deliberately, and a 'fire' was an occurrence within the meaning of the policy.

As far as explosions are concerned, Clause 6.1.2 in the Institute Time Clauses Hulls and Clause 1.1.1 in the Institute Cargo Clauses (B) and (C) cover loss of or damage to the subject-matter insured reasonably attributable to explosion.

Commonwealth Smelting Ltd v Guardian Royal Exchange Assurance Ltd [1984] 2 Lloyd's Rep 608

The plaintiffs operated a smelting complex insured by the defendants. One policy covered destruction of or damage to the buildings, machinery plant and other contents (material damage policy) by fire, lightning or explosion. The other policy covered consequential loss (consequential loss policy) in the event of destruction of or damage to any building or other property used by the plaintiffs at the premises for the purposes of the business by fire, lightning or explosion.

On 11 August 1981, a casualty occurred in the blower which wholly destroyed it. The question for decision was whether the event that caused the damage and destruction was an explosion.

The court held that the word 'explosion' is used in these policies to denote the kind of catastrophe described in Webster, 1961, and Encyclopaedia Britannica: an event that is violent, noisy and caused by a very rapid chemical or nuclear reaction, or the bursting out of gas or vapour under pressure. The damage and destruction in this case were not so caused, or at any rate explosion in that sense was not the predominant cause; it was centrifugal disintegration. Accordingly, the claim failed.

A court may infer on a balance of probabilities that the contents of a container are undeclared cargoes which caused an explosion. See

Northern Shipping Co v Deutsche Seereederei GmbH and Others (The 'Kapitan Sakharov') [2000] 2 Lloyd's Rep 255, CA

The plaintiffs' (NSC's) container vessel Kapitan Sakharov was a 'feeder' vessel for containers shipped to Khor Fakkan. Most of her cargo consisted of DSR's and CYL's laden containers. Shortly before midday a container on deck containing a dangerous cargo exploded, causing a fire on deck which spread below, resulting in the sinking of the vessel shortly after midnight on the next day. Two seamen lost their lives.

The court held that the explosion and fire on deck was caused by an undeclared and dangerous cargo in a DSR container. A large number of DSR containers were stowed on the forepart of hatch 3, one of which was SENU in the bottom tier close to the centre line of the vessel. The explosion occurred at about that point and it resulted from a dangerous cargo which was probably an unstable chemical. Since none of the containers stowed at or near the point of explosion had such declared

> cargo, the cargo responsible for the explosion must have been undeclared. On a balance of probabilities the undeclared cargo responsible for the explosion was in a DSR container and the learned judge was entitled to infer that the contents of SENU or another DSR container caused the explosion.

Theft

The Institute Time Clauses Hulls Clause 6.1.3 provides for the peril of violent theft by persons from outside the vessel, and there is no similar provision in the Institute Cargo Clauses (B) or (C), but the Institute Cargo Clauses (A) covers all risks. The Rules for Construction of Policy Rule 9, on 'Thieves', states:

> The term 'thieves' does not cover clandestine theft or a theft committed by any one of the ship's company, whether crew or passengers.

Rule 9 of the Rules for Construction of Policy provides that 'thieves' does not cover clandestine theft or theft by the ship's company, i.e., crew or passengers. The violence required to support theft does not have to be violence to the person; violence to property would be sufficient. See

> *La Fabrique De Produits Chimiques Societe Anonyme v F. N. Large* [1922] 13 Ll L Rep 269
>
> A policy of insurance dated 21 November 1918, a policy of marine insurance covering risks from warehouse to warehouse, and it is a policy on perishable goods which were contained in three cases. The policy was free of particular average. The two cases of vanillin were stolen from the transporting warehouse on 29 November 1918.
>
> The court held that there was a smashing of two sets of doors by crowbars, which seemed clearly to be a theft by violence. The judge did not think that by violence there must be an assault on some person or other, but that when a person comes along and by crowbars smashes in doors, he breaks in and steals by violence, and that the facts of this case answered the description of a theft by violence.

As to clandestine theft, clandestine means secret, hidden or concealed, and clandestine theft may refer to an act of stealing committed by passenger or crew on board. Regarding clandestine theft followed by violence in order to escape, see

> *Athens Maritime Enterprises Corporation v Hellenic Mutual War Risks Association (Bermuda) Ltd (The 'Andreas Lemos')* [1982] 2 Lloyd's Rep 483
>
> The plaintiff shipowners had insured the ship's hull and machinery on the terms of the standard form of English marine policy with the F.C. &

> S. clause attached. They also entered the vessel in the defendant association. The rules of the association excluded *inter alia* the risks from the Standard Form of English Marine Policy by the F.C. & S. clause in which the assured warranted free from the consequences of piracy and riots or civil commotion.
>
> On 22 June, the vessel was anchored in the Chittagong. That night, a gang of men armed with knives took from the vessel equipment and materials and used force or the threat of force to make good their escape. The dispute was as to whether the loss came within the cover provided by the marine policy or whether it should be borne by the association.
>
> The court held that (1) theft without force or threat of force was not piracy under a policy of marine insurance. (2) the armed men who came on board Andreas Lemos intended and expected to steal without violence but anticipated the possibility of some resistance or interference and intended to use force or threat of force if that possibility materialised; here the case was one of clandestine theft which was discovered and force or a threat of force was used by the men to make good their escape; (3) the act of appropriation had finished when the force or threat of force was first used; the act of appropriation was finished when the goods were thrown into the sea and there was in the circumstances no loss by piracy; (4) a riot did occur but it was not complete until after the loss and the loss was not caused by the riot; clandestine thieves who used or threatened violence in order to escape after the theft had been committed did not give rise to a loss by riots any more than a loss by piracy and the loss was not covered by the association's war risk cover.

Although theft would be covered by Institute Cargo Clauses (A), in the case of Institute Cargo Clauses (B) and (C) it is necessary to have addition of the Institute Theft, Pilferage and Non Delivery Clause (or TPND Clause), subject to the ordinary exclusions in the Institute Cargo Clauses (B) and (C).

Jettison

The risk of jettison is provided for by Clause 6.1.4 of the Institute Time Clauses Hulls. This is included in the Institute Cargo Clauses (A) subject to the exclusion clauses. In Institute Cargo Clauses (B) and (C), jettison is covered in Clause 1.2.2. However, in the Institute Cargo Clauses (B), the peril is that of jettisoning or washing overboard, and in the Institute Cargo Clauses (C), the peril is that of merely jettison. Basically the peril amounts to the casting of the subject matters overboard, i.e., cargo, stores or even bunker, to lighten the vessel provided the cargo must have been properly stowed. See

> *Butler v Wildman* (1820) 3 B & Ald 398
>
> The master threw money into the sea in order to prevent it from falling into the hands of an enemy, and the vessel was subsequently captured. The action was upon a policy upon Spanish property, subscribed by British underwriters who, at the time of effecting the policy, knew that the assured were Spaniards, and that Spain was at war with the State to whom the capturing vessel belonged.
>
> The court held that this was a loss by jettison, signifying any throwing overboard of the cargo for a justifiable cause. It was a loss by enemies. If it was not lost by jettison, in the strictest sense, it was something of the same kind, and therefore came within the words 'all other losses and misfortunes'.

In the Marine Insurance Ordinance, deck cargo is not within the definition of goods unless it is proved there is a custom to the contrary. Rule 17 of the Rules for Construction for Policy on 'Goods' states:

> The term 'goods' means goods in the nature of merchandise, and does not include personal effects or provisions and stores for use on board. In the absence of any usage to the contrary, deck cargo and living animals must be insured specifically, and not under the general denomination of goods.

Jettison in the case of inherent vice is not within the peril. Deck cargo is at the shipowners' risk and there is a need to be specifically insured.

> *Apollinaris v Nord Deutsche Insurance Company* [1904] 1 KB 252
>
> During transit, goods were transshipped on to a Rhine steamer and there stowed on deck. A fire broke out on the steamer and goods were burnt. There was evidence that it was common practice for Rhine steamers to carry goods on deck.
>
> The court held that the general rule, which exempts underwriters on cargo from liability for loss of goods carried on deck during a voyage by sea, does not apply to an inland voyage upon a particular river contemplated by the policy upon which river there is a usage to carry cargoes on deck.

Jettison, although an insured peril, will in many cases fall within the terms of general average as a general average sacrifice as in section 66(4) of the Marine Insurance Ordinance, which on 'General Average Loss' states:

> Subject to any express provision in the policy, where the assured has incurred a general average expenditure, he may recover from the

> insurer in respect of the proportion of the loss which falls upon him; and, in the case of a general average sacrifice, he may recover from the insurer in respect of the whole loss without having enforced his right of contribution from the other parties liable to contribute.

Under the Marine Insurance Ordinance, the assured may recover from the underwriters for the whole loss and does not, as in general average expenditure, have to enforce a right of contribution from the other parties liable to contribute.

The issues of a cargo which is thought to be loaded under deck but is actually loaded on deck are complicated on a contract of carriage and a contract of insurance. The appeal to the Court of Appeal in *The 'Green Island'*[67] arose out of the loss in the course of a voyage of goods sold under a Carriage Insurance Paid (CIP) contract. The issues on which this appeal turns arise as a result of an unusual combination of circumstances.

> *The 'Green Island'*[68]
>
> The claimants (the buyer) (B) agreed to buy on CIP terms Tripoli three ambulances from the defendants (the sellers) (S). S agreed with the freight forwarders (F) to arrange the shipment to and the insurance. The bill of lading was claused on its face with a description of the vehicles as 'unpacked (new) vehicles'. The certificate of marine insurance stated that 'Warranted shipped under deck'. The ambulances were, in fact, shipped on deck, apparently lashed to containers; they were unpacked and unprotected. Two of the three were washed overboard in the course of the voyage.
>
> The Court of Appeal held that the statement in the booking note means that the carriers can only ship vehicles on deck if the relevant bills of lading are claused to that effect. If the bills of lading were claused, appropriate insurance would have been arranged. The insurance requirements of the CIP contract were express and there was no implied obligation to 'match' the insurance with the contract of carriage unless there was an agreement. The Court found that the freight forwarder should have checked to see if the carrier had breached, or was intending to breach its contract by stowing on deck without remarking the bill of lading. However, they also held that this failure was not causative of any loss as the insurance contracted for would never have covered the claim in any event.

67. *Geofizika DD v MMB International Limited, Greenshields Cowie & Co Ltd (The "Green Island")* [2010] 2 Lloyd's Rep 1, CA.
68. ibid.

Piracy

There is no similar provision in Institute Cargo Clauses (B) or (C). The Institute Cargo Clauses (A) covers it as all risks and under Rule 8 of the Rules for Construction of Policy it includes passenger mutiny, or a mob attacking from the shore. The Rules for Construction of Policy Rule 8 on 'Pirates' states:

> The term 'pirates' includes passengers who mutiny and rioters who attack the ship from the shore.

The decision made in *Republic of Bolivia v Indemnity Mutual* [1909] 1 KB 785 was dissented in the following case. See

> *Pan American World Airways Inc v The Aetna Casualty & Surety Co* [1975] 1 Lloyd's Rep 77
>
> The aircraft was hijacked by the Popular Front for the Liberation of Palestine and destroyed at Cairo. The plaintiff claimed against the all-risk insurers. The policy excluded loss resulting from (i) destruction by any military or usurped power; (ii) war, civil war, revolution, rebellion, insurrection or warlike operations, and (iii) riots or civil commotion.
>
> The all risks insurers contended that liability should be borne by the war risk insurers and the US Government because the loss resulted from 'war, invasion, civil war, revolution, rebellion, insurrection' and other causes, occurring in Jordan before and at the time of the hijacking as a result of conflict between the fedayeen and the government in Jordan. The war risk insurers and the US Government denied liability.
>
> The US Court of Appeals held that none of the exclusions covered the loss in question and the all risk insurers were liable.

The peril of piracy is subject to the Clauses Paramount at the end of the Institute Time Clauses Hulls. It is true that Clause 23 (i.e., War Exclusion Clause) of the Institute Time Clauses Hulls excludes the risks of capture and seizure etc, whilst barratry and piracy are excepted from it. It is specifically provided in Clause 24.2 of the Institute Time Clauses Hulls that loss or damage caused by any terrorist or person acting from political motives is excluded. Under Clause 25 (i.e., Malicious Acts Exclusion Clause) of the Institute Time Clauses Hulls, any loss, damage, liability or expense arising from the detonation of an explosive or a weapon of war is excluded if caused by a person acting maliciously or from political motive. These exclusions must be born in mind when dealing in particular with the perils of fire, explosion, violent theft, piracy and barratry. Under Clause 7 of the Institute Cargo Clauses (A), (B) and (C), loss, damage or expense caused by riots or civil commotion and in particular by any terrorist or person acting from political motive is excluded.

Breakdown of or accident to nuclear installations or reactors

The Institute Time Clauses Hulls Clause 6.1.6 provides cover for loss of or damage to the subject-matter insured caused by breakdown of or accident to nuclear installations or reactors. Clause 26 of the Institute Time Clauses Hulls provides that there shall be no liability for damage, etc, arising from any weapon of war using atomic or nuclear fission, fusion or like reaction or radioactive force or matter. The peril, therefore, covered under Clause 6.1.6 in the Institute Time Clauses Hulls must concern only loss or damage arising out of peaceful breakdowns or accidents.

Contact with aircraft, etc

Clause 6.1.7 of the Institute Time Clauses Hulls includes cover for loss of or damage to the subject-matter insured caused by contact with aircraft or similar objects, or objects falling therefrom, or loss or damage from land conveyance, dock or harbour installation. The peril concerns loss or damage to the subject matters insured and not a third party liability clause.

Similar third party liability does not come within Clauses 6 or 8 of the Institute Time Clauses Hulls due to the fact that 3/4ths collision liability covers only collisions between vessels. If the assured wishes to insure for incidents in connection to third party liability in Clause 6.1.7, it must be done so specifically.

Earthquake, volcanic eruption or lightning

The Institute Time Clauses Hulls Clause 6.1.8 includes risks of earthquake, volcanic eruption or lightning. Institute Cargo Clauses (B) Clause 1.1.6 has similar coverage but the Institute Cargo Clause (C) does not. It is necessary to include such perils specifically since they may be interpreted as not being within perils of the seas.

Accidents in loading, discharging or shifting cargo or fuel

Clause 6.2.1 of the Institute Time Clauses Hulls covers loss of or damage to the subject-matter insured caused by accidents in loading, discharging or shifting cargo or fuel.

Latent defect

Clause 6.2.2 in the Institute Time Clauses Hulls provides cover for loss of or damage to the subject-matter insured caused by bursting of boilers, breakage of shafts or latent defect in the machinery or hull. The right of the assured to recover is limited to loss or damage which results from the latent defect and not for the replacement of the part containing the defect. The reason is that a latent defect does not cause a latent defect. See

> *Scindia S.S. v London Assurance* [1937] 1 KB 639
>
> The vessel was insured under a time policy. The vessel dry docked for renewal of the wood lining of stern bush. The tailshaft was withdrawn but broke and it fell into the dock with the propeller attached to it. One blade of the propeller was broken. The insurers admitted liability for the damage to the propeller but not for the cost of replacing the tailshaft which had fractured owing to a latent defect.
>
> The court held that the cost was not recoverable.

Damage caused by a latent defect is covered but not the latent defect itself.

> *Oceanic v Faber* (1907) 13 Com Cas 28, CA
>
> During the currency of a policy, the shaft was condemned due to the discovery of a fracture in 1902. The fracture was the direct result of an imperfect weld in 1891 which had left a latent defect.
>
> The court held that the assured, not having proved that the defect first became patent while the vessel was in port during the currency of the policy, were not entitled to recover from the underwriters the cost of replacing the defective shaft.

Where steel and brass pipes were joined using the wrong mixture to do so, it was held not to be a latent defect.

> *Irwin v Eagle Star Insurance Co Ltd, (The 'Jomie')* [1973] 2 Lloyd's Rep 489
>
> A yacht was insured under a policy containing an Inchmaree Clause covering her in respect of loss through 'any latent defect in the machinery or hull'. An air conditioning firm joined together a steel pipe and a brass pipe which was usually submerged in bilge water. The joint should not have been made, for when it was exposed to air and salt water, electrolysis resulted. Six months after the installation the yacht sank due to a nipple in the piping of the air conditioning system breaking.
>
> The plaintiff claimed an indemnity under the policy on the ground that the loss was caused by a 'latent defect in the hull or machinery'. Evidence was given that (inter alia) he was not aware of the condition of the piping nor had he any reason to suspect the defect.

The US Court of Appeals held that under the law of Florida the assured was not entitled to an indemnity, for there was no defect in the metal but only a mistake by the air conditioning firm in joining steel and brass together. Accordingly, there was no 'latent defect in the machinery or hull' within the meaning of the policy.

As to whether damage suffered by a yacht was caused by external accidental means which could be held to be the equivalent of perils of the seas or latent defect or faulty design, see

J. J. Lloyd Instruments Ltd v Northern Star Insurance Co Ltd (The 'Miss Jay Jay') [1987] 1 Lloyd's Rep 32, CA

The configuration of the hull of the plaintiffs' yacht was such that when driven above a certain speed the boat would plane along the surface of the water, rather than forcing through it, as in the case of an ordinary displacement hull. The skin of the hull was found to be damaged after a trip with moderate sea conditions. The question for decision was whether the damage was caused by external accidental means or a latent defect in the hull.

The Court of Appeal held that (1) the loss had been caused by 'external accidental means'; it was not caused by 'the ordinary action of the wind and waves' but by the frequent and violent impacts of a badly designed hull upon an adverse sea; (2) since the defendants did not exclude unseaworthiness or design defects which contributed to a loss without being the sole cause, the plaintiffs' claim fell within the policy provided that what happened in the sea conditions was a proximate cause of the loss; what had to be decided was whether on the evidence the unseaworthiness due to the design defects was such a dominant cause that a loss caused by the adverse sea could not fairly be considered a proximate cause at all. The evidence did not establish anything of the kind; what it did establish was that but for a combination of unseaworthiness due to design defects and an adverse sea, the loss would not have been sustained; both were proximate causes and the appeal would be dismissed.

Negligence of master, officers, crew or pilots

The Institute Time Clauses Hulls Clause 6.2.3 provides cover for loss of or damage to the subject-matter insured caused by negligence of master, officers, crew or pilots. The only cover provided by the Lloyd's policy regarding negligence of master or crew arose under section 55(2)(a) of the Marine Insurance Ordinance, which on 'Included and Excluded Losses' states:

> the insurer is not liable for any loss attributable to the wilful misconduct of the assured, but, unless the policy otherwise provides, he is liable for any loss proximately caused by a peril insured against, even

though the loss would not have happened but for the misconduct or negligence of the master or crew.

Negligent navigation of a vessel leading to collision is covered.

Lind v Mitchel [1928] 34 Com Cas 81

The vessel was holed by ice. The master felt the ship was a danger to shipping and therefore set fire to it. The shipowner claimed a total loss on the ground of perils of the sea and fire. The underwriter claimed that the loss resulted from the wilful act of the master and crew.

The court held that the loss was caused by the negligence of the master who had panicked and overreacted.

The crew of the salvor tried but failed to hoist a skiff of the vessel being salved. The assured, the vessel being salved, claimed against the underwriter in respect of the total loss of the skiff.

Rosa v Insurance Company of The State of Pennsylvania ('The Belle of Portugal') [1970] 2 Lloyd's Rep 386

The vessel sailed from San Diego and was lost at sea due to an electrical fire. Her crew took to the skiff and were picked up by the motor vessel Port Adelaide. The crew of the Port Adelaide tried to hoist the skiff on board, but their efforts were unsuccessful, and the skiff was lost. The plaintiffs claimed under the policies against the defendant company in respect of the total loss of the vessel, the skiff and the cargo.

The US Court of Appeals held that as to the loss of the vessel fire was one of the perils insured against and the electrician's negligence did not defeat the plaintiffs' right to recover under the policy. As to the loss of the skiff, there was no proof that the crew of the Port Adelaide were negligent in trying to hoist her on board, and even if they were negligent, the 'Inchmaree' clause covered losses due to the negligence of 'mariners'.

As to the negligence of the manager of a ship repair company, see

Baxendale v Fane (The 'Lapwing') [1940] 66 Ll L Rep 174

A yacht was insured under a time policy. The vessel was negligently docked by the manager of a ship-repairing company whilst being placed in a tidal dock to have her bottom cleaned, and suffered damage. The insurer raised the defence that the manager was not the 'master' for the purposes of the clause in the policy.

The court held that the claim succeeded for the manager was the 'master' within the meaning of s 742 of the Merchant Shipping Act 1894, and the loss was therefore covered by the policy.

In *The Lapwing*, 'Master' has been defined in the Merchant Shipping Act 1894, section 742 to include every person (except a pilot) having command or charge of any ship.

Negligence of repairers or charterers

Clause 6.2.4 of the Institute Time Clauses Hulls includes cover for loss of or damage to the subject matter insured caused by negligence of repairers or charterers, provided that such repairers or charterers are not an assured under the policy. It appears that any loss or damage to the subject matters would fall under the perils subject to exceptions in Clauses 21, 22, 24 and 25.

Barratry of master, officers or crew

Clause 6.2.5 of the Institute Time Clauses Hull covers loss of or damage to the subject matter insured caused by barratry of master, officers or crew. Rule 11 of the Rules for Construction of Policy on 'Barratry' states:

> The term 'barratry' includes every wrongful act wilfully committed by the master or crew to the prejudice of the owner, or, as the case may be, the charterer.

A wrongful act done wilfully by a master or crewmember is a barratrous act where it is committed against the interest of the shipowner or demise charterer and also causing damage to cargo. It may be difficult to consider an act to be barratry where a wrongful act is wilfully committed by a master or crewmember solely against a cargo owner. Institute Cargo Clause (A) being all risks would cover those losses or damage but there is no provision for barratry in Institute Cargo Clauses (B) or (C). Under Clause 4.7 of Institute Cargo Clauses (B) and (C), deliberate damage to, or deliberate destruction of the subject matters by the wrongful act of any person or persons is specifically excluded, though malicious damage clauses may be inserted to cover it.

Although it is required that there is a deliberate act of wrong doing by the master or crew against the shipowner or demise charterer, it may still be barratry if it is done with the intention to benefit the shipowners. The important rule is that the shipowner must not be privy to the act of barratry. Barratry includes any crime or fraud causing loss or damage to the shipowners under circumstances which could not reasonably be prevented by the shipowners. In case the shipowner or demise charterer knows about the act of barratry, all reasonable measures should be taken to prevent its occurrence.

As to the best of a man's judgment, though erroneous or through honest incompetence it will not be barratrous, and it has been said that even acts amounting to reckless carelessness will not be barratry as there was no intention to injure. An example is smuggling by the master, which may be barratry. But it may not be so if the shipowner knows about the fact that the master has engaged in such practices before and has not taken any reasonable step to prevent it.

> *Pipon v Cope* [1808] 1 Camp 434
> If through the negligence of the owner of a ship insured, mariners barratrously carry smuggled goods on board, and thereby the ship is seized as forfeited, the underwriters are not liable for the loss.
> If a ship is justly seized as forfeited for smuggling and afterwards restored, the underwriters are not liable for any damage happening to the ship by the perils of the sea, in the interval between the seizure and restoration.

Wilful scuttling by the master without the privity of the assured may amount to barratry.

> *N. Michalos & Sons Maritime SA and Another v Prudential Assurance Co Ltd Public Corporation For Sugar Trade v N. Michalos & Sons Maritime Co Ltd (The 'Zinovia')* [1984] 2 Lloyd's Rep 264
> The owners' vessel Zinovia ran aground in shallow water off the coast of Egypt. After about a month she was refloated and towed to Suez but was found to have suffered such damage as to make her a constructive total loss. The owners contended that the loss was caused by perils of the sea and in the first action they claimed against the defendants under a valued policy of insurance.
> The insurers argued that the vessel had been deliberately cast away in that the owners had recruited a Mr. Kouvaris to join the vessel at Port Said with a view to running her aground and so causing her to become a constructive total loss. In the second action the Public Corporation for Sugar Trade, a Sudanese public body, sued as owners of the cargo of sugar which was on board the vessel or as holders or endorsees of the bills of lading relating to the cargo. They contended that the owners had wilfully cast away the vessel.
> The owners denied that the vessel was deliberately cast away. They further contended that if the vessel was deliberately cast away this occurred without their knowledge or connivance and was contrary to their interest.
> The court held that on the facts and evidence the insurers and cargo-owners had not satisfied the court, according to the high standard of proof required, that the vessel was deliberately cast away; the owners had succeeded in showing that the loss of the vessel was proximately caused by a peril of the sea, namely the grounding of the vessel due to negligent navigation and her subsequent pounding on the bottom.

The court further held that if the vessel had been deliberately run aground by Mr Kouvaris or any other member of the crew, the insurers had not proved that the owners in any way consented or were privy to that action. If the burden of disproving privity lay on the owners then they had discharged it, there would be judgment for the owners in the first action and in the second action the factual issues were as determined in accordance with the judgment. As to the making away with the ship by the crew amounting to barratry, see

> *The Hai Hsuan* [1958] 1 Lloyd's Rep 351
>
> Six vessels at Hong Kong belonging to the Chinese Nationalist Government were ordered to sail to Formosa or to Japan. The master ran up the Red flag, and held them for the benefit of the Chinese Communist Government.
>
> The court held that this constituted 'barratry'.

The Supreme Court in *Tasman Orient Line CV v New Zealand China Clays Ltd (The 'Tasman Pioneer')* [2010] NZSC 37; [2010] 2 Lloyd's Rep 13 held that under the Hague-Visby Rules art IV(2)(a), shipowners and contracting carriers were exempt from liability for the acts or omissions of masters and crew in the navigation and management of the ship unless their actions amounted to barratry, the test for which was whether damage had resulted from an act or omission of the master or crew done with intent to cause damage, or recklessly and with knowledge that damage would probably result. The appellant carrier, Tasman Orient Line CV (T), operated a freight liner service and the respondent shippers (N) contracted with T for the carriage of their goods as deck cargo on the vessel 'Tasman Pioneer' (the Vessel) from Auckland to Busan.

> *Tasman Orient Line CV v New Zealand China Clays Ltd (The 'Tasman Pioneer')* [2010] NZSC 37; [2010] 2 Lloyd's Rep 13
>
> On 1 May 2001 the Vessel left Yokohama bound for Busan. Because the ship was behind schedule, the master decided that instead of taking the usual route, he would take the vessel through a narrow passage which was expected to shorten the journey by 30 to 40 minutes.
>
> Shortly after the master changed course to transit the passage the Vessel grounded, causing significant damage and allowing seawater to penetrate the Vessel. Following the grounding the master did not alert the Japanese coastguard. He continued to steam at full speed through the passage and into the Inland Sea. He did not advise the owners' agents until after he anchored, about two and a half hours later. The master instructed the crew to lie to coastguard investigators. The charts were altered to show a false course.

The Court of First Instance gave judgment in favour of the shippers. The Court found that had the master notified the authorities promptly, salvors would have been engaged and the cargo would have been saved. He held that the post-grounding conduct of the master occurred 'in the navigation of the ship' within the meaning of article IV, rule 2(a), but that T could not rely on the exemption because article IV, rule 2(a) imported an obligation of good faith, and the conduct of the master after the grounding occurred was performed in bad faith.

The New Zealand Court of Appeal, by a majority, dismissed T's appeal, holding that T could not rely on article 4, rule 2(a) because the master's conduct subsequent to the grounding was so outrageous that it was fundamentally at odds with the purpose of the Hague-Visby Rules and so was not conduct 'in the navigation of the ship'.

Tasman Orient appealed to the Supreme Court.

The New Zealand Supreme Court held that the appeal would be allowed. (1) Article IV, rule 2(a) did not import any requirement of good faith. Carriers were exempted from liability for the acts or omissions of masters and crew in the navigation and management of the ship unless their actions amounted to barratry.

(2) Although the actions of the master following the grounding were reprehensible, they were actions in the navigation or the management of the Vessel. The respondents had not pleaded and could not now argue that the actions of the master amounted to barratry. T was accordingly protected by article IV, rule 2(a).

Pollution

Institute Time Clauses Hulls, Clause 7 covers loss of or damage to the vessel caused by any government authority acting under the powers vested in it in order to prevent or mitigate a pollution hazard or threat of such which results directly from damage to the vessel for which the underwriters are liable under the insurance. The government action must not have arisen from want of due diligence on the part of the assured, the owners or managers of the vessel. Master, officers, crew or pilots are not to be considered owners within the meaning of Clause 7 if they hold shares in the vessel.

Running Down Clause (RDC) or 3/4ths Collision Liability

The 3/4ths Collision Liability was formerly known as the Running Down Clause, which is now contained in Clause 8 in the Institute Time Clauses Hulls. The old form of Lloyd's policy on hull and machinery was a policy on the physical loss and damage of the ship and did not extend to third party liability arising out of a collision. An assured may ask on what basis in law to insure for third party liability. The Marine Insurance Ordinance section 3(2)(c), on 'Marine Adventure and Maritime Perils Defined', states:

> Any liability to a third party may be incurred by the owner of, or other person interested in or responsible for, insurable property, by reason of maritime perils.

The Marine Insurance Ordinance section 5, on 'Insurable Interest Defined', states:

> (1) Subject to the provisions of this Ordinance, every person has an insurable interest who is interested in a marine adventure.
> (2) In particular a person is interested in a marine adventure where he stands in any legal or equitable relation to the adventure or to any insurable property at risk therein, in consequence of which he may benefit by the safety or due arrival of insurable property, or may be prejudiced by its loss, or by damage thereto, or by the detention thereof, or may incur liability in respect thereof.

A cargo shipper, ship operator or consignee who would suffer from a loss or damage of a subject matter, e.g., a cargo or ship, or benefit by safe arrival of the same subject matter has insurable interest in the marine adventure. Concerning the measure of third party liability that an assured may insure against, the Marine Insurance Ordinance section 74, on 'Liabilities to Third Parties', states:

> Where the assured has effected an insurance in express terms against any liability to a third party, the measure of indemnity, subject to any express provision in the policy, is the amount paid or payable by him to such third party in respect of such liability.

The nature of the undertaking in Clause 8.1 agrees that the underwriters shall indemnify the assured for 3/4ths of the sum paid to any person or persons by reason of the assured becoming legally liable to pay by way of damages. In Clause 8.2.2 of the Institute Time Clauses Hulls, the liability of the underwriters is restricted to three quarters of the insured value of the vessel insured in respect of any one collision.

Clause 8.2 in the Institute Time Clauses Hulls provides indemnity in addition to the other terms and conditions of the Institute Time Clauses Hulls, though it is subject to particular provisions. An underwriter would pay for the loss or damage of the other vessel involved in a collision and at the same time would pay for the loss or damage of the vessel insured, provided it is a valid claim of the policy.

After a collision where both vessels are to blame, the indemnity is to be decided on the basis of cross liability as if the respective owners had been compelled to pay each other such proportion of each other's damages as may have been properly allowed in determining the sum or balance payable by or to the assured in consequence of such collision. If the liability of one or both for such sum is limited by law then the payment is paid on the basis of single liability. The underwriters also agree to pay out under Clause 8.3 three quarters of the legal costs of the assured or what the assured is compelled to pay in contesting liability or taking action to limit liability provided the prior written consent of the underwriters is obtained.

Liability under clause 8.1 in Institute Time Clauses Hulls: Damages

Clause 8.1.1 relates to loss of or damage to any other vessel or property on any other vessel. Clause 8.1.2 relates to delay or loss of use of other vessel or property on it and Clause 8.1.3 is about general average of, or salvage of, or salvage under contract of any such vessel or property in which such payment arises from the vessel of the assured coming into collision with another vessel. Essentially, collision relates to collisions between vessels; see

> *Chandler v Blogg* [1898] 1 QB 32
> Where a collision with a sunken barge was held to be a collision with a vessel as the barge was not a sunken wreck and since it was subsequently raised continued to exist as a vessel.

As to whether a sunken vessel is a vessel or a wreck, see

> *Maritime Salvors Ltd v Pelton Steamship Company Ltd (The 'Zelo')* [1924] 19 Ll L Rep 9
> The shipowners claimed against a mutual indemnity association in respect of a collision between the vessel Zelo and the wreck of another vessel. The insurer refused liability on the ground that the wreck was no longer a 'vessel', so the association brought an action against the same for the whole of the loss. The association paid one-quarter of the loss and contended that they were not liable for the remaining three-quarters because she was a 'vessel' at the time of collision.
>
> The court held that the claim in respect of the remaining three-quarters failed. The question whether the other vessel was a 'vessel' at the time of collision depended on whether or not any reasonably minded owner would have continued salvage operations on her in the hope of completely recovering her, and by subsequent repair. In the court's opinion, they did have such a reasonable expectation.

Exclusions

The exclusions are provided in clause 8.4 of clause 8 of the Institute Time Clauses Hulls. The proviso of clause 8.4 is that clause 8 always does not extend to any sum which the assured must pay for or in respect of clauses 8.4.1 to 8.4.5.

Clause 8.4.1 excludes any payment which the assured must pay for or in respect of removal or disposal of obstructions, wrecks, cargoes or any other thing whatsoever. Clause 8.4.2 of the Institute Time Clauses Hulls states an exclusion regarding any sum which the assured must pay for or in respect of any real or personal property or thing whatsoever except other vessels or property on other vessels.

Under Clause 8.4.3, the exclusion relates to any sum which the assured must pay for or in respect of the cargo or other property on, or the engagements of the assured vessel. Clause 8.4.4 is about exclusion of any sum which the assured must pay for or in respect of loss of life, personal injury and illness.

Clause 8.4.5 excludes any sum which the assured must pay for or in respect of pollution or contamination of any real or personal property or thing whatsoever except other vessels with which the insured vessel is in collision or property on such other vessels.

The exceptions contained in Clause 8.4 may be taken out by the shipowner or ship operator and insured by means of mutual insurance, i.e., Protection and Indemnity cover, which may be subject to another set of exclusions.

Sistership

Clause 9 of the Institute Time Clauses Hulls is the Sistership Clause. If two vessels which are owned by different shipowners collide with each other and both of them suffer loss or damage, the shipowners of one vessel may claim against the shipowners of another vessel for damages. The shipowners of one vessel may claim for a salvage award against the shipowners of another vessel which has received salvage services from the vessel. The situation becomes unclear if both vessels in a collision or salvage are wholly or partly owned by the same shipowners or under the same management because the same shipowners cannot sue themselves.

A similar difficult situation arises when an underwriter who has paid out on three quarters liability to the shipowners of one vessel takes recourse action against the shipowners of another vessel who in fact are the same shipowners. In order to clarify this, the sistership clause states that in these circumstances the vessels are treated as if in separate ownership. The decision of the sums payable by different parties is referred to a sole arbitrator agreed upon between the underwriters and the assured, as sisterships, may be insured with different underwriters.

Burden of proof in cargo damage

The dispute in *Milan Nigeria Ltd v Angeliki B Maritime Company*[69] arose out of claims by the cargo owners Milan for short delivery and alleged caking/wetting damage to its cargo of bagged rice which had been loaded and carried on board the vessel M/V 'Angeliki B' ('the Vessel'). The Vessel was owned by the owners Angeliki B Maritime Company. The cargo was carried pursuant to bills of lading incorporating the Hague Rules issued by Owners. The bills of lading also incorporated charterparty provisions providing for English law and arbitration in London. The arbitral tribunal ('the Tribunal') rejected Milan's short delivery claim but upheld the damage claim in part on the basis that Milan had not discharged the burden of proving that all of the damage suffered to the cargo was due to Owners' breaches of the contract of carriage. Leave to appeal to court was given to Milan.

> *Milan Nigeria Ltd v Angeliki B Maritime Company*[70]
> The issue is who bears the burden of proof in establishing the cause of the damage ('the burden of proof issue').
>
> There are two separate but related issues on the burden of proof concerning the operation of the exceptions in Article IV rule 2. The first is whether the carrier, who has given short or damaged delivery, need prove merely that the shortage or damage resulted from the operation of one or more of the excepted perils or must he also prove that the loss or damage was not also caused by the operation of other non-excepted perils.
>
> The Court held that it allowed the Milan's appeal on the burden of proof issue. The appropriate course was that the Award should be remitted to the Tribunal for reconsideration in accordance with the law—namely that the burden of proof is upon Owners to establish that some part of the cargo damage fell within an excepted peril.

In *Exportadora Valle de Colina SA and others v A P Moller Maersk A/S (t/a Maersk Line)*,[71] it was held that the damage to grapes carried from Chile to Europe in refrigerated containers was not caused by poor packing, handling or stowage but by excessive periods of power off for which the

69. [2011] EWHC 892 (Comm).
70. ibid.
71. [2010] EWHC 3224 (Comm).

carrier was responsible. The claimants (E) in the case, which were Chilean growers and exporters of grapes, claimed against the defendant shipowners and operators (M) in respect of damage to grapes carried by M in containers to Europe. The grapes were packed at cold stores in refrigerated containers and then carried by road to the terminal where they were delivered into the custody of M before being loaded onto container vessels for Europe. On outturn, problems were encountered with the container shipments. There were certain periods of power off during the carriage of refrigerated containers which were necessary and permissible for operational reasons, but in the case of almost all the containers in respect of which the claims arose the actual periods of power off exceeded what was necessary or permissible. The Court held that under clause 6.1(b), which closely mirrored the terms of the Convention on the Contract for the International Carriage of Goods by Road 1956 (CMR) art 18(2), M as carrier only had to show that one or more of the excluded matters relied upon could plausibly have caused the damage, then the presumption in clause 6.1(b) operated unless E could rebut it. The better view was to rebut the presumption E had to show on a balance of probabilities that the matter relied upon by M did not cause the loss. On that test, E had proved that none of the matters relied on by M was causative of the damage. The grapes were in good order and good condition on stuffing in the containers, but there was damage to the grapes on outturn and none of the exclusions applied; M was liable for that loss and damage. There was no requirement under the clause for E to show what the cause of the loss and damage in fact was, if none of the exclusions applied (see paras 24–30 of judgment).

> *Exportadora Valle de Colina SA and others v A P Moller Maersk A/S (t/a Maersk Line)*[72]
>
> The carriage of the containers by the defendant shipowners and operators (M) involved multimodal transport within clause 6 of M's standard bill of lading. M's case was that any loss or damage to the grapes was caused by one or more of the matters which amounted to exclusions under clause 6.1(a) (iii), (iv) and (v), namely inherent vice, insufficient or defective packing and bad stowage.
>
> The Court held the judgment for the claimants (E). (1) E had established that the grapes were in good order and condition when shipped. (2) E had shown on a balance of probabilities that there was actual damage upon outturn to the grapes the subject of the claims, with two exceptions. (3) E had established on a balance of probabilities that the damage on outturn was not caused by any of the matters relied upon by M, namely inherent vice and poor packing, handling

72. ibid.

and stowage. (4) For nearly all the containers in issue, there were one or more periods of power off which either exceeded any permissible period for a recognised operational power off or the reason for which remained unexplained. Those excessive or unexplained periods of power off constituted breaches by M of its obligations under the contracts of carriage or as bailees. E had established in the case of all those containers that on a balance of probabilities the damage sustained at outturn was caused by those excessive power offs. The burden was on M to show what proportion of the damage was attributable to a cause for which it was not liable, in the instant case damage which would inevitably have occurred as a consequence of permissible power offs, and to the extent that it could not do so, it was liable for the entirety of the damage. In the circumstances M would be liable for the damage sustained even if E had no explanation for the damage on outturn.

10.8 INCLUDED AND EXCLUDED LOSSES

Included losses

In order to recover under the policy, the basic requirement is the assured should have suffered loss or damage caused by perils insured against. These perils appear in Clause 6 of the Institute Time Clauses Hulls and Clause 1 of the Institute Cargo Clauses. It is important to remember in the case of Clauses 23–26 of the Institute Time Clauses Hulls there appear what are known as the Paramount Exceptions. It is specifically stated in heavy type that the clauses which follow should be paramount and override anything in the insurance inconsistent with them.

Clause (A), (B) and (C) of the Institute Cargo Clauses are all subject to exclusions appearing in Clauses 4 to 7. An assured may recover in respect of loss and/or damage proximately caused by a peril insured against. The Marine Insurance Ordinance section 55(1), on 'Included and Excluded Losses', states:

> Subject to the provisions of this Ordinance, and unless the policy otherwise provides, the insurer is liable for any loss proximately caused by a peril insured against, but, subject as aforesaid, he is not liable for any loss which is not proximately caused by a peril insured against.

Under the Institute Cargo Clauses (A), there is an open statement in that the insurance covers all risks except for the provisions of Clauses 4 to 7.

The Institute Cargo Clauses (B) and (C) have two separate requirements in respect of perils listed in Clause 1. The assured may recover in respect of loss or damage 'reasonably attributable' to the perils set out under

Clause 1.1, and the perils under Clause 1.2, which requires that in order to recover for loss or damage it must have been 'caused by' a listed (or named) peril. It is not clear whether this contractual wording was intended to change the old form of burden of proof, but it would seem there ought to be a difference between a test which speaks of 'proximately caused by', 'caused by' and 'reasonably attributable to'. There is no guidance in respect of whatever the test is to be in the future. The same difficulty will appear whether two or more causes operated in respect of the loss or only one of the causes comes within the listed perils.

> *Reischer v Borwick* [1894] 2 QB 548
> The policy was against damage received in collision with any object. The ship ran against a snag which made a hole in her. The vessel was anchored and the hole plugged. A tug was sent to bring her to dock for repairs. Owing to the motion through the water when being towed, the plug came out and the ship sank.
>
> The court held that the loss of the ship was covered by the policy. It is obvious that in that case there was not the intervention of any new cause. The hole occasioned by the collision was the cause of the loss. The fact that ineffectual attempts had been made to stop the hole, and that the plug came out, did not introduce any new element of causation.

> *Leyland Shipping Company, Limited v Norwich Union Fire Insurance Society, Limited* [1918] AC 350, HL
> A ship was insured against, *inter alia*, perils of the sea by a time policy containing a warranty against all consequences of hostilities. The ship was torpedoed and the torpedo struck her well forward and she began to settle down by the head, but with the aid of tugs she reached Havre on the evening of the same day. When she was taken alongside a quay in the outer harbour, a gale sprang up causing her to bump against the quay. The harbour authorities, fearing that she would sink and block the quay, ordered her to a berth inside the outer breakwater. She remained there for two days taking the ground at each ebb tide, but floating again with the flood. Finally her bulkheads gave way and she sank and became a total loss. The shipowners brought an action on the policy claiming to recover as for a loss by perils of the sea.
>
> The court held that the grounding was not a novus causa interveniens, that the torpedoing was the proximate cause of the loss, and that consequently the underwriters were protected by the warranty against all consequences of hostilities. The decision of the Court of Appeal was affirmed.

In *Leyland Shipping Company, Limited v Norwich Union Fire Insurance Society, Limited* [1918] AC 350, HL at p 355 it states that the captain, writing to his owners from Havre, says that on 2 February, the nos 1 and 2 bulkheads gave out with a crash, and goes on to say: 'I am practically certain that the vessel's back is broken in two places between the bridge and the stem, viz., on the fore part of no 2 hatch, and again abreast the no 1 hatch, and the side plating in both those places is very badly buckled, and the sheer of the ship is quite broken from the bridge forward to the forecastle hold, which is plainly visible from the shore.' In a letter of 4 February he says: 'the ship from forward to amidships has absolutely foundered. Apparently the torpedo has wrecked the ribs and keelsons in the forward end, and then her weight being water-borne aft and on the ground forward has broken her back and opened her out forward. Her condition is hopeless.' In his letter of 10 February the following sentence occurs: 'It seems evident from the very sudden way in which the forward end of the vessel crumpled up that her structure was so weakened by the terrific force of the explosion that there was no strength left to resist the additional strain imposed on her, firstly, by the great weight of water against the bulkheads and afterwards by the bursting strain of the cargo as it swelled in the holds.'

Cory v Burr (1883) 8 AC 393, HL
In a time policy including 'barratry of the master', the ship was seized by the Spanish authorities because of the barratrous act of the master in smuggling.

The court held, affirming the decision of the Court of Appeal, that the loss must be imputed to 'capture & seizure' and not to the barratry of the master. The underwriter was not liable.

Canada Rice Mills, Limited v Union Marine and General Insurance Company, Limited [1941] AC 55, PC
The appellants insured with the respondents a cargo of rice under a floating policy. On arrival at the port of discharge it was found that the rice had heated. The special jury found that the rice was damaged by heat caused by the closing of the cowl ventilators and hatches from time to time during the voyage; that such closing was the proximate cause of the damage; and that the weather and sea during the time the ventilation was closed were such as to constitute a peril of the sea.

The court held (1) that where there is an accidental incursion of seawater into a vessel in a manner and at a part of the vessel where seawater is not expected to enter in the ordinary course of things, as through uncovered cowl ventilator openings during stormy weather, and there is consequent damage to the thing insured, there is prima facie a loss by perils of the sea. It is the fortuitous entry of the seawater which is the peril of the sea in such cases, and whether in any particular case there is a loss by such peril is a question of fact for the jury, who are entitled to take a broad commonsense view of the whole

> position; (2) that the damage to the rice which was caused not by the incursion of seawater, but by action necessarily and reasonably taken to prevent the incursion—the peril of the sea—affecting the rice was a loss due to the peril of the sea and was recoverable as such; (3) that the loss was also recoverable as falling within the general words of the policy.

It was said in relation to determining the proximate cause, that the question should be whether the loss or damage becomes inevitable in respect of the cause relied upon by the assured.

The specific focus in *Borealis AB v Geogas Trading SA*[73] was the test for breaking the chain of causation. In order to comprise a *novus actus interveniens*, i.e., breaking the chain of causation the conduct of the claimant 'must constitute an event of such impact that it "obliterates" the wrongdoing' of the defendant. The same test applies in contract. For there to be a break in the chain of causation, the true cause of the loss must be the conduct of the claimant rather than the breach of contract on the part of the defendant. In case the breach of contract by the defendant and the claimant's subsequent conduct are concurrent causes, it must be unlikely that the chain of causation will be broken. In circumstances where the defendant's breach of contract remains *an* effective cause of the loss, at least ordinarily, the chain of causation will not be broken.

The Court continued to state that the claimant's state of knowledge at the time of and following the defendant's breach of contract is likely to be a factor of very great significance. For the chain of causation to be broken, the claimant need not have knowledge of the legal niceties of the breach of contract; nor will the chain of causation *only* be broken if the claimant has actual knowledge a breach of contract has occurred. Otherwise, there would be a premium on ignorance. However, the more the claimant has actual knowledge of the breach, of the danger of the situation which has thus arisen and of the need to take appropriate remedial measures, the greater the likelihood the chain of causation will be broken. Conversely, the less the claimant knows the more likely it is that only recklessness will suffice to break the chain of causation.

> *Borealis AB v Geogas Trading SA*[74]
> The defendant (G) sold a cargo of some 5,200 mt butane to the claimant (B) to be used as feedstock by B in its plastics production plant. The butane was contaminated with fluorides, which cracked during normal processing to produce, amongst other substances, hydrofluoric

73. [2010] EWHC 2789 (Comm).
74. ibid.

acid, which then caused serious and extensive physical damage to the plant and equipment, together with consequential interruption to B's business. B took no action in response to the alarm. B claimed loss of profits from plant down time.

The Court held that the Court was not persuaded that there was unreasonableness, still less unreasonableness breaking the chain of causation. The failure of B to react to the alarm was criticised by the Court. On the evidence, B had no knowledge of, or reason to suspect, the presence of fluorides in the cargo. Even if with hindsight B might have done more, the Court concluded that the action of B at the time did not obliterate the effect of the original breach by G and break the chain of causation.

It is not easy to determine the proximate cause(s) in marine insurance policy involving road and sea, and two insurers.

European Group Ltd v Chartis Insurance UK Ltd[75]
The issue is to determine which of two insurers was liable for losses caused by fatigue stress cracking to tubes in economiser blocks. The claimants and the defendants subscribed to an insurance policy of erection all risks, public liability and delay in start-up in the plant. The defendants also subscribed to a primary marine project cargo/delay in start-up policy. The economiser blocks were manufactured in Romania and transported by road or by road and sea. The policies provided that, if it was not possible to ascertain whether the cause of damage to the insured's property occurred before or after the arrival of the property at the plant, the claimants and the defendants would each contribute 50 per cent of the claim. The marine policy excluded liability for loss or damage caused by inherent vice. After the economisers had been on site for between four and six months fatigue crack damage was discovered in weld joints.

The Court held the judgment for the claimants. (1) The court accepted the evidence of the claimants' expert that resonant vibration by wind buffeting, sufficient to have caused the fatigue cracking to the tubes, could effectively be ruled out. (2) The 50/50 clauses in the insurance policies would be applicable if there was such uncertainty that it was not possible to reach any conclusion as to when the damage occurred; or one theory was so improbable that even if the other theory was ruled out, it could not as a matter of common sense be described as more likely than not to have occurred. (3) On the balance of probabilities the damage occurred prior to arrival of the economiser blocks on site as a result of resonant vibration during transportation from the factory in Romania. (4) The condition of the economiser blocks when they left the factory was such that they could reasonably be expected

75. [2012] 1 Lloyd's Rep 603; [2012] EWHC 1245 (QB).

> to survive the transportation, if properly packed. There was nothing in the inherent condition or design of the economisers which could be described as a proximate cause of the loss. The proximate cause of the loss was resonant vibration during transit resulting from the inadequacy of the packing. That was an external fortuity and in such a case there was no room as a matter of law for inherent vice to be treated as another proximate cause of the loss.

The meaning of the phrase 'arising out of' of a policy was looked into in *British Waterways v Royal & Sun Alliance Insurance Plc*.[76] The meanings of 'the proximate cause' and a wider test that contemplated more remote consequences than those envisaged by the words 'caused by' were found. The Court considered that a stricter test should be applied in the text of a policy exclusion. The Court held that the exclusion did not apply because the proximate cause of the tractor toppling into the canal was its being reversed too close to the bank, not the use of the tractor to cut hedges.

> *British Waterways v Royal & Sun Alliance Insurance Plc*[77]
> The claimant insured (B) claimed on a Policy issued by the defendant insurer (R) in respect of its liability on an insured vehicle. B had engaged two men as independent contractors to cut the hedges along the towpaths of its canal using its tractor and hedge-cutter. The tractor fell into the canal when it was reversing and the two men died. B pleaded guilty to an offence under the Health and Safety at Work legislation. B settled the claims and sought to recover from R.
>
> The Court held that the exception to the tool exclusion did not apply because there was no legislative requirement to insure in respect of liability to the driver of the vehicle and the other contractor was not a passenger but a joint user of the vehicle. The deaths, and the liability for the deaths, did not arise out of the operation of the tractor as a hedge-cutting tool, rather it arose out of the collapse of the bank when, after completing that phase of the operation, the tractor was being reversed. R had not established the tool exclusion, whether the test was one of proximate cause or whether the words 'arising out of' contemplated more remote consequences. Judgment for claimant.

Apart from the general perils in Clause 6 of the Institute Time Clauses Hulls and Clause 1 of the Institute Cargo Clauses, other losses are provided for in respect of which the assured may recover.

The question in the appeal in *New World Harbourview Hotel Co Ltd v ACE Insurance Ltd*[78] concerns the construction of two 'Composite Mercantile

76. [2012] 1 Lloyd's Rep 562; [2012] EWHC 460 (Comm).
77. ibid.
78. [2012] HKEC 264 (Court of Final Appeal).

Policies' issued by the respondent insurers. The appellants (members of the New World Group), who own or operate convention centres, hotels, car parks and other businesses, sued on the policies to recover losses sustained from interruption of their businesses caused by Severe Acute Respiratory Syndrome ('SARS') in 2003.

> *New World Harbourview Hotel Co Ltd v ACE Insurance Ltd*[79]
> Leave was granted to appeal on the ground that the case involved a point of law of great general or public importance namely: 'Whether, in common form and widely issued policies of the type in question, the provision of insurance cover in respect of loss sustained "as a result of notifiable human infectious or contagious disease" is limited to cover losses resulting from infectious diseases which are by statute compulsorily notifiable or whether such cover extends to losses caused by diseases subject to administrative reporting requirements although not backed by statutory sanctions.'
>
> The Court held that the question of construction of the word 'notifiable' is 'not as to the extension of which [the term to be construed] is capable, but of the sense in which it ought to be understood in the particular v context with which it is to be reconciled'. Once due regard is paid to the context, it becomes clear that the word 'notifiable' in the clause calling for construction bears the meaning which the courts attributed to it. That meaning being clear, the *contra proferentem* rule does not enter the picture. The Court dismissed the appeal with costs.

See also *Syarikat Takaful Malaysia Berhad v Global Process Systems Inc and another (The 'Cendor MOPU')*[80] in which the insurers rejected the claim for the loss of the legs and alleged that the loss was the inevitable consequence of the voyage. The Court of First Instance held that the insurers had proved the proximate cause of the loss was inherent vice. The Court of Appeal reversed the decision of the Court of First Instance and concluded the proximate cause of the loss was an insured peril. The insurers appealed to the Supreme Court. The Supreme Court affirmed the decision of the Court of Appeal. Judgment for the assured.

Salvage charges

Under section 65 of the Marine Insurance Ordinance, salvage charges are treated as a loss by perils insured against under an insurance contract. The Marine Insurance Ordinance section 65, on 'Salvage Charges', states:

79. ibid.
80. [2011] 1 Lloyd's Rep 302; [2011] UKSC 5.

> (1) Subject to any express provision in the policy, salvage charges incurred in preventing a loss by perils insured against may be recovered as a loss by those perils.
>
> (2) 'Salvage charges' means the charges recoverable under maritime law by a salvor independently of contract. They do not include the expenses of services in the nature of salvage rendered by the assured or his agents, or any person employed for hire by them, for the purpose of averting a peril insured against. Such expenses, where properly incurred, may be recovered as particular charges or as a general average loss, according to the circumstances under which they were incurred.

Where there is contractual salvage this may amount to particular charges under section 78 of the Marine Insurance Ordinance or general average under section 66 of the Marine Insurance Ordinance.

General average

Under section 66 of the Marine Insurance Ordinance, general average losses brought about by a peril insured against are recoverable in an insurance contract. The Marine Insurance Ordinance section 66, on 'General Average Loss', states:

> (5) Subject to any express provision in the policy, where the assured has paid, or is liable to pay, a general average contribution in respect of the subject insured, he may recover therefor from the insurer.

In the case of general average sacrifice, the assured may recover the entire loss from the underwriters. But in the case of general average expenditure, only the assured's own share of the general average contribution to the expenditure can be recovered. Salvage and general average are dealt with in Clause 11 of the Institute Time Clauses Hulls and Clause 2 of the Institute Cargo Clauses. The Marine Insurance Ordinance section 66, on 'General Average Loss', states:

> (4) Subject to any express provision in the policy, where the assured has incurred a general average expenditure, he may recover from the insurer in respect of the proportion of the loss which falls upon him; and, in the case of a general average sacrifice, he may recover from the insurer in respect of the whole loss without having enforced his right of contribution from the other parties liable to contribute.

> *The 'Abt Rasha'* [2000] 2 Lloyd's Rep 575, CA
>
> The Abt Rasha, which was a ULCC, arrived off Durban to replace two hydraulic steering pumps which had been damaged during the course

of the voyage. Shortly after she got under way again the vessel encountered severe weather. The replacement hydraulic pumps became so heavily damaged that the vessel was no longer navigable. The vessel diverted towards Port Elizabeth as a port of refuge.

The owners and managers of the Abt Rasha entered into a non-separation agreement with cargo interests. This agreement contained *inter alia* a Bigham clause which provided that liability to contribute to general average under the extension provided by the non-separation agreement was limited to a figure not exceeding the costs of transhipment had the cargo-owners taken delivery of their cargo at Port Elizabeth and then paid for the forwarding costs themselves.

The court held that the 'proportion of the loss which falls on' the shipowner within the meaning of s 66(4) [of the Marine Insurance Act 1906] was the proportion so calculated because there was no warrant for giving that sub-section other than its natural meaning. Where a non-separation agreement was entered into in circumstances such as this, 'the vessel's proportion . . . of general average' within the meaning of cl 11.1 of the Institute Clauses was the proportion calculated in accordance with the agreement either par. 2 or par. 3 as the case might be.

Sue and labour

Insofar as the particular charges are concerned, they arise out of the assured seeking to minimize or avert loss or damage from an insured peril and in respect of which the underwriters agree to indemnify the assured as in section 78 of the Marine Insurance Ordinance. The Marine Insurance Ordinance section 78(1), on 'Suing and Labouring' clause, states:

> Where the policy contains a suing and labouring clause, the engagement thereby entered into is deemed to be supplementary to the contract of insurance, and the assured may recover from the insurer any expenses properly incurred pursuant to the clause, notwithstanding that the insurer may have paid for a total loss, or that the subject-matter may have been warranted free from particular average, either wholly or under a certain percentage.

In essence the 'Sue and Labour Clause' represents a supplement to the general contract of insurance, and under section 78(4) of the Marine Insurance Ordinance it is stated that:

> It is the duty of the assured and his agents, in all cases, to take such measures as may be reasonable for the purpose of averting or minimizing a loss.

A Red Line Clause in *Moussi H. Issa NV v Grand Union Insurance Co Ltd* [1984] HKLR 137 stated, *inter alia*, that '[i]t is the duty of the insured and their agents, in all cases, to take such measures as may be reasonable for the purpose of averting or minimising a loss and to ensure that all rights against the carriers, bailees, or other third parties are properly preserved and exercised'. The Bailee Clause was in identical terms to the first sentence of the Red Line Clause above quoted and the policies also contained the Reasonable Despatch Clause from the Institute Cargo Clauses. The defendant underwriter rejected the claims under the policies on the ground the plaintiff failed to comply with the provision in the policies requiring it to ensure that all rights against the carrier were properly preserved and exercised.

> *Moussi H. Issa NV v Grand Union Insurance Co Ltd* [1984] HKLR 137
> The plaintiff was the holder of two Marine Insurance Policies issued by the defendant and claimed for shortage, apparently due to pilferage, of cargo shipped from Hong Kong to Paramaribo in Surinam, South America. The time for bringing suit against the carrier under the Hague Rules expired and shortly afterwards the defendant rejected the claims under the policies on the ground that the plaintiff failed to comply with the provision in the policies requiring it to ensure that all rights against the carrier were properly preserved and exercised.
>
> The court held that generally, a prudent uninsured should preserve his rights against the carrier by issuing a writ within the appropriate time limit but only if such rights are worth preserving. On the particular facts of the case there is no breach either of the Red Line Clause or of the Bailee Clause, since a prudent uninsured would not have incurred the expense of instituting proceedings against the carrier.

Clause 13 of the Institute Time Clauses Hulls and Clause 16 of the Institute Cargo Clauses stipulate that the contractual provision covers the assured, their servants and agents. It is, therefore, not clear whether the master or an officer or other is to be considered as a servant for the purpose of this provision. It is particularly important in the case of master or officer, and it seems to be contradictory with the provisions of section 55(2)(a) of the Marine Insurance Ordinance, under which the assured may recover in respect of a peril insured against even resulting from negligence or misconduct on the part of the master or crew. The Marine Insurance Ordinance section 55(2)(a), on 'Included and Excluded Losses', states:

> The insurer is not liable for any loss attributable to the wilful misconduct of the assured, but, unless the policy otherwise provides, he is liable for any loss proximately caused by a peril insured against, even

though the loss would not have happened but for the misconduct or negligence of the master or crew.

'Both to Blame Collision' Clause:

> The Clause is only mentioned in Clause 3 of the Institute Cargo Clauses, which relates to cargo claims. The effect of the Clause is to extend to cover the assured against the proportion of liability under the contract of affreightment 'Both to Blame Collision' Clause. It further states that the assured in the event of any claim by other shipowners under the 'Both to Blame Collision' Clause has to notify the underwriters, who have the right at their own cost and expense to defend the assured against such a claim.

Excluded losses

Under section 55 of the Marine Insurance Ordinance, certain losses are expressly excluded from ordinary marine insurance contracts unless specific provisions are made for those unusual perils.

The most important exclusion in section 55(2)(a) of the Marine Insurance Ordinance is that the insureds may not recover for any loss attributable to their own wilful misconduct. This is a straight rule for the insurance of marine liability and the basis of all insurance contracts. The Marine Insurance Ordinance section 55(2)(a), on 'Included and Excluded Losses', states:

> (a) The insurer is not liable for any loss attributable to the wilful misconduct of the assured . . .

The term appears in Clause 4.1 of (A), (B) and (C) of the Institute Cargo Clauses. It does not appear as such in the hull clauses, and the most usual form of misconduct on the part of shipowners is the sinking or burning of the vessel under their order. Whilst the Institute Time Clauses Hulls do not make specific provision in respect of wilful misconduct, the perils contained in Clause 6.2 of the Institute Time Clauses Hulls defeat the right of the assured to recover if in fact the loss is attributable to a want of due diligence on the part of the assured, owners or managers.

Further exclusions

Delay

The Marine Insurance Ordinance section 55(2)(b), on 'Included and Excluded Losses', states:

> Unless the policy otherwise provides, the insurer on ship or goods is not liable for any loss proximately caused by delay, although the delay be caused by a peril insured against . . .

Clause 4.5 of the Institute Cargo Clauses (A), (B) and (C) adopts a similar approach to the effect that an assured cannot recover for a loss, damage or expense proximately caused by delay, even though the delay arises because of the perils insured against, unless the policy states otherwise. There is an exception within the meaning of Clause 4.5 when it falls into Clause 2, the General Average Clause, of the Institute Cargo Clauses (A), (B) and (C).

> *Pink v Fleming* (1890) 25 QBD 396
> Goods were insured by a marine policy against among other things damage consequent on collision. The ship in which they were shipped came into collision with another ship and went for repair. A portion of the goods was discharged and subsequently re-shipped. On arrival the goods had been damaged by handling and delay due to their perishable nature.
>
> The court held that the collision was not the proximate cause of the loss, and therefore the plaintiffs could not recover.

Ordinary wear and tear

A loss or damage arising from ordinary wear and tear is not recoverable. The Marine Insurance Ordinance section 55(2)(c), on 'Included and Excluded Losses', states:

> Unless the policy otherwise provides, the insurer is not liable for ordinary wear and tear, ordinary leakage and breakage, inherent vice or nature of the subject-matter insured, or for any loss proximately caused by rats or vermin, or for any injury to machinery not proximately caused by maritime perils.

Insurance markets normally regard ordinary wear and tear as part of the operation costs, because ordinary wear and tear is something occurring inevitably to a subject matter insured like a ship, container, machinery or building. Underwriters do not want to cover events which tend to occur inevitably unless they are unforeseeable. Even if they are foreseeable, say, once every 100 years, it is important nobody can predict precisely when they will happen, or in what magnitude, eg, the strength of a typhoon or earthquake, or how long they will last.

> *Shell UK Ltd (T/A Shell (UK) Exploration & Production) v CLM Engineering Ltd (formerly Land & Marine Engineering Ltd) and Others* [2000] 1 Lloyd's Rep 612
>
> The policy insured against, inter alia, 'All risks of physical loss and/or physical damage to the property insured . . . including physical loss and/or physical damage arising from fault and/or error in design or from faulty and/or defective construction or workmanship or materials which occurs during the period of this insurance or manifests itself or is discovered and reported no later than 24 months from September 1993.' Shell alleged that in November 1993 it was discovered that the arrival temperature of the oil and gas was lower than required by the operating parameters and that the insulated gel broke down into hard gel and liquid monoethylene glycol by reason of a fault in the design of the gel by the contractors or by faulty construction or workmanship.
>
> The court held that (1) there was nothing to indicate in the MAR form that the scope of the policy extended to such matters as errors in design; and the property insured was defined as the works executed in the performance of all contracts relating to the project, including all items of property and equipment but without reference to design or workmanship; (2) for s 55 to apply, the provision must be clear and unambiguous since the presumption must be that the policy only responded to casualties.

It is a necessary part of Shell's case that the policy in the present case does indeed 'otherwise provide' in regard to inherent vice and so on. In my judgment, such provision must be clear and unambiguous since the presumption must be that the policy only responds to casualties. In the classic view of Lord Herschell in *The Xantho* (1887) 12 App Cas 509:

> The purpose of the policy is to secure an indemnity against accidents which may happen, not against events that must happen. If, for instance, a part of the structure of the project contained a latent defect at the inception of the risk, I am unable to accept the submission made on behalf of Shell and the contractors that there is a fortuity if that defect is then discovered during the period of cover: insurance covers fortuities, not losses which have occurred through the ordinary incidents of the operation of the vessel. Similarly the insurance does not cover the costs of maintaining the vessel or running it. As the Judge held to be the case in the present action, the cracking occurred as a result of the ordinary working of the platform at sea and the presence of latent defects in the welds. There was no external accident or cause. Correcting latent defects is, as a matter of principle, an expense to be borne by the shipowners and not by the underwriters. Similarly, the pre-existing defective condition of the subject-matter of the insurance

(be it hull or cargo) can be said to have made the loss something which was bound to happen and therefore not fortuitous . . . But, as section 55 of the Act recognises, if the parties make a specific agreement, a policy can cover such risks. The presence or absence of a latent defect in the hull or machinery of a vessel is, by definition, unknown to the assured and whether or not there is such a defect and whether or not it will during a given period of time or maritime adventure have an impact or cause any damage is fortuitous from the point of view of the assured. As is demonstrated by the Inchmaree clause and other similar clauses, there is both a market need for such cover and a willingness to provide it. However, there are further difficulties. A policy of insurance does not cover matters which already exist at the date when the policy attaches. The assured, if he is to recover an indemnity, has to show that some loss or damage has occurred during the period covered by the policy. If a latent defect has existed at the commencement of the period and all that has happened is that the assured has discovered the existence of that latent defect, then there has been no loss under the policy. (per Lord Justice Hobhouse in *The Nukila*, [1997] 2 Lloyd's Rep 146 at p 15)

Wadsworth Lighterage Co Ltd v Sea Inc (1929) 35 Com Cas 1

A policy on a barge provided: 'the insurance is against the risks of total and/or constructive and/or arranged loss including general average and salvage and damage to such vessel by collision with any other vessel or with any fixed floating or other object or by fire lightning stranding or sinking'. The barge sank through general debility [feebleness].

The court held that the policy did not make the insurers liable for ordinary wear and tear, and therefore they were not liable although the word 'sinking' was used in the policy.

Ordinary leakage and breakage

The risks associated with ordinary leakage and breakage are excluded under section 55(2)(c) of the Marine Insurance Ordinance. In Clause 4.2 of the Institute Cargo Clauses, the insurance does not cover ordinary leakage, ordinary loss in weight or volume, or ordinary wear and tear of the subject matter insured.

In *AMOCO Oil Co v Parpada Shipping Co Ltd (The 'George S.')* [1989] 1 Lloyd's Rep 369, it is stated at p 372 that '[t]he apparent short delivery, however ascertained, included loss by evaporation which the Judge assessed at 0.2 per cent or 1,720 barrels, thus reducing the apparent short delivery for which, if it was a true short delivery, the shipowners would be responsible to 6,783 barrels on shore figures and 4,688 barrels on ship's figures'. The

0.2 per cent loss by evaporation on board which was assessed by the judge was regarded as ordinary loss in weight or volume in the transit.

Inherent vice

Inherent vice of the subject matters is excluded normally under section 55(2)(c) of the Marine Insurance Ordinance. It may be defined as the inability of the subject matters to withstand the contract of carriage as arranged between the parties.

Under the Institute Cargo Clauses, the peril of inherent vice is divided into two parts, i.e., Clause 4.3 as to the insufficient or unsuitability of the packing or preparation of the subject matters, and Clause 4.4 as to the inherent vice or nature of the subject matter insured. Insufficiency of packaging in any event may be deemed to be within inherent vice.

> *Berk v Style* [1955] 2 Lloyd's Rep 382
> The plaintiffs had bought a consignment of cargo on an f.o.b. contract. On arrival, the goods were transferred from the ship's hold to a lighter. A number of bags burst and the plaintiffs incurred expenses in rebagging and landing the goods.
>
> The policies (two) covered 'all risks and loss and/or damage from whatsoever cause arising', and contained a sue and labour clause; and Institute Cargo Clauses (C) (Wartime Extension) Clause 6 provided that insurance should in no case be deemed to cover loss or damage or expense proximately caused 'by . . . inherent vice or nature of the subject-matter insured'.
>
> The court held (1) that the expenses were not recoverable since, being proximately caused by inherent vice in the subject-matter of the insurance, they were excluded by Clause 6, Institute Cargo Clauses (C) (Wartime Extension), which restricted the scope of the risks covered by the words 'all risks of loss or damage' in the policies; (2) that the expenses were also not recoverable under the sue and labour clause because, in order to recover under such a clause, the expenses must be shown to have been due to accident or casualty, and there was no accident or casualty in the present case as the bags were in such a condition at the time of shipment that it was certain that they would not hold their contents in the course of necessary handling and transhipment.

Inherent vice is excluded even in respect of 'all risk' policies as in Clause 4.4 of the Institute Cargo Clauses (A). It is possible to insure against such a loss under the Heat Sweat and Spontaneous Combustion clause (HSSC Clause).

Soya GmbH Mainz Kommanditgesellschaft v White [1983] 1 Lloyd's Rep 122, HL

The plaintiff bought soya beans from a company c.i.f. Antwerp, Rotterdam and the company insured under an open cover with the defendant underwriters including HSSC. Cargoes were damaged in a heated and deteriorated condition. The question for decision was whether the cause of the damage was inherent vice.

The court held that the standard HSSC policy did 'otherwise provide' so as to displace the prima facie rule of construction laid down by s 55(2)(c) that the insurer was not liable for 'inherent vice or nature of the subject-matter insured' to the extent that such inherent vice consisted of a tendency to become hot, to sweat or to combust spontaneously and to hold otherwise would be contrary to commercial common sense.

Rat and vermin

The risks of rat and vermin are excluded under section 55(2)(c) of the Marine Insurance Ordinance.

Hamilton v Pandorf (1887) 12 AC 518, HL

Rice was shipped under a charterparty and bill of lading which excepted 'dangers and accidents of the sea'. During the voyage, rats gnawed a hole in a pipe on board the ship, whereby seawater escaped and damaged the rice, without neglect or default on the part of the shipowners or their servants.

The court held that the damage was within the exception and the shipowners were not liable. Lord Halsbury C. stated: 'it is also necessary to give effect to the words "dangers and accidents of the sea". . . . I think the idea of something fortuitous and unexpected is involved in both words.'

Other exclusions

Under the final part of section 55(2)(c) of the Marine Insurance Ordinance, loss of or damage to machinery not proximately caused by perils insured against is not recoverable. The cover for the perils of machinery damage is provided for in the Institute Time Clauses Hull, but there is exclusion in the case of wear and tear.

Under the Institute Cargo Clauses, there are exclusions which include Clause 4.6 in respect of loss, damage or expense arising from insolvency or financial default of the owner, managers, charterers or operators of the vessel insured.

Under Clause 4.7 of the Institute Cargo Clauses (A), and Clause 4.8 of the Institute Cargo Clauses (B) and (C), loss, damage or expense arising from the use of any weapon of war, employing atomic or nuclear fission and/or fusion or other like reaction or radioactive force or matter, is excluded.

In the Institute Cargo Clauses (B) and (C), Clause 4.7, liability is excluded on the part of the insurer in respect of deliberate damage to or deliberate destruction of the subject matter insured or any part thereof by the wrongful act of a person. This exclusion does not appear in the Institute Cargo Clauses (A), which is an 'all risks' policy and can be covered if required in (B) and (C) by the use of the malicious damage clause.

Infringement of customs regulations

To rely on the exclusion in the Institute War and Strikes Clauses cl 4.1.5, which excluded loss or damage arising from the detention of a vessel by reason of, among other things, an infringement of customs regulations, underwriters did not have to show there was privity or complicity on the part of the insured or its servants or agents in such an infringement.

> *Atlas Navios-Navegacao LDA v Navigators Insurance Co Ltd (The 'B Atlantic')*[81]
>
> The court was asked to determine preliminary issues in an action in which the claimant (C) claimed under its war risks insurance for the constructive total loss of a vessel owned by it. A pre-departure inspection by the Venezuelan authorities discovered three bags of cocaine strapped to her hull. The vessel had been detained in Venezuela ever since. It was common ground that C knew nothing about the drugs and had no involvement in any attempted drug trafficking; it was C's case that the crew had no involvement with the drugs either. The defendant underwriters (N) relied on the exclusion contained in cl 4.1.5 of the Institute War and Strikes Clauses, which had been incorporated in the policy. Clause 4.1.5 excluded loss, damage, liability or expense arising from 'arrest restraint detainment confiscation or expropriation . . . by reason of infringement of any customs . . . regulations'. The issues were (i) whether, for N to be able to rely on the exclusion in cl 4.1.5, they had to show that there was privity or complicity on C's part in any infringement of customs regulations; (ii) if not, whether they had to show that there was privity or complicity on the part of C's servants or agents in any such infringement.
>
> Preliminary issues determined in favour of defendants. (1) The answer to the first question was no, C having decided not to pursue its contention that N had to show that there was privity or complicity

81. [2012] 1 Lloyd's Rep 363; [2012] EWHC 802 (Comm).

on the part of the insured in any infringement of customs regulations. (2) To rely on the exclusion in cl 4.1.5, N did not have to show that there was privity or complicity on the part of C's servants or agents in any infringement of customs regulations. The words of cl 4.1.5 were general and unqualified and the clause did not say that the infringement of customs regulations was one to which owners, their servants or agents had to be privy, or in which they had to be complicit. Had the draughtsmen of the Institute War and Strikes Clauses intended that cl 4.1.5 should apply only to infringements of customs regulations by particular persons, they could easily have said so, but they did not.

Shipment of dangerous goods

A bill of lading and/or charterparty will generally contain an express or implied term prohibiting the shipment of dangerous cargo. No cargo of a dangerous nature should be shipped onboard without the agreement of the shipowner. The cargo interest should provide for the shipowner the information in respect of the nature and character of the cargo which can damage directly or indirectly the vessel or other cargo on the vessel. The Court in *The 'Aconcagua'*[82] considered whether an undisclosed dangerous goods or a rogue batch of calcium hypochlorite constituted a dangerous cargo and would, therefore, permit shipowners to recover their losses from the charterers or shippers.

Compania Sud Americana De Vapores SA v Sinochem Tianjin Import and Export Corporation (The 'Aconcagua')[83]

CSAV (C), as disponent owners, claimed against Sinochem (S), the shippers of the calcium hypochlorite, following an explosion and a fire onboard the vessel. The fire severely damaged the vessel and other cargo. The cause of the fire was the self-ignition of 334 kegs of calcium hypochlorite cargo. Calcium hypochlorite is a known dangerous cargo. At the time of shipment, Sinochem had declared the cargo as calcium hypochlorite.

The Court held that on the basis of the expert evidence it was found that the cargo shipped was a rogue batch which would have ignited at a temperature of 25°C to 30°C rather than 60°C as would ordinarily be expected. Accordingly, it was found that C had not agreed to the carriage of this cargo with knowledge of its characteristics and C was able to recover from Sinochem.

C had been at fault as, in breach of the IMDG Code, they had stowed the container adjacent to bunker tanks which were heated

82. [2011] 1 Lloyd's Rep 683; [2010] EWCA Civ 1403.
83. ibid.

> during the course of the voyage. It was found however that, had C complied with the IMDG Code in this respect, the cargo would still have been stowed below deck where temperatures would have exceeded 30° C. In the circumstances, it was found that the fault in relation to the stowage of the container was not causative of the loss (i.e. the cargo would have caught fire anyway). The appeal was dismissed.

The Court further pointed out that calcium hypochlorite is not a cargo whose nature is 'such that even a strict compliance with the accepted methods of carriage will not suffice to eliminate the possibility of an accident'. It has been carried safely for decades and if the carriage by the shipowner (or, as in this case, the time charterer) cannot be faulted, the likelihood must be, both in common sense and in law, that the claim by the owner/charterer for breach of contract in shipping dangerous cargo is likely to succeed.

Deductible

The deductible is an amount specified in the policy either in monetary terms or a percentage of a sum insured, which must be exceeded before a claim is payable. When the deductible is exceeded, only the amount or percentage which is in excess of the deductible is recoverable under the policy.

A Deductible Clause incorporated in Clause 12 of the Institute Time Clauses Hulls provides that no claim arising from perils insured against payable under the insurance is to be payable unless the aggregate of all such claims arising out of each separate accident or occurrence exceeds a stated sum, in which case the sum should be deducted. There is no similar provision in respect of the Institute Cargo Clauses. Parties to a cargo insurance contract may agree to include a deductible in a policy. It is common to include a deductible in a cargo policy which covers bulk cargo, e.g., crude oil or sugar in bulk.

Under Clause 12.2 of the Institute Time Clauses Hulls, claims for damage by heavy weather occurring in a single sea passage between two successive ports are treated as being due to one accident. The effect is that an assured is only liable for one deductible for more than one claim in a single sea passage between two successive ports, but the aggregate of all claims then reaches the limit of a policy more easily. If the aggregate exceeds the limit, the assured may have to be liable for any amount above the limit. On the other hand, if claims which are caused by heavy weather, e.g., five claims, are considered separately in a single sea passage between two successive ports, the assured has to pay five times the deductible but can claim the maximum amount up to the limit of the policy in each claim case.

Successive losses

The general principle in connection to a partial loss is that the underwriter cannot be liable in respect of any one partial loss for more than the sum insured in the policy. The Marine Insurance Ordinance section 77(1), on 'Successive Losses', states:

> Unless the policy otherwise provides, and subject to the provisions of this Ordinance, the insurer is liable for successive losses, even though the total amount of such losses may exceed the sum insured.

This principle applies whenever the partial losses are repaired, or if not repaired, the subject matter is still in existence when the policy lapses. This means there has been no constructive total loss or actual total loss before the termination of the policy.

An assured is entitled to treat a constructive total loss as a partial loss and repair the subject matter if wishing to do so. The Marine Insurance Ordinance section 61, on 'Effect of Constructive Total Loss', states:

> Where there is a constructive total loss the assured may either treat the loss as a partial loss, or abandon the subject-matter insured to the insurer and treat the loss as if it were an actual total loss.

Where there is a partial loss of the subject matter, which is not repaired, and this is followed by a total loss by a peril insured against under the policy during the currency of the policy, then the assured can only recover for the total loss. The Marine Insurance Ordinance section 77(2), on 'Successive Losses', states:

> Where, under the same policy, a partial loss, which has not been repaired or otherwise made good, is followed by a total loss, the assured can only recover in respect of the total loss: Provided that nothing in this section shall affect the liability of the insurer under the suing and labouring clause.

If the total loss is not caused by a peril insured against, and it occurs during the currency of the policy, then the assured will be unable to recover for either the partial loss or the total loss. See

> *Livie v Janson* (1810) 12 East 648; 104 ER 253
>
> An American ship insured from New York to London, warranted free from American condemnation, having, for the purpose of eluding her national embargo, slipped away in the night, was by force of the ice, wind, and tide, driven on shore, where she sustained only partial damage, but was seized the next day, and afterwards, with great

> difficulty and expense, got off and finally condemned by the American government for breach of the embargo.
>
> The court held that as there was ultimately a total loss by a peril excepted out of the policy, the assured could neither recover for a total loss, nor for any previous partial loss arising from the stranding, etc., which in the event became wholly immaterial to the assured.

In *Livie v Janson,* it is stated at p 654 that if a ship is insured against marine perils but not against fire, then if a partial loss is caused by marine perils and this is not repaired, and later the ship, during the currency of the policy, is destroyed by fire, the assured will not be able to recover under the policy.

Principle of indemnity

The contract of marine insurance is based on the principles of indemnity in which an insurer agrees to indemnify an assured against all losses or damage to the subject matter insured proximately caused by perils insured against within the period of a policy. Therefore, the assured should not be indemnified if not actually suffering any loss or damage. If there is loss or damage, the compensation recovered should not exceed the loss or damage. The Marine Insurance Ordinance section 2(1), on 'Marine insurance defined', states:

> A contract of marine insurance is a contract whereby the insurer undertakes to indemnify the assured, in manner and to the extent thereby agreed, against marine losses, that is to say, the losses incident to marine adventure.

The general principles of indemnity in marine insurance can be seen from the following.

Subrogation

By section 79(1) of the Marine Insurance Ordinance, it is provided that where an underwriter pays out on an actual or constructive total loss either of the whole of the subject matter or in the case of goods, of any apportionable part of the subject matter, the underwriter may take over the rights of the assured in whatever may remain. The underwriter is subrogated to all rights or remedies of the assured in respect of the subject matter as from the time of the casualty or loss. The Marine Insurance Ordinance section 79(1), on 'Right of Subrogation', states:

> Rights of Insurer on Payment
>
> Where the insurer pays for a total loss, either of the whole, or in the case of goods of any apportionable part, of the subject-matter insured, he thereupon becomes entitled to take over the interest of the assured in whatever may remain of the subject-matter so paid for, and he is thereby subrogated to all the rights and remedies of the assured in and in respect of that subject-matter as from the time of the casualty causing the loss.

As to a partial loss, the underwriter does not acquire any interest in the subject matter upon making payment to the assured, but the underwriter is subrogated to only the rights and remedies of the assured in respect of the partial loss. The Marine Insurance Ordinance section 79(2), on 'Right of Subrogation', states:

> Subject to the foregoing provisions, where the insurer pays for a partial loss, he acquires no title to the subject-matter insured, or such part of it as may remain, but he is thereupon subrogated to all rights and remedies of the assured in and in respect of the subject-matter insured as from the time of the casualty causing the loss, in so far as the assured has been indemnified, according to this Ordinance, by such payment for the loss.

The right of subrogation in relation to an underwriter arises only when the underwriter actually pays for the loss. It is reasonable to say that in case the settlement of a claim is postponed or delayed for some reason or other, the right of subrogation is also postponed or delayed. The right of the underwriter to enforce a right of action in its own name rather than the name of the assured arises only when that right has in fact been conferred on the underwriter by statute or where the assured has made a formal assignment of the right of action to the underwriter.

> *M. H. Smith (Plant Hire) Ltd v D. L. Mainwaring (T/A Inshore)* [1986] 2 Lloyd's Rep 244
>
> The plaintiff hired a boat from the defendants in order to transport a dumper. When the dumper was loaded into the boat it sank. The underwriter paid out the claim and alleged that the boat had sunk as a result of the negligence of the defendants. In July 1985, they began an action against the defendants in the name of the assured.
>
> There was an authority to the insurers to take proceedings in the name of the assured dated July 1984, but neither the insurers nor their solicitors were aware that in March 1985, the assured had been wound up and finally dissolved and did not therefore exist at the date of the particulars of claim. In November 1985 the insurers discovered that the assured had been wound up and they applied to the court for leave to substitute the insurers as plaintiffs.

The court held that subrogation entitled the insurers to bring an action in the name of the assured against the wrongdoer to recover anything that was recoverable since the right of action was vested in the assured. Action could be brought by the insurers in its own name where it had taken a legal assignment of the cause of action from the assured. The insurers were entitled to instruct solicitors to bring this action in the name of their assured as long as their assured existed. In March 1985 the assured had ceased to exist when the company was dissolved and a non-existent party could not be party to an action.

Sisterships

Where two vessels belonging to the same shipowners are involved in a collision, then the underwriters of the same or different insurance companies who pay out for damage to one of the ships cannot bring an action against the same or different underwriter on the other ship. The reason for this is that in subrogation the underwriter stands in the shoes of the assured and the assured cannot in English law sue themselves.

Clause 9 in the Institute Time Clauses Hulls about sister ships provides that when the vessel insured comes into collision with or receives salvage services from another vessel belonging wholly or in part to the same owners or under the same management, the assured has the same rights under the insurance as they would have were the other vessel entirely the property of owners not interested in the vessel insured. However, in such cases, the liability for the collision or the amount payable for the services rendered is referred to a sole arbitrator to be agreed upon between the underwriters and the assured.

Double insurance

Double insurance may arise where the assured takes out a second or further policy in respect of the same subject matter and the same risks. The Marine Insurance Ordinance section 32(1), on 'Double Insurance', states:

> Where two or more policies are effected by or on behalf of the assured on the same adventure and interest or any part thereof, and the sums insured exceed the indemnity allowed by this Ordinance, the assured is said to be over-insured by double insurance.

Where the assured over-insures through double insurance, only up to the value of the subject matter insured may be recovered from any underwriter up to the limit insured under the particular policy in any order the assured pleases from any of the underwriters concerned in the double insurance. The Marine Insurance Ordinance section 32(2)(a), on 'Double Insurance', states:

> Where the assured is over-insured by double insurance– the assured, unless the policy otherwise provides, may claim payment from the insurers in such order as he may think fit, provided that he is not entitled to receive any sum in excess of the indemnity allowed by this Ordinance

The right of contribution

The right of contribution between underwriters, where an assured has over-insured by double insurance, is governed by section 80 of the Marine Insurance Ordinance. By this section the underwriters who have paid out or are later found to have paid out more than their proportion in respect of the loss may recover rateably from those who have not paid their proper share. Each underwriter is bound to contribute his proper share in relation to the loss. The Marine Insurance Ordinance section 80, on 'Right of Contribution', states:

> (1) Where the assured is over-insured by double insurance, each insurer is bound, as between himself and the other insurers, to contribute rateably to the loss in proportion to the amount for which he is liable under his contract.
>
> (2) If any insurer pays more than his proportion of the loss, he is entitled to maintain an action for contribution against the other insurers, and is entitled to the like remedies as a surety who has paid more than his proportion of the debt.

The right of contribution only arises in those cases where policies are effected on behalf of the same assured on the same adventure and interest. It does not apply where different persons take out two or more policies on the same subject matter against the same risks but in respect of different interests.

The case *National Farmers Union Mutual Insurance Society Ltd v HSBC Insurance (UK) Ltd*[84] is not a case in relation to marine insurance but the issue of double insurance is relevant to circumstances where a buyer purchases a vessel from an owner. It is found in the above case that on its proper construction, a policy of buildings insurance taken out by owners of a property did not extend to the buyers of the property pending completion of the sale. Insurers may limit their exposure of contribution in case it is held to be

84. [2011] 1 Lloyd's Rep 86; [2010] EWHC 773 (Comm).

a double insurance by designing their cover as an excess layer which has no direct relationship with policy of another layer, specifying that they are not subject to liability to a rateable proportion of any loss, or excluding liability of indemnity.

The property in the above case was covered only by a policy of insurance taken out by the buyers and no question of double insurance arose, so that the buyers' insurers were solely responsible for settling the resulting claim after fire damaged the property shortly before the sale was completed, and the owners' insurers were not liable to make a contribution. The defendant insurance company (H) had provided buildings cover to owners of a property under a policy of insurance and extended to anyone buying the property until the sale was completed, although it stated that it would not pay if the buildings were insured under any other insurance. The owners subsequently agreed to sell the property. Just prior to the exchange of contracts, the buyers took out a policy of insurance with the claimant insurance company (N) in respect of the property. That policy stated that if, when the buyers claimed, there was other insurance covering the same damage, N would only pay its share.

> *National Farmers Union Mutual Insurance Society Ltd v HSBC Insurance (UK) Ltd*[85]
>
> Between the dates of exchange of contracts for sale and completion, when the property was at the risk of the buyers and the subject matter of the buildings insurance taken out with both N and H, a fire broke out at the property which caused extensive damage. The sale still took place, at the full purchase price, so the owners suffered no loss and therefore made no claim under their policy with H. The buyers made a claim on their buildings cover with N in respect of the fire damage and were paid a sum in settlement of their insurance claim. N then sought a contribution from H on the basis that, at the time of the fire, H also provided buildings cover to the owners against the risk of damage to the property, which resulted in a case of double insurance.
>
> The Court held that the preliminary issues were determined in favour of the defendant. On its proper construction, the grant by H of insurance to any buyers of the property was subject to the exception that an indemnity would not be provided to them in respect of any physical loss or damage occurring up to the time of completion if those buyers had, themselves, taken out buildings insurance covering the same risk. At the time of the fire, the buyers had taken out insurance covering the same risks with another insurer, N. As a result, H did not cover the buyers in respect of the fire damage to the property. N's policy contained no provision excluding coverage in the event that

85. ibid.

the buyers were otherwise insured in respect of the same risk. The only policy covering the buyers for physical loss to the property was therefore N's policy. They were not insured with both H and N, and the instant case was not, therefore, one of double insurance. That analysis was sufficient to dispose of the preliminary issues, which would be answered accordingly (i) on the true construction of H's policy, it did not cover the buyers; (ii) on the true construction of N's policy, it contained only a pro rata clause where there was another insurance covering the same damage; (iii) in light of those conclusions, the third issue did not arise.

10.9 PARTIAL LOSS AND TOTAL LOSS

Partial losses

In the Marine Insurance Ordinance, a partial loss of the subject matter insured caused by a peril insured against is described as a particular average loss. The Marine Insurance Ordinance section 64(1), on 'Particular Average Loss', states:

> A particular average loss is a partial loss of the subject-matter insured, caused by a peril insured against, and which is not a general average loss.

The partial losses provided for in section 64 of the Marine Insurance Ordinance include salvage and general average and particular charges. Any expense which is incurred by the assured in keeping safe or preserving the subject matter insured, other than general average or pure maritime salvage, comes under the heading of 'particular charges' and is not included in particular average. The measure of indemnity, which is the amount payable by the underwriters in the event of a particular average loss in the case of a ship, freight or cargo, is dealt with by sections 69, 70 and 71 of the Marine Insurance Ordinance respectively.

The Court of Appeal in *The 'Front Comor'*[86] held that the court had power under the Arbitration Act 1996 section 66 to order judgment to be entered in the terms of an arbitral award in a case where the award was in the form of a negative declaration.

> *West Tankers Inc v Allianz SpA (The 'Front Comor')*[87]
>
> The respondent tanker owner (W) chartered the vessel to an Italian oil company (E). The vessel collided with a pier owned by E and caused

86. [2012] 1 Lloyd's Rep 398.
87. ibid.

extensive damage. The appellant insurers (G) were E's insurers and were subrogated to their claims against W. The arbitrators appointed under the charterparty made an award declaring that W were under no liability to E and G in respect of the collision. While the arbitration was pending G had brought proceedings against W in the Italian court in respect of the same incident. W obtained permission to enforce the award under s 66(1) and to enter judgment in terms of the award under s 66(2), on the basis that once the award had been entered as a judgment any subsequent Italian judgment obtained by G would not be recognised in England by virtue of Regulation 44/2001 art 34.

The Court held that the appeal dismissed. The phrase 'enforced in the same manner as a judgment to the same effect' in s 66 was not confined to enforcement by one of the normal forms of execution of a judgment but could include other means of giving judicial force to the award on the same footing as a judgment. Ultimately the efficacy of any award by an arbitral body depended on the assistance of the judicial system.

Particular average loss of a ship

Particular average loss of a ship may fall into several categories: where the ship has been repaired, where the ship is only partially repaired, where the ship is not repaired at all nor sold during the period of the policy, and where the ship is not repaired but is sold during the currency of the risks.

Where the ship has been repaired

Where the ship has been repaired, the assured can recover the reasonable cost of the repairs less customary deductions. The assured may not recover more than the sum insured in respect of any one particular average loss. The Marine Insurance Ordinance section 69(a), on 'Partial Loss of Ship', states:

> Where a ship is damaged, but is not totally lost, the measure of indemnity, subject to any express provision in the policy, is as follows–
>
> (a) where the ship has been repaired, the assured is entitled to the reasonable cost of the repairs, less the customary deductions, but not exceeding the sum insured in respect of any one casualty;

Where the ship is only partially repaired

When a ship is damaged but has a tight schedule to meet, the shipowners may arrange temporary repair or partial permanent repair subject to the recommendations of the surveyor of the classification or the flag state. In some cases, the shipowners may want to carry out permanent repairs but there is no appropriate repair facility in the vicinity. The shipowners may carry out temporary repairs and then the vessel sails or is towed to a nearby shipyard for permanent repair. Where the ship is only partially repaired the assured can recover a reasonable cost of the repairs less customary deductions, and they are also entitled to a sum for reasonable depreciation in respect of the unrepaired damage. The total amount of the cost of repair and depreciation of unrepaired damage may not exceed the cost for repairing the entire damage suffered. The Marine Insurance Ordinance section 69(b), on 'Partial Loss of Ship', states:

> Where a ship is damaged, but is not totally lost, the measure of indemnity, subject to any express provision in the policy, is as follows–
>
> (b) where the ship has been only partially repaired, the assured is entitled to the reasonable cost of such repairs, computed as above, and also to be indemnified for the reasonable depreciation, if any, arising from the unrepaired damage, provided that the aggregate amount shall not exceed the cost of repairing the whole damage, computed as above.

Where the ship is not repaired at all NOR sold during the period of the policy

Where the ship is not repaired at all nor sold during the period of the policy the assured may recover a reasonable depreciation arising from the unrepaired damage. The sum recoverable may not exceed the reasonable cost of repairing the whole of the damage less customary deductions. In the case of a time policy, the measure of the depreciation is to be taken at the time of the expiration of the time policy. The Marine Insurance Ordinance section 69(c), on 'Partial Loss of Ship', states:

> *Helmville, Ltd v Yorkshire Insurance Company, Ltd (The 'Medina Princess')* [1965] 1 Lloyd's Rep 361
>
> Damage was sustained by the plaintiffs' ship during the voyage from Bremen to China laden with a cargo of flour, and the ship was insured under a valued policy containing the Institute Time Clauses Hulls. The damaged value of the ship was nil.
>
> The court held that 'reasonable cost of repairs' in Sect. 69 was not confined to the reasonable cost of permanent repairs, and that it was a question of fact in every case as to what the phrase included; that Sect. 69 (3) required the measure of indemnity to be quantified on the basis of what it would have cost to repair the Medina Princess if the repairs had been carried out; that, in order to repair her, it would

> have been necessary to tow her to Karachi; and that, therefore, the cost of towing (£14,560), although not incurred, was recoverable by the plaintiffs as part of their partial loss claim.

The Institute Time Clauses Hulls Clause 18.1 provides that the measure of indemnity in respect of claims for unrepaired damage is the reasonable depreciation in the market value of the vessel at the time the insurance terminates arising from such unrepaired damage, but not exceeding the reasonable cost of repairs. In Clause 18.2 of the Institute Time Clauses Hulls, the underwriters are not liable for unrepaired damage in the event of a subsequent total loss, caused by a peril insured against or not, sustained during the period covered by the policy or any extension.

> *British and Foreign Insurance Company, Limited v Wilson Shipping Company, Limited* [1921] 1 AC 188, HL
>
> By a time policy a vessel was insured against marine risks only, but including particular average. The vessel was under charter to the Admiralty, under which charter the Admiralty contracted to pay for loss by war risks, the value to be ascertained at the date of the loss. During the currency of the policy the vessel sustained damage by marine risks, but, to the extent of 1770l, this damage was not repaired. On a subsequent voyage, but during the currency of the policy, the vessel was torpedoed by an enemy submarine and became a total loss. The Admiralty in consequence of the unrepaired damage paid the owners 1770l. less than they would otherwise have done. The owners claimed to recover from some of the underwriters their proportion of the 1770l.
>
> The court held that as the unrepaired damage was followed by a total loss during the currency of the same policy, the smaller loss merged in the larger, and therefore the underwriters were not liable for the unrepaired damage, notwithstanding the fact that the liability for the total loss did not fall upon them.
>
> Under Clause 18.3 of the Institute Time Clauses Hulls, the underwriters are not liable in respect of unrepaired damage for more than the insured value at the time the policy terminates.

Where the ship is not repaired but is sold during the currency of the risks

Where the ship suffers a particular average which is not repaired but the ship is sold during the currency of the risks, the assured may recover the reasonable cost of repairing such damage but the sum recovered is not to exceed the actual depreciation in the value of the ship as ascertained by the sale and not more than the insured value.

Particular average loss of freight

Part or whole of the freight may be lost when part or whole of the cargo has not been properly delivered in accordance with the terms of a contract of carriage. The measure of indemnity of particular average loss of freight depends on whether the policy is a valued policy or an unvalued one. The Marine Insurance Ordinance section 70, on 'Partial Loss of Freight', states:

> Subject to any express provision in the policy, where there is a partial loss of freight, the measure of indemnity is such proportion of the sum fixed by the policy in the case of a valued policy, or of the insurable value in the case of an unvalued policy, as the proportion of freight lost by the assured bears to the whole freight at the risk of the assured under the policy.

Particular average loss of goods

The measure of indemnity is dealt with by section 71 of the Marine Insurance Ordinance. Measurement of the loss is provided for both valued and unvalued policies and also for measuring the loss depending upon whether some part of the goods have been actually lost or merely arrived in a damaged condition. The Marine Insurance Ordinance section 71, on 'Partial Loss of Goods, Merchandise, etc.', states:

> Where there is a partial loss of goods, merchandise, or other movables, the measure of indemnity, subject to any express provision in the policy, is as follows–
> (a) Where part of the goods, merchandise or other movables insured by a valued policy is totally lost, the measure of indemnity is such proportion of the sum fixed by the policy as the insurable value of the part lost bears to the insurable value of the whole, ascertained as in the case of an unvalued policy;
> (b) Where part of the goods, merchandise, or other movables insured by an unvalued policy is totally lost, the measure of indemnity is the insurable value of the part lost, ascertained as in the case of total loss;
> (c) Where the whole or any part of the goods or merchandise insured has been delivered damaged at its destination, the measure of indemnity is such proportion of the sum fixed by the policy in the case of a valued policy, or of the insurable value in the case of an unvalued policy, as the difference between the gross sound and damaged values at the place of arrival bears to the gross sound value;

(d) 'Gross value' means the wholesale price or, if there be no such price, the estimated value, with, in either case, freight, landing charges, and duty paid beforehand; provided that, in the case of goods or merchandise customarily sold in bond, the bonded price is deemed to be the gross value. 'Gross proceeds' means the actual price obtained at a sale where all charges on sale are paid by the sellers.

Salvage charges

Salvage charges, which are charges incurred in a voluntary salvage in preventing a loss caused by perils insured against, are recoverable as a loss by the perils. The Marine Insurance Ordinance section 65, on 'Salvage Charges', states:

(1) Subject to any express provision in the policy, salvage charges incurred in preventing a loss by perils insured against may be recovered as a loss by those perils.

A salvor may recover charges under maritime law incurred by the salvor who is not bound to carry out the service of salvage under a contract. Expenses of services in salvage provided by the assured, their agent or any persons employed by the assured for preventing a loss from perils insured against are not recoverable. The Marine Insurance Ordinance s 65, on 'Salvage charges', states:

(2) 'Salvage charges' means the charges recoverable under maritime law by a salvor independently of contract. They do not include the expenses of services in the nature of salvage rendered by the assured or his agents, or any person employed for hire by them, for the purpose of averting a peril insured against. Such expenses, where properly incurred, may be recovered as particular charges or as a general average loss, according to the circumstances under which they were incurred.

David Aitchison and A. F. Brandt v Haagen Alfsen Lohre 4 App Cas 755, HL
Lord Blackburn at p 765 stated that salvage charges were not recoverable under the sue and labour clause because the owners of the ship doing the services did so as salvors acting on the maritime law . . . quite independently of whether there was insurance or not, or whether, if there was a policy of insurance, it contained the suing and labouring clause or not.

In *David Aitchison and A. F. Brandt v Haagen Alfsen Lohre*, it was held that salvage did not come within the suing and labouring clause of a policy of marine insurance. Lord Cairns LC stated at pp 766–67: 'I will only make one observation with regard to salvage expenses. It appears to me to be quite clear that if any expenses were to be recoverable under the suing and labouring clause, they must be expenses assessed upon the quantum meruit principle. Now salvage expenses are not assessed upon the quantum meruit principle; they are assessed upon the general principle of maritime law, which gives to the persons who bring in the ship a sum quite out of proportion to the actual expense incurred and the actual service rendered, the largeness of the sum being based upon this consideration—that if the effort to save the ship (however laborious in itself, and dangerous in its circumstances) had not been successful, nothing whatever would have been paid. If the payment were to be assessed and made under the suing and labouring clause, it would be payment for service rendered, whether the service had succeeded in bringing the ship into port or not.'

In *Royal Boskalis Westminster NV and Others v Mountain and Others* [1999] QB 974, the court held that the fact that a payment cannot be valued as a quantum meruit does not prevent a claim under the sue and labour clause. Neither does the fact that the expense is incurred by way of waiving a claim rather than making a payment. Where a salvage award is made, which includes an element of life salvage, the assured may nevertheless recover the entire sum/amount of that award from the underwriter even though it is not solely in respect of the subject matter insured, i.e., master, officer, crew, passenger or supernumerary. It is all because of public policy it is not regarded as a 'perfect' indemnity in marine insurance.

> *Grand Union (Shipping), Ltd v London Steam-ship Owners' Mutual Insurance Association, Ltd (The 'Bosworth' (No. 3))* [1962] 1 Lloyd's Rep 483
>
> The plaintiffs' motor vessel Bosworth entered for protection in the defendant P. & I. Club. Salvage services were rendered to the Bosworth by the steamship Finnmerchant, the trawler Wolverhampton Wanderers and the trawler Faraday. The Finnmerchant poured oil onto the sea to assist the saving of Bosworth's crew by the Wolverhampton Wanderers.
>
> The court held (1) that life salvage was not a form of common-law maritime salvage, but a species of salvage created by an Act of Parliament; that Wolverhampton Wanderers' award was an award against ship, cargo and freight for services rendered to ship, cargo and freight, enhanced by services rendered for saving life; and that, therefore, it was recoverable under the Lloyd's policy and not from the defendants; (2) that, by the practice of the Admiralty Court, an award made in these circumstances was treated as an award for services rendered to ship and cargo and was within s 65(1) of the Marine Insurance Act 1906; (3) that the sum paid to the Finnmerchant was life salvage

and would have been recoverable from the defendants, but more than that amount had already been paid by the defendants to the plaintiffs. Judgment was given for the defendants.

Particular charges

The Marine Insurance Ordinance section 64, on 'Particular average loss', states:

> (1) A particular average loss is a partial loss of the subject-matter insured, caused by a peril insured against, and which is not a general average loss.
>
> (2) Expenses incurred by or on behalf of the assured for the safety or preservation of the subject-matter insured, other than general average and salvage charges, are called particular charges. Particular charges are not included in particular average.

Particular charges are dealt with in section 78 of the Marine Insurance Ordinance and arise in respect of the 'sue and labour clause' as supplementary to the policy. The insurer may have to pay sue and labour claims even after a total loss is paid to the assured. General average losses, general average contributions and salvage charges are not recoverable under the sue and labour clause. The Marine Insurance Ordinance section 78, on 'Suing and Labouring' clause, states:

> (1) Where the policy contains a suing and labouring clause, the engagement thereby entered into is deemed to be supplementary to the contract of insurance, and the assured may recover from the insurer any expenses properly incurred pursuant to the clause, notwithstanding that the insurer may have paid for a total loss, or that the subject-matter may have been warranted free from particular average, either wholly or under a certain percentage.
>
> (2) General average losses and contributions and salvage charges, as defined by this Ordinance, are not recoverable under the suing and labouring clause.

Without a sue and labour clause, an assured who has incurred expense or liability in attempting to mitigate or minimize risks in connection to the subject matter insured from the perils insured against would be unable to recover in respect of such expense or liability from the underwriter. Subsequently, the assured would be unwilling to incur expense or liability for the purpose of averting or diminishing loss to the subject matter by perils insured against. The underwriters would be likely to have a greater number of claims for higher particular average losses. Under the terms of the sue and labour clause, it is provided that the assured and their servants, agents, etc may sue and labour and travel for, in and about, the defence, safeguards and recovery of the subject matter insured.

Clause 13.2 of the Institute Time Clauses Hulls stipulates that, subject to the provisions and to Clause 12, the underwriters will contribute to charges properly and reasonably incurred by the assured, their servants or agents for such measures. General average, salvage charges, except as provided for in Clause 13.5, and collision defence or attack costs are not recoverable under the sue and labour clause. The assured may only recover in respect of charges incurred for the purpose of 'averting or diminishing loss to the subject matter' by perils insured against. If the assured incurs charges in respect of perils not insured against, then they will be unable to recover such charges from the underwriter. The Marine Insurance Ordinance section 78, on 'Suing and Labouring' clause, states:

> (3) Expenses incurred for the purpose of averting or diminishing any loss not covered by the policy are not recoverable under the suing and labouring clause.

An assured may be in a difficult situation in not taking any proactive action to avert the loss to the subject matter insured because the underwriter may claim that it is the duty of the assured to take appropriate measures to avert the loss or damage. The Marine Insurance Ordinance section 78, on 'Suing and Labouring' clause, states:

> (4) It is the duty of the assured and his agents, in all cases, to take such measures as may be reasonable for the purpose of averting or minimizing a loss.

In section 78(4) of the Marine Insurance Ordinance, it is stated to be the duty of the assured and their agents in all cases to take such measures as may be reasonable for the purpose of averting or minimizing the loss. There is no apparent conflict between this provision and the provisions of section 55(2)(a) of the Marine Insurance Ordinance, which states that the assured may recover for a loss proximately caused by a peril insured against, even though the loss would not have occurred but for the misconduct or negligence of the master or crew. The provision in section 78(4) does not qualify section 55(2)(a) and must be read as applying solely to the sue and labour clause.

The provision of section 78(4) of the Marine Insurance Ordinance relates to the assured and their agents; it would seem reasonable that 'agents' in this case should relate only to the management of the ship and that it should not be read as including the master, officer or crew of the ship. It is not completely clear whether a failure by the assured to sue and labour as required by the Ordinance would incur a recourse action against the assured by the underwriters. If the provision of section 78(4) of the Marine Insurance Ordinance is regarded as a part of a contract, then it would appear that an action should lie. However, the sue and labour clause itself merely says that it shall be lawful for the assured to sue and labour and not that it is a must.

Clause 13.1 of the Institute Time Clauses Hulls stipulates that in case of any loss or misfortune it is the duty of the assured and their servants and agents to take such measures as may be reasonable for the purpose of averting or minimizing a loss which would be recoverable under the policy. The provision of clause 13.1 includes the wording 'it is the duty of the Assured and their servants and agents'. Clause 16 of the Institute Cargo Clauses (A), (B) and (C) provides that it is the duty of the assured and their servants and agents in respect of loss recoverable in the policy to take such measures as may be reasonable for the purpose of averting or minimizing such loss, and to ensure that all rights against carriers, bailees or other third parties are properly preserved and exercised and that the underwriters will, in addition to any loss recoverable in the policy, reimburse the assured for any charges properly and reasonably incurred in pursuance of these duties. It is again stated that 'it is the duty of the assured and their servants and agents' to avert or minimize a loss.

A sue and labour clause usually contains a 'waiver clause' which states that any measure either by the assured or the underwriter for the purpose of protecting, recovering, preserving or saving the subject matter insured is not considered as a waiver or acceptance of the rights of either party in relation to the giving of or accepting a notice of abandonment. The waiver clause is important for the underwriter because of the provision of section 62(5) of the Marine Insurance Ordinance, which on 'Notice of Abandonment' states:

> The acceptance of an abandonment may be either express or implied from the conduct of the insurer. The mere silence of the insurer after notice is not an acceptance.

Whether or not expenses incurred in respect of the subject matter have been properly incurred and are, therefore, recoverable as particular charges is a question of fact. See

> *Wilson Brothers Bobbin Company, Limited v Green* [1917] 1 KB 860
>
> By a policy of marine insurance underwritten by the defendant, the plaintiffs were insured in respect of a wood cargo laden on a Norwegian

ship for a voyage from a Baltic port to an English port. The policy, which contained the usual suing and labouring clause, was against war risk only, and excluded all claims arising from delay. Shortly after sailing in November 1914, the ship was stopped by German war vessels, and the master was told that as wood had been declared contraband by the German Government the ship would not be allowed to pass the Sound. The ship put into a Norwegian port, where the cargo was discharged and stored for some time, but was subsequently reshipped and forwarded to its destination in England.

The court held that under the suing and labouring clause the plaintiffs were entitled to recover the cost of storage for a reasonable time and the proper cost of forwarding the cargo to its port of destination at the expiration of that time.

General average loss

General average loss is defined in section 66(1) of the Marine Insurance Ordinance, which on 'General Average Loss' states:

> A general average loss is a loss caused by or directly consequential on a general average act. It includes a general average expenditure as well as a general average sacrifice.

As to the meaning of a general average act, the Marine Insurance Ordinance section 66(2), on 'General Average Loss', states:

> There is a general average act where any extraordinary sacrifice or expenditure is voluntarily and reasonably made or incurred in time of peril for the purpose of preserving the property imperilled in the common adventure.

In order for an assured to recover in respect of a general average loss, it is necessary that the general average act is made in respect of a real peril and not a merely imagined peril.

> *Joseph Watson and Son, Limited v Firemen's Fund Insurance Company of San Francisco* [1922] 2 KB 355
>
> Under a mistaken assumption of fire the captain of a ship caused steam to be turned into the hold to extinguish the supposed fire, and so damaged the plaintiffs' goods. On a claim by them for a general average loss against the defendants as insurers, the court held that the 'peril' being in fact non-existent there was no general average loss, and, secondly, even if there were, that the defendants were not liable, as the loss had not been caused through a peril insured against, such a peril having never existed. Further, that the words 'loss . . . incurred for the

purpose of avoiding, or in connexion with the avoidance of, a peril insured against' in s 66 (6) of the Marine Insurance Act 1906 did not operate to cover such a loss as this, but only losses collateral to the main process of avoiding a peril insured against.

The right of the assured to recover in respect of a general average loss arises at the time when the loss occurs and not at the time when the amount payable by the assured is determined by an average adjusters. See

> *Chandris v Argo Insurance Company, Ltd and Others* [1963] 2 Lloyd's Rep 65
>
> General average and particular average losses were incurred by the plaintiff shipowners in respect of their vessels, insured by the defendants under policies containing Institute Time Clauses Hulls. The actions were commenced by the shipowners, claiming to be indemnified by the insurers, more than six years after the dates of the losses or terminations of adventures, but less than six years after the relevant average adjustments were completed and issued.
>
> The court held (a) that, by authority and by s 66 of the Marine Insurance Act 1906, the assured's right to recover from his insurer in respect of a general average loss arose when the loss occurred; and that, accordingly, the cause of action arose at the same time; (b) that Clause 8 prescribed the method of computing liability and did not result in there being no liability on the insurers or time not running until the amount had been computed; that the Rules provided that an average adjustment should be produced, but that did not mean that that was a condition precedent to a cause of action arising; and (c) that the plaintiffs had failed to establish that, on those policies, a cause of action arose and time began to run only when an average statement had been prepared and published to parties to the adventure.

Where a general average loss is not incurred for the avoidance of a peril insured against, the insurer is not liable for general average loss or contribution. The Marine Insurance Ordinance section 66(6), on 'General Average Loss', states:

> In the absence of express stipulation, the insurer is not liable for any general average loss or contribution where the loss was not incurred for the purpose of avoiding, or in connection with the avoidance of, a peril insured against.

In the case of a general average expenditure, the assured may only recover from the underwriter that proportion of the expenditure which is payable by the assured. Where there is a general average sacrifice, the assured need not have enforced their rights of contribution from the other parties in the adventure. The Marine Insurance Ordinance section 66(4), on 'General Average Loss', states:

> Subject to any express provision in the policy, where the assured has incurred a general average expenditure, he may recover from the insurer in respect of the proportion of the loss which falls upon him; and, in the case of a general average sacrifice, he may recover from the insurer in respect of the whole loss without having enforced his right of contribution from the other parties liable to contribute.

If the assured has paid or is liable to pay a general average contribution in respect of the subject matter insured, then they may recover this from the underwriter. In *Castle Insurance Co Ltd and Others v Hong Kong Islands Shipping Co Ltd* [1984] AC 226, it is stated that at common law the liability to make a general average contribution arises at the time the general average sacrifice is made or the expenditure incurred, subject to such liability's being defeated by non-arrival at the port of destination. A shipowner's right of lien on cargo against payment of contribution or the provision of adequate security depends on the coming into existence of an accrued liability to pay general average. This must not be later than the vessel's safe arrival at the port of destination. (As to the time of accrual of the cause of action, see *Tate & Lyle Ltd v Hain Steamship Co Ltd* (1934) 49 Ll L Rep 123, 135; (1936) 55 Ll L Rep 159, 174; *Morrison Steamship Co Ltd v Greystoke Castle Cargo Owners* [1947] AC 265 and *Chandris v Argo Insurance Co Ltd* [1963] 2 Lloyd's Rep 65, 79. See also *Schothorst and Schuitema v Franz Dauter GmbH. (The Nimrod)* [1973] 2 Lloyd's Rep 91, 97, per Kerr J.)

A general average contribution arises where there is a general average loss and the party suffering the loss is entitled to recover a rateable contribution from the other parties interested in the adventure.

> *Castle Insurance Co Ltd and Others v Hong Kong Islands Shipping Co Ltd* [1984] AC 226, PC
>
> Between 30 October and 27 November 1972, salvage operations which included jettison of cargo were carried out on the ship Potoi Chau. The ship was on a voyage carrying general cargo consigned under bills of lading containing a clause which provided general average. Between November 1972 and the end of February 1973 the ship managers released the preserved cargo to its several consignees on each of the consignees signing a Lloyd's standard form average bond supported by an insurer's letter of guarantee from the insurer of the particular consignment as security.
>
> The court held (1) that the effect of the general average clause in the bill of lading was to transfer to the consignee as endorsee of the bill

> of lading the common law liability to contribute to general average of whoever had been the owner of the cargo at the time the sacrifice was made; that, therefore, since by its terms the clause itself did not postpone the date when the cause of action accrued and, since at common law a cargo owner's liability to contribute to general average accrued at the time the sacrifice was made or the expense incurred, the shipowners' cause of action against the consignees under the general average clause had been time-barred at the time when the ship managers had applied to join the shipowners as additional plaintiffs.

In *Castle Insurance Co Ltd and Others v Hong Kong Islands Shipping Co Ltd*, the Court stated that where a consignee of cargo executed a Lloyd's standard form average bond in return for release of cargo, the consignee undertook a fresh contractual obligation to contribute to general average. Since by their terms the bonds provided that the obligation to contribute was not to arise until the adjusters had completed the general average statement, it was the earliest date at which the shipowners' cause of action against the consignees for payment of contribution arose under the general average bonds. Therefore, the shipowners' claims under the bonds against the consignees had not been time-barred when the ship managers applied to join the shipowners as plaintiffs. The Court further held that by the terms of the letters of guarantee used, each of the insurers had assumed a primary liability to pay a sum of money either explicitly or implicitly expressed to be the general average contribution which might properly be found to be due on completion of the average statement by the adjusters. Therefore, at the time when the ship managers applied to join the shipowners as plaintiffs, the shipowners' right of action against the insurers had not been time-barred.

If the ship, the freight, the cargo or any two of them are owned by the same assured, it would not be an issue of same ownership. The Marine Insurance Ordinance section 66(7), on 'General Average Loss', states:

> Where ship, freight, and cargo, or any two of those interests, are owned by the same assured, the liability of the insurer in respect of general average losses or contributions is to be determined as if those subjects were owned by different persons.

Where the underwriter pays out under a general average sacrifice in respect of the entire loss, then the underwriter may enforce the assured's rights for contribution against the other owners of the sums involved.

> *Dickenson and Others v Jardine and Others* [LR] 3 CP 639
>
> A insured goods at Canton by a policy which included jettison among the perils insured against. The goods were jettisoned under

> circumstances which entitled A to a general average contribution from the owners of the ship and of the rest of the cargo, which arrived safely at London, the port of discharge. A having sued the underwriters for the whole amount insured, without having first collected the contributions to which he was entitled from the other owners of the ship and cargo, the court held that he was entitled to recover and that the underwriters having paid him would then be entitled to stand in his place with respect to the general average contribution. The liability of the underwriters under the policy could not be varied by a custom, alleged to exist in the port of London between merchants and underwriters, to hold the latter liable only for the share of the loss cast upon the owner of jettisoned goods in the general average statement.

Clause 11.1 of the Institute Time Clauses Hulls provides that the insurance covers the insured vessel's proportion of salvage, salvage charges and/or general average, reduced in respect of any under-insurance, but in case of general average sacrifice of the vessel insured the assured may recover in respect of the whole loss without first enforcing its right of contribution from other parties. In Clause 11.2 of the Institute Time Clauses Hulls, it is stated that adjustments are to be according to the law and practice obtaining at the place where the adventure ends, as if the contract of affreightment contained no special terms upon the subject; but where the contract of affreightment so provides, the adjustment must be according to the York-Antwerp Rules.

Notice of claim and tender

There is no provision dealing with the issue of notice of claim or tender in the Marine Insurance Ordinance except in the case of a constructive total loss. The Marine Insurance Ordinance section 62(3), on 'Notice of Abandonment', states:

> Notice of abandonment must be given with reasonable diligence after the receipt of reliable information of the loss, but where the information is of a doubtful character the assured is entitled to a reasonable time to make inquiry.

Clause 10.1 of the Institute Time Clauses Hulls provides that in the event of accident whereby loss or damage may result in a claim under the policy, notice must be given to the underwriters prior to the survey and also, if the vessel insured is abroad, to the nearest Lloyd's Agent so that a surveyor may be appointed to represent the underwriters should they so desire. Clause 10.3 of the Institute Time Clauses Hulls states that the underwriters may take tenders or may require further tenders to be taken for the repair of the vessel insured. In 10.4 of the Institute Time Clauses Hulls, it is provided that in the event of failure to comply with the conditions of Clause 10 a deduction of 15 per cent is to be made from the amount of the ascertained claim.

Management of claim

An insured cannot recover any part of a claim under a policy if the claim has been fraudulently exaggerated or claim has been supported by dishonest mechanism. On the other hand, a person cannot be deprived of a judgment for damages unless it is proved to have an abuse of process.

The principal issues in this appeal for the Supreme Court are whether a civil court ('the court') has power to strike out a statement of case as an abuse of process after a trial at which the court has held that the defendant is liable in damages to the claimant in an ascertained sum and, if so, in what circumstances such a power should be exercised. The driving force behind the appeal is the defendant's liability insurers, who say that fraudulent claims of the kind found to exist are rife and should in principle be struck out as an abuse of the court's process under CPR 3.4(2) or under the inherent jurisdiction of the court.

> *Fairclough Homes Limited v Summers*[88]
> The claimant (C) was accidentally injured by falling from a stacker truck whilst working for the defendant. C claimed damages of £838,000 from his employer and the claim amount was later found by the Court to be grossly exaggerated. An amount of £88,716 was finally awarded for the genuine part of C's claim.
>
> The Court held that although the Court had accepted the defendant's submission that the court has power under the CPR and under its inherent jurisdiction to strike out a statement of case at any stage of the proceedings, even when it has already determined that the claimant is in principle entitled to damages in an ascertained sum, the Court had concluded that that power should in principle only be exercised where it is just and proportionate to do so, which is likely to be only in very exceptional circumstances. The Court had concluded that this not such a case.

88. [2013] 1 Lloyd's Rep 159, SC; [2012] UKSC 26.

PART III

Dispute Resolution

11

Admiralty Jurisdiction

11.1 INTRODUCTION

Hong Kong was a British Dependent Territory before 1 July 1997. The legal system was based on that in England and the laws of both jurisdictions were similar and often identical. The People's Republic of China resumed the exercise of sovereignty over Hong Kong on 1 July 1997. Hong Kong has become a Special Administrative Region ('SAR') directly under the Central People's Government and is recognized as an inalienable part of the People's Republic of China. The Basic Law is the constitutional document for the Hong Kong SAR. It sets out the new constitutional order of 'one country, two systems', 'a high degree of autonomy' and 'Hong Kong People ruling Hong Kong'.

Sources of maritime law in Hong Kong

Article 8 of the Basic Law states that the laws previously in force in Hong Kong, that is, the common law, rules of equity, ordinances, subordinate legislation and customary law shall be maintained, except for any that contravene the Basic Law, and subject to any amendment by the legislature of the Hong Kong SAR. Maritime law is the law being applied by the Admiralty Jurisdiction of the High Court in Hong Kong by legislation, or has been adopted by the court by decision, tradition and principle. There are three main sources:

- legislation (ie, ordinances)
- case law
- international conventions

Legislation (Ordinances)

The Hong Kong SAR has the power to make new ordinances through the legislature under article 17 of the Basic Law. The law-making body in the SAR is the Legislative Council. It has wide law-making powers except in the areas of defence and foreign affairs. All ordinances passed by the Hong Kong Legislative Council must be reported to the Standing Committee of the National People's Congress. The following is a summary of the legislative procedure:

Instructions from the policy branches of the Hong Kong government

↓

Bill drafting by the Law Drafting Division of the Department of Justice

↓

Consideration by the Chief Executive and the Executive Council

↓

Publication of the bill

↓

First reading, second reading (consideration by
the various bills committees)
and third reading

↓

Assent by the Chief Executive

↓

Publication of the Ordinance

↓

Reporting of the Ordinance to Standing Committee
of the National People's Congress

The Hong Kong Legislative Council is entitled to delegate some of its law-making powers to other bodies. 'Delegated legislation' is the description given to the vast body of rules, orders, regulations and by-laws created by subordinate bodies under specific powers delegated to those bodies by the

Legislative Council. The advantage of delegated legislation is that it enables regulations to be made and amended quickly without the need for placing them before the Legislative Council. Also, it must be noted that delegated legislation is valid only if it is within the legislative powers conferred by the Legislative Council. For example, the Bills of Lading Analogous Shipping Documents Ordinance (Cap 440, Laws of Hong Kong) came into force on 1 March 1994 replacing the old Bill of Lading Ordinance. It governs the transfer of rights and liabilities under bills of lading, and the representations in the bills of lading. The Ordinance provides that the Secretary for Trade and Industry may by regulation make provision for the application of the Ordinance to cases where a telecommunication system or any other information technology is used for effecting transactions relating to bills of lading.

Case law

Case law is a major source of the law in Hong Kong. It refers to the creation and refinement of law in the course of judicial decisions. It is also called the 'judge-made' law. The doctrine of *stare decisis*, that is, the doctrine of binding precedent, is fundamental to the Hong Kong legal system. Within the hierarchical structure of the Hong Kong courts, a decision of a higher court will be binding on a court lower than it in that hierarchy.

Although Hong Kong has continued to apply common law since the sovereignty reverted to China in 1997, judicial decisions in England are unlikely to be binding on local courts in Hong Kong. The English decisions, however, may well continue to be persuasive in Hong Kong.

International conventions

Article 13 of the Basic Law stipulates that the Central People's Government is responsible for foreign affairs relating to the HKSAR, but it authorizes the HKSAR to conduct the relevant external affairs in accordance with the Basic Law. Article 153 of the Basic Law deals with the application of international agreements to the HKSAR and their implementation. The views of the HKSAR government have to be sought before international agreements to which the PRC is a party (or becomes a party) are extended to the HKSAR. Article 153 further stipulates that international agreements to which China is not a party but which are implemented in Hong Kong may continue to be implemented in the HKSAR. On the other hand, article 151 of the Basic Law provides that, the HKSAR, using the name 'Hong Kong, China' may maintain and develop relations and conclude and implement agreements on its own, with foreign states and regions and international organizations, in such matters as economic affairs, trade, finance and monetary affairs, shipping, communications, tourism, culture and sports.

On the Bilingual Laws Information System (BLIS) website (http://www.justice.gov.hk), the Hong Kong government provides a list of all the multilateral international agreements applying to the HKSAR. The list identifies a number of multilateral international agreements not applying to the rest of China.

11.2 ADMIRALTY JURISDICTION

Admiralty jurisdiction in Hong Kong refers to the jurisdiction conferred by the High Court Ordinance (Cap 4, Laws of Hong Kong). In other words, the jurisdiction of the Admiralty jurisdiction of the Court of First Instance in Hong Kong is statutory. It is governed by s 12A to s 12E of the High Court Ordinance. The Admiralty jurisdiction of the Court of First Instance shall consist of jurisdiction to hear and determine any of the following questions and claims:

(a) any claim to the possession or ownership of a ship or to the ownership of any share therein;
(b) any question arising between the co-owners of a ship as to possession, employment or earnings of that ship;
(c) any claim in respect of a mortgage of or charge on a ship or any share therein;
(d) any claim for damage received by a ship;
(e) any claim for damage done by a ship;
(f) any claim for loss of life or personal injury sustained in consequence of any defect in a ship or in her apparel or equipment, or in consequence of the wrongful act, neglect or default of:
    (i) the owners, charterers or persons in possession or control of a ship; or
    (ii) the master or crew of a ship, or any other person for whose wrongful acts, neglects or defaults the owners, charterers or persons in possession or control of a ship are responsible, being an act, neglect or default in the navigation or management of the ship, in the loading, carriage or discharge of goods on, in or from the ship, or in the embarkation, carriage or disembarkation of persons on, in or from the ship;
(g) any claim for loss of or damage to goods carried in a ship;

(h) any claim arising out of any agreement relating to the carriage of goods in a ship or to the use or hire of a ship;
(i) any claim–
  (i) under the Salvage Convention 1989;
  (ii) under any contract for or in relation to salvage services; or
  (iii) in the nature of salvage not falling within subparagraph (i) or (ii);
  or any corresponding claim in connection with an aircraft;
(j) any claim in the nature of towage in respect of a ship or an aircraft;
(k) any claim in the nature of pilotage in respect of a ship or an aircraft;
(l) any claim in respect of goods or materials supplied to a ship for her operation or maintenance;
(m) any claim in respect of the construction, repair or equipment of a ship or in respect of dock charges or dues;
(n) any claim by a master or member of the crew of a ship for wages (including any sum allotted out of wages or adjudged by a superintendent to be due by way of wages);
(o) any claim by a master, shipper, charterer or agent in respect of disbursements made on account of a ship;
(p) any claim arising out of an act which is or is claimed to be a general average act;
(q) any claim arising out of bottomry;
(r) any claim for the forfeiture or condemnation of a ship or of goods which are being or have been carried, or have been attempted to be carried, in a ship, or for the restoration of a ship or any such goods after seizure, or for droits of Admiralty;
(s) any claim arising under section 7 of the Merchant Shipping (Prevention and Control of Pollution) Ordinance (Cap 413).

The Jupiter [1925][1]
This was a dispute which required the determination of the meaning of 'any claim to the possession of a ship' under the English equivalent of s 12A(2)(a) High Court Ordinance.

The vessel 'Jupiter' was owned by a Russian company. Because of the Russian Revolution, all private property in shipping was abolished by Government decree in 1918. An English company acting as agents for the Soviet Government purported to sell the ship to an Italian buyer. The Russian company brought proceedings against the ship claiming possession. However, the Italian buyer protested that the English court had no jurisdiction to hear the claim.

1. [1925] All ER Rep 203.

The Court of Appeal held that the court had jurisdiction to deal with the claim for possession, even though the parties were foreigners.

Atkin LJ: 'The only question that is left is whether or not there is a discretion in the court to decline to exercise jurisdiction in such cases, and, if so, whether that jurisdiction ought to be so exercised in this case. As to that, the law seems to . . . [be] that the court, in such a case, has a discretion whether it will exercise its jurisdiction or not, and in cases where the parties both belonging to a foreign State have merely taken the occasion of the ship being temporarily here to get a question of title, which depends on the municipal laws of another country, determined by the courts of this country, the court may, in the exercise of its discretion, decline to do so. But, on the facts of this case, there seems to me to be no reason why the court should not exercise its discretion and entertain the suit. The vessel has been in this country for a period of years, and the question arises in respect of her disposition by a contract entered into in this country by a limited company of this country . . . and, although questions may arise as to the right of title of the vendors to the defendants, yet it appears to me to be a case which can properly be tried in this country.'

The Acrux [1965][2]

By s 12A(2)(c) of the High Court Ordinance in Hong Kong, the Admiralty jurisdiction includes the determination of 'any claim in respect of a mortgage of or charge on a ship'.

In this English case, action was brought against the proceeds of sale of the vessel 'Acrux', which was registered in Italy for unpaid seamen's social insurance contributions. Under Italian law, the unpaid social insurance contributions could give rise to a right to arrest the vessel.

The court held that the word 'charge' meant a charge in the nature of a mortgage, and not a charge for unpaid social insurance contributions.

Hewson J: '"Other charges", in my view, means any charge on a vessel given under the law of any nation to secure claims similar to those recognised by this court as carrying a maritime lien such as wages, damage, salvage and bottomry. The categories of maritime lien as recognised by this court cannot, in my view, be extended except by the legislature . . . This court had no original jurisdiction to entertain claims *in rem* for insurance contributions, or anything like them, because they were not thought of.'

2. [1965] 1 Lloyd's Rep 565.

United Africa Co Ltd v Owners of m.v. 'Tolten' [1946][3]

By s 12A(2)(e) of the High Court Ordinance, the Admiralty jurisdiction includes jurisdiction to hear and determine 'any claim for damage done by a ship'.

A British ship caused damage to a wharf in Nigeria. The owner of the wharf brought proceedings against the ship in the Admiralty court, claiming damages for a trespass to land situated abroad. The shipowner argued that the Admiralty court had no jurisdiction to entertain the claim.

The Court of Appeal held that it had the jurisdiction to hear the claim being for the enforcement of a maritime lien by proceedings *in rem*. Such jurisdiction is based on the presence within the jurisdiction of the offending ship.

Somervell LJ: 'The ultimate issue in this case is whether a claim *in rem* in respect of which a maritime lien is exercisable in which the plaintiff claims for damage to foreign land . . . Speaking broadly I should have thought that it was in the interests of nations *inter se* that where damage is done which gives rise to a maritime lien, that lien should be enforceable in the courts of any country to which the ship may proceed. I have assumed and think I am entitled to assume that the law of other countries or at any rate most other countries is the same as ours on the matter in question here.'

Corps and Corps v The '*Queen of the South*' [1968][4]

S 12A(2)(h) of the High Court Ordinance gives the Admiralty court jurisdiction to hear and determine 'any claim arising out of any agreement relating to the carriage of goods in a ship or to the use or hire of a ship'.

The watermen claimed a sum of about £290 for services provided to the defendant's peddle steamer, in that they had moored and unmoored the peddle steamer through the use of the watermen's motor boats. The watermen argued that the agreement between the watermen and the peddle steamer's manager, under which the service claimed was rendered, was an agreement relating to the use or hire of a ship.

The court held that it had jurisdiction to hear the claim.

Brandon J: 'In the present case, however, it seems to me clear, on the written and oral evidence before me, that the whole of the services rendered by the plaintiffs were based on the use of motor boats owned and operated by them. It is true that in some cases the men engaged in mooring and unmooring did their work on a quay or on a buoy. But they were landed on the quay or on the buoy from a motor boat, and taken off again by the same means . . . I hold that the court has jurisdiction to entertain it on that ground.'

---

3. [1946] 2 All ER 372.
4. [1968] 1 Lloyd's Rep 182.

Petrofina SA v AOT Limited (The Maersk Nimrod) [1991][5]

Under an international contract for the sale of goods (CIF contract), the sellers sold the buyers 50,000 tonnes of fuel oil. The contract provided that demurrage calculated according to the provision in the charterparty would be payable in the event of certain delays. The sellers were not party to the charterparty and subsequently claimed for loss pursuant to the CIF contract. One of the issues raised was whether the seller's claim fell within the jurisdiction of the Admiralty court as 'relating to the carriage of goods in a ship'.

The court held that the seller's claim did not arise out of the charterparty. It was only related to the CIF contract. The words 'any agreement relating to the carriage of goods in a ship' did not embrace a CIF contract for the sale of goods. The charterparty was independent of the CIF sale of goods contract and therefore the claim did not fall within the Admiralty jurisdiction of the court.

Phillips J: 'I agree . . . that the authorities do not provide a definition of "arising out of" that can be applied to determine whether the necessary degree of proximity exists between claim and agreement. My task is simply to ask whether, giving the English language its ordinary and natural meaning, the sellers' claim arises out of the charterparty. In my judgment it does not. It arises out of the CIF contract under which the claim is brought and, more particularly, out of the agreement to pay demurrage which forms part of that contract. The claim is independent of the charterparty, which is only relevant in that it provides the demurrage rate and the other terms, if any, that bear on the computation of demurrage. For these reasons my conclusion is that the seller's claim is not one which falls within the Admiralty jurisdiction.'

The 'Conoco Britannia' [1972][6]

By virtue of s 12A(2)(j) of the High Court Ordinance, the admiralty court has jurisdiction to hear and determine 'any claim in the nature of towage in respect of a ship'.

The tug owner supplied the shipowner of a vessel with a tug under a towage contract. The shipowner, under the towage contract, agreed to indemnify the tug owner against any loss of or damage to the tug. The tug collided with the vessel and as a result sunk after the collision.

The court held that the court had jurisdiction to hear the dispute in relation to the towage contract.

5. [1991] 3 All ER 161.
6. [1972] 2 All ER 238.

The 'Fairport' (No. 5) [1967][7]

S 12A(2)(l) of the High Court Ordinance confers jurisdiction to the Admiralty court to hear and determine 'any claim in respect of goods or materials supplied to a ship for her operation or maintenance'. S 12A(2)(o) concerns 'any claim by a master, shipper, charterer or agent in respect of disbursements made on account of a ship'.

The ship's chandlers advanced two loans to the vessel for the payment of the crew's wages. The chandlers claimed repayment of the loan.

The court held that the Admiralty jurisdiction included the jurisdiction to hear the dispute in connection with payments made by way of advances to enable the wages of the crew to be settled.

The 'D'Vora' [1952][8]

Under s 12A(2)(l), the Admiralty court can entertain 'any claim in respect of the construction, repair or equipment of a ship'.

The fuel supplier supplied oil to a vessel, but the shipowner refused to pay for the oil.

The Admiralty court held that the court had no jurisdiction to hear and determine the claim because the claim was not one in respect of the 'equipment' of a ship.

Willmer J: 'In my judgment, there is an important difference between "equip" and "supply", "supply" being a word which is appropriate for use in connection with consumable stores, such as fuel oil whereas "equip", to my mind, connotes something of a more permanent nature than consumable stores. I can well understand that anchors, cables, hawsers, sails, ropes, and such things, may be said to be part of a ship's equipment, and that, nonetheless, although they may have to be renewed from time to time; but such things as fuel oil, coal, boiler water and food—consumable stores—seem to me to be in quite a different category.'

In addition, the Admiralty jurisdiction of the Court of First Instance also has jurisdiction in relation to any of the following proceedings:

(a) any application to the Court of First Instance under
    (i) the Merchant Shipping Acts 1894 to 1979 in their application to Hong Kong;
    (ii) the Merchant Shipping Ordinance (Cap 281);
    (iii) the Merchant Shipping (Safety) Ordinance (Cap 369);
    (iv) the Merchant Shipping (Liability and Compensation for Oil Pollution) Ordinance (Cap 414);

7. [1967] 2 Lloyd's Rep 162.
8. [1952] 2 Lloyd's Rep 404.

(v) the Merchant Shipping (Registration) Ordinance (Cap 415);
(vi) the Merchant Shipping (Limitation of Shipowners Liability) Ordinance (Cap 434);

(b) any action to enforce a claim for damage, loss of life or personal injury arising out of:
(i) a collision between ships;
(ii) the carrying out of or omission to carry out a manoeuvre in the case of 1 or more of 2 or more ships; or
(iii) non-compliance, on the part of 1 or more of 2 or more ships, with the collision regulations;

(c) any action by shipowners or other persons under–
(i) the Merchant Shipping Acts 1894 to 1979 in their application to Hong Kong;
(ii) the Merchant Shipping Ordinance (Cap 281);
(iii) the Merchant Shipping (Safety) Ordinance (Cap 369);
(iv) the Merchant Shipping (Liability and Compensation for Oil Pollution) Ordinance (Cap 414); or
(iv) the Merchant Shipping (Limitation of Shipowners Liability) Ordinance (Cap 434) for the limitation of the amount of their liability in connection with a ship or other property.

The jurisdiction of the Admiralty jurisdiction of the Court of First Instance in connection with 'any question arising between the co-owners of a ship as to possession, employment or earnings of that ship' includes power to settle any account outstanding and unsettled between the parties in relation to the ship, and to direct that the ship, or any share thereof, shall be sold, and to make such other order as the court thinks fit. Furthermore, 'any claim for damage done by a ship' extends to:

(a) any claim in respect of a liability incurred under Part II of the Merchant Shipping (Liability and Compensation for Oil Pollution) Ordinance (Cap 414); and
(b) any claim in respect of a liability incurred by the International Oil Pollution Compensation Fund under Part III of that Ordinance.

In respect of 'claim to salvage', the 'Salvage Convention 1989' means the International Convention on Salvage 1989 as it has effect under section 9 of the Merchant Shipping (Collision Damage Liability and Salvage) Ordinance (Cap 508). The reference to salvage services includes services rendered in saving life from a ship and the reference to any claim under any contract for or in relation to salvage services includes any claim arising out of such a contract, whether or not arising during the provision of the services.

11.3 MAIN CLASSIFICATION OF ADMIRALTY ACTIONS: IN *REM* OR IN *PERSONAM*

Section 12B of the High Court Ordinance states that:

- An action in *rem* may be brought in the Court of First Instance against the ship or property in connection with which the claim or question arises and such action shall be deemed to be brought and upon the issue of the writ in *rem*.
- An action in *personam* may be brought in the Court of First Instance in all cases within the Admiralty jurisdiction of that court.

What is the difference between admiralty actions in *rem* or in *personam*? The maritime law in Hong Kong allows a plaintiff to proceed not only in *personam* (eg, against the shipowner or the charterer), but also in *rem* against the thing itself (eg, the ship). Once a plaintiff has obtained a writ in the proceedings in *rem*, the plaintiff can apply for a warrant to arrest the ship in order to obtain security before judgment for the claim. The whole point of the action in *rem* is that it is an action against the ship itself or, in appropriate circumstances, cargo or freight where the cargo or freight is subject to a maritime lien. An action in *rem* is not a personal action against the shipowner. The classic statement of the nature of the action in *rem* can be found in *The Burns*,[9] where Lord Justice Fletcher Moulton of the English Court of Appeal stated that 'the action in *rem* is an action against the ship itself. It is an action in which the shipowners may take part, if they think proper, in defence of their property, but whether or not they will do so is a matter for them to decide, and if they do not decide to make themselves parties to the suit in order to defend their property, no personal liability can be established against them in that action. It is perfectly true that the action indirectly affects them'. The 'fundamental nature of the action in *rem* is to obtain security against a ship', observed Waung J in *The Chong Bong*[10] who went on to state 'that ability to obtain security depends on a number of factors chief of which are the visits if any of a ship into the jurisdiction and the change of ownership of that ship or its sister ship after the cause of action has arisen'.

An action in *personam*, on the other hand, is a personal action against the shipowner or the relevant party (eg, charterer) liable for the maritime claim.

9. [1907] P 137.
10. [1997] 3 HKC 579. In *Indian Grace (No. 2)* [1998] AC 878, Lord Stern of the UK House of Lords stated that 'a more realistic view of the nature of actions in *rem* . . . stripped away the form and revealed that in substance the owners were the parties to the action in *rem*.' This view was widely criticized and is not binding authority in Hong Kong.

In practice, the shipowners are usually domiciled outside the Hong Kong jurisdiction. For example, the shipowners may be companies incorporated in Panama or Greece. To bring actions in *personam* against these companies, leave (ie, special permission) of the court is needed to serve a writ of summons outside the jurisdiction.

As regards an action in *rem*, the plaintiff only needs to establish that the claim falls within the Admiralty jurisdiction of the Hong Kong Court of First Instance under s 12A of the High Court Ordinance (as discussed above), and that the ship is within Hong Kong waters. It is possible to issue the writ in *rem* in advance before the entry of the vessel into Hong Kong waters. If the court grants the arrest order, the vessel subject to the arrest will be placed on the 'watch list' kept by the Hong Kong Admiralty court. Once the vessel enters Hong Kong waters, she will be arrested.

11.4 MODE OF EXERCISE OF ADMIRALTY JURISDICTION: ARREST OF VESSELS

What objects may be arrested?

Section 12E of the High Court Ordinance stipulates that a ship is 'any description of vessel used in navigation and . . . includes, subject to any regulations made by the Hong Kong government, a hovercraft'. A 'hovercraft' is defined as 'a vehicle designed to be supported when in motion wholly, or partly, by air expelled from the vehicle to form a cushion of which the boundaries include the ground, water or other surface beneath the vehicle'. At common law, barges intended for use in tidal waters, requiring a tug to tow them, have been held by the court to be ships: *Harlow*.[11] An Admiralty action in *rem* is most typically brought against a ship. However, it is also possible to bring an Admiralty action in *rem* against other property in a very limited category of claims. In a salvage claim, a salvor acquires, at common law, a maritime lien in connection with not only the ship which the salvor salves but also any cargo, bunkers and stores on board, and freight.

Sister ships

Section 12B(4) of the High Court Ordinance allows the arrest of 'sister ships'. In the case of any such claim as is mentioned in s 12A(2)(e) to (q), where:

11. [1992] P 175.

- the claim arises in connection with a ship ('Ship A'); and
- the person who would be liable on the claim in an action in *personam* ('the relevant person') was, when the cause of action arose, the owner or charterer of, or in possession or in control of, the ship ('Ship A').

An action in *rem* may (whether or not the claim gives rise to a maritime lien on that ship) be brought in the Court of First Instance against:

- that ship ('Ship A'), if at the time when the action is brought the relevant person is either the beneficial owner of that ship as respects all the shares in it or the charterer of it under a charter by demise; or
- any other ship ('Ship B' or 'Ship C') of which, at the time when the action is brought, the relevant person is the beneficial owner as respects all the shares in it.

In *Tian Sheng No. 8*,[12] the Hong Kong Court of Final Appeal held that in the context of s 12B(4)(b) of the High Court Ordinance, the word 'owner' means legal owner (as opposed to beneficial owner) where a registered ship is concerned. In *Decurion*,[13] while the time charterer was in possession or in control of 'Ship A', the time charterer's associated company (registered owner of 'Ship B') was not regarded as having the right to exercise control over 'Ship A'. The court declined to lift the corporate veil by ignoring the separate corporate identities of the time charterer and its associated company. Hence, the arrest of 'Ship B' in relation to a maritime claim concerning 'Ship A' was held to be unjustifiable.

In the case of a claim in the nature of towage or pilotage in respect of an aircraft, an action in *rem* may be brought in the Court of First Instance against that aircraft if, at the time when the action is brought, it is beneficially owned by the person who would be liable on the claim in an action in *personam*. Where, in the exercise of its Admiralty jurisdiction, the Court of First Instance orders any ship, aircraft or other property to be sold, the court shall have jurisdiction to hear and determine any question arising as to the title to the proceeds of sale.

The Ordinance does not provide a precise definition of the phrase 'beneficially owned'. It was held to mean 'the right to sell or dispose of a ship'.[14] In appropriate cases, the court may look behind the 'corporate veil' to determine beneficial ownership.

12. [2000] HKCFA 105.
13. [2013] HKCA 180.
14. *The Convenience Container* [2007] 3 HKLRD 575; *The Decurion* [2013] HKCA 180.

Rung Ra Do [1994] 3 HKC 621

The plaintiff time-chartered a vessel to the government of the Democratic People's Republic of Korea (North Korea). However, the North Korean government failed to pay freight as claimed by the plaintiff. The plaintiff obtained the arrest of 'Rung Ra Do' in Hong Kong and alleged that the vessel was beneficially owned by the North Korean government. The North Korean government argued that it was not the beneficial owner of 'Rung Ra Do' and alleged that the vessel was owned by KRRD, a North Korean co-operative organisation.

The Hong Kong court examined the expert evidence provided by a number of university professors specialising in North Korean law and economics. The court held that the vessel 'Rung Ra Do' was owned by the North Korean government. A North Korean co-operative organisation, according to the North Korean Constitution, could not own an international trading vessel although it might operate or manage the vessel on behalf of the North Korean government.

Barnett J: 'It is not in dispute that the court may look behind the registered owner of a vessel to see whether that owner is the beneficial owner or merely a nominal owner, the beneficial owner being another. Authority for this proposition can be found in *I Congreso Del Partido* [1978] 1 QB 500 and *The Aventicum* [1978] 1 Lloyd's Rep 184.'

Aventicum [1978] 1 Lloyd's Rep 184

The plaintiff was the consignee of a cargo shipped from Finland to Japan. During the voyage, damage to the cargo was caused by wetting. The defendant shipowner applied to set aside the writ and all subsequent proceedings on the ground that the court had no jurisdiction over the ship or the owner of the ship in that at the time the action was brought, the ship was not beneficially owned by the same person who was the owner at the time when the cause of action arose.

Slynn J had the following observations: '. . . where damages are claimed by cargo-owners and there is a dispute as to the beneficial ownership of the ship, the court in all cases can and in some cases should look behind the registered owner to determine the true beneficial ownership . . . the court shall not be limited to a consideration of who is the registered owner or who is the person having legal ownership of the shares in the ship; the directions are to look at the beneficial ownership. Certainly in a case where there is a suggestion of a trusteeship or a nominee holding, there is no doubt that the court can investigate it.'

Arrest procedure

Fundamental to the Admiralty procedure is the ability to issue and serve the writ of summons in *rem* on a vessel in order to arrest her at the beginning of an action. The arrest procedure enables a plaintiff to obtain security for his claim at the very start of the proceedings. An issued writ of summons is only valid for service within one year from the date of issue, unless leave is granted by the court to extend the validity of the writ. An affidavit leading to a warrant of arrest of a ship must be sworn and filed. It should contain the following particulars and information:

- the solicitor's authority to commence action in *rem* and to swear the affidavit on behalf of the plaintiff;
- the law firm's undertaking to pay the bailiff's costs of arrest on demand;
- the port of registry of the ship (check Lloyd's Register of Ships);
- the particulars of and the circumstances surrounding the nature of the maritime claim which help explain why the Admiralty jurisdiction of the Court of First Instance can entertain the claim;
- the documentary evidence substantiating the claim, which can be attached to the affidavit as exhibits;
- evidence of the beneficial ownership or identity of the demise charterer of the vessel to be arrested;
- the vessel in respect of which the claim arose, if the vessel to be arrested is not the vessel in respect of which the claim arose.

If the Admiralty judge is satisfied with the affidavit leading to the warrant of arrest, he will order a warrant of arrest of the ship be issued. From the moment the ship is arrested, the bailiff is responsible for the custody of the ship. The following is part of a sample warrant of arrest and undertaking:

WARRANT OF ARREST

To the Bailiff:

We hereby command you to arrest the ship or vessel "ABC" of the port of St. Vincent and Grenadines Flag and to keep the same under safe arrest until you shall receive further orders from us.

The plaintiff as owner and/or shipper and/or holder and/or indorsee of bill of lading No. 1234 and/or as person otherwise interested in a cargo lately laden on board the Defendant's ship "ABC" claims damages sustained by reason of the Defendant's breach of contract evidenced in writing by bill of lading No. 1234 dated [date] under which the said cargo was shipped and/or for breach of duty as bailees for reward and/or conversion and/or negligence in or about the carriage of the said cargo.

Witness the Honourable Mr./Madam Chief Justice [. . .] of Hong Kong, the [date]

[. . .]
Registrar

UNDERTAKING

An undertaking is hereby given that the Bailiff shall be indemnified in respect of all fees and expenses that may be incurred by him in consequence of the arrest of the ship or vessel "ABC" of the port of St. Vincent and Grenadines Flag.

This undertaking will be implemented forthwith on the Solicitors for the Plaintiffs being called upon to make payment of such fees and expenses.

[Date]

[. . .]
Solicitors for the Plaintiffs

The Bailiff of the court will arrange for the service of the warrant of arrest. Service of the warrant is effected physically on the ship or cargo in compliance with order 75 rule 11 of the Rules of High Court. Service of a warrant of arrest or writ in an action in *rem* against a ship freight or cargo shall be effected by:

- affixing the warrant or writ for a short time on any mast of the ship or on the outside of any suitable part of the ship's superstructure, and
- on removing the warrant or writ, leaving a copy of it affixed (in the case of the warrant) in its place or (in the case of the writ) on a sheltered, conspicuous part of the ship.

Service of a warrant of arrest or writ in an action in *rem* against freight or cargo or both shall, if the cargo has been landed or transhipped, be effected:

- by placing the warrant or writ for a short time on the cargo and, on removing the warrant or writ, leaving a copy of it on the cargo, or
- if the cargo is in the custody of a person who will not permit access to it, by leaving a copy of the warrant or writ with that person.

Bail bond or the provision of security by the shipowner

As stated in the judgment of *The Cap. Bon,*[15] the primary function and object of the admiralty action in *rem* is to 'provide security for a plaintiff in respect of any judgment which he may obtain as a result of the hearing and determination of the claim'. One of the forms of security to be used is a 'bail bond' as prescribed in order 75 rule 16 of the Rules of High Court:

(1) Bail on behalf of a party to an action in *rem* must be given by bond in Form No. 11 in Appendix B; and the sureties to the bond must enter into the bond before a commissioner for oaths or a solicitor exercising the powers of a commissioner for oaths under section 7A of the Legal Practitioners Ordinance (Cap 159) not being a solicitor who, or whose partner, is acting as solicitor or agent for the party on whose behalf the bail is to be given.
(2) Subject to paragraph (3), a surety to a bail bond must make an affidavit stating he is able to pay the sum for which the bond is given.
(3) Where a corporation is a surety to a bail bond given on behalf of a party, no affidavit shall be made under paragraph (2) on behalf of the corporation unless the opposite party requires it, but where such an affidavit is required it must be made by a director, manager, secretary or other similar officer of the corporation.
(4) The party on whose behalf bail is given must serve on the opposite party a notice of bail containing the names and addresses of the persons who have given bail on his behalf and of the commissioner before whom the bail bond was entered into. After the expiration of 24 hours from the service of the notice (or sooner with the consent of the opposite party) he may file the bond and must at the same time file the affidavits (if any) made under paragraph (2) and an affidavit proving due service of the notice of bail to which a copy of that notice must be exhibited.

A sample bail bond and affidavit of surety to bail bond are now shown as follows:

15. [1967] 1 Lloyd's Rep 547.

BAIL BOND

Whereas this Admiralty action in *rem* against the above-mentioned property is pending in the Court of First Instance and the parties to the said action are the above-mentioned plaintiff and defendant:

Now, therefore, we, XYZ Insurance Company Limited of No.123 King's Road, Hong Kong hereby submit ourselves to the jurisdiction of the said court and consent that if they, the above mentioned defendant, do not pay what may be adjudged against them in this action, with costs, or do not pay any sum due to be paid by them in consequence of any admission of liability therein or under any agreement by which this action is settled before judgment and which is filed in the said court, execution may issue against us or our goods and chattels, for the amount unpaid or an amount of [. . .] whichever is the less.

The bail is put up without prejudice to the right of the defendant to apply for an order that these proceedings be stayed on the ground that the plaintiff has agreed that all disputes arising under or in connection with [Bill of Lading No.1234] be determined in the courts of, or by arbitration in, [the United Kingdom], and/or that in all circumstances the courts of [the United Kingdom] are the natural and appropriate forum for the trial of this dispute.

AFFIDAVIT OF SURETY TO BAIL BOND

I, [Chan Siu Ming], of No. 255 Queen's Road Central, Hong Kong, Vice President of XYZ Insurance Company Limited of No. 123 King's Road, Hong Kong, the proposed surety for the defendant in this action, make oath and say that XYZ Insurance Company is worth more than the sum of [US$800,000.00] after payment of all its debts, and is therefore able to pay the sum for which the bond is given.

Very often, a shipowner provides contractual security in lieu of a formal bail bond, which the plaintiff accepts. The contractual security is usually in the form of a letter of undertaking given by a P & I Club (Shipowner's

Protection and Indemnity Club) or a bank. Upon the provision of a satisfactory bail bond or contractual security, the shipowner will immediately request the court that the arrested vessel be released from the arrest. The defendant will normally give consent to the release of the ship after receiving the bail bond or security which adequately secures the amount of the claim in question. The court will then grant the order of release:

RELEASE

To the Bailiff:

Whereas in this action we did command you to arrest the ship or vessel "ABC" of the port of St. Vincent and Grenadines Flag, and to keep the same under safe arrest until you should receive further orders from us.

NOW WE DO HEREBY COMMAND you to release the said ship or vessel "ABC" from the arrest effected by virtue of our warrant in this action.

Under the common law, the court has the inherent jurisdiction to prevent the abuse of the judicial process or the use of court procedure in an oppressive manner by ordering the release of the arrested vessel and to control the amount of security (*The Polot II*[16] and *The Hua Tian Long*).[17]

Sale of the arrested vessel

If the defendant has failed to acknowledge service of the writ of summons in *rem*, or if the defendant has failed to provide a satisfactory bail bond or contractual security securing the maritime claim, the plaintiff may apply to the court at any time after arrest by way of 'notice of motion' for an order that the vessel be appraised and sold either on judgment or *pendente lite* (ie, prior to judgment). The notice of motion is supported by an affidavit explaining that the ship is a wasting asset, the value of which is being reduced continually by the expense of maintaining it under arrest. In the light of the plaintiff's claim, the plaintiff may be unable to satisfy his judgment at the conclusion of the proceedings. If the defendant has filed a defence to dispute

16. [1977] 2 Lloyd's Rep 115.
17. [2008] 4 HKLRD 719.

liability, the affidavit needs to show that the defence is unlikely to succeed, and that any judgment is unlikely to be satisfied. Therefore, the ship shall be sold in line with the interests of both parties to obtain the highest possible value for the ship. A sample of the notice of motion is now shown below:

NOTICE OF MOTION

Take notice that the court will be moved on [. . .] at 9:30 a.m. for orders that:

1. The Defendant's vessel "ABC" be sold *pendente lite*.
2. The vessel be sold by the Admiralty Bailiff by private treaty or public tender provided that:
   (a) Such private treaty or public tender conforms to the Admiralty Bailiff's usual terms and conditions for sale;
   (b) It exceeds a sum being the average of two valuations carried out by [Ship Surveyor A] and [Ship Surveyor B]; and
   (c) The balance of the tender price (assuming the relevant tender is accepted) be received by this court within 5 banking days of the acceptance of the tender.
3. In the event of a tender being made which satisfies the requirements of paragraphs (2)(a) and (b) herein, the Admiralty Bailiff shall accept the same within 2 days and so advise the prospective buyers or its representatives and [the solicitors for the plaintiff].
4. Costs of this application be to the plaintiff.

[Date]
This Notice of Motion was taken out by [the solicitors for the plaintiff].

Appraisement is the official evaluation of the property by court-appointed valuers in order to prevent the property from being sold at too low a price. The Admiralty court has the inherent jurisdiction to make an order for appraisement and sale of the ship prior to judgment. In *Castrique v Imrie*,[18] Justice Blackburn said:

18. (1869) LR 4 HLC 414.

> It is not essential that there should be an actual adjudication on the status of the thing. Our courts of Admiralty, when property is attached and in their hands, on a proper case being shown that it is perishable, order (for the benefit of all parties concerned) that it shall be sold and the proceeds paid into court to abide the result of the litigation. It is almost essential to justice that such a power should exist in every case where property, at all events perishable property, is detained.

In addition, order 29 rule 4(1) of the Rules of High Court provides that: 'The Court may, on the application of any party to a cause or matter, make an order for the sale by such person, in such manner and on such terms (if any) as may be specified in the order of any property (other than land) which is the subject-matter of the cause or matter or as to which any question arises therein and which is of a perishable nature or likely to deteriorate if kept or which for any other good reason it is desirable to sell forthwith.'

In *The Margo L*,[19] the plaintiff bank, mortgagee of the arrested vessel 'Margo L', applied for a private sale of the vessel after the defendant's failure to acknowledge service of the writ of summons in *rem*. The plaintiff adduced evidence that a named purchaser had proposed a price of US$3.4 million. In an attempt to convince the court that the ship should be sold by private sale to that particular purchaser, the bank relied on three valuation certificates, the highest of which valued the vessel at US$3.2 million. The court held that to ensure protection for all maritime claimants, the court should normally order an appraisement of the vessel so as to determine the minimum selling price. Further, even if valuation was done in the most convincing way, the court would still order a sale by public tender in order to obtain the best price possible for the vessel. Justice Waung laid down the following principles and guidelines:

- The unique feature of the Admiralty jurisdiction of Hong Kong and the United Kingdom is that all claims in admiralty in *rem* are made against one ship and when that ship is sold, against the proceeds of that ship. Admiralty claims are of numerous varieties and the claimants could be from any part of the world where the ship had sailed, therefore it may take time for claims to surface but that does not matter too much so long as the following two essential steps are taken by the Admiralty court:
  (1) the ship must be publicly sold at the best possible price;
  (2) the public sale of the ship must be made known to the maritime world so that potential claimants can come to the Admiralty court to make claims against the proceeds of sale.

19. [1998] 1 HKC 271.

- The best possible price realized by a public sale of the ship will ensure that no admiralty claimant is prejudiced, because the proceeds of that public sale will replace the ship against which claims would otherwise be made.
- Another unusual feature of the Admiralty court is the ranking of different claims by a determination of priorities, so that those claimants ranking high in priority (such as crew and salvors) will have the first right to the proceeds of sale before those ranking low in priority (such as cargo claimants). But the practice of the Admiralty court is that the determination of priority takes place a long time after the ship is arrested or sold, and very often it is postponed until such time as all the claims from various parts of the world have come in. To ensure fairness to all claimants, the Admiralty court had developed a system of claimants entering caveats, which will prevent any arrested ship being released or any proceeds of sale in court being paid out without prior notice being given to the caveators of such application for release or payment.
- It is fundamental to any developed admiralty jurisdiction such as that in Hong Kong, that litigation in the Admiralty court does not merely consist of matters between two parties, the plaintiff and the defendant, as in ordinary High Court actions. Every sale of a ship by the Admiralty court intimately affects the interests of other potential claimants in admiralty, many of whom might not even know that ship has been arrested or sold by the Admiralty court in Hong Kong. The role of the Admiralty court is to ensure that any sale of the vessel is effected in such a way as to protect all admiralty claimants, not merely the plaintiff who arrested the ship or who requested the Admiralty court to sell the ship, or who has obtained judgment, or who has a high priority claim. The best way, which is also the normal way, in which the Admiralty court ensures protection for all admiralty claimants is by insisting upon the sale of a ship being done by the well-tried method of appraisement and sale by public tender.
- To ensure the Admiralty court sells arrested ships properly and that a ship is not sold at an undervalue, the Admiralty court adopts a system of appraisement of a ship before it is sold. The Admiralty Registrar and the Chief Bailiff obtain from court appointed experts, the proper valuation of a ship and very often two or three experts would be involved in the appraisement of a ship. But this exercise is only what might be called 'floor protection', namely to ensure a ship is not sold below its appraised price. An appraisement cannot accurately predict what might be the best possible price for a particular ship which is unique.

- The best possible price can only be achieved by the market and by an appropriate exposure of a particular ship to the market interests in that ship. This is readily achieved by Hong Kong and worldwide shipping brokers immediately sending out news of the order for sale made by the Hong Kong Admiralty court, and by the international maritime community reading the order of public sale in a shipping newspaper such as the *Lloyd's List of Shipping*. All these can only be achieved and be meaningful if the Admiralty court makes the usual order of sale by public tender.

In England, Justice Sheen also dismissed an application to effect a private sale of the arrested ship in *The 'APJ Shalin'*.[20] He took the view that while a ship is under arrest, it is in the custody of the Admiralty Bailiff. It is immaterial who the owners are. When an order for sale by the court has been made, there cannot be a private sale because that would be open to abuse. All offers to purchase the ship must be made to the Admiralty Bailiff, who must realize the highest possible price obtainable. Private negotiations could adversely affect the market, because they could have the result that potential bids would be withheld.

The sale of the arrested ship to a purchaser is carried out by the execution of a 'bill of sale'. The bill of sale recites that the vessel is sold 'free from liens and encumbrances and debts whatsoever' up to the date of execution of the bill of sale. Such a bill of sale gives clean title to the purchaser. In *The Acrux*,[21] the court stated that 'it would be intolerable, inequitable and an affront to the court if any party who invoked the process of this court and received its aid and, by implication, assented to the sale to an innocent purchaser, should thereafter proceed or was able to proceed elsewhere against the ship under her new and innocent ownership. This court recognises proper sales by competent Courts of Admiralty . . . it is part of the comity of nations, as well as a contribution to the general well-being of international maritime trade.' The court then ordered the claimant to give an undertaking that they would not proceed elsewhere against the ship in respect of the unsatisfied balance of their claim, nor institute proceedings in *rem*, or equivalent proceedings, against the ship anywhere in respect of their claim.

20. [1991] 2 Lloyd's Rep 62.
21. [1962] 1 Lloyd's Rep 405.

Distribution of proceeds of sale and maritime liens

The priorities in distribution of proceeds of sale are as follows:

(1) The Admiralty Bailiff's expenses and costs of arrest
(2) The recoverable legal costs of arresting and preserving the vessel and proceeding to sell the vessel
(3) Maritime liens and possessory lien of a ship repairer
(4) Mortgages, in their respective order of priorities
(5) The claims of others entitled to proceed by Admiralty action in *rem*
(6) The claims of in *personam* creditors of the shipowner
(7) The shipowner is entitled to the balance remaining, if any

Maritime liens

An action in *rem* may be pursued against a ship subject to a maritime lien under s 12B(3) of the High Court Ordinance: 'In any case in which there is a maritime lien or other charge on any ship, aircraft or other property for the amount claimed, an action in *rem* may be brought in the Court of First Instance against that ship, aircraft or property'. The court in *The 'Bold Buccleugh'*[22] gave a definition of a maritime lien as follows: 'Having its origin in the rule of the civil law, a maritime lien is well defined . . . to mean a claim or privilege upon a thing to be carried into effect by legal process and Mr. Justice Story explains that process to be a proceeding in *rem* and adds, that wherever a lien of claim is given upon the thing, then the Admiralty enforces it by a proceeding in *rem*, and indeed is the only court competent to enforce it. This claim or privilege travels with the thing into whosoever's possession it may come. It is inchoate from the moment the claim or privilege attaches, and, when carried into effect by legal process by a proceeding in *rem*, relates back to the period when it first attached.'

At common law, four categories of maritime liens are recognized:

(1) damage done by a ship;
(2) salvage;
(3) seamen's wages; and
(4) bottomry and respondentia.

The maritime lien also attaches to the master's wages and disbursements under the Merchant Shipping (Seafarers) Ordinance (Cap 478). In respect of the remedies of the master for remuneration and disbursements, the section reads: 'The master of a Hong Kong ship shall have the same lien for his

22. (1851) 7 Moo PC 267.

remuneration, and all disbursements or liabilities properly made or incurred by him on account of the ship, as a seafarer has for his wages'. According to *The 'Orienta'*,[23] the real meaning of the word 'disbursement' in Admiralty practice is disbursements by the master, which he makes himself liable for in respect of necessary things for the ship, for the purposes of navigation, which he, as master of the ship, is there to carry out. The expenses are necessary in the sense they have to be incurred immediately, and when the shipowner is not there able to give the order, and the shipowner is not so near to the master that the master can ask for his authority.

Maritime liens are discharged by the sale of the ship by court in proceedings in *rem* to a third party purchaser. Also, maritime liens are lost if there is a total and permanent destruction of the vessel.

Common law possessory lien of the ship repairer

A ship repairer, in possession of the ship at the shipyard, will have a valid possessory lien on the vessel for repairs done and materials supplied if the shipowner fails to pay the repairer the repair charges. At common law, the possessory lien holder has priority over all other claims save for maritime liens already in existence at the time the possessory lien was exercised.

On the other hand, it is a well-settled principle that the ship repairer is not to contend with the Admiralty Bailiff. The repairer should co-operate with the Admiralty Bailiff to surrender the ship to the officer of the court if a warrant of arrest is issued by the Admiralty court. Under the order of the court, the Admiralty Bailiff will remove the ship and sell her. However, the ship repairer is protected in the sense that by arresting, a ship repairer will not lose his possessory lien.

In *The 'Arantzazu Mendi'*,[24] Lord Atkin said: 'The ship arrested does not by the mere fact of arrest pass from the possession of its then possessors to a new possession of the [Bailiff]. [The bailiff's] right is not possession, but custody. Any interference with his custody will be properly punished as a contempt of the court which ordered arrest, but subject to the complete control of the custody, all the possessory rights which previously existed continue to exist, including all the remedies which are based on possession'.

In *The 'Acacia'*,[25] Justice Townsend also refused to hold that by arresting the ship, the repairer lost his possessory lien. He said: 'I am reluctant to decide for the first time that the effect of an Admiralty arrest is to destroy

23. [1895] P 49.
24. [1939] AC 256.
25. (1880) 23 ChD 330.

the lien for the active enforcement of which it was sued out, or that a party having a valid claim up to that moment can be deemed to forgo it by asking the statutory aid of the court to make it effectual. In the absence, therefore, of authority to show that the taking out of the Admiralty warrant would discharge the possessory lien, I cannot in reason or in justice hold that [the repairer] is to lose the fruit of their expenditure.'

Ship mortgages

The present Hong Kong shipping register was established in December 1990 under the Merchant Shipping (Registration) Ordinance, which sets out the procedure for registering and deleting ships in Hong Kong, creating and discharging mortgages on such ships, and other related matters. Ships registered on the register established under the Merchant Shipping (Registration) Ordinance remained British ships until 1 July 1997. From that date they became Chinese ships. Until 1 July, ships registered in Hong Kong flew the British red ensign directly above the blue ensign that showed Hong Kong's armorial bearings. Following 1 July 1997, the flag of the PRC is now to be flown above the flag of the Hong Kong SAR.

Broadly speaking, a mortgage could be said to be 'any charge by way of lien on any property for securing money or money's worth'. A ship registered in Hong Kong may be made security for any obligation by way of a mortgage. Foreign mortgages are frequently enforced in Hong Kong's Admiralty court in the same way as mortgages registered in Hong Kong. For example, in *The Maule*,[26] the Hong Kong Admiralty court enforced a foreign mortgage. Both the ship and mortgage were registered at the port of Limassol in Cyprus. The vessel was arrested in Hong Kong pursuant to the mortgage granted by the defendant over the ship as security in favour of the plaintiff. The foreign law of the mortgage must give the mortgagee the status of a secured lender, and the particular mortgage must comply with the requirements of the foreign law in question and be valid and binding thereunder.

Even if the mortgage is not registered, the court may nevertheless give effect to it in circumstances where equitable principles so require. The court may still consider that the unregistered mortgage is valid between the original parties as an equitable mortgage and is good against the mortgagor. Registration, however, gives a mortgage priority over earlier unregistered mortgages and later registered or unregistered mortgages. A registered mortgage does not have priority over mortgages registered earlier.

26. [1997] 1 HKC 231.

11.5 STAY OF HONG KONG PROCEEDINGS: *FORUM NON CONVENIENS*

General principles: Conflict of laws

Once it has been established that the Hong Kong court has jurisdiction to hear the case (for example, a writ of summons in *rem* has been served on the ship for one of the maritime claims under s 12A of the High Court Ordinance), it does not automatically follow that the court will, or will be able to, exercise that jurisdiction. There are occasions on which the court will, in its discretion, refuse to exercise its jurisdiction and stay the Hong Kong proceedings. One of the purposes of the conflict of laws is to discourage 'forum shopping', a process by which a plaintiff commences proceedings in an inappropriate forum, in the belief that some advantages will be gained by the plaintiff before that forum.

The basic principle is that a stay will only be granted on the ground of *forum non conveniens*, where the court is satisfied that there is some other available forum, having jurisdiction, which is the appropriate forum for trial of the action. In other words, the other forum is the forum in which the case may be tried more suitably for the interests of all the parties and the ends of justice. If another forum is available, which prima facie is clearly more appropriate for the trial of the action, the Hong Kong court will ordinarily grant a stay unless there are circumstances by reason of which justice requires that a stay should nevertheless not be granted. In contrast, if there is no clearly more appropriate forum abroad, the Hong Kong court will ordinarily refuse a stay of proceedings.

In *Spiliada Maritime Corporation v Cansulex Ltd*,[27] a cargo of sulphur was shipped from British Columbia to India on board the ship 'Spiliada'. Severe corrosion was caused to the vessel, allegedly because the cargo was wet when loaded. The shipowner (a Liberian company) decided to sue the shipper (a British Columbian firm) in England, and applied for leave to serve the writ of summons on the shipper outside the jurisdiction. The bill of lading contained an express choice of English law. In an extensive review of the applicable law, the court decided that it should exercise discretion to hear the case in England in the interests of the parties and for the ends of justice. Leave to serve the writ outside jurisdiction was granted. Speaking in the context of stay of actions, Lord Goff summarized the relevant principles in the following six propositions:

---

27. [1986] 3 WLR 972.

(1) The basic principle is that a stay will only be granted on the ground of *forum non conveniens* where the court is satisfied that there is some other available forum, having competent jurisdiction, which is the appropriate forum for the trial of the action, ie, in which the case may be tried more suitably for the interests of all the parties and the end of justice.

(2) In general the burden of proof rests on the defendant to persuade the court to exercise its discretion to grant a stay, but if the court is satisfied there is another available forum which is prima facie the appropriate forum for the trial of the action, the burden will then shift to the plaintiff to show special circumstances by reason of which justice requires that the trial should nevertheless take place in this country.

(3) The court sees no reason why the court should not refuse to grant a stay where no particular forum can be described as the appropriate forum. It is significant that in all the leading cases where a stay has been granted there has been another clearly more appropriate forum. The burden resting on the defendant is not just to show that England is not the natural or appropriate forum, but to establish that there is another available forum which is clearly or distinctly more appropriate. If the connection of the defendant with the English forum is a fragile one (for example, if he is served with proceedings during a short visit to England), it should be all the easier for him to prove that there is another clearly more appropriate forum for the trial overseas.

(4) Since the question is whether some other forum exists which is clearly more appropriate for the trial of the action, the court will look first to see what factors there are which point in the direction of another forum. Circumstances may indicate that justice can be done in the other forum at 'substantially less inconvenience or expense'. But it may be more desirable to adopt the expression used by Lord Keith in *The Abidin Daver*[28] when he referred to the 'natural forum' as being 'that with which the action has the most real and substantial connection'. So it is for connecting factors in this sense that the court must first look; and these will include not only factors affecting convenience or expense (such as availability of witnesses), but also other factors such as the law governing the relevant transaction and the places where the parties respectively reside or carry on business.

(5) If the court concludes at that stage there is no other available forum which is clearly more appropriate for the trial of the action, it will ordinarily refuse a stay. It is difficult to imagine circumstances when, in such case, a stay may be granted.

28. [1984] AC 398.

(6) If, however, the court concludes some other forum is available which prima facie is clearly more appropriate for the trial of the action, it will ordinarily grant a stay unless there are circumstances by reason of which justice requires that a stay should nevertheless not be granted. In this inquiry, the court will consider all the circumstances of the case, including circumstances going beyond those taken into account when considering connecting factors with the other jurisdiction. One such factor can be the fact, if established objectively by cogent evidence, that the plaintiff will not obtain justice in the foreign jurisdiction. But then the burden of proof shifts to the plaintiff.

Foreign jurisdiction clauses

Sometimes, the parties may, by agreement contained in the bill of lading or charterparty, submit their disputes to the exclusive jurisdiction of a foreign court. For example, a charterparty may provide that:

- This contract shall be construed and the relations between the parties determined in accordance with the laws of England.
- Any dispute arising under this charter shall be decided by the English Courts to whose jurisdiction the parties hereby agree.

In *The Eleftheria*,[29] Justice Brandon stated the prima facie rule that an action commenced in the local court in breach of an agreement to submit to a foreign jurisdiction will be stayed:

(1) Where the plaintiff sues in the local court in breach of an agreement to refer disputes to a foreign court, and the defendant applies for a stay, the local court, assuming the claim to be otherwise within its jurisdiction, is not bound to grant a stay but has a discretion whether to do so or not.
(2) The discretion should be exercised by granting a stay unless strong cause for not doing so is shown.
(3) The burden of proving such a strong cause is on the plaintiff.
(4) In exercising its discretion the court should take into account all circumstances of the particular case. For example: (a) In what country the evidence on the issues of fact is situated, or more readily available, and the effect of that on the relative convenience and expense of trial as between the local and foreign courts; (b) Whether the law of the foreign court applies and, if so, whether it differs from the law of the local court in any material aspects; (c) With what country either party

29. [1970] P 94.

is connected, and how closely; (d) Whether the defendants genuinely desire trial in the foreign country, or are only seeking procedural advantages; (e) Whether the plaintiffs would be prejudiced by having to sue in the foreign court because they would: (i) be deprived of security for their claim; (ii) be unable to enforce any judgment obtained; (iii) be faced with a time bar not applicable in the local jurisdiction; or (iv) for political, racial, religious or other reasons be unlikely to get a fair trial.

The following Hong Kong case can illustrate the above principles:

> *Kanematsu-Gosho Ltd and Others v The Owners of the Ship 'Ocean Friend'*[30]
>
> The holder of bill of lading commenced an action in *rem* against the defendant for damages in respect of damage to and non-delivery of goods. The bill of lading stated that the contract be governed by English law and any dispute be determined in England, to the exclusion of any other court. The plaintiff argued that the Hong Kong court should not grant a stay of proceedings because:
>
> (a) the plaintiff would lose their security for the action;
> (b) the right of action in England was time-barred;
> (c) the defendant did not genuinely wish for trial in England;
> (d) the plaintiff and the defendant both had connections with Hong Kong, while the plaintiffs had very little connection with England and the defendant had none;
> (e) the evidence could more conveniently be heard in Hong Kong; and
> (f) the proposed action by the defendant against the plaintiff on counter guarantees would be heard in Hong Kong.
>
> The court held that the matters put forward on behalf of the plaintiff did not show good reason why a stay should not be granted, apart from the matters of security and the time bar. If a stay without terms is granted, the plaintiff would be left without a remedy and that would not be just. The court granted the stay of proceedings on two terms:
>
> (a) the defendant would provide acceptable security to satisfy any judgment that the plaintiff might obtain in an English court and;
> (b) the defendant had to undertake not to rely on the time bar in English proceedings.
>
> O'Conner J: 'The effect of a time bar in respect of proceedings in the foreign jurisdiction was . . . in my view . . . a neutral factor, although circumstances of a particular case may make it a factor pointing towards or against a stay. For example, where the failure to bring proceedings before the time bar became effective was due to a deliberate decision

30. [1981] HKLR 253.

> made for tactical reasons in order to provide grounds for opposing a stay, in the hope of forcing the court's hand. In such circumstances, which I do not consider to be the circumstances of the present case, a time bar might well be viewed as a factor in favour of a stay.'

Arbitration agreements

Section 20 of the Arbitration Ordinance states that a stay of court proceedings will be granted unless the court is satisfied the arbitration agreement is null and void, inoperative or incapable of being performed. This is a mandatory provision. The proper test is well-established: is there a plainly arguable case that the parties were bound by an arbitration clause?[31]

In a Hong Kong case, *Tai Hing Cotton Mill Ltd v Glencore Grain Rotterdam BV & Anor*,[32] a Dutch company had a dispute with a Hong Kong company in respect of the sale and purchase of a consignment of cotton. The contract contained an arbitration clause providing for arbitration in Liverpool, England in accordance with the rules of the Liverpool Cotton Association Ltd, and that the disputes should be resolved by the application of English law. The Court of Appeal held that a stay of the court proceedings should be granted for the following reasons:

- It is not the function of the court to investigate whether the defendant to the legal proceedings (ie, the applicant for the stay of court action) has an arguable basis for disputing the claim. The merits are for the arbitration tribunal alone to decide.
- It is also not the function of the court to consider in detail whether an arbitrator is likely to have jurisdiction to determine the dispute, as this is a question for the arbitration tribunal to determine.
- If a claim is made against the defendant and the defendant did not admit the claim, then there was a dispute within the meaning of the Arbitration Ordinance. If the defendant sought to apply for a stay of action, the court must grant a stay unless the plaintiff could show the arbitration agreement was null and void, inoperative or incapable of being performed.
- Since the plaintiff failed to show the arbitration agreement was null and void, inoperative or incapable of being performed, the judge stayed the action and refused the plaintiff any relief therein.

31. *Pacific Crown Engineering Ltd v Hyundai Engineering and Construction Co Ltd* [2003] 3 HKC 659.
32. [1996] 1 HKC 363.

Relationship between stay of proceedings and arrest of vessel to obtain security

Order 12 rule 18 of the Rules of High Court provides a mechanism enabling the defendant to contend that the Hong Kong Court has no jurisdiction over the claim. The Hong Kong proceedings may be stayed if there is a legally binding clause providing for the exclusive jurisdiction of a *foreign court*.

Under s 20 of the Arbitration Ordinance, the court is obliged to order a stay of court proceedings unless it finds that the arbitration agreement is null and void, inoperative or incapable of being performed.

The Hong Kong Admiralty court has jurisdiction to maintain the arrest of the vessel in the face of a mandatory stay of proceedings.

> *The Britannia:*[33]
> In considering whether the court had jurisdiction to maintain the arrest of the vessel despite the mandatory stay of proceedings under the Arbitration Ordinance, Waung J stated: 'I have no doubt that the law does not prohibit any arrest of a ship in admiralty notwithstanding the mandatory stay of proceedings. . . . In a situation of mandatory stay of proceedings for arbitration, neither the pre-June 1997 Arbitration Ordinance nor the amended new Arbitration Ordinance compels the court to release an arrested ship or takes away the power of the court to arrest a ship or to maintain the arrest of the ship.'

It is necessary for the arrest applicant to show that an arbitration award in his favour is unlikely to be satisfied by a defendant. The arrest applicant should therefore disclose evidence that the defendant may well be unable to satisfy any arbitration or foreign court award made against him in order to maintain the arrest. Then the security available in the action in *rem* could be ordered to stand so as to remain available to enable a plaintiff to pursue his action. Waung J in *The Britannia* said:

> I accept of course that it is not legitimate if the arrest was to provide security for the possible foreign arbitration award, but it is perfectly legitimate and proper if the arrest is for the purpose of this action when the stay might be lifted because of the defendant's failure to honour the arbitration award or to redeliver the vessel back to the owners. This is precisely the situation here. We have a defendant with no known financial assets, which has not paid either hire, or its crew or now possibly its lawyers.

33. [1998] 1 HKC 221.

12

Arbitration and Mediation

12.1 ARBITRATION

Generally speaking, an arbitration is the reference of a dispute between two (or more) persons/parties for determination, after hearing both sides in a judicial manner by another person (or persons), other than a court of competent jurisdiction. The decision maker(s), that is, the arbitrator(s), is/are selected by the parties involved in the dispute. The parties agree on the rules and procedures of the arbitration.

The goal of arbitration is to obtain the fair resolution of disputes by an impartial tribunal without unnecessary delay or expenses. The parties should be free to agree how their disputes are resolved, subject only to such safeguards as are necessary in the public interest. Also, intervention by the courts should be restricted, and the grounds for appealing against an arbitration award should be limited.

Advantages of arbitration

There are several advantages for the parties in referring a dispute to arbitration rather than commencing an action in the courts. Arbitration can be much more flexible both in time and procedure. As a result, the arbitration process may be speedier than a court case, and there can be a saving in legal costs. Unlike court proceedings, arbitrations are conducted in private so that confidentiality is preserved. When the shipping dispute concerns a technical matter such as ship collision, general average, tug and towage problem or marine insurance claim, the arbitrators appointed to arbitrate often possess the appropriate special expertise and qualifications.

Differences between arbitration and mediation

Mediation is different from arbitration. Mediation is about disputing parties appointing a skilled third party to assist them in finding a mutually acceptable solution to their differences. Mediation is a process in which a neutral person, the mediator (can be one, two, or a panel), helps parties involved in a dispute to negotiate effectively with each other in order to solve the problem cooperatively and successfully. Mediators perform a number of functions. They manage the process of meeting and negotiating, assist the parties to move forward towards settlement, open and maintain the pathways for communications, encourage full expression of information and complete exploration of the merits with regard to the dispute, generate options and alternatives for the parties to consider, and maintain fairness during the mediation procedure. A successful mediation may lead to the best discovery of the participants' interests, as well as the path to an effective settlement of the dispute between the parties.

It is possible to combine mediation and arbitration, beginning with mediation. If there is a determination by the parties or mediators that negotiated settlement is not feasible, then the matter will be referred to arbitration.

12.2 ARBITRATION ORDINANCE (CAP 609)

Arbitration agreement

According to section 19 of the Arbitration Ordinance, an arbitration agreement shall be in writing. An arbitration agreement is in writing if its content is recorded in any form, whether or not the arbitration agreement or contract has been concluded orally, by conduct, or by other means.

The requirement that an arbitration agreement be in writing is met by an electronic communication, if the information contained therein is accessible so as to be useable for subsequent reference. Furthermore, an arbitration agreement is in writing if the agreement is in a document, whether or not the document is signed by the parties to the agreement.

This is a sample arbitration clause recommended by the Hong Kong International Arbitration Centre (HKIAC):

> Any dispute, controversy or claim arising out of or relating to this contract, including the validity, invalidity, breach or termination thereof,

> shall be settled by arbitration in Hong Kong under the Hong Kong International Arbitration Centre Administered Arbitration Rules in force when the Notice of Arbitration is submitted in accordance with these Rules. The number of arbitrators shall be . . . (*one or three*). The arbitration proceedings shall be conducted in . . . (*insert language*).

Default appointment

Default appointment is provided for in sections 13 and 24 of the Arbitration Ordinance. In an arbitration with a sole arbitrator, if the parties are unable to agree on the arbitrator, the arbitrator shall be appointed, upon request of a party, by the Hong Kong International Arbitration Centre (HKIAC).

In an arbitration with three arbitrators, each party shall appoint one arbitrator, and the two arbitrators thus appointed shall appoint the third arbitrator. If a party fails to appoint the arbitrator within thirty days of receipt of a request to do so from the other party, or if the two arbitrators fail to agree on the third arbitrator within thirty days of their appointment, the appointment shall be made, upon request of a party, by HKIAC.

Commencement of arbitration

Generally speaking, a claimant must commence arbitration proceedings within the time limit. For example, arbitration in connection with a dispute under a bill of lading governed by the Hague-Visby Rules must be brought within one year of the delivery of the cargo or of the date when it should have been delivered.

Section 58 of the Arbitration Ordinance empowers the arbitration tribunal to extend time for arbitration proceedings. Two alternative tests apply to such applications:

- whether the circumstances were outside the reasonable contemplation of the parties when they made the arbitration agreement and that it would be just to extend the period; and
- in any other case, whether the conduct of one party makes it unjust to hold the other to the strict terms of the agreement.

Intervention by court

Section 12 of the Arbitration Ordinance is drafted on the basis of Article 5 of the UNCITRAL (United Nations Commission on International Trade Law) Model Law on International Commercial Arbitration, which provides that 'in matters governed by this law, no court shall intervene except where so provided in this law'. Generally speaking, the award shall be treated as final, and the Court does not have jurisdiction to set aside an arbitral award on the ground of errors of fact or law on the face of the award.[1] Bringing a question of law to the Court for consideration can only be done with the agreement of all the other parties to the arbitral proceedings, or with the leave (ie, special permission) of the Court.[2]

In *Re UDL Contracting Limited*,[3] arbitration proceedings were commenced in September 1998 against UDL by ADPL. UDL had made a counterclaim which substantially exceeded ADPL's claim. Later, UDL applied to the court under the Companies Ordinance for an order restraining all further proceedings in the arbitration until the conclusion of the winding up proceedings against UDL, or until further order. Section 181 of the Companies Ordinance provides that 'at any time after the presentation of a winding-up petition and before a winding-up order has been made, the company or any creditor or contributory may, where any action or proceedings against the company is pending in any court or tribunal other than the Court of First Instance or the Court of Appeal, apply to the Court of First Instance to restrain further proceedings in the action or proceeding, and the court to which application is so made may, as the case may be, stay or restrain the proceedings accordingly on such terms as it thinks fit'. The court held that the words 'action or proceeding' and 'court or tribunal' were not confined to court proceedings. Arbitrations were encompassed, and therefore the court had the power to grant the stay of the arbitration proceedings. The object of the Arbitration Ordinance is to facilitate the fair and speedy resolution of disputes. Whilst the parties to a dispute are to be accorded freedom to decide how the dispute is to be resolved, that freedom is nevertheless subject to public interest safeguards. The principle that the court should only interfere as expressly provided by the Arbitration Ordinance must be read in that context. Hence the court reached the conclusion that the court does have jurisdiction to restrain further proceedings in the arbitration. The court then exercised its discretion to stay the arbitration proceedings. Justice Le Pichon said:

> In the exercise of the court's discretion . . . the test is whether substantial injustice will result if the arbitration is not stayed. It is inarguably in the interest of the company's creditors and contributories that ADPL's claim be properly defended and the counterclaim asserted. Against that

1. Section 81(3) of the Arbitration Ordinance.
2. Sections 4 and 5 of the Second Schedule, Arbitration Ordinance.
3. [2000] 1 HKC 390.

> I have to balance any potential prejudice to ADPL. A stay will not have the effect of eliminating ADPL's claim: that will be adjudicated sooner or later since the stay that is being sought is not permanent but is of limited duration. It is presently anticipated that the restructuring process would be completed by March or April. In practical terms, this means an adjournment of no more than two to three months since it is accepted that the hearing scheduled for January is likely to go part-heard. If the scheme is sanctioned by the court, the stay would cease. The arbitration can then resume. In this connection, it ought to be mentioned that under the draft scheme of arrangement, a sum of $2m out of the interim financing obtained of $5m has been set aside for arbitration expenses involving the subsidiaries . . . If the scheme either does not proceed as anticipated or is not sanctioned, then a winding-up order would seem inevitable. In that eventuality, if the liquidator were to consider that the company does have a sound counterclaim, he may seek funding from the creditors to resolve the dispute. In all the circumstances, it would be appropriate for the court to exercise its discretion in favour of the company.

12.3 ENFORCEMENT OF ARBITRATION AWARDS

Hong Kong arbitration awards

An award made in arbitration proceedings by an arbitral tribunal is enforceable in the same way as a judgment of the court of Hong Kong, but only with the leave of the Hong Kong court. If that leave is given, the Hong Kong court may enter judgment in terms of the arbitration award.[4]

Mainland Chinese arbitration awards

On 21 June 1999, agreement was reached between the HKSAR and Mainland Chinese governments on the reciprocal enforcement of arbitration awards. The agreement was based on the provisions of and practice under the New York Convention. It allowed the arbitration awards made by recognized mainland arbitral authorities[5] to be enforced in the HKSAR. HKSAR awards will also be enforceable in Mainland China. The Arbitration (Amendment) Ordinance 2000, which implements the agreement reached, came into operation on 1 February 2000. Sections 92–98 of the Arbitration Ordinance deal with the recognition and enforcement of mainland arbitration awards.

---

4. Section 84 of the Arbitration Ordinance.
5. These include the China International Economic Trade Arbitration Commission (CIETAC) and the China Maritime Arbitration Commission (CMAC).

Foreign arbitration awards

The recognition and enforcement in Hong Kong of arbitral awards made overseas pursuant to the New York Convention are governed by sections 87–91 of the Arbitration Ordinance. The New York Convention currently applies to the recognition and enforcement of international arbitration awards in about 146 states and territories.[6] A New York Convention award may, by leave of the Hong Kong court, be enforced in the same manner as a judgment or order of the Hong Kong court to the same effect. Where leave is so granted by the Hong Kong court, judgment may be entered in terms of the award.

Taiwanese arbitration awards

Owing to Taiwan's lack of legal statehood, the New York Convention has never applied to Taiwanese arbitration awards. The Arbitration (Amendment) (No 2) Ordinance took effect on 23 June 2000. It avails the enforcement in Hong Kong of arbitration awards made in Taiwan.

12.4 LONDON MARITIME ARBITRATORS ASSOCIATION (LMAA)

London is one of the world leaders in the field of settling international commercial and maritime disputes through arbitration. The London Maritime Arbitrators Association (the LMAA) is an association of practising maritime arbitrators. One of the objects of which is to advance and encourage the professional knowledge of London maritime arbitrators and, by recommendation and advice, to assist the expeditious procedure and disposal of disputes.

The Association was founded on 12 February 1960 at a meeting of the arbitrators on the Baltic Exchange Approved List. In 1972 it was decided to create a new category of members known as supporting members, who are described in the rules of the Association as those who not as a general rule practice as umpires or arbitrators, but who wish to lend their support to, the achievement of the objects of the Association.

The LMAA has also instituted its Small Claims Procedure and the Mediation Terms, which together with the LMAA Terms offer the shipping

6. The Chief Executive in Council may, by order in the Gazette, declare that any state or territory is a party to the New York Convention (section 90 of the Arbitration Ordinance). Alternatively, a regular update of the list of states and territories can be found in the website of The United Nations Commission on International Trade Law <www.uncitral.org>.

community a range of choices for the resolution of commercial disputes. Maritime arbitrations in London are normally conducted on the basis of the LMAA Terms. LMAA arbitrators will accept appointment subject to the Terms. These have been introduced after extensive consultation with those interested in maritime arbitration, and are intended to assist in the speedier and more economical resolution of disputes.

12.5 CHINA MARITIME ARBITRATION COMMISSION (CMAC)

The PRC promulgated its first Arbitration Law on 31 August 1994, and this law came into force on 1 September 1995. The new law requires that arbitration be carried out independently according to law and without interference by any administrative organ, social organization, or individual. It also requires the existence of a written arbitration agreement voluntarily entered into by the parties describing the matters to be arbitrated and selecting the arbitration commission to arbitrate the dispute. Apart from these changes, the new legislation has detailed provisions regarding arbitration proceedings, applications to set aside arbitral awards, the enforcement of arbitral awards, and procedural matters concerning foreign-related arbitration.

Arbitration Rules for the China Maritime Arbitrations Commission (CMAC) were enacted on 1 October 1995. These rules provide for the independence of arbitration agreements from their underlying contracts. Thus, if the validity of a contract is challenged, CMAC continues to wield the power to decide on the existence and validity of arbitration agreements, and the scope of arbitral jurisdiction. However, where the parties dispute the validity of the arbitration agreement, and one party requests that CMAC decide the matter while the other party requests that the People's Court issue a ruling, the power to make a final decision will be vested in the People's Court. The CMAC Arbitration Rules allow the arbitration tribunal to decide the matter under the principles of mediation if the parties agree. As a practical matter, most Chinese enterprises prefer 'friendly consultation and mediation'. Many arbitration clauses in Chinese commercial contracts provide that friendly consultation or mediation must be attempted before arbitration. Even after formal proceedings have started, the parties are still expected to continue to attempt to find a resolution to the dispute.

13

Conflict of Laws in Carriage

At common law, a contract of carriage is governed by its 'proper law'. The 'proper law' is the law which governs the contract and the parties' obligations under it. It also normally determines the contract's validity and legality, its construction and effect, and the conditions of its discharge. The classic definition of the term 'proper law of a contract' can be found in *Bonuthon v Commonwealth of Australia*,[1] where Lord Simonds stated that the proper law of the contract is:

- the system of law expressly chosen by the parties, by which the parties intended the contract to be governed; or
- in the absence of express choice of the proper law, the law impliedly selected by the parties if such intention can be clearly inferred from the terms of the contract or the relevant surrounding circumstances; or
- where the intention of the contractual parties is neither expressed nor to be inferred from the circumstances, the system of law with which the transaction has its closest and most real connection.

13.1 EXPRESS CHOICE OF LAW

The parties may expressly choose the law to govern their contract of carriage. The expressed intention of the parties determines the proper law of the contract, and thus relieves the court of the onerous task of ascertaining the proper law of the contract. The leading case establishing this

1. [1951] AC 201.

principle is *Vita Food Products Inc v Unus Shipping Company Limited.*[2] Unus Shipping, a Nova Scotian company, owned the vessel 'Hurry On', and agreed with a New York company Vita Food Products in bills of lading signed in Newfoundland to carry a cargo of herrings in the 'Hurry On' from Middle Arm (a Newfoundland port) to New York. Through the negligence of the captain the vessel ran aground and the cargo was damaged. The herrings had to be 'reconditioned' before they could be sold. Vita Foods sued Unus Shipping in the Nova Scotian courts and, on appeal, came before the Privy Council. The bills of lading declared that English law was the proper law. The Privy Council held that the parties' choice of English law was effective. Lord Wright said:

> It will be convenient at this point to determine what is the proper law of the contract. In their Lordships' opinion, the express words of the bill of lading must receive effect, with the result that the contract is governed by English law. It is now well settled that by English law . . . the proper law of the contract is the law which the parties intended to apply. That intention is objectively ascertained, and, if not expressed, will be presumed from the terms of the contract and the relevant surrounding circumstances . . . But where the English rule that intention is the test applies, and where there is an express statement by the parties of their intention to select the law of the contract, it is difficult to see what qualifications are possible, provided the intention expressed is bona fide and legal, and provided there is no reason for avoiding the choice on the ground of public policy . . . There is, in their Lordships' opinion, no ground for refusing to give effect to the express selection of English law as the proper law in the bills of lading.

Palace Hotel Ltd v Owners of the Ship or Vessel Happy Pioneer[3]

The cargo owner sued the carrier for failure to deliver goods to the cargo owner at Tientsin. The bill of lading stated that 'all actions under the contract of carriage shall be brought before the court at Amsterdam, and no other court shall have jurisdiction with regard to any such action unless the carrier appeals to another jurisdiction or voluntarily submits himself thereto'. The bill of lading also stated that 'the law of the Netherlands shall apply to this contract'.

The Hong Kong court stayed the proceedings.

Justice Power: 'I am satisfied, as I have indicated, that the law of the Netherlands is the proper law of the contract and this would clearly be best applied by the court at Amsterdam.'

2. [1939] AC 277.
3. [1982] HKC 641.

13.2 IMPLIED CHOICE OF LAW

Where the parties do not choose the law to govern their contract at all, the courts may nevertheless be able to infer from the terms of the contract and the relevant surrounding circumstances that the parties have made an implied selection of the proper law. According to *Jacobs v Credit Lyonnais*,[4] the court has to apply 'sound ideas of business, convenience, and sense to the language of the contract itself, with a view to discovering from it the true intention of the parties'.

In *Amin Rasheed Shipping Corporation v Kuwait Insurance Company*,[5] a Liberian company resident in Dubai (the plaintiff) had insured a ship with the defendants, a Kuwaiti insurance company. A claim made by the plaintiff was rejected by the defendant whereupon the plaintiff sued in England. The terms of the marine insurance policy adopted the obsolete language of the Lloyd's SG policy as schedules to the English Marine Insurance Act 1906. Leave to serve the writ outside the jurisdiction was required under the Rules of Supreme Court in England. Under the rules, leave would only be granted if the contract was 'by its terms, or by implication, governed by English law'. The House of Lords held that the intention of the parties, although not expressed in the contract, was that English law should govern the marine insurance contract. Lord Diplock made the following observations:

> The crucial surrounding circumstance, however, is that it was common ground between the expert witnesses on Kuwaiti law that at the time the policy was entered into there was no indigenous law of marine insurance in Kuwait. Kuwait is a country in which the practice since 1961, when it began to develop as a thriving financial and commercial centre, has been to follow the example of the civil law countries and to embody the law dealing with commercial matters, at any rate, in written codes. In Kuwait there had been in existence since 1961 a Commercial Code dealing generally with commercial contracts but not specifically with contracts of marine insurance. The contract of marine insurance is highly idiosyncratic; it involves juristic concepts that are peculiar to itself such as sue and labour, subrogation, abandonment and constructive total loss; to give but a few examples. The general law of contract is able to throw but little light upon the rights and obligations under a policy of marine insurance in the multifarious contingencies that may occur while the contract is in force. The lacuna in the Kuwaiti commercial law has since been filled in 1980 by the promulgation for the first time of a code of marine insurance law. This code does not simply adopt the English law of

4. (1884) 12 QBD 589.
5. [1984] 1 AC 50.

> marine insurance, there are significant differences. However, it did not come into operation until August 15, 1980, and it is without retrospective effect. It does not therefore apply to the policy which was entered into at a time before there was any indigenous law of marine insurance in Kuwait . . . Except by reference to English statute . . . it is not possible to interpret the policy or to determine what those legal rights and obligations are. So, applying, as one must in deciding the jurisdiction point, English rules of conflict of laws, the proper law of the contract embodied in the policy is English law.

13.3 THE CLOSEST AND MOST REAL CONNECTION

Where the intention of the contractual parties is neither expressed nor to be inferred from the circumstances, the system of law with which the transaction has its closest and most real connection is the proper law of the contract. Different factors will be taken into consideration by the court, including:

- the place of contracting, ie, the country in which the contract is concluded;
- the place of performance of the contractual obligations (eg, the place of loading, the place of discharge, the place of payment of freight and demurrage . . . etc.);
- the principal places of business of the parties, namely, the carrier, the shipper and the consignee . . . etc.

In *Assunzione*,[6] a charterparty that contained no express choice of law clause had been signed in Paris by the agents of French shippers and Italian shipowners. The charterparty was for the carriage of wheat from Dunkirk, France to Venice, Italy in the 'Assunzione', an Italian ship, and the wheat was being shipped as part of an exchange agreement between the French and Italian governments. The charterparty was in English with additional clauses in French. A great proportion of the freight and demurrage were payable in lire in Italy. The bills of lading were in French but endorsed by the Italian consignees. Out of the diverse but balanced facts, the English Court of Appeal had to determine whether the contract was governed by French law or Italian law. It was unanimously held that Italian law was the proper law of the contract, and the decisive factor was that both parties had to perform their contractual obligations in Italy. Singleton LJ:

> With regard to the circumstances which support the defendant's contention that Italian law should be applied, I mention these: the ship

6. [1954] P 150.

> was an Italian ship owned by two Italians in partnership, and a ship wearing the Italian flag; the owners were Italians, the master was an Italian; the contract was for carriage from a French port to an Italian port; the cargo was to be delivered at an Italian port, and one may appear to cancel the other, but there are further considerations; the charterparty provided that freight and demurrage should be paid in Italian currency. I have read clause 16 of the charterparty; I bear in mind that 80% of the freight had to be paid in Italy before the ship arrived at an Italian port, and the balance had to be paid at the discharging port. Clause 7 as to demurrage payable in Italian lire, too, is of importance on this part of the case. The next point is that the bills of lading were indorsed by Italian consignees before the arrival of the ship at Venice. The judge thought that that must have been in the contemplation of the parties as the bills of lading were made out to order, by which it must be assumed that indorsement was contemplated . . . Although I believe it to be impossible to state any rule of general application, I feel that matters of very considerable importance are the form of, and place of, payment. In this case I regard it as a very important feature, coupled as it is with the facts that the ship was an Italian ship and that the destination was an Italian port . . . Applying the rule which I have stated and weighing all the facts to which attention was directed, I am satisfied that the scale comes down in favour of the application of Italian law.

13.4 SPECIAL CASES

In some cases, the local legislation of the forum may override an express choice of law. The *Hollandia*[7] provides an example of a situation in which the plaintiff was able to ask the court to strike down the exclusive jurisdiction clause. A piece of machine had been shipped from Leith in Scotland to Bonaire in the Dutch West Indies aboard a Dutch vessel. It sustained £20,000 damage while being unloaded and the shippers sued the carriers in England. The bill of lading contained an express choice of the law of the Netherlands and the Dutch law incorporated the Hague Rules limiting the carrier's liability to £250 per packet. The bill of lading also contained a clause giving the Amsterdam courts exclusive jurisdiction over disputes arising under the contract of carriage. The Hague-Visby Rules, on the other hand, were part of English law by virtue of the English Carriage of Goods by Sea Act 1971. Under the Hague-Visby Rules, recovery of £11,000 would be allowed, which was why the plaintiff sued in England rather than the Netherlands. The defendant carrier sought a stay of the English proceedings because of the

7. [1983] 1 AC 565.

exclusive jurisdiction clause. The House of Lords held that the Hague-Visby Rules, which had the force of law in England, stipulated at Article III r 8 that any clause in a contract of carriage relieving the carrier from liability otherwise than as provided in those rules would be null and void and of no effect. In these circumstances, the exclusive jurisdiction clause was rendered null and void and of no effect. The House of Lords refused to grant the stay of proceedings. Lord Diplock:

> It is, in my view, most consistent with the achievement of the purpose of the Act of 1971 that the time at which to ascertain whether a choice of forum clause will have an effect that is proscribed by Art III r 8 should be when the condition subsequent is fulfilled and the obligations of the carrier or ship that are referred to in that rule and it is established as a fact (either by evidence or as in the instant case by the common agreement of the parties) that the foreign court chosen as the exclusive forum would apply a domestic substantive law which would result in limiting the carrier's liability to a sum lower than that to which he would be entitled . . . then an English court is in my view commanded by the Act of 1971 to treat the choice of forum clause as of no effect.

As demonstrated in the above case, the Hong Kong court will not enforce a contract that constitutes a violation of local legislation. Furthermore, the Hong Kong court will not enforce a contract if its performance involves doing an act in a foreign and friendly state which violates the law of that state. In *Regazzoni v KC Sethia Ltd*,[8] the buyer and the seller of certain cargo entered into a contract of sale governed by English law. The cargo was to be shipped from India to Genoa, although both parties knew that the cargo would be transshipped to South Africa. The relevant Indian statute banned the export of cargo 'destined for any port or place in the Union of South Africa or in respect of which the Chief Customs Officer is satisfied . . . although destined for a port or place outside the Union of South Africa are intended to be taken to the Union of South Africa'. The House of Lords held that the contract would not be enforced against the seller. Lord Reid:

> To my mind, the question whether this contract is enforceable by English courts is not, properly speaking, a question of international law. The real question is one of public policy in English law: but in considering this question we must have in mind the background of international law and international relationships often referred to as the comity of nations. This is not a case of a contract being made in good faith but one party thereafter finding that he cannot perform his

8. [1958] AC 301.

part of the contract without committing a breach of foreign law in the territory of the foreign country. If this contract is held to be unenforceable, it should, in my opinion, be because from the beginning the contract was tainted so that the courts of this country will not assist either party to enforce it.

References

Baatz Y and others, *The Rotterdam Rules: A Practical Annotation* (Maritime and Transport Law Library, 2009)

Bates J, *Marine Environment Law* (Lloyd's of London Press Limited, 1993)

Baughen S, *Shipping Law* (Cavendish Publishing Limited, 2015)

Bonnick S, *Gram on Chartering Documents* (Lloyd's of London Press Limited, 1999)

Brodie P R, *Dictionary of Shipping Terms* (Lloyd's of London Press Limited, 2013)

———, *Illustrated Dictionary of Cargo Handling* (Lloyd's of London Press Limited, 2013)

Bureau Veritas, *Ship Safety Handbook* (Lloyd's of London Press Limited, 1989)

Cashmore C, *Parties to a Contract of Carriage* (Lloyd's of London Press Limited, 1990)

Chan Felix W H, Ng Jimmy, Tai S K and others, *Halsbury's Laws of Hong Kong: Maritime Law*, vol 18(1) (Butterworths, 2013)

Curtis S, *Law of Shipbuilding Contracts* (Lloyd's of London Press Limited, 2012)

Darling G and Smith C, *LOF 90 and the New Salvage Convention* (Lloyd's of London Press Limited, 1991)

Davies D, *Commencement of Laytime* (Lloyd's of London Press Limited, 2006)

Davison R and Snelson A, *The Law of Towage* (Lloyd's of London Press Limited, 1990)

de la Rue C, *Liability for Damage to the Marine Environment* (Lloyd's of London Press Limited, 1993)

Dicey and Morris, *Conflict of Laws* (Sweet and Maxwell, 2012)

Douglas R P and Green G K, *Law of Harbours and Pilotage* (Lloyd's of London Press Limited, 1993)

Eder B and others, *Scrutton on Charterparties and Bills of Lading* (Sweet and Maxwell, 2014)

Fogarty A, *Merchant Shipping Legislation* (Lloyd's of London Press Limited, 2004)
Forsyth C F, *Conflict of Laws* (HTL Publications, 1995)
Ganado M and Kindred H, *Marine Cargo Delays* (Lloyd's of London Press Limited, 1990)
Golann D, *Mediating Legal Disputes* (Aspen Publishers Inc, 2009)
Goldrein I, *Ship Sale and Purchase* (Lloyd's of London Press Limited, 2013)
Goode R, *Commercial Law* (Penguin Group, 2010)
Gorton L and Ihre R, *A Practical Guide to Contracts of Affreightment and Hybrid Contracts* (Lloyd's of London Press Limited, 1990)
Gorton L, Ihre R and Sandevarn A, *Shipbroking and Chartering Practice* (Lloyd's of London Press Limited, 2009)
Griggs P and Williams R, *Limitation of Liability for Maritime Claims* (Lloyd's of London Press Limited, 1998)
Grime R, *Shipping Law* (Sweet and Maxwell, 2012)
Hazelwood S, *P and I Clubs—The Law and Practice* (Lloyd's of London Press Limited, 2010)
Hodges S, *Law of Marine Insurance* (Cavendish Publishing Limited, 1996)
Hudson N G, *The York-Antwerp Rules* (Lloyd's of London Press Limited, 1996)
Hudson N G and Allen J C, *The Institute Clauses Handbook* (Lloyd's of London Press Limited, 1995)
——— and Allen J C, *The Marine Claims Handbook* (Lloyd's of London Press Limited, 1996)
Institute of Maritime Law, *The Ratification of Maritime Conventions* (Lloyd's of London Press Limited, 1990)
Lower-Hill B, *Lloyd's Survey Handbook* (Lloyd's of London Press Limited, 1996)
Luddeke C, *Marine Claims* (Lloyd's of London Press Limited, 1996)
Meeson N, *Admiralty Jurisdiction and Practice* (Lloyd's of London Press Limited, 2011)
Miller M D, *Marine War Risks* (Lloyd's of London Press Limited, 2005)
Ready N P, *Ship Registration* (Lloyd's of London Press Limited, 1995)
Richardson J W, *A Guide to the "BOXTIME" Charterparty* (Lloyd's of London Press Limited, 1990)
———, *A Guide to the Hague and Hague-Visby Rules* (Lloyd's of London Press Limited, 1994)
Rogers P, Strange J and Studd B, *Coal: Carriage by Sea* (Lloyd's of London Press Limited, 1997)
Schmitthoff C, *Schmitthoff's Export Trade* (Sweet and Maxwell, 2012)
Schofield J, *Laytime and Demurrage* (Lloyd's of London Press Limited, 2011)
Sparks A, *Steel: Carriage by Sea* (Lloyd's of London Press Limited, 2009)
Spruyt J, *Ship Management* (Lloyd's of London Press Limited, 1990)
Sturt R, *Collision Regulations* (Lloyd's of London Press Limited, 1984)
Sullivan E, *Marine Encyclopaedic Dictionary* (Lloyd's of London Press Limited, 1999)

Sutherland V J and Cooper C L, *Man and Accidents Offshore* (Lloyd's of London Press Limited, 1986)
Sutton J, Kendall J and Gill J, *Russell on Arbitration* (Sweet and Maxwell, 2007)
Todd P, *Bills of Lading and Bankers' Documentary Credits* (Lloyd's of London Press Limited, 2007)
Treitel G, *Carver on Bills of Lading* (Sweet and Maxwell, 2005)
Vincenzini E, *International Salvage Law* (Lloyd's of London Press Limited, 1992)
Wild P and Dearing J, *The Reefer Market* (Lloyd's of London Press Limited, 1994)
Wilson J F, *Carriage of Goods By Sea* (Pitman Publishing, 2010)
Yates and others, *Contracts for the Carriage of Goods by Land, Sea and Air* (Lloyd's of London Press Limited, 2000)

Index